Off-Road Vehicle Engineering Principles

Carroll E. Goering
Professor Emeritus
University of Illinois

Marvin L. Stone
Professor
Oklahoma State University

David W. Smith
Research Engineer
Deere & Company

Paul K. Turnquist
Professor & Head Emeritus
Auburn University

ASAE is an educational and scientific organization
dedicated to the advancement of engineering
applicable to agricultural, food, and biological systems.

Copy editing by Peg McCann and Denise Sicking
Cover design by Melissa Miller

Cover Photographs
Courtesy of Caterpillar, Inc. Reprinted with permission.

For information, contact the
American Society of Agricultural Engineers
2950 Niles Road, St. Joseph, MI 49085-9659 USA
Phone: 269-429-0300 Fax: 269-429-3852 E-mail: hq@asae.org

Library of Congress Card Number (LCCN) 2003106933
International Standard Book Number (ISBN) 1-892769-26-3
ASAE Publication 801M0103

CONTENTS

FOREWORD

Off-Road Vehicle Engineering Principles is a successor to *Tractors and Their Power Units*. The earlier text had gone through four editions before going out of print. While users of the earlier text will recognize a considerable amount of material that has been carried over to this new text, much of the material is new and the arrangement of chapters has been completely reorganized. For example, material on performance has been moved to a much earlier point in the new text to aid instructors in capturing student interest; undergraduate students are typically more interested in engine and vehicle performance than in other topics in the book. Also, the chapter on electrical systems has been completely rewritten to reflect the use of computer-like architecture in the design of vehicle electrical systems. The chapters on traction and on chassis mechanics are more tightly integrated in the new book than they were in the earlier book. Every chapter was rewritten to reflect current technology. Other changes were made to enhance student learning; each chapter now includes an introduction and a summary. Also, the new book includes many more homework problems than were in the earlier book. There is more material than can be included in a typical one-semester course; however, the instructor can use those chapters that best fit the needs of his or her own course.

Much of the book is accessible to those with knowledge of undergraduate physics and college algebra. A full understanding of Chapter 14 on Chassis Mechanics requires a prerequisite course in differential equations. Although the level of difficulty of the book is best suited for upper-class undergraduates and for graduate students, much of the material could also be taught to students in their first or second year of college.

The authors owe a debt of gratitude to numerous persons who assisted in the preparation of *Off-Road Vehicle Engineering Principles*. As previously mentioned, this book is built on the legacy of *Tractors and Their Power Units*. The current authors acknowledge the distinguished authors of the various editions of the earlier book; these include W.M. Carleton, Roy Bainer, J. Bruce Liljedahl, E.L. Barger, E.G. McKibben (all deceased), Makoto Hoki, David W. Smith, and Paul K. Turnquist. The latter two persons continue as authors of the current book.

A number of persons with special expertise provided valuable assistance by reviewing various chapters. These persons helped find and eliminate errors and also made valuable suggestions for improving the respective chapters. Rex Hanson of the John Deere Dubuque Works reviewed Chapter 9 on Cooling Systems. John Solie, from Oklahoma State University, reviewed Chapter 10 on Electrical and Electronic Systems. Gordon Wiegardt, retired from the John Deere Product Engineering Center, reviewed Chapter 11 on Hydraulic Systems. Dick Michael, retired from the John

Deere Product Engineering Center, and Bernie Romig of the John Deere Technical Center reviewed Chapter 12 on Drive Trains. Claude Laguë of the University of Saskatchewan, Andrew Shorter of Goodyear, Thomas Way of the National Soil Dynamics Laboratory, Shrini Upadhyaya of the University of California at Davis, and Frank Zoz, Robert Wismer, Jack Wiley, and Carol Plouffe of John Deere reviewed Chapter 13 on Traction and Transport Devices. The authors are grateful to each of these persons for their valuable assistance.

The authors are especially grateful to Al Hansen of the University of Illinois. He was the first to use the new book as a text for a course. In preparing to teach his AgE 346 course in the Fall 2000 semester, he digitized all of the illustrations, arranged the various chapter drafts into a bound book and had the book reproduced for his students. In using the book, he and his students found numerous errors that were then corrected by the authors. Also, Kevin Shinners used a pre-publication version of the book in a course at the University of Wisconsin in the Spring 2002 semester. Based on that experience, he provided valuable suggestions for improving the book.

Despite the best efforts of the authors and the publisher, it is likely that some errors still remain in the book. If and when users of the book encounter such errors, the authors will be grateful if you will call them to our attention so that they can be removed before succeeding editions of the book are published.

Finally, the authors are grateful to their employers and to their families for understanding the time commitment that is required to produce a new book and for allowing each of us to make that commitment.

Carroll E. Goering
Marvin L. Stone
David W. Smith
Paul K. Turnquist

CHAPTER 1

DEVELOPMENT OF OFF-ROAD VEHICLES

1.1 Introduction

The National Academy of Engineering named the 20 greatest engineering achievements of the 20th century (Inside ASAE, 2000). Agricultural mechanization was ranked seventh ahead of computers, the telephone and spacecraft. Development and use of off-road vehicles, primarily the agricultural tractor, played a major role in the agricultural mechanization achievements of the past century.

It was recognized more than a century ago that human beings were inefficient as a primary source of power, about 0.1 kW continuous output. A steam-powered tractor plowing sod was demonstrated in 1858, but tractors got their first impetus during World War I and really came into their own during World War II, in each case because of the enormous increase in demand for food and fiber production, while experiencing a severe decrease in available agricultural labor.

Evolution of the tractor has accompanied changes in farm technology and sizes of farms. The tractor has progressed from its original primary use as a substitute for animal power to the present units designed for multiple uses. Traction and power, belt power, power take-off drives, mounted tools, hydraulic remote-control units, electronic sensing and control of machine performance, and ergonomically designed operator workplace and climate-controlled cabs have extended the usefulness and efficiency of the tractor for agricultural and other off-road uses.

The word *tractor* has been attributed to various sources (*Farm Implement News*, 1952, 1953), however, according to the *Oxford Dictionary*, the word was first used in 1856 in England as a synonym for traction engine. The term "tractor" appears in an 1890 U.S. patent for a track-laying steamtraction engine. The rapid increase in numbers and versatility of tractors has been accompanied by a similar decrease in numbers of horses and mules on farms in the United States (USDA, 1960). Depending upon local and domestic conditions, various types and sizes of tractors have been developed and used worldwide.

The early steam engines furnished belt power, but they had to be pulled from place to place by horses or oxen. The next step in the evolution in farm power was the conversion of the steam engine into a self-propelled traction engine. Successful steam plows were developed in the 1850s, and improvements continued until 1900.

Inadequate traction plagued the inventors of the large heavy tractors, who tried to solve the problem by making the driving wheels wider and wider. One big-wheeled tractor made in 1900 for use in California had two wood-covered drive wheels, each 5 m wide and 3 m in diameter. The tractor weighed about 36 tons. Other attempts to solve the problem of traction resulted in the development of track-type agricultural tractors about 1900.

Early attempts to develop gasoline tractors were stimulated by the need for reduction in the number of workers required to attend the steam tractors, both when plowing and operating threshing machines. Early gasoline tractors resembled steam tractors. Considerable development was needed before a reasonably successful internal combustion engine was available for a tractor. The internal combustion engine did not assume much importance until after the expiration of the Otto patents in 1890.

The first Winnipeg tractor trials were held in 1908, giving the public an opportunity to compare field operations of steam and gas tractors. Succeeding trials were held each year through 1912, when they were discontinued. The first U.S. tractor demonstration, held at Omaha, Nebraska, in 1911, was conducted as an exhibition and not as a competition between machines.

The Nebraska Tractor Test Law, passed in 1919, specified that every tractor sold in the state of Nebraska should be tested and the results published. In addition, the manufacturer was required to maintain an adequate supply of parts for repairs. The tests, which attained worldwide recognition, have provided standards for rating tractors, speeded improvements, and eliminated many types that were inferior in design and performance.

1.2 History of Vehicle Development

Some of the important highlights in tractor development are listed here.

1858-1919

Steam plowing engine by J. W. Fawkes drew eight plows at 4.8 km/h in virgin sod in 1858.

The Parvins steamer was probably the first U.S. attempt at a track-laying device in 1873, although the U.S. Patent Office recorded crawler tractor developments in the early 1850s.

Otto patents were issued for an internal combustion engine in 1876.

At least one company built a tractor with an internal combustion engine in 1889.

The first Winnipeg tractor trials were held in 1908.

The first tractor demonstration was held in the United States at Omaha, Nebraska, in 1911.

Smaller, lightweight tractors were introduced.

The frameless-type tractor was introduced.

Cast iron was first used in chassis on the 1917 Fordson tractor (Miller, 2000). This promoted frame rigidity and accuracy of drive train alignment.

Oil cooled clutch facings were first used on 1917 Fordson tractor (Miller, 2000). The oil used was from the engine sump.

The Nebraska Tractor Test Law was passed in 1919.

The power take-off was introduced on the International 8-16 tractor in 1919, which expanded the versatility by providing engine power for drawn equipment (Miller, 2000).

1920-1929

The first Nebraska tractor test was completed on a John Deere Waterloo Boy in 1920 (Bashford, 2000).

A highly successful row-crop or all-purpose farm tractor was developed and marketed as the McCormick-Deering Farmall Regular in 1924 (Miller, 2000).

The mechanical rockshaft to lift mounted equipment was introduced in 1927 on the Deere model GP tractor (Miller, 2000).

The power take-off was gradually adopted.

1930-1939

Caterpillar introduced the diesel engine in 1931 in one of their crawler tractors. Nebraska Test No. 208 reported it developed 83.86 belt horsepower (Miller, 2000).

Rubber tires by Firestone Rubber and Allis-Chalmers were first offered in 1932 (an earlier option was available on the Model U). Rubber tires replaced steel wheels during the next 20 years. Rubber tires enabled higher field and transport speeds, increased tractive power and provided a more comfortable ride (Miller, 2000).

First wheeled diesel tractor tested at Nebraska.

Hydraulic rockshaft replaced the mechanical rockshaft in 1936 on the Deere Model A tractor (Miller, 2000). The need for powered lift prompted all manufacturers to provide this feature.

Interest in high-compression engines increased and in 1937 the Oliver 70 HC was the first tractor available with a higher compression ratio that could utilize the 70-octane fuel. This resulted in higher power for the same engine size and better fuel economy (Miller, 2000).

Outside styling consultants were hired to enhance tractor appearance in 1938 by International Harvester and Deere. IH used Raymond Loewy and Deere used Henry Dreyfuss Associates (Miller, 2000).

The weight-transfer hitch was introduced in 1938 on the Ford 9N tractor. It used the hydraulically operated rockshaft to increase or decrease traction by lifting or lowering the hitch (Miller, 2000).

Full electric equipment was adopted.

1940-1949

Live power take-off was introduced.

LPG engines burned petroleum gases consisting of butane and propane. Minneapolis-Moline marketed the first LPG tractor engine in about 1941. LPG is a clean burning fuel and extended engine life. Use continued through the 1960s (Miller, 2000).

Remote hydraulic cylinders for mounted and drawn implements were adopted and standardized by SAE and ASAE, accelerating their adoption and use (Miller, 2000).

First LPG tractor tested at Nebraska.

Number of lawn and garden tractors expanded rapidly.

1950-1959

Power of tractors increased rapidly.

Percentage of diesel tractors increased.

Large four-wheel drive tractors were developed for the wheatland areas of North America.

Last LPG tractor tested at Nebraska.

Refinements such as power steering, automatic transmissions, and transmissions with greater speed selections became widely available.

1960-1969

Power of tractors continued to increase.

Except for the smallest tractors in the United States, all tractors now have diesel engines.

Much more emphasis was placed on operator comfort and safety.

Full power-shift transmissions became available.

Planetary final drives for the final transmission reduction in large tractors (Miller, 2000).

Ergonomic seat and suspension designs resulted in improved comfort and adjustability for operators (Miller, 2000).

Radial-ply tractor tires became available.

The number of power tillers started to expand in Japan.

1970-1979

Turbocharger and intercoolers were added to diesel engines.

Roll-over protective structures (ROPS) became available and, beginning October 15, 1976, were required on all new tractors sold for use by employees in the United States.

Most large tractors were equipped with cabs.

Nebraska tractor test included sound level measurements in 1970.

Four-wheel drive increased in popularity.

Last gasoline tractor tested at Nebraska in 1978; since then, only diesel tractors have been tested.

Percentage tractors of more than 75 kW continued to increase in North America.

1980-2000

Tractors equipped with electronic sensing and control systems for improved performance

Nebraska tractor tests start testing to meet requirements of the Organization for Economic Cooperation and Development (OECD) codes that are internationally accepted (Bashford, 2000).

First Nebraska lift test was published in 1984 (Bashford, 2000).

Number of maintenance, landscape and garden tractors expanded rapidly.

Introduction of rubber-belted track agricultural vehicles.

Tractor size and power have appeared to reach upper limit of 300 kW.

1.3 Types of Off-Road Vehicles

Early tractors, essentially a substitute for draft horses, were first powered by a steam engine. The internal combustion engine came later. The first tractors were large and cumbersome and were suited mainly for plowing and threshing. As time passed, the tractor was developed for many more purposes. Adaptations were made for its use as a motor cultivator, and later came the general-purpose tractor to perform major farm operations.

The many uses, adaptations, and refinements of the tractor have resulted in the gradual evolution of several recognized classifications. One descriptive classification, listed below, is based on steering method, the arrangement of the frame, and traction members (see Figures 1.1 to 1.9).

- Crawler
- Standard row-crop
- High-clearance
- Utility
- Orchard
- Lawn and garden
- Multipurpose
- Power tiller
- Tree skidder
- Skid-steer loader
- Four-wheel drive with smaller front steering wheels
- Four-wheel drive with equal-sized wheels and articulated frame steering

The *crawler* tractor originally introduced in the 1930s had steel tracks. In the early 1990s Caterpillar introduced the Challenger series with a rubber-belted track drive (Figure 1.1). It is designed for row-crop and tillage operations.

The *standard* tractor (Figure 1.2) was developed primarily for traction. It is characterized by a drive through the two rear wheels, with center of gravity located at approximately one-third the wheelbase ahead of the rear axles. *Row-crop* tractors, in addition to being suitable for traction work, are especially adapted for use on row crops, allowing quick attachment of various tillage and cultivation tools as well as implements for such tasks as mowing and handling forages.

Figure 1.1. Crawler tractor. CAT® Challenger Model 75E with rubber belted tracks. Power ranges from 131 to 306 kW. (Reprinted with permission from Caterpillar, Inc.)

Figure 1.2. Standard, or row-crop tractor. Power ranges from 56 to 82 kW. (Reprinted with permission from New Holland North America, Inc.)

Figure 1.3. Utility or light industrial tractor. Power ranges from 24 to 54 kW. (Reprinted with permission from Kubota Tractor Corporation.)

Figure 1.4. Lawn and garden tractor. Power ranges from 5 to 15 kW. (Reprinted with permission from Areins Company.)

High-clearance tractors are usually modified row-crop tractors designed especially for crops that require extra clearance, such as cane.

Utility tractors (Figure 1.3) generally have less clearance than standard or row-crop tractors. They are used for many jobs on the farm or ranch and are often equipped with a front loader. A typical use for such a tractor is cleaning a feedlot.

An *orchard* tractor differs little from a utility tractor except that it is a few centimeters lower. The ROPS standard by the Occupational Safety and Health Administration (OSHA) does not apply, and therefore the tractor is not equipped with a rollover protective structure.

Lawn and garden tractors (Figure 1.4) have a power output of less than 15kW and are primarily designed for the care of large lawns. They can tow or carry a lawn mower, a sweeper, a snow blower, a bulldozer blade, and many other attachments. Because these tractors have very low clearance and are designed for light-duty use, they are not generally used for agricultural production.

A *power tiller* is a two-wheeled tractor guided by hand and used commonly in the rice-growing region of Japan and Southeast Asia. The power tiller is equipped with a rotary tiller and powered by a horizontal single-cylinder engine of 5 to 12 kW.

The *tree skidder* (Figure 1.5) is a four-wheel-drive tractor developed especially for moving tree trunks out of the forest to an area where they can be loaded onto trucks. The tree skidder is a combination of an agricultural and an industrial tractor.

The *skid-steer* tractor (Figure 1.6) was designed especially for industrial use, but it is often used in agriculture when turning room is restricted, such as in a dairy or fruit-storage building. The name "skid-steer" comes from the method of turning, which is exactly like that of a crawler. The wheels on one side are braked or reversed, causing the tractor to skid in order to turn.

A *four-wheel-drive tractor with smaller front wheels* (Figure 1.7) is simply a standard or a row-crop tractor with the front wheels also being driven. In regard to price and traction, this type of tractor falls between the standard and the four-wheel-drive tractor with equal-sized wheels (Figure 1.9). This tractor has become particularly popular in Japan because of its excellent steering and traction characteristics in soft, wet rice fields.

Four-wheel-drive, or simply 4WD, tractors have been developed to produce more drawbar power. The size of 4WD tractors varies in the United States and Canada from 100 kW to more than 300 kW. In Europe, 4WD tractors may be as small as 15 kW and are used especially in vineyards. Four-wheel-drive tractors can be steered by pivoting the tractor in the center (frame steer) or steering the wheels (as in Figure 1.8).

The complexity of a tractor is illustrated by Figure 1.9, which shows such features as power steering, epicyclic gear reduction in the front wheels, wet disk brakes, power-shift transmission, and a ROPS cab.

Figure 1.5. Tree skidder. This vehicle has four-wheel drive. Power ranges from 75 to 150 kW. (Reprinted with permission from Caterpillar, Inc.)

Figure 1.6. Skid-steer tractor. This type of tractor has four-wheel drive. Power ranges from 27 to 82 kW. (Reprinted with permission from GEHL Company.)

Figure 1.7. Four-wheel drive tractor with smaller steering wheels in front. Power ranges from 70 to 190 kW. (Reprinted with permission from Deere and Co.)

Figure 1.8. Four-wheel drive tractor with equal-sized wheels, articulation and front-wheel steering. Power ranges from 20 to 300 kW. (Reprinted with permission from Case IH.)

Figure 1.9. Phantom view of a off-road vehicle showing the design complexity. (Reprinted with permission from Deere and Co.)

1.4 Some Global Variations in Tractors

Diverse geographical, climate and use factors require some variations, so that:

- Tractors made and used in Japan are usually equipped with riceland tires.
- The power-to-mass ratio is greater for Japanese tractors than for others.
- European tractors more commonly use radial-ply tires for traction.
- Tractors made outside of North America may have as many as four different speeds on the power take-off.
- In North America, 4WD is used mainly on tractors over 100 kW; in Japan, however, tractors, tractors as small as 10 kW may have front-wheel-assisted 4WD.
- In Europe, the United Kingdom, and Japan, 4WD with front wheels smaller than the rear is common, whereas almost all 4WD tractors in North America have equal-sized wheels.
- Crawler tractors are more popular in Europe and the United Kingdom.
- Power tillers are quite popular in Japan and Southeast Asia, primarily because of the small rice farms and soft, wetland conditions found there.

1.5 Functions of Off-Road Vehicles

Today's off-road vehicle is a complex machine used to propel and power a large variety of implements for agricultural production. Therefore, implement applications have considerable effect on vehicle design. The engineer must be aware of implement requirements to properly weigh such design considerations as crop clearance, wheelbase, width of tractor, power-to-mass ratio and mass distribution, as well as to provide placement of controls to satisfy the majority of operations. Every design will necessarily include some compromises.

Implements are generally attached to and operated by tractors in one of four ways:

- Towed, single-point hitch
- Mounted, three-point hitch
- Semi-mounted, three-point hitch
- Frame mounted

Power may be transmitted to the implement by a power take-off shaft or by oil-hydraulic hoses. Each new tractor provides many engineering problems related to proper drawbar height, provision for sufficient horizontal and vertical angular movement, and implement clearance.

The widespread adoption of hitch-mounted and frame-mounted equipment is probably the result of one or both of two primary factors: decreased investment cost to the farmer and increased versatility. In addition, the extra weight transfer of mounted implements adds significantly to the tractive ability of the tractor.

1.6 Sources of Energy

One of the characteristics of a technically advancing society, at least before 1978, is that the rate of growth of energy use is considerably higher than the rate of population growth and closely parallels the growth of economic output (Sporn, 1957). Possession of adequate energy is a requisite for any kind of advanced civilization, for if human beings must depend strictly on their own muscles, it will take all of their mental and physical strength to supply the bare necessities of life. The difference in per capita energy consumption is closely related to differences in standards of living. The important sources of energy are:

- Solar energy, direct
- Solar energy, indirect

    Fossil fuels: oil, natural gas, coal, peat, shale and tar sands

    Biomass: wood, corncobs, etc.

    Wind

    Tides

    Water

- Nuclear energy
- Geothermal energy

Our present age may sometime be called the fossil fuel age since most of our present power use is coming from stored fossil fuels–oil, gas, and coal. How long these supplies will last is debatable, but they are nonrenewable. Farming and other off-road operations in the 21st century depend on petroleum fuels, so it is of interest to know when the supply will begin to diminish. The date is uncertain, but in the United States, for example, production of petroleum has not met demand since 1947 (Scarlott, 1987). In 1973, the United States imported about 40% of its oil needs (Ray, 1973), and in 2000, more than 50% was imported. It has also been predicted that we may have nearly reached the year of peak production, after which crude oil production will begin to decline.

Oil shales, from which it is possible to obtain nearly a barrel of oil per ton, are available in considerable but indeterminate amounts. For power plant use, some attention is being given to recovery of the oil by methods other than mining.

Solar energy is our one inexhaustible energy source. A rough measure of energy falling on the earth is a maximum of approximately 940 W/m^2 and an average of 630 W/m^2. Although this energy source will likely become very useful for space heating and for agricultural operations such as crop drying, at present it is doubtful that it will be useful for operating tractors because of the low concentration of energy and a lack of suitable means of collecting and concentrating it. Another problem is the large difference in the amount of available solar energy at different times of the year or day and at different locations.

Nuclear energy could ease the load on fossil fuels, as it is well adapted to large power-plant operations where radiation shielding can be employed. It is a concentrated, clean, and easily controlled energy source; the energy in 4 ml of uranium 235 being equivalent to the energy stored in 140 m^3 of coal ("Nuclear Energy," 1961).

Controlled fusion, the process of combining the lighter elements rather than breaking up the heavy atoms as in the fission process, may offer some hope for future power plants, even the farm tractor. Fusion atomic energy does not require heavy shielding since it is not radioactive.

1.7 Energy Conversion Devices

Common techniques for the conversion of energy include:

- Human and animal energy
- Piston engines
- Compression ignition (diesel)
- Spark ignition (gasoline)
- Gas turbine
- Rotary (Wankel) engine
- Free-piston engine
- Sterling engine
- Steam piston engine
- Steam turbine engine
- Thermoelectricity
- Fuel cell
- Solar cell
- Electric motor and electric generator
- Storage battery
- MHD (magnetohydrodynamic) devices

The second law of thermodynamics states that conversion of heat to work is limited by the temperatures at which conversion occurs. For a cycle with a high temperature T_1 and a low temperature T_2 the maximum efficiency, usually called the *Carnot cycle efficiency*, is

$$e = \frac{T_1 - T_2}{T_1}$$

Thus the Carnot cycle establishes an inherent limit to the maximum useful conversion of energy that may be obtained from a fuel used in a power plant, such as in a tractor engine. In the best tractor engines only about one-third of the heating value of the fuel is converted into useful work.

Conventional piston engines have hundreds of parts, and the reciprocating units must be stopped and started thousands of times per minute. Nevertheless, they are highly successful units.

Rotary engines are of interest because of their simple design. Inside the combustion chamber is a three-lobed rotor with curved sides. The shape of the rotor and the chamber causes the rotor to move in an eccentric orbit as it rotates. All three corners of the rotor stay in contact with the walls of the chamber at all times. Vanes, which correspond to piston rings in the conventional engine, seal the chambers from loss of gas. As the rotor turns, its corners form three moving cavities, which in turn go through the usual cycle of intake, compression, power, and exhaust. The output shaft turns at three times the speed of the rotor in this example. The power-to-mass ratio is greater than 1.0, which is not an advantage on a tractor. The major mechanical problem, the rotating seals, has been solved well enough to include the Wankel in production line automobiles but not yet in industrial-type diesel engines. The Wankel engine so far is less efficient than conventional spark-ignition engines used in tractors.

The simplicity and reliability of a gas-turbine engine plus its high power-to-weight ratio makes it a desirable power unit for aircraft. Its high power-to-weight ratio is, however, not an asset on a tractor used for the usual agricultural or industrial purposes. The efficiency of gas-turbine engines, when the power is transmitted through a shaft, is roughly one-half to two-thirds that of diesel engines. The efficiency compares more closely with gasoline-burning spark-ignition engines.

The principle of operation of a gas-turbine engine has much in common with the conventional piston-type internal combustion engine. The events that take place in a piston engine (intake, compression, ignition, power, and exhaust) also take place in a gas-turbine engine. Gas turbine engines run at a relatively constant rpm, which also restricts their use on a tractor in which variations in both speed and torque are required.

1.8 The Future Engine

The piston engine, especially the diesel is well established (Amann, 1974) and is not likely to be replaced immediately with other types of energy conversion systems. However, development of new engine materials has progressed considerably. Ceramic materials, which have high heat resistance, as well as high anticorrosion and antiwear properties, have indicated their potential applications to piston engines, permitting operations at higher temperatures and thus higher thermal efficiency compared to conventional metal-based engines. Although their brittleness has been a limitation to their wide application for engine structures, recent development of ultra-plastic fine ceramics may open a new era for wide usage in piston engines. Complete replacement of engine parts with ceramics is not likely to occur soon, but applications of ceramics to specific engine parts including turbochargers will. Tractor and engine manufactures continue to use better engine materials and to develop new types of energy conversion devices.

1.9 Trends in Tractor Design

The design and performance of tractors have changed considerably since the first Nebraska tractor tests. The future tractor design greatly depends on technological advancement and agricultural and economic development, including energy and environmental systems (Cragle, 1983). From Figure 1.10, it is clear that tractors continue to have a greater power-to-mass ratio, which results in the tractors traveling faster, since improvements in traction have not keep pace with the increase in power. It is also clear that the power output of tractors has continued to increase. The growth, however, in power and weight of tractors will probably not continue at the same rate as in the past. One reason is the failure of farm sizes to grow because of the decrease in the availability of land, as well as market limitations and the instability of export opportunities. The trend of farm size stabilization is now clearly noted in most European countries (Gohlich, 1984). Another reason for limiting the increase in size of future tractors is increasing soil compaction that affects soil conservation as well as plant growth. Detailed discussion on the future trends in tractor development will not be made here since it is beyond the scope of this book. However, Gohlich (1984) presents an excellent discussion on this subject. He concludes that:

- More emphasis will be concentrated on reducing the total weight of the tractor-implement system in order to minimize the ground pressure and soil compaction, particularly during seedbed preparation.
- Front- and rear-mounted implements will contribute increasingly to better, more convenient, and more economical field operations.
- Instead of the standard heavy tractors today, for most of the required field operations a lighter, second tractor power unit with low-pressure tires, which would transfer the majority of its power through the power take-off (PTO) rather than through the ground drive, may be a desirable approach on the future farm.
- The optimization of power transfer will be facilitated by driver information displays as well as by control circuits that may eventually take over some present manual functions.
- Components for the driver's comfort and safety will be further improved; thereby helping to ensure that farming remains a satisfying job for the farm worker.

1.10 Engineering Standards

Design engineers utilize engineering standards in vehicle design, development, and evaluation. Sources of engineering standards, engineering practices and data are: the American Society of Agricultural Engineers (ASAE), the Society of Automotive Engineers (SAE), and the International Standards Organization (ISO). Standards are cited and used in several of the chapters in this book.

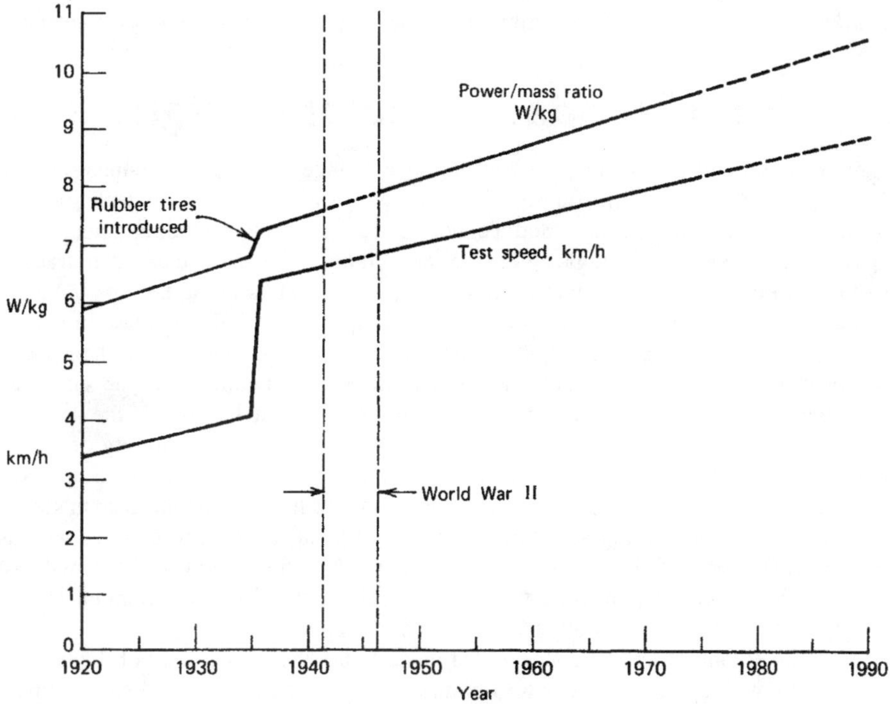

Figure 1.10. Trend of changes in tractor sped and power (Yahva and Goering, 1977).

1.11 Chapter Summary

Development and use of off-road vehicles, primarily the agricultural tractor, played a major role in the agricultural mechanization achievements of the 20th century. The Nebraska Tractor Test Law, passed in 1919, specified that every tractor sold in the state of Nebraska should be tested and the results published. The tests, which attained worldwide recognition, have provided standards for rating tractors.

Traction power, belt power, power take-off drives, mounted tools, hydraulic remote-control units, electronic sensing and control of machine performance, and ergonomically designed operator workplace and climate-controlled cabs have extend the usefulness and efficiency of the tractor for agricultural and other off-road uses.

Today's off-road vehicle is complex and is being used to propel and power a large variety of implements for agricultural production. Implement applications, therefore, have considerable effect on vehicle design, and the engineer must be aware of implement requirements to weigh properly such design considerations as crop clearance, wheelbase, width of tractor, power-to-mass ratio and mass distribution, as well as to provide placement of controls to satisfy the majority of operations.

Agricultural and other off-road operations in the 21st century depend on petroleum fuels, so it is of interest to know when the supply will begin to diminish. The date is uncertain, but in 1973 the United States imported about 40% of its oil needs (Ray, 1973), and in 2000, more than 50% was imported. *Solar energy* is our one inexhaustible energy source, but it is doubtful that it will be useful for operating tractors because of the low concentration of energy and a lack of suitable means of collecting and concentrating. *Controlled fusion* may offer some hope for future power plants, even those as small as a farm tractor, since fusion atomic energy does not require heavy shielding because it is not radioactive.

The piston engine, especially the diesel is well established (Amann, 1974) and is not likely to be replaced immediately with other types of energy conversion systems. However, development of new engine materials has progressed considerably. Ceramic materials, which have high heat resistance, as well as high anticorrosion and antiwear properties, have indicated their potential applications to piston engines, permitting operations at higher temperatures and thus higher thermal efficiency compared to conventional metal-based engines. Future tractor design greatly depends on technological advancement and agricultural and economic development, including energy and environmental systems (Cragle, 1983).

Design engineers utilize engineering standards, engineering practices and data in vehicle design, development, and evaluation.

References and Suggested Readings

Amann, C.A. 1974. Why the piston engine lives on. *Machine Design* (21 February).

Bashford, L.L. 2000. Tractor tests. *Resource* 7(4): 9-10.

Baumheckel, R., and K. Borghoff. 1997. *International Harvester Farm Equipment Product History 1831-1985.* St. Joseph, MI: ASAE.

Caterpillar Inc., Product Research and Agricultural Business Unit Engineers. 1995. In pursuit of the ideal tractor: Development of the rubber belt tractor. ASAE Distinguished Lecture Series No. 20. St. Joseph, MI: ASAE.

Cragle, R.G. 1983. Trends which will impact agricultural technology in the next decade. SAE Technical Paper Series No. 831268. International Off-Highway Meeting and Exposition of the Society of Automotive Engineers, 12-15 September. Warrendale, PA: SAE.

Erb, D., and E. Brunbaugh. 1993. *Full Steam Ahead J. I. Case Tractors and Equipment 1842-1955.* St. Joseph, MI: ASAE.

Farm Implement News, 25 Dec. 1952; 25 Sept. 1953; 25 Nov. 25 1953. Origin of the term of tractor.

Gay, L. 1995. *Farm Tractors 1975-1995.* St. Joseph, MI: ASAE.

Göhlich, H. 1984. The development of tractors and other agricultural vehicles. *Journal of Agricultural Engineering Research* 29(1): 3-16.

Inside ASAE. 2000. Agricultural mechanization named among top 20 engineering achievements. *Resource* 7(4): 17.

Larsen, L. 1981. *Farm Tractors 1950-1975.* St. Joseph, MI: ASAE.

Leffingwell, R. 2000. *Farm Tractor Milestones.* Osceola, WI: MBI Publishing Company.

Macmillan, D., and R. Jones. 1988. *John Deere Tractors and Equipment Vol. 1, 1837-1959.* St. Joseph, MI: ASAE.

Macmillan, D., and R. Harrington. 1990. *John Deere Tractors and Equipment Vol. 2, 1960-1990.* St. Joseph, MI: ASAE.

Miller, M.L. 1999. *Designing the New Generation John Deere Tractors.* St. Joseph, MI: ASAE.

Miller, M.L. 2000. A history of tractor development. ASAE Agricultural Equipment Technology Conference. Kansas City, MO. 23-25 February.

Power Engineering, January 1961. Nuclear energy.

Ray, D.L. 1973. The nation's energy future, WASH-1282. Washington, D.C.: U.S. Atomic Energy Commission.

Scarlott, C.A. 1987. Changing energy science. *Science Monthly* (May): 221ff.

Sporn, P. 1957. Energy requirements and the role of energy in an expanding economy. *Agricultural Engineering* (September): 657ff.

Stonehouse, T., and E. Brumbaugh. 1996. *JI Case Agricultural & Constructions Equipment 1956-1994, Vol. 2.* St. Joseph, MI: ASAE.

Swinford, N. 1994. *Allis-Chalmers Farm Equipment 1914-1985.* St. Joseph, MI: ASAE.

Swinford, N. 1999. *The Proud Heritage of AGCO Tractors.* St. Joseph, MI: ASAE.

Swinford, N. 2000. *A Century of Ford and New Holland Farm Equipment.* St. Joseph, MI: ASAE.

USDA. 1960. The Yearbook of Agriculture. Washington, D.C.: U.S. Department of Agriculture.

Yahva, R.K., and C.E. Goering. 1977. Some trends in fifty-five years of Nebraska tractor test data. ASAE Paper No. MC77503. St. Joseph, MI: ASAE.

CHAPTER 2

ENGINE PERFORMANCE MEASURES

2.1 Introduction

In this chapter, you will learn some important terminology and relationships for describing and calculating engine performance. These performance measures include torque, power, mean effective pressures, fuel-use efficiencies, and specific fuel consumption. Terminology of governed engines will be presented as a lead-in to a discussion of engine maps and how they can be used to maximize engine efficiency.

2.2 Power Flows in an Engine

Engines may be rated by power, but power varies greatly depending on where it is measured. Tractor manufacturers usually rate tractors by their PTO (power take-off) power but, since some large, 4WD (four-wheel drive) tractors do not have a PTO, the drawbar power may be given instead. Figure 2.1 illustrates power flows through an engine. In the discussion of engine performance below, lower case p is used to indicate pressure, while upper case P is used to indicate power.

The fuel is the source of the engine power. The power embodied in the fuel is called *fuel equivalent power*, P_{fe}, and is given by

$$P_{fe} = \frac{\dot{m}_f \, H_g}{3600} \tag{2.1A}$$

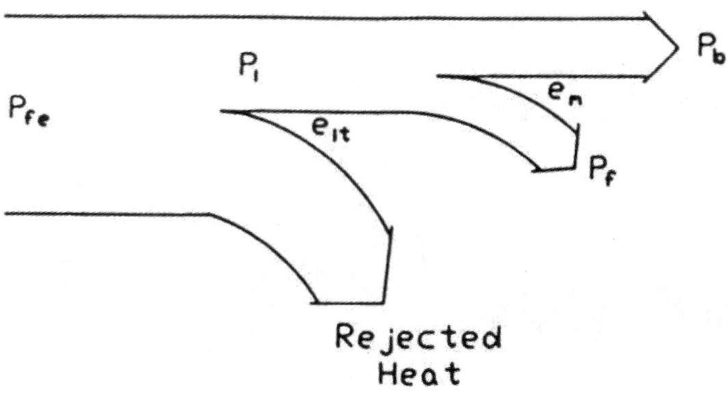

Figure 2.1. Power flows in an engine.

where

P_{fe} = fuel equivalent power, kW

$\dot{m}_f$ = fuel consumption rate, kg/h

H_g = gross (higher) heating value of the fuel, kJ/kg

Equation 2.1A is used when the fuel consumption is measured on a mass basis. If the fuel consumption is measured volumetrically, the following equation is used to calculate fuel equivalent power:

$$P_{fe} = q_f \rho_f H_g / 3600 \tag{2.1B}$$

where

q_f = fuel consumption rate, L/h

ρ_f = fuel density, kg/L

For example, an engine consuming 30 kg/h of No. 2 diesel (H_g = 45,000 kJ/kg) would have a P_{fe} of 375 kW. The heating value of the fuel is measured by a calorimeter, as will be discussed in the chapter on fuels. More than half of the fuel equivalent power is lost and unavailable for useful work. In the paragraphs below, we trace the power through the engine to identify the losses.

Burning of the fuel produces high pressure on each piston which, when multiplied by the piston area, generates a force. Although the pressure varies throughout the piston stroke, it is possible to calculate an *indicated mean effective pressure*, p_{ime} (sometimes also shown as *imep*). The procedure will be explained a little later. Multiplying the force by the piston stroke gives the work per engine cycle (in the typical, four-stroke cycle, an engine cycle occurs each two revolutions of the crankshaft). Finally, multiplying by the number of cycles per unit time, i.e., the crankshaft speed, gives the *indicated power*, as

$$P_i = \frac{p_{ime} D_e N_e}{120{,}000} \tag{2.2}$$

where

P_i = indicated power, kW

p_{ime} = indicated mean effective pressure, kPa

D_e = engine displacement, L

N_e = crankshaft speed, rpm

The number in the denominator is the product of 2 revs/cycle times 60 sec/min times 1000 W per kW. Equation 2.2 also works for a two-stroke cycle engine, except that the number in the denominator would be 60,000 instead of 120,000. The engine displacement, D_e, of the engine is calculated as

$$D_e = \frac{A_p\, L\, n}{1000} \tag{2.3}$$

where

A_p = top area of piston (cm^2) = $\pi\, d^2/4$ (d = piston diameter, cm)

L = stroke length, cm

n = number of cylinders in the engine

Continuing the above example, suppose the engine that consumed 375 kW of fuel equivalent power was a six-cylinder engine with a bore of 11.67 cm, a stroke of 12.0 cm, and the crankshaft speed was 2200 rev/min. Also, the p_{ime} was 1200 kPa. Using Equation 2.3, the displacement would be 7.7 L. Using Equation 2.2, the indicated power would be 169.4 kW.

Equation 2.2 shows that, for any engine, there are only three possible ways to increase the indicated power, i.e., the power at the head of the pistons. These are:

1. Increase the engine size (D_e), or
2. Increase the engine speed (N_e), or
3. Increase the internal pressure and consequently the stress on the engine.

The indicated power is not immediately available for useful work; rather, it must first be transmitted to the flywheel of the engine. Therefore, interest is in the *flywheel power* of the engine that, in the early days of engine testing, was measured with a device called a *prony brake.* Thus, flywheel power is commonly referred to as *brake power*, P_b. The derivation of the brake-power equation begins with a consideration of how work per crankshaft revolution relates to crankshaft torque.

Consider applying a force, F, to a wrench at a radius of R from the center of a shaft that is to be turned by the wrench. The torque applied to the shaft thus becomes FR. In traveling one full revolution, the force travels a distance equal to the circumference of the circle, 2πR. The work per revolution then is 2πRF, or 2πT, where T is the shaft torque. Thus, the work per revolution is simply 2π times the shaft torque. Multiplying the work per revolution by the number of revolutions per unit of time (i.e., the shaft speed) gives the shaft power. Thus, the brake power is given by the following equation:

$$P_b = \frac{2\pi\, T_b\, N_e}{60{,}000} \tag{2.4}$$

where

P_b = brake (flywheel) power, kW

T_b = engine brake torque, N·m

N_e = engine speed, rpm

The significance of the term, *brake torque*, will be explained a little later. The 60,000 in the denominator is simply a units constant. Continuing the above example, assume the brake torque was 625 N·m. Then Equation 2.4 shows that the brake power of the example engine was 144 kW.

For the example, the engine consumed 375 kW of fuel equivalent power but had only 169.4 kW of indicated power at the head of the pistons and only 144 kW of brake power at the flywheel. The loss, 375 – 169.4 = 205.6 kW, is primarily caused by thermodynamic limitations imposed by the second law of thermodynamics. However, the loss, 169.4 – 144 = 25.4 kW, is due to friction losses in the engine. *Friction power* is given by

$$P_f = P_i - P_b \tag{2.5}$$

where P_f = friction power, kW.

Thus, by definition, friction power includes any part of the indicated power that is not delivered to the flywheel. In our example engine, the friction power would be 25.4 kW. What is included in friction power? It includes friction in the rings, bearing, etc. However, it also includes power to run the oil pump, cooling fan, alternator, and, for an air-conditioned vehicle, power to run the compressor. A practical consequence of Equation 2.5 is that, during an official test, turning off the air conditioner reduces the friction power and adds to the brake power.

2.3 Mean Effective Pressures

The term, indicated mean effective pressure, came about at follows. In the early days of engine testing, a cylindrical chart was attached to the front of the engine and driven at crankshaft speed. An inked stylus, driven by a pressure transducer in the engine cylinder, produced a graph of cylinder pressure versus crankshaft rotation. Such a diagram was called an "indicator diagram." By knowing the ratio of connecting rod length over crank throw radius, it was possible to convert the indicator diagram to a p-v diagram similar to the one in Figure 2.2. The area within the p-v diagram, divided by the base width of the diagram, gives the indicated mean effective pressure. The base width of the diagram is simply the displacement of a single cylinder, D_c. More modern methods have replaced the use of the indicator diagram in preparing p-v diagrams, but the term *indicated mean effective pressure* is still used in engine theory. In modern testing, a pressure transducer continuously measures the cylinder pressure while an encoder measures the crankshaft rotational position. The data are fed into a computer that calculates the p_{ime}.

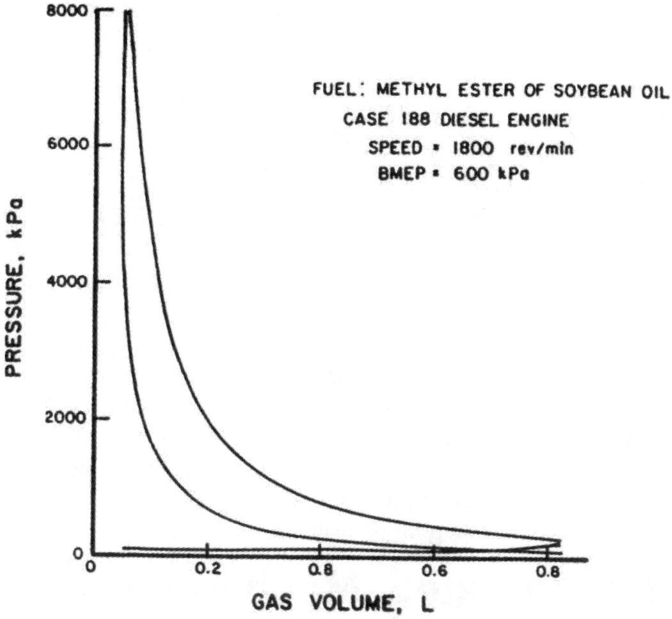

Figure 2.2. Engine p-v diagram and indicated mean effective pressure.

For an engine of displacement D_e running at a given speed, N_e, Equation 2.2 shows that the indicated power varies proportionally with the indicated mean effective pressure. Engine analysts have used a variation of Equation 2.2 to develop other mean effective pressures. Thus, a *brake mean effective pressure*, p_{bme} or bmep, can be calculated by

$$p_{bme} = \frac{120{,}000\, P_b}{D_e\, N_e} \tag{2.6}$$

where p_{bme} = brake mean effective pressure, kPa.

The p_{bme} does not exist as an actual pressure in the engine and thus cannot be measured. It can be calculated using Equation 2.6 if the engine displacement, speed and brake power are known. By combining Equations 2.4 and 2.6 and simplifying, the following equation can be derived:

$$T_b = \frac{D_e\, p_{bme}}{4\pi} \tag{2.7}$$

The denominator in Equation 2.7 would be 2π for a two-stroke cycle engine. Brake mean effective pressure is often referred to as *specific torque*; it is proportional to brake torque but is independent of the engine displacement. For our example engine, either Equation 2.6 or Equation 2.7 shows the bmep would be 1020 kPa.

A variation of Equation 2.2 is used to calculate *friction mean effective pressure*, i.e.,

$$p_{fme} = \frac{120{,}000\,P_f}{D_e\,N_e} \tag{2.8}$$

where p_{fme} = friction mean effective pressure, kPa.

For a compression-ignition (CI) engine, the p_{fme} is primarily a function of engine speed, i.e.,

$$p_{fme} = A_0 + A_1\left(\frac{N_e}{1000}\right) + A_3\left(\frac{N_e}{1000}\right)^2 \tag{2.9}$$

where A_0, A_1, and A_2 are constants for a specific engine.

For most CI engines, A_1 is negative, while A_0 and A_2 are positive. Thus, p_{fme} tends to increase with the square of engine speed and, from Equation 2.8, friction power tends to increase with the cube of engine speed. For our example engine, Equation 2.8 shows the friction mean effective pressure would be 180 kPa. For throttled, spark-ignition (SI) engines, the equation for p_{fme} is more complex.

2.4 Indicated, Mechanical, and Brake Thermal Efficiencies

We have seen how power is lost at various points in an engine. Several efficiency equations have been developed to determine how well engines avoid such power losses. The *indicated thermal efficiency* is defined as

$$e_{it} = \frac{P_i}{P_{fe}} \tag{2.10}$$

where e_{it} = indicated thermal efficiency, decimal.

The *mechanical efficiency* is defined as

$$e_m = \frac{P_b}{P_i} \tag{2.11}$$

where e_m = mechanical efficiency, decimal.

The overall, or *brake thermal efficiency* is defined as

$$e_{bt} = \frac{P_b}{P_{fe}} \tag{2.12}$$

As the names show, the indicated thermal efficiency is a measure of the combustion process in converting the power in the fuel to mechanical power at the head of the pistons. The mechanical efficiency is a measure of what fraction of that mechanical power is transmitted to the flywheel. In the earlier example, turning off the vehicle air conditioner would increase the mechanical efficiency. Finally, to have a high overall or brake thermal efficiency, the engine must have both a high indicated thermal efficiency and a high mechanical efficiency. For our example engine, e_{it} = 0.452, e_m = 0.85 and e_{bt} = 0.384.

2.5 Specific Fuel Consumption

Fuel consumption, either in kg/h or L/h, is not a good measure of engine performance; such fuel consumption depends not only on the engine efficiency, but also on the engine displacement and load. Thus, *specific fuel consumption* (SFC) has been developed as a measure of engine performance. Specific fuel consumption is defined as

$$XSFC = \frac{\dot{m}_f}{P_X} \tag{2.13}$$

where XSFC = specific fuel consumption, kg/kWh.

The X must always be specified when using the term, specific fuel consumption. Thus, ISFC, or *indicated specific fuel consumption*, is calculated by dividing $\dot{m}_f$ by the indicated power. BSFC, or *brake specific fuel consumption*, is calculated by dividing $\dot{m}_f$ by brake power. Both ISFC and BSFC are used widely in the literature on engine theory. For a tractor, it would also be useful to have specific fuel consumption terms associated with PTO power and drawbar power. Such terms are rarely seen in the literature but, in this book, we will use PSFC to indicate *PTO specific fuel consumption* and DSFC to indicate *drawbar specific fuel consumption.* In general, XSFC can be interpreted as the amount of fuel required per unit of work accomplished. Lower XSFC values correspond to more efficient operation. The XSFC is related to efficiency by

$$XSFC = \frac{3600}{H_g\, e_{xt}} \tag{2.14}$$

where I, B, D, or P is substituted for X. Thus, as Equation 2.14 shows, the higher the efficiency, the lower the specific fuel consumption. For our example engine, ISFC = 0.177 kg/kWh and BSFC = 0.208 kg/kWh.

2.6 Engine Speed Control

Virtually all CI engines and some SI engines are equipped with governors to automatically control engine speed. As early as 1769, James Watt's steam engine included a mechanical governor of the type illustrated in Figure 2.3c. Use of mechanical governors continued on some engines, but electronic governors offer advantages for larger engines. Much of the terminology relating to governed engines came from mechanical governors but also applies to engines with electronic governors. Figure 2.3 will be used to explain governor action.

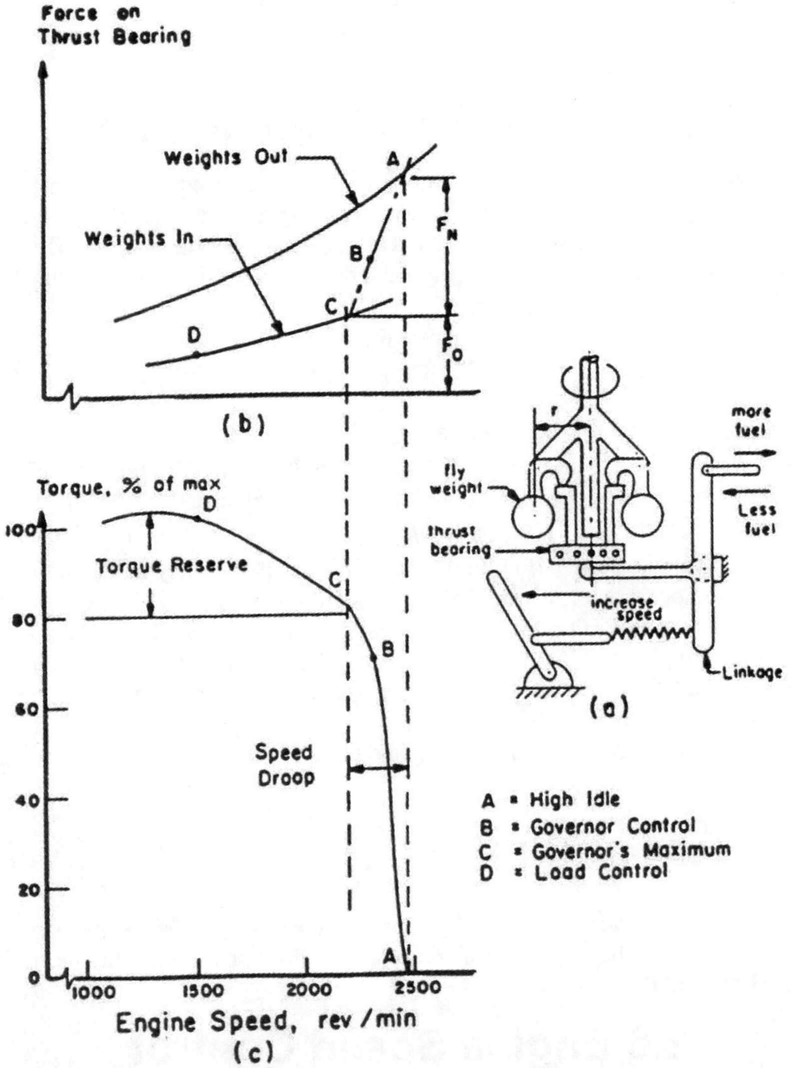

Figure 2.3. Illustration of governor action.

Mechanical governors of the type shown in Figure 2.3a can be a separate unit or, on CI engines, can be included in the fuel injector pump housing. In either case, the governor shaft speed is proportional to the crankshaft speed of the engine. As engine speed increases, increased centrifugal force on the hinged flyweights causes them to swing outward, thus pushing downward on the thrust bearing and causing the linkage to rotate counterclockwise and stretch the governor spring. The fuel control rod then moves to decrease fuel delivery to the engine. Conversely, if the engine speed falls, the spring causes the linkage to rotate clockwise, pushing the thrust bearing upward and forcing the flyweights to move inward. The rotating linkage also moves the fuel control rod to increase fuel delivery to the engine. All mechanical governors include a thrust bearing because the governor flyweights are attached to a rotating governor shaft, while the pivot of the linkage is stationary; therefore, the bearing accommodates the relative motion between the rotating and non-rotating parts of the governor. The fuel control rod could change fuel delivery in a number of ways. In some SI engines, the rod would control fueling by opening or closing the throttle plate in a carburetor. In CI engines, rod movement would change the delivery of a fuel injection pump. Fuel injection pumps are discussed in Chapter 7.

When the governor is in equilibrium, the force imposed on the thrust bearing by the flyweights is equal to the force imposed by the governor spring. It is helpful in understanding governor action to calculate the force on the thrust bearing. From Newton's second law, the force on the thrust bearing is

$$F_{tb} = K\,r\,N_g^2 \tag{2.15}$$

where

K = a units constant also incorporating mechanical advantage of flyweight linkage

r = radius from center of governor shaft to center of flyweights

Plotting thrust-bearing force versus engine speed (Figure 2.3b) leads to the problem that the radius, r, is changing as speed changes. Two limiting cases have been plotted in the figure, one for weights in (minimum r) and one for weights out (maximum r).

Now consider an engine running without load but with substantial fuel delivery. The governor prevents the engine from accelerating until it destroys itself; as speed increases, the flyweights swing to the full-out position to reduce the fuel delivery and limit the engine speed to that shown at Point A of Figures 2.3b and 2.3c. Point A is called the *high idle point* because the speed is high and the engine is idling, i.e., not doing any work. Now imagine that increasing torque load is placed on the engine; the resulting decline in speed causes the flyweights to move inward, thus increasing fuel delivery to the engine to accommodate the increased load. When the torque reaches and exceeds that shown at Point C in Figure 2.3c, the governor flyweights would be at their innermost position and the linkage would not be able to move the fuel control rod any farther. Point C is called the *governor's maximum point* because the governor is unable to affect fuel delivery beyond that point. To the left of Point C in Figure 2.3c, the speed is controlled entirely by the amount of torque load placed on the engine, i.e., higher loads cause lower engine speeds. Figure 2.3c is called the *engine map*. The area to the right of Point C on the map is called the *governor-control region*, while the area to the left is called the *load-control region*. Although a perfect governor would maintain constant speed in the governor-control region, actual governors permit some speed variation. *Governor regulation*, as calculated using Equation 2.16, is a measure of how closely the governor can maintain constant speed.

$$\text{Reg} = 200\left(\frac{N_{HI} - N_{GM}}{N_{HI} + N_{GM}}\right) \tag{2.16}$$

where

Reg = governor regulation, %

N_{HI} = high idle speed, rev/min

N_{GM} = speed at governor's maximum, rev/min

The lower the regulation value, the more closely the governor is able to control the engine speed. High-quality mechanical governors can achieve regulation of 6% or less, while electronic governors can achieve regulation as low as 1%.

Figure 2.3 shows a mechanical governor holding the engine speed within the range identified as speed droop in Figure 2.3c. However, the operator could select a different speed-control range by use of the lever in Figure 2.3a. For example, moving the lever to the left would further stretch the governor spring, increasing the force on the thrust bearing and requiring a higher engine speed to move the weights outward. The result would be that the governor control range would move to the right on Figure 2.3c, i.e., the engine would operate at higher speeds in the governor-controlled range. Conversely, moving the speed-control lever to the right would reduce the spring tension and the engine would operate at lower speeds in the governor-controlled range.

One other feature in Figure 2.3c is of interest; the increase in torque between torque at governor's maximum and that at peak torque is known as *torque reserve*. Torque reserve is normally expressed as a percentage of the *rated torque*. Normally, governed engines are rated at the governor's maximum point when the speed control lever is set for maximum engine speed. Torque reserve of CI engines ranges from less than 10% up to 50% but is typically higher than 25%.

2.7 Engine Torque Generation

Equations 2.1A, 2.2, 2.4, 2.5, 2.7 and 2.11 can be combined and simplified to form the following set of equations:

$$T_b = T_i - T_f \tag{2.17}$$

where

T_b = brake torque, N·m

T_i = indicated torque, N·m

T_f = friction torque, N·m

T_i and T_f are defined as

$$T_i = \left(\frac{H_g\, e_{it}}{0.06\pi}\right)\left(\frac{\dot{m}_f}{2N_e}\right) \tag{2.18}$$

and

$$T_f = \frac{D_e\, p_{fme}}{4\pi} \tag{2.19}$$

For a two-cycle engine, the denominator in Equation 2.19 would be 2π. In Equation 2.18, note that the second quantity in parentheses can be interpreted as the mass of fuel entering the engine per engine cycle, while the first quantity in parentheses has little variation over the engine map. Thus, the indicated torque is nearly proportional to the amount of fuel used per engine cycle. The friction torque varies proportionally with the friction mean effective pressure that, according to Equation 2.9, is a function of engine speed. In the governor-controlled range, the governor controls the amount of fuel delivered to the engine per engine cycle and thus controls the indicated torque; the engine speed and thus the p_{fme} and friction torque are nearly constant in the governor-controlled range. As the engine moves into the load-controlled range, the governor can no longer change the amount of fuel per engine cycle. As the engine speed falls in the load-controlled range, the friction torque drops and the brake torque increases accordingly. Thus, accessories that add to friction power at high speeds actually add to torque reserve as speed falls! However, if the fuel per cycle remained constant throughout the load-controlled range, the engine would have insufficient torque reserve. In fact, due to features of injection-system design, including increasing volumetric efficiency of the fuel-injection pump, the fuel per cycle actually increases gradually as the engine speed falls in the load-controlled range. The combination of decreasing friction torque plus some increase in fuel per cycle provides the total torque reserve of the engine.

2.8 Engine Performance Maps

Engine manufacturers plot engine performance maps as an aid to vehicle manufacturers in choosing suitable engines for their vehicles. Figure 2.4 is an example of an engine performance map. The torque-speed curve on the map is for the engine running with the governor set for maximum speed, i.e., it is the *torque envelope* that defines the top and right margins of the map. Within the envelope is a family of *constant power curves*; these are plotted by solving Equation 2.4 for brake torque. Then, the equation can be used to calculate the torque required at each of a series of speeds to attain the desired constant power level of each curve.

Also shown on the map of Figure 2.4 are a set of *constant BSFC contours*, i.e., along each contour, the BSFC is constant at the value labeled on the curve. There is always one torque-speed combination at which the BSFC is at a minimum. On Figure 2.4, this minimum is 278 g/kWh and occurs below the torque envelope; as torque increases at any given speed, the BSFC decreases until the minimum is reached and then starts to increase again. The engine is said to be *over-fueled* in the area in which the BSFC increases with increasing torque. On some engines, the minimum BSFC point is above the torque envelope and the engine is never over-fueled.

Calculating constant BSFC contours can be a formidable undertaking. One way to do it is to measure torque, speed and fuel consumption at many places on the map. The BSFC is then calculated at each point and constant BSFC contours are sketched in by hand. Hundreds of data points must be taken to get accurate contours using this method. Use of theory can greatly simplify the plotting of constant BSFC contours. By combining Equations 2.1A, 2.4, 2.7, 2.10, 2.11, 2.12, and 2.13, the following equation can be derived:

$$BSFC = \left(\frac{3600}{H_g\, e_{it}}\right)\left(1 + \frac{p_{fme}}{p_{bme}}\right) \qquad (2.20)$$

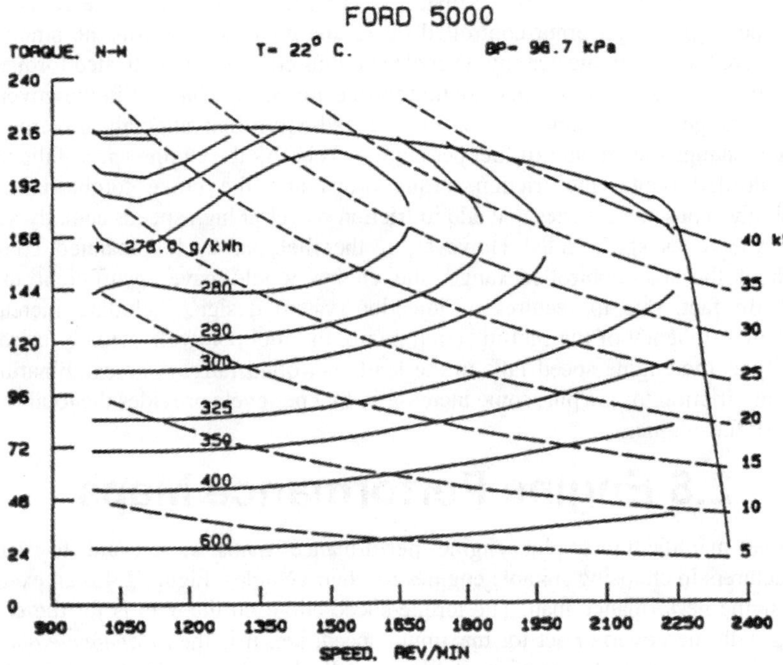

Figure 2.4. A performance map of an over-fueled engine.

Equation 2.20 is valid for any engine. For any brake torque on the map of any engine, the p_{bme} in Equation 2.20 can be calculated using Equation 2.7. For a given diesel engine, the p_{fme} can be calculated from the speed if the constants in Equation 2.9 are known. Finally, an equation is needed to relate e_{it} to engine torque and speed. For a given CI engine, approximates value for e_{it} can be calculated using

$$e_{it} = \frac{e_{ito}}{1 + T_b^n \, f(N_e)} \tag{2.21}$$

where

e_{ito} = indicated thermal efficiency at low torque, decimal
n = a constant specific to a given engine
$f(N_e)$ = a function of engine speed

Reasonable approximations of e_{it} can be obtained when the following function is used:

$$f(N_e) = B_0 + B_1 \left(\frac{N_e}{1000}\right)^{-1} + B_2 \left(\frac{N_e}{1000}\right)^{-2} + B_3 \left(\frac{N_e}{1000}\right)^{-3} + B_4 \left(\frac{N_e}{1000}\right)^{-4} \tag{2.22}$$

where $B_0 - B_4$ = constants specific to a given engine and can vary somewhat even between engines of the same model. All speeds in Equation 2.22 were divided by 1000 to provide B constants of reasonable size.

For a CI engine, e_{ito} is essentially constant in the lower half of the torque-speed map. By measuring torque, speed and fuel consumption at approximately 50 points on the map, it is possible to fit a model to the engine, i.e., to find values for A_0, A_1, A_2, B_0, B_1, B_2, B_3, B_4, e_{ito}, and n. Then the model can be used to find the BSFC at any point on the map. Computer programs based on the model can plot the entire engine map.

2.9 Optimizing Engine Efficiency

Inspection of the engine map in Figure 2.4 reveals several points regarding engine efficiency. Engines are most efficient at or near full load and become increasingly inefficient as torque load declines. At zero brake torque, Equation 2.4 shows the brake power would be zero; then Equations 2.10 and 2.12 show the brake thermal efficiency would also be zero. Finally, Equation 2.14 shows the BSFC would rise to infinity. At zero brake torque, all of the fuel energy is expended in engine friction. Figure 2.4 also shows that BSFC falls and the engine becomes more efficient as engine speed declines. Studying Equations 2.8 and 2.9 will reveal that friction power decreases markedly as engine speed declines, improving the brake thermal efficiency and reducing the BSFC. The designer's selection of engine-rated speed is thus a compromise. Lower rated speeds provide lower BSFC, but also reduce the torque reserve of the engine.

The variation of BSFC with load, described above, is further illustrated in Figure 2.5. The data were plotted as the brake torque was increased from low torque, through governor's maximum and to peak torque. Data points with extremely high BSFC, i.e., those near the high idle point, were not plotted. Note the steep drop in BSFC with initial power increases, then a leveling of the curve approaching rated power. The BSFC falls farther as the engine reaches the maximum power point in the load-controlled range, and then remains low as power falls with further movement into the load-controlled range. The shape of the curve in Figure 2.5 is typical for a governed engine.

Part-load fuel economy can be improved by appropriate management of the engine. If the engine of Figure 2.4 was loaded to 20 kW, for example, the BSFC would be approximately 400 g/kWh at an engine speed of 2250 rev/min. By shifting the transmission to a higher gear and lowering the engine speed to 1850 rev/min, the brake power could be kept at 20 kW while maintaining constant travel speed and lowering the BSFC to less than 325 g/kWh for a fuel savings of more than 18%. As will be discussed in a later chapter, one of the objectives of transmission design is to permit engines to continually operate in more efficient areas of the engine map.

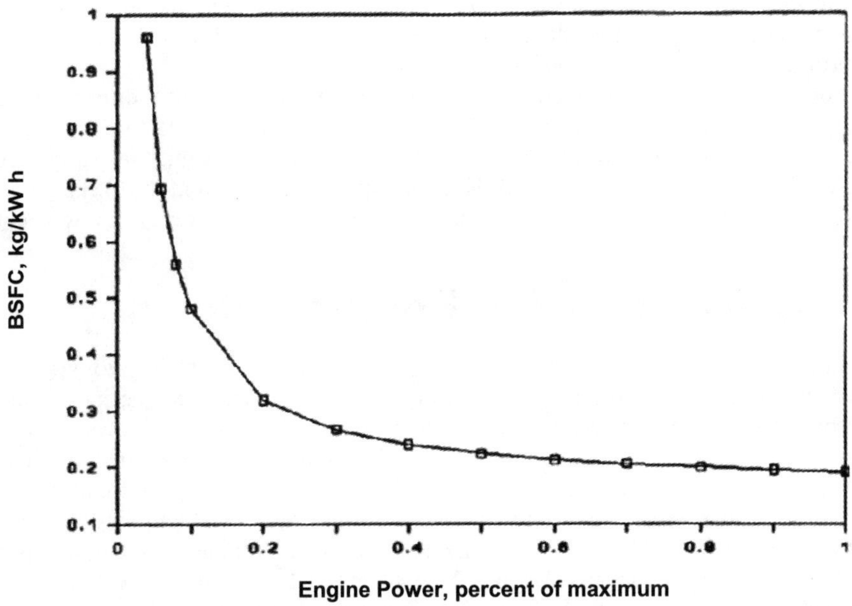

Figure 2.5. Typical shape of a BSFC versus power curve for a CI engine.

2.10 Chapter Summary

Fuel is the source of all engine power. Because of the second law of thermodynamics, less than half of the fuel equivalent power is converted into indicated power (power at the piston) and the rest is lost as rejected heat. The portion of the indicated power that is not delivered as brake power to the flywheel is the friction power used to run engine accessories and overcome friction. There are only three possible ways to increase the power output of an engine; these are to make a bigger engine (larger displacement), run it faster, or increase the mean effective pressure and thus the stress on the engine.

Indicated mean effective pressure is calculated as the area within the p-v diagram of an engine divided by the displacement of a single cylinder. The term "indicated" is used because an indicator diagram was the original method used to develop p-v diagrams. Engine designers have found it useful to develop other mean effective pressures, including brake and friction mean effective pressures. Brake mean effective pressure is proportional to brake torque and is sometimes called specific torque. Friction mean effective pressure increases with engine speed and, for a CI engine, is determined almost entirely by engine speed.

The indicated thermal efficiency is a measure of the combustion efficiency of the engine, while the mechanical efficiency indicates the efficiency in converting indicated power to brake power. The brake thermal efficiency is the product of the indicated thermal efficiency and the mechanical efficiency. Thus, for high overall efficiency, an engine must have both an efficient combustion process and must deliver a high portion of the indicated power to the flywheel.

Specific fuel consumption (SFC) indicates the amount of fuel consumed per unit of work accomplished by the engine. Adjectives must be used with the SFC to indicate where the power is measured. For example, BSFC is appropriate when the brake power is measured.

Engine load has a large effect on engine efficiency and SFC. At zero engine load, both the mechanical and the brake thermal efficiencies are zero, while the SFC is infinite! As load increases, the efficiency approaches its maximum value asymptotically, while the SFC approaches its minimum value asymptotically.

In the governor-controlled range of operation, speed is nearly constant; both indicated and brake torque thus increase as the governor increases the amount of fuel delivered to the engine per engine cycle. In the load-controlled range, the governor cannot increase the fuel delivered per cycle. However, the design of the injection system does allow some increase in the fuel per cycle as the engine slows, thus providing torque reserve. A smaller contribution to torque reserve is provided by the decline in friction torque as the engine slows.

Engine performance maps display a torque-speed envelope, a set of constant-power contours, and a set of constant BSFC contours. The constant-power contours are easily plotted through use of the brake-power equation. Constant BSFC contours require more data and effort to plot, but plotting is made easier through calibration of an engine model. The BSFC contours show areas of efficient engine operation. On some maps, the point of minimum BSFC is below the torque envelope. On such engines, BSFC decreases with increasing load until the minimum BSFC is reached and then increases in a phenomenon called over-fueling. Other engines have the minimum BSFC point above the torque envelope and cannot be over-fueled.

Homework Problems

2.1. A certain CI engine consumes No. 2 diesel fuel ($H_g = 45{,}000$ kJ/kg) at the rate of 26.3 kg/h while running at a rated speed of 2200 rpm and producing 595 N·m of torque. Assuming the mechanical efficiency is 0.85,

(a) Calculate the fuel equivalent power.

(b) Calculate the brake power.

(c) Calculate the indicated power.

(d) Calculate the friction power.

(e) Calculate the indicated thermal efficiency.

(f) Calculate the brake thermal efficiency.

(g) Calculate the BSFC.

2.2. A certain SI engine consumes gasoline ($H_g = 47{,}600$ kJ/kg) at the rate of 29.5 kg/hr while running at a rated speed of 2200 rpm and producing 508 N·m of torque. Assuming the mechanical efficiency is 0.85,

(a) Calculate the fuel equivalent power.

(b) Calculate the brake power.

(c) Calculate the indicated power.

(d) Calculate the friction power.

(e) Calculate the indicated thermal efficiency.

(f) Calculate the brake thermal efficiency.

(g) Calculate the BSFC.

2.3. The engine of Problem 2.1 is a six-cylinder engine with a 115.8 mm bore and a 120.7 mm stroke. Calculate the engine displacement in liters.

2.4. The engine of Problem 2.2 is a six-cylinder engine with a 112.5 mm bore and a 118.4 mm stroke. Calculate the engine displacement in liters.

2.5. Using data from Problems 2.1 and 2.3, calculate the mean effective pressures, p_{fme}, p_{ime}, and p_{bme} for the engine.

2.6. Using data from Problems 2.2 and 2.4, calculate the mean effective pressures, p_{fme}, p_{ime}, and p_{bme} for the engine.

2.7. Making use of theory in Chapter 2, derive the equation $p_{fme} = p_{ime} - p_{bme}$. Then use answers from Problem 2.5 to check the accuracy of your derived equation.

2.8. Making use of theory in Chapter 2, derive the equation $p_{fme} = p_{ime} - p_{bme}$. Then use answers from Problem 2.6 to check the accuracy of your derived equation.

2.9. Making use of theory in Chapter 2, derive equations to calculate e_m as a function of certain mean effective pressures.

2.10. The engine of Problem 2.1 has high idle speed of 2363 rpm. The torque reserve is 31.4% and peak torque occurs at 1000 rpm.

(a) Calculate the governor regulation.

(b) Calculate the peak torque in N·m.

(c) Sketch the torque and power curves versus engine speed.

2.11. The engine of Problem 2.2 has high idle speed of 2385 rpm. The torque reserve is 25.4% and peak torque occurs at 1200 rpm.

(a) Calculate the governor regulation.

(b) Calculate the peak torque in N·m.

(c) Sketch the torque and power curves versus engine speed.

2.12. The engine of Problem 2.2 has high idle speed of 2392 rev/min. The torque reserve is 50.0% and peak torque occurs at 1200 rpm.

(a) Calculate the governor regulation.

(b) Calculate the peak torque in N·m.

(c) Sketch the torque and power curves versus engine speed.

2.13. Using theory from Chapter 2, derive Equation 2.20.

2.14. In a test of a certain CI engine, the speed is held constant while the torque is increased from 10% of rated torque to full-rated torque. The heating value of the fuel is 45,500 kg/kJ. Assume that the indicated thermal efficiency remains constant at 0.45 and that the BSFC is 0.2 kg/kWh at full rated torque. Use Equation 2.20 to plot BSFC versus percent of maximum torque. Note that the p_{fme} remains constant because the engine speed is constant. This is an excellent spreadsheet problem. (*Hint:* Note that you can solve Equation 2.20 for the ratio, p_{fme}/p_{fme} at full-rated torque, and then adjust it for various torque percentages.)

2.15. In a test of a certain CI engine, the speed is held constant while the torque is increased from 8% of rated torque to full-rated torque. The heating value of the fuel is 45,500 kg/kJ. Assume that the indicated thermal efficiency remains constant at 0.45 and that the BSFC is 0.19 kg/kWh at full rated-torque. Use Equation 2.20 to plot BSFC versus percent of maximum torque. Note that the p_{fme} remains constant because the engine speed is constant. This is an excellent spreadsheet problem. (*Hint:* Note that you can solve Equation 2.20 for the ratio, p_{fme}/p_{fme} at full-rated torque, and then adjust it for various torque percentages.)

References and Suggested Readings

Goering, C.E., and H. Cho. 1988. Engine model for mapping BSFC contours. *Mathematical and Computer Modeling* 11: 514-518.

Gui, X.O., C.E. Goering and N. L. Buck. 1989. Simulation of a fuel-efficient augmented engine. *Transactions of the ASAE* 32(6): 1875-1881.

SAE. 1992a. Engine power test code–Spark ignition and compression ignition–Gross power rating. SAE Standard J1995. *SAE Handbook,* vol. 3. Warrendale, PA: SAE.

SAE. 1992b. Engine power test code–Spark ignition and compression ignition–Net power rating. SAE Standard J1349. *SAE Handbook,* vol. 3. Warrendale, PA: SAE.

SAE. 1992c. Procedure for mapping engine performance–Spark-ignition and compression-ignition engines. SAE Standard J1312. *SAE Handbook,* vol. 3. Warrendale, PA: SAE

CHAPTER 3

ENGINE AND VEHICLE TESTING

3.1 Introduction

In this chapter, you will be introduced to equipment used for testing engines, including dynamometers, fuel flow meters and air flow meters. You will also be introduced to methods for measuring cylinder pressures and gas volumes in a running engine and will learn techniques for analyzing and interpreting such data. You will be introduced to tractor testing as conducted according to SAE/ASAE/ISO standards and according to OECD procedures. Finally, you will learn how to correct observed power output of an engine to standard atmospheric conditions.

3.2 Dynamometers

A *dynamometer* for testing engines must include the following four essential elements:

- A means for controlling torque
- A means for measuring torque
- A means for measuring speed
- A means for dissipating power

A *prony brake* (see Figure 3.1) is the simplest type of dynamometer. The rotor (a) is connected to the crankshaft of the engine and rotates at engine speed. Surrounding the rotor are brake shoes (b) that create a torque drag on the rotor. The torque is controlled by tightening or loosening the hand lever (c). The torque is measured by a scale (d) located at the end of a lever arm (e). A tachometer (not shown) must be provided for measuring the rotor speed. After the torque and speed are measured, the power can be calculated using Equation 2.4. The brake shoes in the prony brake convert all of the engine power to heat. The brake shoes would soon burn up unless a means is provided for circulating water or some other coolant to carry off the heat.

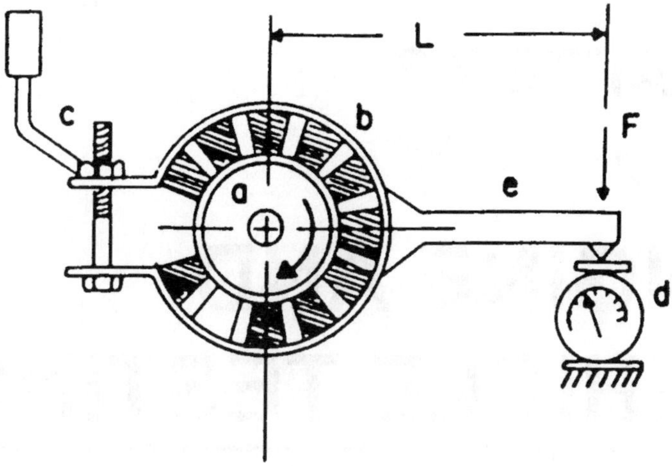

Figure 3.1. A prony brake dynamometer.

The prony brake is an example of an *absorption-type dynamometer*. *Electric dynamometers* are also available. An electric generator is cradled in a set of trunion bearings that would permit the entire generator housing to spin when the generator rotor is driven by an engine. However, a load arm attached to the generator housing prevents the generator from spinning. The end of the load arm is attached to a scale or some other load-measuring device to measure the load and thus the torque. With electric generators, the engine power is converted into electricity, which can be dissipated into a resistor bank or delivered to an electric power grid. The flow of electricity is controlled as a means of controlling the torque load on the engine. An advantage of electric dynamometers is that the generator can function as an electric motor when supplied with an electric current. As a motor, the dynamometer can be used to start and/or to *motor the engine*, i.e., to cause it to spin at a normal operating speed when no fuel is supplied. Motoring an engine permits measurement of the friction torque or friction mean effective pressure of an engine. The importance of friction torque and friction mean effective pressure were discussed in Chapter 2.

3.3 Measurement of Fuel Consumption

3.3.1 Volume-Based Fuel Flowmeters

Fuel consumption is frequently measured during an engine test; the measurement can be on a volume or mass basis. Figure 3.2 is an example of a *rotameter* for measuring fuel consumption on a volume basis. The transparent tube has a slight taper, i.e., its diameter increases toward the top. Fuel enters at the bottom of the tube, flows upward around the float in the tube, and exits at the top. The higher the flow rate, the higher the flow must rise to allow sufficient space around the edges of the float to

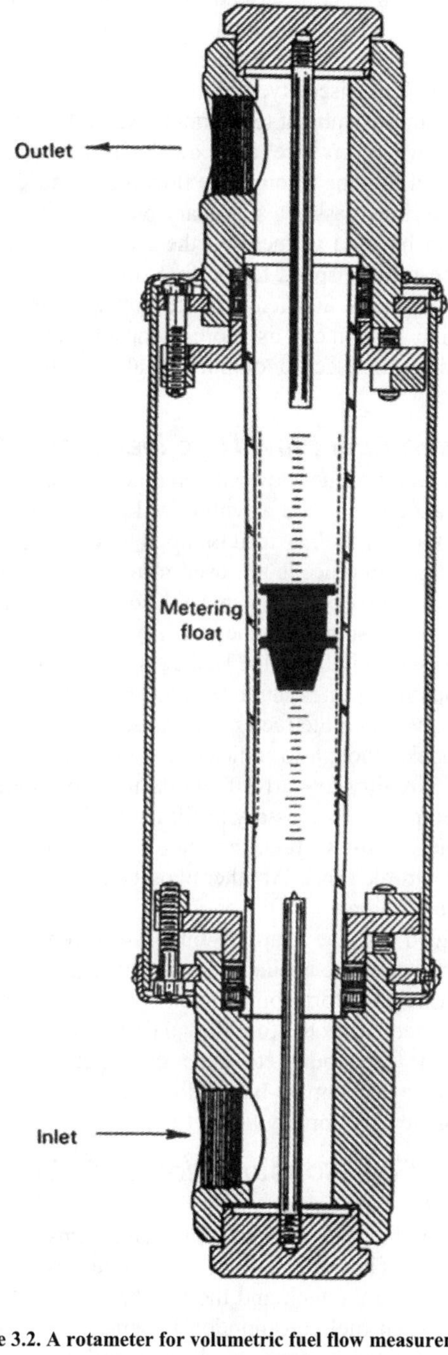

Figure 3.2. A rotameter for volumetric fuel flow measurement. (Courtesy of Fisher-Porter Company.)

allow fuel to pass. Thus, a scale inscribed directly on the outside of the tube can be calibrated directly in terms of volumetric flow rate. The rotameter reading is dependent upon the viscosity and the temperature of the fuel. Usually, the rotameter is selected to match the fuel viscosity, but the reading can be corrected if the fuel temperature differs from the ambient temperature when the rotameter was calibrated.

Turbine wheel flow meters are also available for measuring fuel flow rate volumetrically. The fuel flowing through the flow meter causes the turbine to rotate. A small magnet on the turbine passes a stationary pick-up coil, one each revolution. An electronic counter can be used to measure the turbine speed. A calibration curve is provided to relate the turbine speed to the fuel flow rate. The turbine flow meter is subject to the same viscosity and temperature concerns as were mentioned for the rotameter. The turbine flow meter is more compact than a rotameter and has the advantage of providing an electric output signal for automatic recording or data processing.

3.3.2 Mass-Based Fuel Flow Rate Measurement

A scale is used to measure fuel flow rate on a mass basis. A small beaker is placed on one pan of a balance scale, and a withdrawal tube is immersed in the beaker to allow the engine to draw its fuel. A feed pump is used to supply fuel to the container from a fuel storage tank. Photocells are used to detect the vertical movement of the scale pan. Electronic control circuitry must also be provided. To run the engine without measuring fuel consumption, the circuitry turns on the feed pump whenever the scale pan lifts enough to indicate the beaker is almost empty. Since the feed pump supplies fuel at a much faster rate than the engine can withdraw it from the beaker, the scale pan begins sinking when the feed pump runs. The circuitry stops the feed pump when the scale pan sinks enough to indicate that the beaker is nearly full. When the operator triggers the circuitry to start a fuel consumption measurement, the circuitry must first run the feed pump if necessary to fill the beaker. Then the fuel pump is shut off and, as the engine withdraws fuel, the scale pan begins to sink. At some point, a photocell starts an electronic timer. Another photocell stops the timer when a specified mass of fuel has been withdrawn by the engine. The mass flow rate of fuel is then the specified mass divided by the elapsed time measured by the electronic timer. Typically, several photocells are available for stopping the timer. These photocells are arranged at different heights corresponding to specified amounts of fuel withdrawn after the timer was started. After the fuel flow measurement is completed, the circuitry resumes control of the feed pump to keep the beaker from becoming empty or overfilled. An advantage of the mass-based flow meter is that it is independent of fuel viscosity and thus can be used for any liquid fuel.

3.3.3 Measuring Fuel Consumption of CI Engines

Measurement of the fuel consumption rate of a CI engine presents a special problem, since these engines have a return line that returns leakage flow from the fuel injectors to the fuel tank. Thus, the net fuel consumption is the difference between the fuel being withdrawn from the tank and the fuel being returned to it. The return fuel presents no problem when fuel consumption is being measured on a mass basis; the return fuel is simply directed into the small container on the scale pan and the timer thus measures the time required for the net fuel consumption of the engine. When a volumetric measurement is used, the normal procedure is to tee the return fuel back into the main supply line at a point between the flow meter and the engine. The flow meter thus measures the net fuel consumption. One difficulty is that the return fuel is usually hot and teeing it into the main fuel line eliminates the opportunity for the fuel to return to the fuel tank for cooling. Sometimes a small heat exchanger is installed in the return line to cool the return fuel.

3.4 Measurement of Air Consumption

Measurement of air consumption is important because the power output of any engine is limited by its ability to consume air. To maintain a combustible mixture, air must be supplied in massively larger amounts than the fuel. Thus, although the power is derived from the fuel, it is the limitations on the air-consuming capacity of the engine that actually limits the engine power.

The most common method of measuring air consumption is to insert a calibrated orifice or nozzle in a line supplying air to the engine. A number of manufacturers of engine test equipment can supply airflow meters. The air pressure falls when passing through the orifice or nozzle; the pressure drop increases with the airflow rate. Thus, a differential pressure transducer can be used to measure the airflow rate. The manufacturer calibrates the airflow meter under controlled conditions and supplies calibration data with the flow meter. Note that, when air entering the flow meter is at atmospheric pressure, the pressure drop across the airflow meter causes the engine to receive its air at sub-atmospheric pressure. Later in Chapter 3, a method will be presented for correcting the engine power output to standard pressure conditions. In some cases, an externally driven blower is placed in the air supply line to offset the pressure drop across the orifice or nozzle in the airflow meter.

3.5 Combustion Data Acquisition and Interpretation

The development of electronic instrumentation and personal computers (PCs) has reached the point at which pressure-volume conditions within the combustion chamber of a running engine can be measured easily. Figure 3.3 is a schematic diagram of one system for obtaining such data. An optical, two-channel shaft encoder is coupled to the engine crankshaft; one channel produces a pulse when the piston (usually in the No. 1 cylinder in a multi-cylinder engine) reaches *head dead center* (HDC; sometimes also called TDC for *top dead center*). The other channel produces pulses at periodic crankshaft intervals, which should not exceed 0.25° of crankshaft rotation for accurate combustion studies. A piezoelectric pressure transducer is installed in the engine with its sensor slightly recessed in the top of the combustion chamber. Typically, the transducer is water cooled to prevent it from being damaged by the high combustion temperatures. A special coating is applied to the face of the transducer to limit radiant energy transfer to the transducer. The transducer produces a signal proportional to the

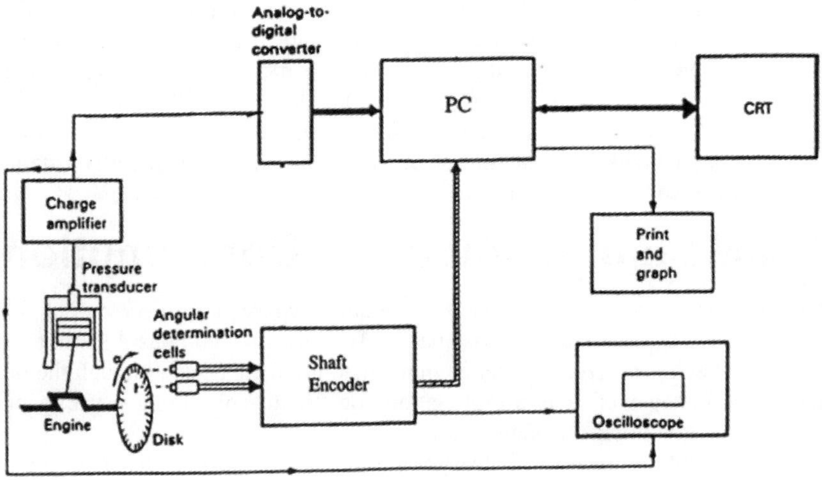

Figure 3.3. A system used for measuring cylinder pressure and gas volume in a running engine.

cylinder pressure, and an analog to digital (A/D) converter converts the signal to digital before it enters the PC. The piezoelectric pressure transducer has very high output impedance and a charge amplifier is used to produce a useable voltage signal.

When a test is initiated, the HDC marker pulse signals the PC to begin recording pressures, e.g., at 0.25° intervals, while the crankshaft turns through two full revolutions for a four-cycle engine. Only one revolution of data would be needed for a two-cycle engine. Normally, many consecutive cycles are recorded and averaged, point by point, to remove cycle-to-cycle variation before the data are analyzed. The measurement equipment must be very fast. If the engine is running at 2400 rpm and pressure measurements are made at 0.25° intervals, for example, the system must be able to record 57,600 pressure readings per second. The high data rate is necessary to capture the steep pressure waveform that occurs in a running engine.

One use of pressure data is to plot a pressure-volume (p-v) diagram for the engine. Such diagrams were used to determine indicated mean effective pressure in Chapter 2 and will be discussed in more detail in Chapter 4. The PC can compute the instantaneous combustion chamber (gas) volume at any time by

$$\frac{V_g}{R\,A_p} = \frac{V_c}{R\,A_p} + (1 - \cos\theta) + \frac{L}{R}\left(1 - \sqrt{1 - \left(\frac{R}{L}\sin\theta\right)^2}\,\right) \tag{3.1}$$

where

V_g = gas volume, cm^3
V_c = clearance volume, cm^3
R = crank throw radius, cm
L = connecting rod length, cm
A_p = piston top area, cm^2
θ = crankshaft angle, measured from HDC

The clearance volume, i.e., the gas volume when the piston is at HDC, is further discussed in Chapter 4. The PC counts pulses from the encoder to determine θ and then uses Equation 3.1 to determine V_g each time the cylinder pressure is measured. The PC continues to measure and calculate through a full engine cycle, i.e., through two complete revolutions of the crankshaft. Then a complete p-v diagram can be plotted.

Correct phasing of the encoder is critical when pressure measurements are being taken, i.e., the HDC marker must be triggered exactly when the piston is at HDC. A good technique is to take pressure measurements while the engine is being motored, as discussed in Section 3.2. Then the PC is used to calculate and plot the logarithm of the pressure versus the logarithm of the gas volume. As will be discussed in Chapter 4, the plot of the compression and expansion strokes should be superimposed straight lines if the encoder is phased correctly. Incorrect phasing can be corrected by mechanically adjusting the connection of the encoder to the crankshaft. Alternatively, a phase offset angle can be used to correct the value of θ in Equation 3.1 when the data are being reduced.

A useful technique is to use the recorded pressure-volume data to infer the instantaneous rate of energy release from the fuel in a running engine. The following equation relates the rate of energy release to pressures and volumes:

$$\frac{dQ}{d\theta} = \frac{V\frac{dp}{d\theta} + \gamma p \frac{dV}{d\theta}}{\gamma - 1} - \frac{dQ_w}{d\theta} \tag{3.2}$$

where

$dQ/d\theta$ = rate of energy release, J/degree

θ = crank angle, degrees

γ = ratio of specific heats

$dQ_w/d\theta$ = rate of heat transfer from the walls, J/degree

The calculations can be done in a spreadsheet. The Suggested Readings and References by Goering at the end of this chapter gives details as to how to compute γ, $dQ_w/d\theta$, and the derivatives in Equation 3.2. In the equation, the term involving the volume derivative relates to work done by the combustion gases on the piston. The term involving the pressure derivative relates to the change in internal energy of the combustion gases. The combination of these three energy flows is assumed to equal the rate of energy from the combustion process. The model is called a zero-dimensional model because it ignores the spatial aspects of combustion and treats the contents of the combustion chamber as homogeneous. Energy release diagrams provide valuable insights into combustion in CI engines, as explained in Chapter 5.

3.6 Standardized Tractor Tests

Data from standardized tests may be used in comparing the performances of various makes and models of tractors. When comparative tests are administered by agencies independent of the tractor manufacturers, the resulting competition among the manufacturers also tends to promote improvements in tractor design. The original independent agency to do tractor testing was the University of Nebraska Tractor Testing Laboratory (NTTL). It was authorized by the 1919 session of the Nebraska legislature, which passed a bill requiring testing and maintenance of service stations for all models of tractors sold within the state. The bill was prompted by the presence of inferior tractors on the market at that time. The provisions of the current law do not apply to tractors whose engines develop less than 30 kW, nor to tractors manufactured or sold as nonagricultural tractors. Standard tests conducted at the NTTL were developed and are periodically updated by Society for Automotive Engineers (SAE) working in cooperation with the American Society of Agricultural Engineers (ASAE). The SAE/ASAE tractor tests are also in good agreement with International Standards Organization (ISO) test standards. Thus, these tests will be referred to as SAE/ASAE/ISO tests.

The NTTL served as a uniquely independent tractor-testing agency for many years. As tractor manufacturing and marketing became global in the 1980s, the Organization for Economic Cooperation and Development (OECD) test codes became the commonly accepted official test procedures for worldwide marketing. These test codes were developed in Europe by 1960. The United States was not a full participant during the development of the OECD test codes, but the Farm and Industrial Equipment Institute (FIEI – later renamed EMI, the Equipment Manufacturer's Institute and more recently known as AEM, the Association of Equipment Manufacturers) attended as a U.S. observer. In 1986, the Nebraska legislature changed its tractor test law to accept OECD tests as well as those conducted by the NTTL as a prerequisite to selling tractors in the state. By 1988, the NTTL had been designated as an official OECD testing station for the United States. In addition, the NTTL began producing summaries of OECD tests conducted elsewhere if the tractor required a Nebraska sales permit. The summaries are similar in format to the reports of the SAE/ASAE/ISO tests and are distributed to Nebraska farmers and other interested parties.

Three standard test procedures are now widely recognized for testing tractors. They include the standard ASAE/SAE/ISO test, the OECD restricted test, and the OECD long test. The OECD tests are described by code designations, as follows:

Code I: OECD Standard Code for the Official Testing of Agricultural Tractor Performance

Compulsory tests:

1. Main PTO
2. Hydraulic power and lifting force
3. Drawbar power, ballasted tractor
4. Turning area and turning circle
5. Position of center of gravity
6. Braking (wheeled tractors only)
7. External (bystander) noise (wheeled tractors only)

Tests performed and reported at manufacturer's option:

8. Engine
9. Performance at the belt or belt pulley shaft
10. Performance in a hot atmosphere
11. Low temperature starting

12. Drawbar power and fuel consumption for unballasted tractor

Code II: OECD Restricted Standard Code for the Official Testing of Agricultural Tractor Performance

Compulsory tests:

1. Main PTO
2. Hydraulic power and lifting force
3. Drawbar power and fuel consumption for unballasted tractor

(The drawbar tests are conducted on a concrete track. In addition, Tests 3 through 11 from the Standard Code may be performed and reported at the manufacturer's option.)

Code III: OECD Standard Code for the Official Testing of Protective Structures on Agricultural Tractors (Dynamic Test).

Code IV: OECD Standard Code for the Official Testing of Protective Structures on Agricultural Tractors (Static Test).

Code V: OECD Standard Code for the Official Measurement of Noise in Protective Structures on Agricultural Tractors.

The SAE/ASAE/ISO test and the OECD restricted test are similar; the ASAE/SAE/ISO procedure includes a 10-hour drawbar test at 75% load, whereas the OECD test requires only a 5-hour test. The OECD test report requires more detailed tractor specifications, including more design data on the engine, power train, brakes, etc. In some countries, either of the OECD tests can be conducted using the manufacturer's test facilities provided an OECD-authorized supervisor monitors the tests. Conversely, the SAE/ASAE/ISO tests are all performed at the NTTL by NTTL personnel. An OECD test can be carried out at more than one facility. For example, an SAE/ASAE/ISO test conducted at the NTTL can be supplemented by tests carried out in other countries to obtain an OECD test. NTTL is not equipped to conduct OECD Code III or Code IV tests, but NTTL personnel witness such tests conducted at the manufacturer's facility.

Manufacturing tolerances are an inherent part of mass production of tractors but, because of such tolerances, there is variation between tractors produced on the same assembly line. Thus, some procedure must be followed to select the specific tractor used for testing. The procedure used for the SAE/ASAE/ISO test or either of the OECD tests is to permit the manufacturer to select a tractor from the assembly line and pre-test it before submitting it for an official test. In addition to pre-testing tractors prior to an official test, manufacturers routinely test tractors and their components for durability and to gain other useful information. For example, an engine may be tested to reveal which combinations of speed and torque produce the highest fuel economy.

3.7 Interpretation of Test Reports

Official OECD reports are too lengthy for distribution to the public. However, for every tractor model sold in the state of Nebraska that has an OECD test, the NTTL publishes a summary of the test in a format similar to that of the earlier Nebraska Tractor Test reports. An example of such a report is shown in Figure 3.4. The purpose

NEBRASKA OECD TRACTOR TEST 1678—SUMMARY 150 JOHN DEERE 7200 SYNCROPLUS DIESEL 12 SPEED

POWER TAKE-OFF PERFORMANCE

Power HP *(kW)*	Crank shaft speed rpm	Gal/hr *(l/h)*	lb/hp.hr *(kg/kW.h)*	Hp.hr/gal *(kW.h/l)*	Mean Atmospheric Conditions
MAXIMUM POWER AND FUEL CONSUMPTION					
Rated Engine Speed—(PTO speed—1003 rpm)					
94.02 *(70.11)*	2099	5.40 *(20.44)*	0.402 *(0.244)*	17.42 *(3.43)*	
Maximum Power (2 hours)					
95.06 *(70.88)*	1900	5.21 *(19.71)*	0.383 *(0.233)*	18.26 *(3.60)*	
VARYING POWER AND FUEL CONSUMPTION					
94.02 *(70.11)*	2099	5.40 *(20.44)*	0.402 *(0.244)*	17.42 *(3.43)*	Air temperature
82.93 *(61.84)*	2184	5.02 *(19.00)*	0.423 *(0.257)*	16.52 *(3.26)*	80°F *(27°C)*
63.10 *(47.06)*	2209	4.29 *(16.24)*	0.475 *(0.289)*	14.71 *(2.90)*	Relative humidity
42.07 *(31.37)*	2227	3.39 *(12.83)*	0.563 *(0.343)*	12.41 *(2.45)*	65%
21.16 *(15.78)*	2245	2.53 *(9.58)*	0.836 *(0.509)*	8.36 *(1.65)*	Barometer
0.90 *(0.67)*	2260	1.72 *(6.49)*	13.266 *(8.070)*	0.53 *(0.10)*	29.06" Hg *(98.44 kPa)*

Maximum Torque 311 lb.-ft. *(422 Nm)* at 1400 rpm
Maximum Torque Rise 32.5%
Torque rise at 1700 engine rpm 20%

DRAWBAR PERFORMANCE

FUEL CONSUMPTION CHARACTERISTICS

Power Hp *(kW)*	Drawbar pull lbs *(kN)*	Speed mph *(km/h)*	Crank-shaft speed rpm	Slip %	Fuel Consumption lb/hp.hr *(kg/kW.h)*	Fuel Consumption Hp.hr/gal *(kW.h/l)*	Temp.°F (°C) cool-ing med	Temp.°F (°C) Air dry bulb	Barom. inch Hg *(kPa)*
Maximum Power—6th (C1) Gear									
82.70 *(61.67)*	6456 *(28.72)*	4.80 *(7.73)*	2101	3.97	0.457 *(0.278)*	15.31 *(3.02)*	196 *(91)*	70 *(21)*	28.73 *(97.29)*
75% of Pull at Maximum Power—6th (C1) Gear									
65.16 *(48.59)*	4838 *(21.52)*	5.05 *(8.13)*	2198	3.49	0.505 *(0.307)*	13.86 *(2.73)*	190 *(88)*	70 *(21)*	28.74 *(97.32)*
50% of Pull at Maximum Power—6th (C1) Gear									
44.39 *(33.10)*	3229 *(14.36)*	5.16 *(8.30)*	2221	2.33	0.603 *(0.367)*	11.60 *(2.29)*	186 *(86)*	70 *(21)*	28.74 *(97.32)*
75% of Pull at Reduced Engine Speed—8th (C2) Gear									
65.26 *(48.67)*	4843 *(21.54)*	5.05 *(8.13)*	1572	3.49	0.440 *(0.268)*	15.90 *(3.13)*	192 *(89)*	70 *(21)*	28.74 *(97.32)*
50% of Pull at Reduced Engine Speed—8th (C2) Gear									
44.34 *(33.06)*	3223 *(14.34)*	5.16 *(8.30)*	1589	2.41	0.491 *(0.299)*	14.23 *(2.80)*	184 *(84)*	70 *(21)*	28.74 *(97.32)*

Location of Test: Tractor Testing Laboratory, University of Nebraska, Lincoln, Nebraska 68583-0832

Dates of Test: May 10 to June 1, 1994

Manufacturer: John Deere Tractor Works, P.O. Box 270, Waterloo, Iowa 50704

FUEL OIL and TIME: Fuel No. 2 Diesel **Cetane No.** 53.9 **Specific gravity converted to 60°/60° F *(15°/15°C)*** 0.8400 **Fuel weight** 6.994 lbs/gal *(0.838 kg/l)* **Oil SAE** 15W-40 **API service classification** SG/CE **To motor** 4.251 gal *(16.093 l)* **Drained from motor** 4.178 gal *(15.817 l)* **Transmission and hydraulic lubricant** John Deere Hy-Gard fluid **Front axle lubricant** John Deere GL-5 Gear Lubricant **Total time engine was operated** 20.0 hours.

ENGINE: Make John Deere Diesel **Type** six cylinder vertical with turbocharger **Serial No.** *TO6059T431193* **Crankshaft** lengthwise **Rated engine speed** 2100 **Bore and stroke** (as specified) 4.19"×4.331" *(106.5 mm×110.0 mm)* **Compression ratio** 17.8 to 1 **Displacement** 359 cu in *(5880 ml)* **Starting system** 12 volt **Lubrication** pressure **Air cleaner** two paper elements and aspirator **Oil filter** one full flow cartridge **Oil cooler** engine coolant heat exchanger for crankcase oil, radiator for hydraulic and transmission oil **Fuel filter** one paper element and prestrainer **Fuel cooler** radiator for inlet fuel **Muffler** underhood **Exhaust** vertical **Cooling medium temperature control** two thermostats and variable speed fan

ENGINE OPERATING PARAMETERS: Fuel rate: 37.3-40.3 lb/h *(16.9-18.3 kg/h)* **High idle:** 2225-2325 rpm **Turbo boost** nominal 8.7-10.2 psi *(60-70 kPa)* as measured 9.0 psi *(65 kPa)*

CHASSIS: Type front wheel assist **Serial No.** *RW7200S001437* **Tread width** rear 60.0" *(1524 mm)* to 100.3" *(2548 mm)* front 60.0" *(1524 mm)* to 88.0" *(2235 mm)* **Wheel base** 103.3" *(2625 mm)* **Hydraulic control system** direct engine drive **Transmission** selective gear fixed ratio **Nominal travel speeds mph *(km/h)*** first 1.43 *(2.30)* second 2.00 *(3.22)* third 2.64 *(4.25)* fourth 3.03 *(4.87)* fifth 4.24 *(6.82)* sixth 4.82 *(7.75)* seventh 5.59 *(9.00)* eighth 6.74 *(10.85)* ninth 8.90 *(14.33)* tenth 9.99 *(16.08)* eleventh 13.98 *(22.50)* twelfth 18.46 *(29.71)* reverse 1.75 *(2.81)*, 3.70 *(5.96)*, 5.89 *(9.48)*, 12.21 *(19.65)* **Clutch** multiple wet disc hydraulically actuated by foot pedal **Brakes** wet multiple disc hydraulically actuated by two foot pedals which can be locked together **Steering** hydrostatic **Power take-off** 540 rpm at 2080 engine rpm and 1000 rpm at 2093 engine rpm **Unladen tractor mass** 12522 lb *(5680 kg)*

REPAIRS AND ADJUSTMENTS: No repairs or adjustments

Figure 3.4 (continued on next pages).

DRAWBAR PERFORMANCE
MAXIMUM POWER IN SELECTED GEARS

Power Hp (kW)	Drawbar pull lbs (kN)	Speed mph (km/h)	Crank-shaft speed rpm	Slip %	Fuel Consumption lb/hp.hr (kg/kW.h)	Fuel Consumption Hp.hr/gal (kW.h/l)	Temp. °F (°C) cool-ing med	Temp. °F (°C) Air dry bulb	Barom. inch Hg (kPa)
					2nd (A2) Gear				
60.51 (45.13)	12254 (54.51)	1.85 (2.98)	2201	14.66	0.553 (0.336)	12.65 (2.49)	186 (85)	65 (18)	28.77 (97.43)
					3rd (A3) Gear				
78.25 (58.35)	11598 (51.59)	2.53 (4.07)	2156	9.67	0.480 (0.292)	14.58 (2.87)	190 (88)	66 (19)	28.76 (97.39)
					4th (B1) Gear				
81.84 (61.03)	10856 (48.29)	2.83 (4.55)	2060	8.14	0.460 (0.280)	15.22 (3.00)	193 (89)	69 (21)	28.74 (97.32)
					5th (B2) Gear				
83.16 (62.01)	8260 (36.74)	3.78 (6.08)	1903	5.25	0.438 (0.267)	15.95 (3.14)	197 (92)	70 (21)	28.73 (97.29)
					6th (C1) Gear				
83.22 (62.06)	7209 (32.07)	4.33 (6.97)	1903	4.45	0.439 (0.267)	15.93 (3.14)	197 (91)	70 (21)	28.73 (97.29)
					7th (B3) Gear				
83.14 (62.00)	6171 (27.45)	5.05 (8.13)	1901	3.81	0.440 (0.268)	15.90 (3.13)	198 (92)	75 (24)	28.68 (97.12)
					8th (C2) Gear				
79.95 (59.62)	4876 (21.69)	6.15 (9.90)	1905	2.99	0.452 (0.275)	15.46 (3.05)	197 (91)	71 (22)	28.71 (97.22)
					9th (C3) Gear				
79.14 (59.02)	3642 (16.20)	8.15 (13.11)	1903	2.50	0.452 (0.275)	15.46 (3.05)	196 (91)	71 (22)	28.73 (97.29)

DRAWBAR PERFORMANCE AT 2100 RPM (Front Drive Engaged)
MAXIMUM POWER IN SELECTED GEARS

Power Hp (kW)	Drawbar pull lbs (kN)	Speed mph (km/h)	Crank-shaft speed rpm	Slip %	Fuel Consumption lb/hp.hr (kg/kW.h)	Fuel Consumption Hp.hr/gal (kW.h/l)	Temp. °F (°C) cool-ing med	Temp. °F (°C) Air dry bulb	Barom. inch Hg (kPa)
					3rd (A3) Gear				
76.19 (56.81)	12039 (53.55)	2.37 (3.82)	2130	14.34	0.499 (0.304)	14.01 (2.76)	190 (88)	66 (19)	28.77 (97.43)
					4th (B1) Gear				
82.36 (61.42)	10678 (47.50)	2.89 (4.66)	2103	7.84	0.461 (0.280)	15.18 (2.99)	192 (89)	68 (20)	28.74 (97.32)
					5th (B2) Gear				
83.04 (61.92)	7419 (33.00)	4.20 (6.75)	2101	4.45	0.457 (0.278)	15.32 (3.02)	196 (91)	70 (21)	28.73 (97.29)
					6th (C1) Gear				
82.70 (61.67)	6456 (28.72)	4.80 (7.73)	2101	3.97	0.457 (0.278)	15.31 (3.02)	196 (91)	70 (21)	28.73 (97.29)
					7th (B3) Gear				
82.65 (61.63)	5529 (24.59)	5.61 (9.02)	2100	3.49	0.460 (0.280)	15.21 (3.00)	196 (91)	74 (23)	28.69 (97.16)
					8th (C2) Gear				
79.64 (59.39)	4392 (19.54)	6.80 (10.94)	2099	2.83	0.472 (0.287)	14.82 (2.92)	196 (91)	71 (22)	28.72 (97.26)
					9th (C3) Gear				
78.94 (58.87)	3293 (14.65)	8.99 (14.47)	2096	2.41	0.476 (0.290)	14.69 (2.89)	194 (90)	71 (22)	28.73 (97.29)

REMARKS: All test results were determined from observed data obtained in accordance with official OECD, SAE and Nebraska test procedures. For the maximum power tests, the fuel temperature at the injection pump return was maintained at 156° F *(69°C)*. This tractor did not meet manufacturers claim of 72.0 dB(A) cab sound level. The performance results on this summary were taken from OECD tests conducted under the Code II Restricted Standard Test Code procedure.

We, the undersigned, certify that this is a true and correct report of official Tractor Test No. **1678**, Summary 150, July 21, 1994.

LOUIS I. LEVITICUS
Engineer-in-Charge

R.D. GRISSO
M.F. KOCHER
K. VON BARGEN
Board of Tractor Test Engineers

Figure 3.4 (continued on next pages).

DRAWBAR PERFORMANCE (FRONT DRIVE DISENGAGED)

FUEL CONSUMPTION CHARACTERISTICS

Power Hp (kW)	Drawbar pull lbs (kN)	Speed mph (km/h)	Crank-shaft speed rpm	Slip %	Fuel Consumption lb/hp.hr (kg/kW.h)	Fuel Consumption Hp.hr/gal (kW.h/l)	Temp. °F (°C) cool-ing med	Temp. °F (°C) Air dry bulb	Barom. inch Hg (kPa)
Maximum Power—7th (B3) Gear									
83.81	5708	5.51	2099	3.57	0.454	15.40	197	72	28.70
(62.50)	*(25.39)*	*(8.86)*			*(0.276)*	*(3.03)*	*(91)*	*(22)*	*(97.19)*
75% of Pull at Maximum Power—7th (B3) Gear									
66.36	4274	5.82	2198	2.68	0.505	13.85	192	76	28.67
(49.48)	*(19.01)*	*(9.37)*			*(0.307)*	*(2.73)*	*(89)*	*(24)*	*(97.09)*
50% of Pull at Maximum Power—7th (B3) Gear									
45.01	28.54	5.91	2220	2.11	0.592	11.82	186	76	28.67
(33.57)	*(12.70)*	*(9.52)*			*(0.360)*	*(2.33)*	*(86)*	*(24)*	*(97.09)*
75% of Pull at Reduced Engine Speed—8th (C2) Gear									
66.11	4272	5.80	1825	3.09	0.456	15.35	190	76	28.67
(49.30)	*(19.00)*	*(9.34)*			*(0.277)*	*(3.02)*	*(88)*	*(24)*	*(97.09)*
50% of Pull at Reduced Engine Speed—8th (C2) Gear									
44.87	2847	5.91	1840	2.02	0.512	13.66	185	76	28.67
(33.46)	*(12.66)*	*(9.51)*			*(0.311)*	*(2.69)*	*(85)*	*(24)*	*(97.09)*
MAXIMUM POWER IN SELECTED GEARS									
4th (B1) Gear									
70.38	9612	2.75	2181	14.20	0.519	13.47	190	67	28.75
(52.48)	*(42.75)*	*(4.42)*			*(0.316)*	*(2.65)*	*(88)*	*(19)*	*(97.36)*
5th (B2) Gear									
83.21	7603	4.10	2103	5.38	0.455	15.36	195	70	28.73
(62.05)	*(33.82)*	*(6.60)*			*(0.277)*	*(3.02)*	*(91)*	*(21)*	*(97.29)*
6th (C1) Gear									
83.31	6628	4.71	2102	4.37	0.454	15.41	197	70	28.73
(62.12)	*(29.48)*	*(7.59)*			*(0.276)*	*(3.04)*	*(92)*	*(21)*	*(97.29)*
7th (B3) Gear									
83.81	5708	5.51	2099	3.57	0.454	15.40	197	72	28.70
(62.50)	*(25.39)*	*(8.86)*			*(0.276)*	*(3.03)*	*(91)*	*(22)*	*(97.19)*
8th (C2) Gear									
81.41	4557	6.70	2101	2.76	0.463	15.10	196	71	28.72
(60.71)	*(20.27)*	*(10.78)*			*(0.282)*	*(2.97)*	*(91)*	*(22)*	*(97.26)*
9th (C3) Gear									
80.07	3374	8.90	2099	2.11	0.469	14.92	197	71	28.73
(59.71)	*(15.01)*	*(14.32)*			*(0.285)*	*(2.94)*	*(92)*	*(22)*	*(97.29)*

TRACTOR SOUND LEVEL WITH CAB	**Front Wheel Drive Disengaged dB(A)**	**Front Wheel Drive Engaged dB(A)**
At 75% load in7th (B3) Gear	73.0	73.5
Bystander in 12th (D3) Gear	82.5	—

TIRES, BALLAST AND WEIGHT	**Tested Without Ballast**
Rear Tires—No., size, ply & psi *(kPa)*	Two 18.4R38; *; 16 *(110)*
Front Tires—No., size, ply & psi *(kPa)*	Two 13.6R28;***; 24 *(165)*
Height of Drawbar	21.5 in *(545 mm)*
Static Weight with Operator—Rear	8336 lb *(3781 kg)*
—Front	4352 lb *(1974 kg)*
—Total	12688 lb *(5755 kg)*

Figure 3.4 (continued on next page).

THREE POINT HITCH PERFORMANCE (SAE Static Test)

Observed Maximum Pressure psi. *(bar)*	2950 *(203)*
Location	lift cylinders
Hydraulic oil temperature °F *(°C)*	140 *(60)*
Location	hydraulic sump
Category	II
Quick attach	No

With lift cylinders—1 × 70 mm and 1 × 80 mm					
Hitch point distance to ground level in. *(mm)*	8.0 *(203)*	15.0 *(381)*	22.0 *(559)*	29.0 *(737)*	36.0 *(914)*
Lift force on frame lb	11178	10092	9491	8851	8061
Lift force on frame *(kN)*	*(49.7)*	*(44.9)*	*(42.2)*	*(39.4)*	*(35.9)*

With lift cylinders - 2 × 80 mm					
Hitch point distance to ground level in. *(mm)*	8.0 *(203)*	15.0 *(381)*	22.0 *(559)*	29.0 *(737)*	36.0 *(914)*
Lift force on frame lb.	12739	11605	10926	10131	9239
Lift force on frame *(kN)*	*(56.7)*	*(51.6)*	*(48.6)*	*(45.1)*	*(41.1)*

THREE POINT HITCH PERFORMANCE (OECD Static Test)

CATEGORY: II		
Quick Attach: none		
Maximum Force Exerted Through Whole Range:	6858 lbs	*(30.5 kN)*
	7790 lbs	*(34.6 kN)* with 80 mm lift cylinders
i) Opening pressure of relief valve:	NA	
Sustained pressure with pump stalled:	2880 psi	*(199 bar)*
ii) Pump delivery rate at minimum pressure and rated engine speed:	25.9 GPM	*(98.0 l/min)*
iii) Pump delivery rate at maximum hydraulic power:	23.7 GPM	*(89.7 l/min)*
Delivery pressure:	2640 psi	*(182 bar)*
Power:	36.5 HP	*(27.2 kW)*

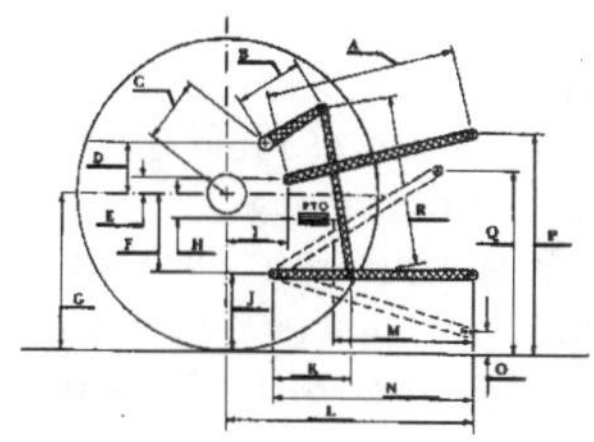

HITCH DIMENSIONS AS TESTED—NO LOAD

	inch	*mm*
A	27.8	*705*
B	15.7	*400*
C	21.8	*554*
D	20.6	*523*
E	4.9	*125*
F	9.8	*250*
G	32.3	*820*
H	3.1	*80*
I	18.0	*456*
J	22.5	*570*
K	21.1	*537*
L	44.2	*1122*
M	19.8	*502*
N	37.9	*962*
O	8.0	*203*
P	46.5	*1180*
Q	35.9	*911*
R	35.0	*889*

JOHN DEERE 7200 SYNCROPLUS DIESEL

Agricultural Research Division
Institute of Agriculture and Natural Resources
University of Nebraska–Lincoln
Darrell Nelson, Dean and Director

Figure 3.4 (continued from previous pages). Nebraska OECD tractor test 1678 – Summary 150 of a John Deere 7200 Syncroplus Diesel 12 Speed. (Courtesy of University of Nebraska Tractor Test Laboratory.)

of the summary reports is to allow prospective tractor buyers to compare tractors. Performance at the PTO is directly comparable by using either the OECD summaries or reports of SAE/ASAE/ISO tests. Care must be used in comparing drawbar performance because of differences in test procedures. For example, SAE/ASAE/ISO tests are run at rated engine speed, while some OECD tests are run at the maximum power point.

If the tractor has a PTO shaft, a two-hour PTO test is run at the speed that produces maximum power. Short tests are also run at rated engine speed and at standard PTO speed if these differ from the speed at maximum power. In the report of Figure 3.4, for example, rated engine speed was 2100 rpm but maximum power was at 1900 rpm. A series of short, part-load PTO tests are run in the governor-controlled range. The engine is also loaded into the load-controlled range so that the torque reserve can be reported. In addition to power and speed, fuel consumption is measured during each PTO test. By neglecting power losses between the engine and PTO shaft, the approximate engine torque for each test can be calculated from power and engine speed by using the brake power equation. The ambient temperature and barometric pressure affect the mass air consumption of an engine and thus the atmospheric conditions prevailing during the PTO tests are reported. In SAE/ASAE/ISO testing, ambient conditions are recorded for every PTO test. As shown in Figure 3.4, however, the average ambient conditions during all PTO tests are recorded in OECD tests. Note that, in the Varying Power tests of Figure 3.4, the SFC (in kg/kWh) increases substantially as engine load decreases, in accordance with our discussions in Chapter 2.

In OECD tests, the tractor is tested in each gear provided that slip and speed limits are not exceeded. The first drawbar test reported in the NTTL summary (Figure 3.4) is for the best pulling gear, i.e., the gear that produces the most drawbar power. Although not mentioned in the report, the front-wheel drive was engaged for this series of drawbar tests. Comparing the engine speed in the maximum power – 6th gear test with engine speeds in the PTO tests shows that this drawbar test was run at rated engine speed. Next, the NTTL reports results of a test at 75% of the pull that produces maximum power. Listed next are results of a test at 50% of the pull that produces maximum power. The next two tests are again at 75% and 50% of pull, but at reduced engine speed. In the latter two tests, the tractor was shifted to a higher gear (from 6th to 8th gear in this case) and the engine speed was reduced to give approximately the same travel speed as in the earlier tests. The latter two tests demonstrate the fuel savings that can be achieved by operating the engine in more favorable areas of the engine map, as was discussed in Chapter 2. The fuel savings can be calculated by comparing specific fuel consumption across tests. For example, the reader is encouraged to compare the SFC for the two tests at 50% load, one with the tractor in 6th gear and the engine running at 2221 rpm; the other with the tractor in 8th gear and the engine running at 1589 rpm.

The next sections of the report are on series of drawbar tests in selected gears. The first series was intended to show drawbar performance of the tractor at maximum power, i.e., with the engine running at 1900 rpm, and with the front-wheel drive engaged. Note that, for the engine to produce maximum power, the drawbar load must be increased as the tractor is shifted to lower gears. Also note the increase in reported wheel slip as the drawbar pull was increased. First gear was not tested because the pull would have produced wheel slip greater than 15%. The engine was at maximum power during the tests in gears 5 through 9. In gears 2, 3 and 4, however, the drawbar load was reduced to prevent excessive slip and the engine produced less than maximum power. Gears 10, 11 and 12 were not tested to avoid use of excessive speeds on the concrete test track.

The second series of tests of various gears was intended to show performance with the engine at rated speed (2100 rpm) and with the front-wheel drive engaged. A third and fourth series of drawbar tests were run for this test report; they were similar to earlier series except that the front-wheel drive was disengaged.

The test report contains details about the fuel and lubricating oil used, engine specifications and operating parameters, and information about the chassis. The latter includes the gear selections available and advertised speeds in each gear. Data are given as to the center of gravity location, axle loads, drawbar height and tires used, which would be useful in calculating tractive performance. The lifting performance of the 3-point hitch is measured and reported in accordance with ASAE Standard S349.2. Finally, sound levels inside the cab were measured and reported.

NTTL or OECD tests have another use in addition to providing a comparison of different tractors. The data in the tests can be further analyzed to gain some valuable insights into the performance of engines and tractors. In this chapter, for example, some of the homework problems use test report data to show the fuel savings that can be achieved by shifting a partially loaded tractor to a higher gear and reducing the engine speed. Homework problems in other chapters will make use of the test data to illustrate other points concerning engine or tractor performance.

3.8 Correcting Power for Atmospheric Conditions

The density of air changes with temperature and barometric pressure. Such density changes affect the rate at which an engine can consume air and, since it is air consumption that limits engine power, the power output varies with atmospheric conditions. SAE Standard J1349 provides a method for correcting engine power to standard conditions. The theory below is based on that standard.

Power correction begins with use of the ideal gas law to correct for atmospheric conditions. From the ideal gas law, pV= MRT. Here, p and V, are the absolute pressure and the gas volume in the combustion chamber that contains mass M of gas. The absolute temperature of the gas is T, and R is an ideal gas constant. The ideal gas law is further explained in Chapter 4. Because mass density is the mass per unit volume, the ideal gas law can be restated as

$$\rho = \frac{M}{V} = \frac{BP}{RT} \tag{3.3}$$

where

ρ = air density, kg/m^3
BP = barometric pressure, kPa
T = absolute temperature of air, K
R = ideal gas constant

Equation 3.3 can be used to obtain the ratio of air densities at observed and standard conditions. The gas constant cancels out in obtaining the ratio, with the following result:

$$\frac{\rho_s}{\rho_o} = \frac{BP_s}{BP_o}\frac{T_o}{T_s} \tag{3.4}$$

where

ρ_o = air density at observed conditions, kg/m^3
ρ_s = air density at standard conditions, kg/m^3
BP_o = barometric pressure at observed conditions, kPa
BP_s = barometric pressure at standard conditions, kPa
T_o = absolute temperature of the air at observed conditions, K
T_s = absolute temperature of the air at standard conditions, K

Note that increased ambient air temperature reduces the air density and thus reduces the mass of air that is present in the displacement volume of each cylinder. However, the engine is hotter than the ambient air and transfers heat to it as the air enters the engine, i.e., the air increases in temperature. The higher the temperature of the incoming air, the less heat is transferred to it and thus, the air density in the combustion chamber doesn't vary with T^{-1}, but with some lesser power. Therefore, for SI engines, the power correction factor is

$$f_a = \frac{BP_s}{BP_o}\left(\frac{T_o}{T_s}\right)^n \tag{3.5}$$

where exponent n = 0.5 for typical, naturally aspirated SI engines.

Theoretically, the correction should be applied to the indicated power, then

$$P_{bs} + P_{fs} = f_a\,(P_{bo} + P_{fo}) \tag{3.6}$$

But $P_{fs} = P_{fo}$, because friction power is not affected by atmospheric conditions. Thus, Equation 3.6 reduces to

$$P_{bs} = f_a\,P_{bo} + P_f(f_a - 1) \tag{3.7}$$

Because P_f is often much smaller than P_b and f_a is close to 1.0, the last term is usually dropped. The final power correction equation is then

$$P_{bs} = f_a\,P_{bo}a \tag{3.8}$$

That is, the correction is applied directly to the brake power. The usual choice of standard conditions is BP_s = 99 kPa and T_s = 298 K (25 °C).

The power correction for CI engines is more complex; it is

$$P_{bs} = f_a^{fm} f_f P_{bo} \tag{3.9}$$

where f_a is defined as

$$f_a = \left(\frac{99}{BP_o}\right)^m \left(\frac{T_o}{298}\right)^n \tag{3.10}$$

where values for m and n are given in Table 3.1 and

$$f_f = f_d f_v \tag{3.11}$$

$$f_d = 1 + 0.7\left(\frac{0.85 - SG}{SG}\right) \tag{3.12}$$

For conventional, pump/line/nozzle systems,

$$f_v = 0.945 + 0.142/\nu_f \tag{3.13}$$

where

ν_f = kinematic viscosity of fuel, mm^2/s
SG = specific gravity of fuel
f_v = 1 for unit injector systems

Values for exponents m and n are given in Table 3.1.

Table 3.1. Values for m and n.

Pressure Charging	Air Cooling	m	n
Naturally aspirated	None	1	0.7
Supercharged	All	1	0.7
Turbocharged	None	0.7	1.2
Turbocharged	Air to air	0.7	1.2
Turbocharged	Air to water	0.7	0.7

Also,

$$fm = \begin{cases} 0.2 & q/r < 37.2 \\ 0.036q/r - 1.14 & 37.2 \le q/r \le 65 \\ 1.2 & q/r > 65 \end{cases}$$

where

q = 16,667 FX/(D_e N_e) (q is proportional to ISFC)
F = fuel consumption in kg/h
D_e = engine displacement in liters
N_e = engine speed in rpm
X = 1 for a two-cycle engine or 2 for a four-cycle engine
r = 1 + boost/BP_o (boost = turbocharger boost; see Chapter 8)
r = 1 for naturally aspirated engines

Power values reported in Nebraska or OECD tests are not corrected to standard atmospheric conditions. The observed conditions are reported, however, so the report user can make the corrections if desired. When turbocharged tractor engines are tested, the boost is measured and reported with the test results.

3.9 Chapter Summary

An engine dynamometer must be able to control the torque load on the engine, measure the torque and speed, and dissipate the power. Two categories of dynamometers are available. Absorption dynamometers can only absorb the engine power and must be provided with a means for dissipating the heat. Electric dynamometers are generators that convert the engine power to electric power for transmission into an electrical circuit. Also, electric generators are capable of motoring an engine to start it or measure its friction characteristics.

Engine fuel consumption can be measured volumetrically by using a rotameter or a turbine flow meter. Fuel flow can be measured on a mass basis by allowing the engine to withdraw fuel from a beaker on a scale. With CI engines, there is return flow from the fuel injectors and care must be taken to measure the net fuel flow to the engine.

Measuring air consumption is important because maximum power output of an engine is limited by its air-handling ability. Air consumption is measured by inserting a calibrated orifice or nozzle into the air intake system.

Through use of a piezoelectric pressure transducer tapped into the combustion chamber and a shaft encoder attached to the crankshaft, it is possible to measure cylinder pressures and gas volumes while an engine is running. The data can be used to plot a pressure-volume diagram of the type discussed in Chapter 2. Through further analysis, the data can be used to infer the rate of energy release from the combustion process in a running engine. Interpretation of energy release diagrams was deferred to Chapter 5.

For many years, the Nebraska Tractor Testing Laboratory performed tractor tests. These tests were run according to standards developed by SAE, ASAE and ISO and could be used to compare various tractors. More recently, the NTTL tests are being supplanted by OECD tests, which are more varied than the NTTL tests. However, personnel at the NTTL produce summaries of OECD tests that are similar in format to NTTL test reports. An example NTTL summary of an OECD test was included in Chapter 3; by reviewing the test report, the reader can verify some of the concepts that were presented in Chapter 2.

Engine performance is affected by atmospheric conditions and it is seldom possible to perform all tests at standard conditions. However, the chapter presents equations for correcting the observed power to standard conditions. The power outputs of tractors tested by the NTTL or in OECD tests are not corrected to standard conditions. However, the ambient temperatures and barometric pressures during the test are reported, thus allowing the report user to make the corrections.

Homework Problems

3.1 During the test of a certain CI engine, the torque was 360 N·m, the speed was 2100 rpm and the engine consumed 200 grams of diesel fuel in 42.0 seconds. The air consumption was measured at 480 kg/h.

(a) Calculate the brake power.

(b) Calculate the fuel consumption in kg/h.

(c) Calculate the BSFC in g/kWh and (d) the air-to-fuel consumption ratio.

3.2 During the test of a certain CI engine, the torque was 75.8 N·m, the speed was 2245 rpm and the engine consumed 100 grams of diesel fuel in 44.8 seconds. The air consumption was measured at 640 kg/h.

(a) Calculate the brake power.

(b) Calculate the fuel consumption in kg/h.

(c) Calculate the BSFC in g/kWh and (d) the air-to-fuel consumption ratio.

3.3 An engine has a 106.4 mm bore and a 110.0 mm stroke. The connecting rod length is 3 times the crank throw radius and the clearance volume is 58.3 cm^3.

(a) Calculate the crank throw radius, R.

(b) Calculate the area of the piston, A_p.

(c) Then, using Equation 3.1, plot gas volume, V_g, for one full revolution of the crankshaft, beginning at HDC.

(d) On the same graph, also plot the gas volume if the last term in Equation 3.1 was dropped, i.e., if the piston movement was purely sinusoidal. Even if the difference between the curves is small, the full equation for gas volume is used in engine performance calculations.

3.4 An engine has a 115.8 mm bore and a 120.7 mm stroke. The connecting rod length is 3 times the crank throw radius and the clearance volume is 84.8 cm^3.

(a) Calculate the crank throw radius, R.

(b) Calculate the area of the piston, A_p.

(c) Then, using Equation 3.1, plot gas volume, V_g, for one full revolution of the crankshaft, beginning at HDC.

(d) On the same graph, also plot the gas volume if the last term in Equation 3.1 was dropped, i.e., if the piston movement was purely sinusoidal. Even if the difference between the curves is small, the full equation for gas volume is used in engine performance calculations.

3.5 Starting with a slider-crank diagram, derive Equation 3.1.

3.6 Using data from the "varying power and fuel consumption" power take-off tests of Figure 3.4, plot specific fuel consumption versus power. Discuss how the shape of the curve is or is not similar to what was suggested in Chapter 2.

3.7 Using the drawbar performance data from Figure 3.4 in which the front wheel drive was engaged, calculate the percentage reduction in specific fuel consumption at 75% pull when the tractor was shifted to a higher gear and the engine speed was reduced.

3.8 Same as Problem 3.7, except use the data at 50% pull.

3.9 Same as Problem 3.7, except use the data for the front drive disengaged.

3.10 Same as Problem 3.7, except use the data for 50% pull when the front drive was disengaged.

3.11 Using theory in Chapter 3 and data from Figure 3.4, correct the power measured on the power take-off test at rated engine speed to standard conditions. Assume the fuel viscosity is 2.6 mm^2/s. Note that the correction method applies to PTO power as well as brake power. Unless ambient conditions are extreme, the correction is usually very small.

3.12 Same as Problem 3.11, except correct the power measured during the 2-hour maximum power PTO test.

3.13 Using the data from Problem 3.11, plot the value of P_{bs}/P_{bo} for ambient temperatures ranging from 270 to 310°K. Plot two curves on the same graph, one with $BP_o = 97$ kPa and one with $BP_o = 100$ kPa. The curves should indicate larger corrections at higher temperatures.

3.14 Using the data from Problem 3.11, except using r = 1, plot the value of P_{bs}/P_{bo} for ambient temperatures ranging from 270 to 310°K. Plot two curves on the same graph, one with $BP_o = 97$ kPa and one with $BP_o = 100$ kPa. The curves should indicate larger corrections at higher temperatures.

3.15 Using theory in Chapter 3 and data from Figure 3.4, correct the power measured on the power take-off test at rated engine speed to conditions measured on a certain day in Denver, Colorado when the barometric pressure was 85 kPa and the ambient temperature was 20°C. Assume the fuel viscosity is 2.6 mm^2/s. Note that the correction method applies to PTO power as well as brake power. This problem illustrates the derating that is necessary if a naturally aspirated engine is operated at high elevations.

References and Suggested Readings

Faletti, J.J., S.C. Sorenson and C.E. Goering. 1984. Energy release rates from hybrid fuels. *Transactions of the ASAE* 27(2): 322-325.

Goering, C.E. 1998. Engine heat release rates via spreadsheet. *Transactions of the ASAE* 41(5): 1249-1253.

SAE, 1992a. Measurement of intake air or exhaust gas flow of diesel engines. SAE Standard J244. *SAE Handbook,* vol. 3. Warrendale, PA: SAE.

SAE, 1992b. Agricultural tractor test code. SAE Standard J208. *SAE Handbook,* vol. 4. Warrendale, PA: SAE.

SAE, 1992c. Agricultural tractor test code (OECD). SAE Standard J2708. *SAE Handbook*, vol. 4. Warrendale, PA: SAE.

CHAPTER 4

THERMODYNAMICS AND ENGINE CYCLES

4.1 Introduction

In this chapter, a brief engine history is presented to trace some of the thermodynamic ideas that are used in modern engines. The ideal gas law and polytropic compression/expansion law are reviewed as lead-ins to cycle analysis. Then the Otto cycle is presented as the ideal model for four-cycle SI engines. The dual cycle is presented as the ideal model for four-cycle CI engines. These models provide upper limits against which certain performance parameters of actual four-cycle engines can be compared. Finally, two-cycle SI and CI engines are discussed.

4.2 A Brief Engine History

Internal combustion (IC) engines, as the name implies, produce power through a combustion process occurring within the piston chambers. In contrast, combustion occurs outside the piston chambers in a steam engine–the combustion heats a boiler to produce steam and the steam is delivered to the piston chambers to produce power. One of the earliest experiments with IC engines occurred when Abbe Hautefeuille, a Frenchman, built a closed chamber in which he exploded gunpowder. The resulting high pressure was used to raise a water column that was in a connecting chamber. By 1680, a Dutch physicist named Huyghens replaced the water column with a piston, which would move when the gunpowder was exploded. Such engines of the exploding type were not very efficient, primarily because the gases in the combustion chamber were not compressed before ignition occurred.

More than 50 years later, in 1838, a scientist named Burnett Pointed out the advantages of compressing the gases in the combustion chamber before ignition. Then, in 1862, *Beau de Rochas* set forth his famous four principles of operation for an efficient IC engine. They were:

1. The combustion chamber should have the smallest possible surface-to-volume ratio,
2. The expansion should take place as rapidly as possible,
3. The piston should have the longest possible stroke, and
4. The pressure should be as high as possible at the start of expansion.

The first two principles were aimed at reducing the heat loss to a minimum, while the second two were aimed at obtaining the maximum possible work from each power stroke.

Nikolaus Otto built on Beau de Rochas' principles, patenting his famous Otto-cycle engine in 1876. Two years later he built a successful IC engine. Otto was the first to use the four-stroke cycle, i.e., the intake, compression, power, and exhaust strokes that are still used in most IC engines today. With the expiration of the Otto patent in 1890, there was a spurt in development and commercialization of IC engines.

Otto's contemporaries thought that having only one power stroke in two crankshaft revolutions was a serious disadvantage and experimenters turned their attention back to two-stroke cycles of the explosive type–that is, without compression. In 1881, three years after Otto patented his engine, Sir Dugald Clerk built a two-stroke cycle engine with compression, but he abandoned it due to mechanical difficulties. Joseph Day simplified the design and, in 1891, he patented a two-stroke cycle engine that used a gas-tight crankcase as a pumping cylinder. Day's design is still used on modern two-stroke cycle engines of the spark-ignition type. By 1906, the Cushman Company was producing successful two-cycle, two-cylinder spark-ignition engines for use in farm tractors.

Dr. *Rudolph Diesel* patented his compression-ignition engine in 1892. The first diesel engine in the United States was built for the Busch brewery in St. Louis under license from Rudolph Diesel.

4.3 Four-Cycle Engine Analysis

As was mentioned in the previous section, the intake, compression, expansion and exhaust processes can be carried out in two or four piston strokes. Thus, there are two types of engines, those with *two-stroke cycles* and those with *four-stroke cycles*. For brevity, these are usually called two-cycle and four-cycle engines, respectively. *Four-cycle engines* have become much more popular than *two-cycle engines*. Thus, the reader can assume that all discussions in this textbook refer to four-cycle engines unless two-cycle engines are specifically mentioned.

4.3.1 Basic Thermodynamic Equations

Two thermodynamic laws are used in cycle analysis, the ideal gas law and the polytropic compression law. The ideal gas law is

$$pV = MRT \tag{4.1}$$

where

p = absolute gas pressure, kPa
V = gas volume, m^3
M = mass of trapped gas, kg
T = absolute temperature of gas, K
R = ideal gas constant = 8.314/mole wt of gas

In cycle analysis, changes in the condition of a gas are calculated rather than the absolute condition, and the gas constant drops out of the analysis.

The mass of gas in a combustion chamber remains essentially constant during the compression process. If the temperature also remained constant, Equation 4.1 indicates that the pressure would vary inversely with the volume. That is, the pressure would double if the volume was halved. Such a process is called an *isothermal* process. In an engine, the temperature increases considerably during the compression stroke; therefore, the pressure increases more than in an isothermal process. A *polytropic* process follows

$$pV^n = C_p \tag{4.2}$$

where C_p = a constant and n = an exponent between 1 and 1.4.

If the combustion chamber were perfectly insulated, so that there was no energy loss, then $n = k = 1.4$ and the process is called an *adiabatic* process. Thus, the use of k instead of n in Equation 4.2 indicates the process is an adiabatic process. For an isothermal process, $n = 1$. The value of n is always between 1 and 1.4 in an IC engine. It is usually a little above 1.3 on the compression stroke, but can fall to 1.26 or lower on the power stroke.

Suppose that a gas changed from state 1 at the beginning of a compression to state 2 at the end of the compression. Then, the ideal gas law can be used to show that

$$\frac{p_1 V_1}{T_1} = \frac{p_2 V_2}{T_2} \tag{4.3}$$

The polytropic law can be used to show that

$$\frac{p_2}{p_1} = \left(\frac{V_1}{V_2}\right)^n \tag{4.4A}$$

The pressure rises from p_1 to p_2 as the gas is compressed from volume V_1 to smaller volume, V_2. The ratio, V_1/V_2, is thus referred to as the *compression ratio*, r, i.e.,

$$r = \frac{V_1}{V_2} \tag{4.5}$$

Then Equation 4.4A can also be rewritten as

$$\frac{p_2}{p_1} = r^n \tag{4.4B}$$

By combining the ideal gas law and the polytropic law, it can be shown that

$$\frac{T_2}{T_1} = r^{n-1} \tag{4.6}$$

4.3.2 The Otto Cycle

The theoretical Otto cycle is shown in Figure 4.1. It includes intake from Point 0 to Point 1, adiabatic compression from Point 1 to 2, constant-volume heat input from Point 2 to 3, adiabatic expansion from Point 3 to 4, blow down from Point 4 to 1, and exhaust from Point 1 to 0. The volume, V_2, is called the *clearance volume*; it is the cylinder volume when the piston is at HDC. The cylinder *displacement* is the change in volume from Point 1 to 2, i.e.,

$$D_c = V_1 - V_2 \tag{4.7}$$

where D_c = displacement of a single cylinder, L.

Although Equation 4.1 shows volumes measured in m^3, use of liters is permitted since we are working with ratios of volumes and engine displacements are generally measured in liters.

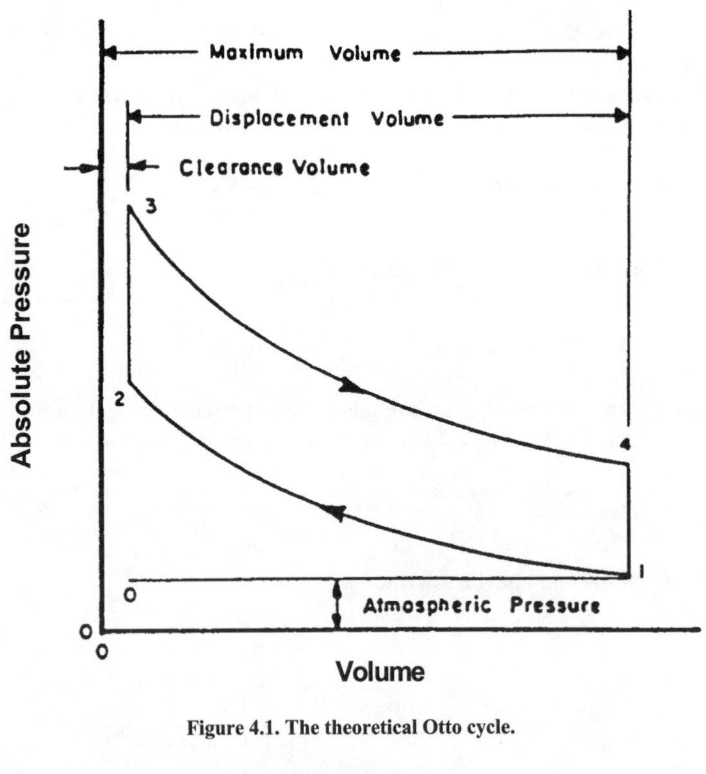

Figure 4.1. The theoretical Otto cycle.

In analyzing the Otto cycle, it is assumed that the conditions at Point 1 are atmospheric and thus p_1 and T_1 are given. It is also assumed that cylinder displacement, D_c, and compression ratio, r, are given. It can then be shown that

$$V_2 = \frac{D_c}{r-1} \tag{4.8}$$

and that

$$V_1 = \frac{r\,D_c}{r-1} \tag{4.9}$$

Finally, it can be shown that T_3 is the maximum temperature occurring in the cycle. Thus, T_3 is also specified as one of the givens for the cycle. The *adiabatic flame temperature*, T_3 = 2700 K, is usually specified as the given value for T_3. It is then possible to calculate the conditions at all Points in the Otto cycle, as shown in Table 4.1.

The work accomplished in the Otto cycle can be calculated by integrating the quantity p dv through the cycle. The results can be simplified to

$$W = p_1 D_c [\frac{(r - r^k) + \frac{T_3}{T_1}(r - r^{2-k})}{(k-1)(r-1)}] \tag{4.10}$$

where

W = work, J/cycle
p_1 = initial pressure, kPa
D_c = displacement volume, L
k = 1.4

By definition, the *cycle mean effective pressure*, p_{cme}, is

$$p_{cme} = \frac{W}{D_c} \tag{4.11}$$

where p_{cme} = cycle mean effective pressure, kPa.

The cycle power is

$$P_c = \frac{p_{cme} D_c N_e}{2(60{,}000)} \tag{4.12}$$

where

P_c = cycle power, kW
N_e = crankshaft speed of the engine in which the cycle is implemented, rpm

Table 4.1 Otto cycle summary.

Quantity	Point 1	Point 2	Point 3	Point 4
P, kPa	p_1	$p_1\, r^k$	$(T_3 / T_1)\, p_1\, r$	$(T_3 / T_1)\, p_1\, r^{1-k}$
V, L	$r\, D_c / (r-1)$	$D_c / (r-1)$	$D_c/(r-1)$	$r\, D_c / (r-1)$
T, °K	T_1	$T_1\, r^{k-1}$	T_3	$T_3\, r^{1-k}$

The 2 in the denominator would become a 1 for a two-cycle engine. Finally, it can be shown that the cycle efficiency is

$$e_{Otto} = 1 - r^{1-k} \tag{4.13}$$

where e_{Otto} = Otto-cycle efficiency, decimal.

The energy flows into and out of the cycle are

$$Q_{in} = \frac{W}{e_{Otto}} \tag{4.14}$$

and

$$Q_{out} = Q_{in} - W \tag{4.15}$$

where

Q_{in} = energy into cycle, J/cycle

Q_{out} = energy rejected from cycle, J/cycle

The Otto-cycle efficiency serves as a goal against which to measure indicated thermal efficiencies attained in actual spark-ignition (SI) engines running at full load. Also, p_{cme} serves as a goal against which to measure achieved values of indicated mean effective pressure in SI engines running at full load. There are a number of simplifying assumptions that underlie the theoretical Otto cycle. These include zero friction, air as the only working fluid, zero heat transfer, constant-volume heat addition and constant-volume heat rejection. Also, as shown in Figure 4.2, an actual pV diagram includes pumping losses during the intake and exhaust stroke that are not included in the theoretical Otto cycle. All of these assumptions are violated to some extent in a running engine and thus the e_{Otto} and p_{cme} values calculated for the Otto cycle can never be achieved in practice.

4.3.3 The Classic Diesel Cycle

Rudolph Diesel's original idea was to develop a compression-ignition engine that would burn the coal dust that was a waste product of that time. He soon found the coal dust to be unsuitable, but continued to develop his engine to burn liquid fuel. At the time that Diesel was developing his engine, he was well aware of the Otto cycle. The Diesel cycle differs from the Otto cycle only in that the process from Point 2 to 3 is a constant-pressure process rather than a constant-volume process. Let's try to reconstruct Diesel's rationale for substituting the constant-pressure process. Diesel realized his engine needed a compression ratio high enough to self-ignite the fuel at the end of the compression stroke. From Equation 4.6, by setting T_2 = 750 K (the self-ignition temperature of diesel fuel), taking T_1 = 300 K (a reasonable ambient temperature) and assuming n = 1.33, the compression ratio would have to be at least 16:1 to self-ignite the fuel. An even higher r would be required for lower values of T_1. From Equation 4.4B, assuming typical barometric pressure of 100 kPa, the pressure, p_2, at the end of the compression stroke would be 3994 kPa or higher. Then, if energy was released into the combustion chamber in a constant-volume process, the pressure

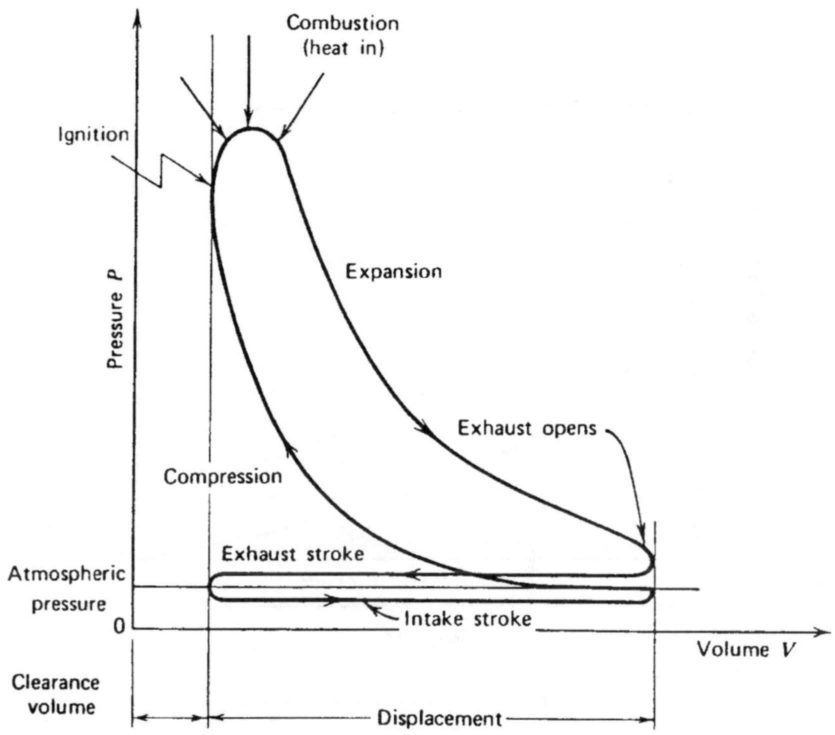

Figure 4.2. Actual pV diagram of a spark-ignition engine.

would rise much higher than 3994 kPa, i.e., higher than the engine could withstand. Diesel's solution was to release the energy in such a way that the pressure would rise no higher than the pressure p_2.

We will not analyze the theoretical diesel cycle because it is not a good model of the modern diesel engine. The dual cycle is a better model, because modern diesel engines have sufficient strength to allow some energy input at nearly constant volume. The theoretical dual cycle provides for part of the energy input at constant volume and the remainder at constant pressure. The dual cycle is a more general cycle, i.e., it includes the Otto cycle and the theoretical diesel cycle as special cases.

4.3.4 The Dual Cycle

The dual cycle is illustrated in Figure 4.3. It is similar to the Otto cycle, except the process between Points 2 and 2a is constant-volume heat input, while 2a to 3 is constant-pressure heat input. Equations 4.4A through 4.9, 4.11, and 4.12 are all valid for the dual cycle. In analyzing the dual cycle, it is assumed that conditions at Point 1 are atmospheric and thus p_1 and T_1 are given. As with the Otto cycle, r, D_c and T_3 are also given. In addition, it is necessary to specify the fraction of energy input that occurs at constant pressure, i.e.,

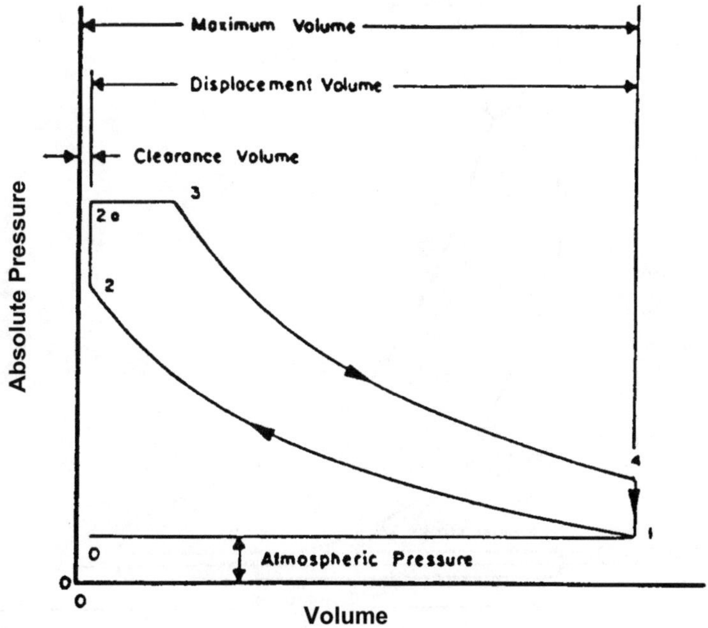

Figure 4.3. The theoretical dual cycle.

$$\beta_{dc} = \frac{Q_{inp}}{Q_{inp} + Q_{inv}} \tag{4.16}$$

where

β_{dc} = fraction of energy input at constant pressure
Q_{inp} = heat input at constant pressure, J/cycle
Q_{inv} = heat input at constant volume, J/cycle

The *fuel cutoff ratio*, r_{co}, defines the volume at which energy input ceases and can be calculated as

$$r_{co} = \frac{V_3}{V_2} = \frac{1 + \frac{\beta_{dc}}{k}}{1 + (\frac{\beta_{dc}}{k})(\frac{T_1}{T_3})r^{k-1}} \tag{4.17}$$

The quantity, u, defined below, is needed in calculating certain pressures.

$$u = k(\frac{1}{\beta_{dc}} - 1) \quad \text{for } \beta_{dc} > 0 \tag{4.18}$$

In calculating conditions at all Points in the dual cycle, note that the pressures, volumes and pressures at Points 1 and 2 are identical to those in the Otto cycle. The other dual-cycle pressures are

$$\frac{p_{2a}}{p_1} = \frac{p_3}{p_1} = \frac{r^k + u\,r(\frac{T_3}{T_1})}{u+1} \tag{4.19}$$

and

$$\frac{p_4}{p_1} = \frac{r_{co}[1 + u\,r^{1-k}(\frac{T_3}{T_1})]}{u+1} \tag{4.20}$$

The volumes at Points 1, 2 and 4 are identical to those in the Otto cycle. The remaining volumes are

$$V_{2a} = V_2 \tag{4.21}$$

and

$$V_3 = r_{co}\,V_2 \tag{4.22}$$

The temperatures at Points 1, 2 and 3 are identical to those in the Otto cycle. The temperatures at the remaining Points are:

$$T_{2a} = T_3 r_{co}^{-1} \tag{4.23}$$

and

$$T_4 = T_3(\frac{r_{co}}{r})^{k-1} \tag{4.24}$$

The cycle mean effective pressure is

$$\frac{p_{cme}}{p_1} = \frac{r - r^k + T_r(r - r^{2-k} r_{co}^{k-1} + r\,r_{co}^{-1}(k-1)(r_{co}-1))}{(k-1)(r-1)} \tag{4.25}$$

where

$$T_r = \frac{T_3}{T_1}$$

The dual-cycle efficiency is

$$e_{dual} = 1 - r^{1-k}[\frac{\beta_{dc}(r_{co}^k - 1) + k(r_{co}-1)(1-\beta_{dc})}{k(r_{co}-1)}] \tag{4.26}$$

The dual-cycle efficiency increases from that of the theoretical Diesel cycle at $\beta_{dc} = 1$ to that of the theoretical Otto cycle at $\beta_{dc} = 0$, as shown in Figure 4.4. The fourth principle of Beau de Rochas is consistent with the efficiency being highest when $\beta_{dc} = 0$, i.e., when all of the energy is input at constant volume. The smaller the volume into which a given amount of energy is released, the higher the pressure resulting from that energy input, and thus, according to Beau de Rochas, the higher the efficiency. This change in cycle efficiency has practical implications for diesel engines, as will be explained in the next section.

The total heat into the dual cycle is

$$Q_{inp} + Q_{inv} = \frac{p_{cme} D_c}{e_{dual}} \tag{4.27}$$

and

$$Q_{out} = (Q_{inp} + Q_{inv}) - (p_{cme} D_c) \tag{4.28}$$

The Otto cycle and original Diesel cycle are special cases of the dual cycle. If β_{dc} is nearly zero (Equation 4.18 fails if $\beta_{dc} = 0$), then Point 3 of Figure 4.3 moves to the left to Point 2a, $r_{co} = 1$, and $T_r = T_3/T_1$. Then the cycle mean effective pressure and the cycle efficiency become those of the Otto cycle. If $\beta = 1$, then Point 2 of Figure 4.3 moves upward to Point 2a and the cycle mean effective pressure and cycle efficiency become those of the original diesel cycle.

The theoretical dual cycle is based on many of the same simplifying assumptions that applied to the Otto cycle. All of the simplifying assumptions are violated to some extent in a running compression-ignition engine and thus the pressure and temperature values and the cycle efficiency can never be achieved in practice. However, these values serve as goals against which achieved values can be compared.

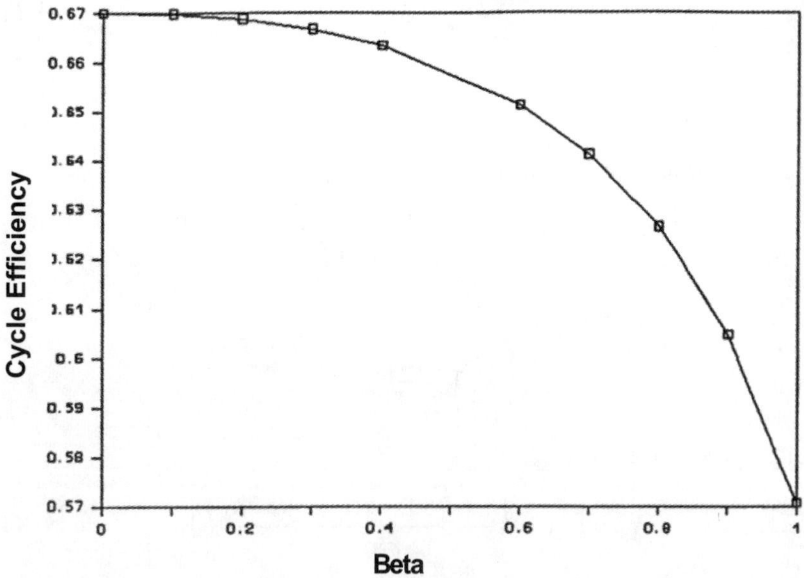

Figure 4.4. Dual-cycle efficiency versus fraction of energy input at constant pressure ($r = 16$, $T_3/T_1 = 9$).

4.4 Part-Load Efficiency of SI and CI Engines

To help provide a theoretical basis for the comparison of SI and CI engines, let us consider the brake thermal efficiency of engines, i.e., the fraction of the fuel equivalent power that is converted into power at the flywheel. By manipulation of equations in Chapter 2, the following equation for brake thermal efficiency can be derived:

$$e_{bt} = e_{it}(1+\frac{p_{fme}}{p_{bme}})^{-1} \tag{4.29}$$

Let us consider how the two engines respond to changes in load when the speed of both is held constant. With constant speed, note that the p_{fme} is constant for both engines, and that the p_{bme} varies in direct proportion to the torque load on each engine. As p_{bme} increases with increasing torque load on either an SI or a CI engine, the second term in Equation 4.29 increases, thus improving the brake thermal efficiency because of rising mechanical efficiency. Both engines have higher mechanical efficiency at full load and lower mechanical efficiency at part load. How does the indicated thermal efficiency vary in each of the engines as the load changes? To answer that question, we study the dual-cycle efficiency.

The best theoretical model for the indicated thermal efficiency is the cycle efficiency. The dual-cycle efficiency is given by Equation 4.26 and is plotted in Figure 4.4. As previously noted, β_{dc} is the fraction of the fuel energy that enters the cycle at constant pressure while fraction ($1-\beta_{dc}$) enters at constant volume. The design of CI fuel-injection systems is such that injection begins at a fixed Point before HDC and the *rate* of injection cannot be changed when the engine load changes. Instead, the *duration* of injections must change to suit the load on the engine; the heavier the load, the later in the cycle the injections must terminate and the larger the value of β_{dc}. Conversely, smaller values of β_{dc} correspond to higher values of cycle efficiency and, at $\beta_{dc} = 0$, the cycle efficiency rises to that of the Otto cycle. The dual-cycle efficiency equation thus predicts that CI engines increase their indicated thermal efficiencies as torque load declines. On the other hand, the theoretical Otto cycle predicts no change in indicated thermal efficiency when the load on an SI engine changes. Another prediction of Figure 4.4 is that the cycle efficiency of an SI engine would be higher than that of a CI engine if both were operated at the same compression ratio. As discussed in Chapter 5, however, engine knock considerations limit the compression ratios of SI engines to much lower values than those for CI engines. Thus, CI engines operate more efficiently than SI engines, both at full load and at part load. Part-load efficiencies are discussed in more detail in Chapter 7.

4.5 Two-Cycle Engines

As with four-cycle engines, two-cycle engines can be of either the SI or the CI type. Although there are many more SI than CI engines of the two-cycle type, both types will be discussed in the following sections.

4.5.1 Two-Cycle SI Engines

The two-cycle engine shown in Figure 4.5 uses the crankcase as an air pump. Upward piston movement creates a vacuum in the crankcase, causing the spring-loaded check valve to open to admit air-fuel mixture into the crankcase. The check valve closes and the mixture in the crankcase is compressed as the engine moves on its down stroke. The piston uncovers the intake and exhaust ports in the cylinder walls during the down stroke. When the ports are uncovered, the compressed mixture in the crankcase rushes into the combustion chamber. A deflector on the piston top directs the incoming mixture towards the top of the cylinder. The incoming mixture helps to push the spent exhaust gases from the previous cycle towards the exhaust port. These flow processes continue while the piston moves to CDC (Crank Dead Center, also called Bottom Dead Center or BDC), reverses direction and starts upward. After the rising piston covers the intake and exhaust ports, the trapped mixture is compressed as the piston continues to rise. Near HDC, a spark plug fires to ignite the mixture and the resulting pressure increase forces the piston downward. The cycle is completed when the intake and exhaust ports are again uncovered. Use of the crankcase as an air pump interferes with its use as a sump for lubricating oil. Lubrication of the moving parts in this two-cycle SI engine is accomplished by mixing oil with the fuel.

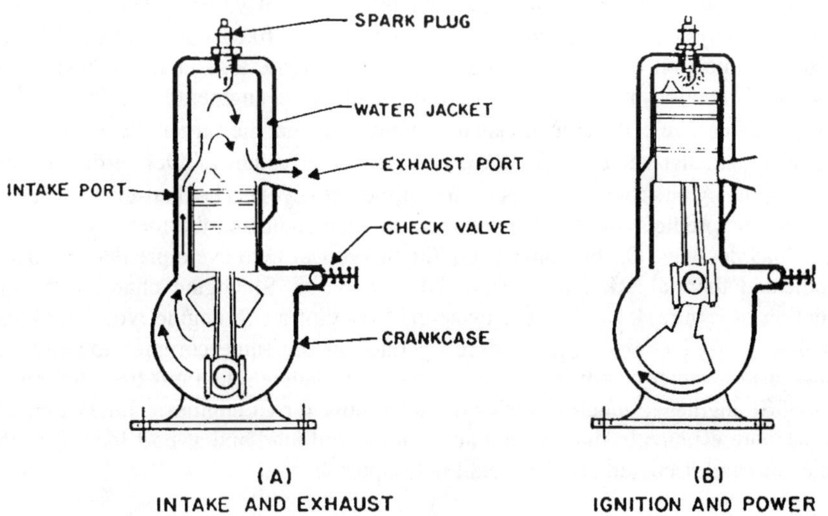

Figure 4.5. Two-cycle spark-ignition engine strokes.

The Otto cycle is not a good theoretical model for the two-cycle SI engine. The intake and exhaust ports restrict a portion of the piston travel to flow processes, rather than to the compression and expansion of a trapped fluid. The concepts of compression ratio and displacement involve the issue of what volume to use for V_1. One approach is to set V_1 equal to the trapped volume at the instant that the intake and exhaust ports are both covered by the rising piston. The definition of clearance volume is the same for two-cycle engines as for four-cycle engines. The ideal gas law and polytropic law are valid for any type of engine.

Two-cycle SI engines have a high power-to-size ratio. By producing a power stroke every crankshaft revolution, a two-cycle engine can produce more power than a four-cycle engine of the same physical size. On the other hand, two-cycle engines idle more erratically, have poorer fuel economy and can be more difficult to start than their four-cycle counterparts. Thus, two-cycle engines have been used primarily in applications in which their power-to-size ratio is an advantage, e.g., in chain saws, string trimmers, some lawn mowers, and outboard engines for boats. Some automobile manufacturers have developed prototype, two-cycle SI engines in which a mechanically-driven blower delivers air to the combustion chambers through conventional, cam-driven intake valves and exhaust is eliminated through conventional exhaust valves. Fuel is injected directly into the combustion chamber. These engines can use a conventional lubrication system, since the crankcase is available for use as an oil sump. The camshaft runs at crankshaft speed, while each spark plug fires once per crankshaft revolution. The reduced weight of these two-cycle SI engines is intended to help improve the fuel economy of automobiles. To date, however, these two-cycle automotive engines have not entered mass production.

4.5.2 Two-Cycle CI Engines

A two-cycle CI engine is shown in Figure 4.6. A mechanically driven blower delivers air through an intake port in the cylinder wall. A conventional, cam-driven valve is used to expel exhaust gases. The crankcase is available as an oil sump and thus the engine can have a conventional lubrication system. Typical timing events are shown in Figure 4.7. When the piston is about halfway through its down stroke, the exhaust valve opens to release the exhaust gases and then the intake port is uncovered. Compressed air from the blower enters the combustion chamber and helps to sweep the residual exhaust gases from the chamber. The exhaust valve begins closing after CDC and, while it is closing, the rising piston covers the intake port. The trapped air is further compressed by the continued upward movement of the piston and fuel injection begins just before HDC. The injected fuel is ignited by compression and the resulting pressure increase forces the piston downward to complete the cycle.

Like its SI counterpart, the two-cycle CI engine has a high power-to-size ratio. The reduced weight is less of an advantage in the heavy-duty vehicles in which two-cycle CI engines are used. These engines have been popular as bus engines, where their small size permits them to fit in a compact engine compartment at the rear of the bus. They have also been used in some farm tractors and heavy-duty trucks. The added cost of the required air blower is a disadvantage of two-cycle CI engines, and they have been available in far fewer numbers than their four-cycle counterparts.

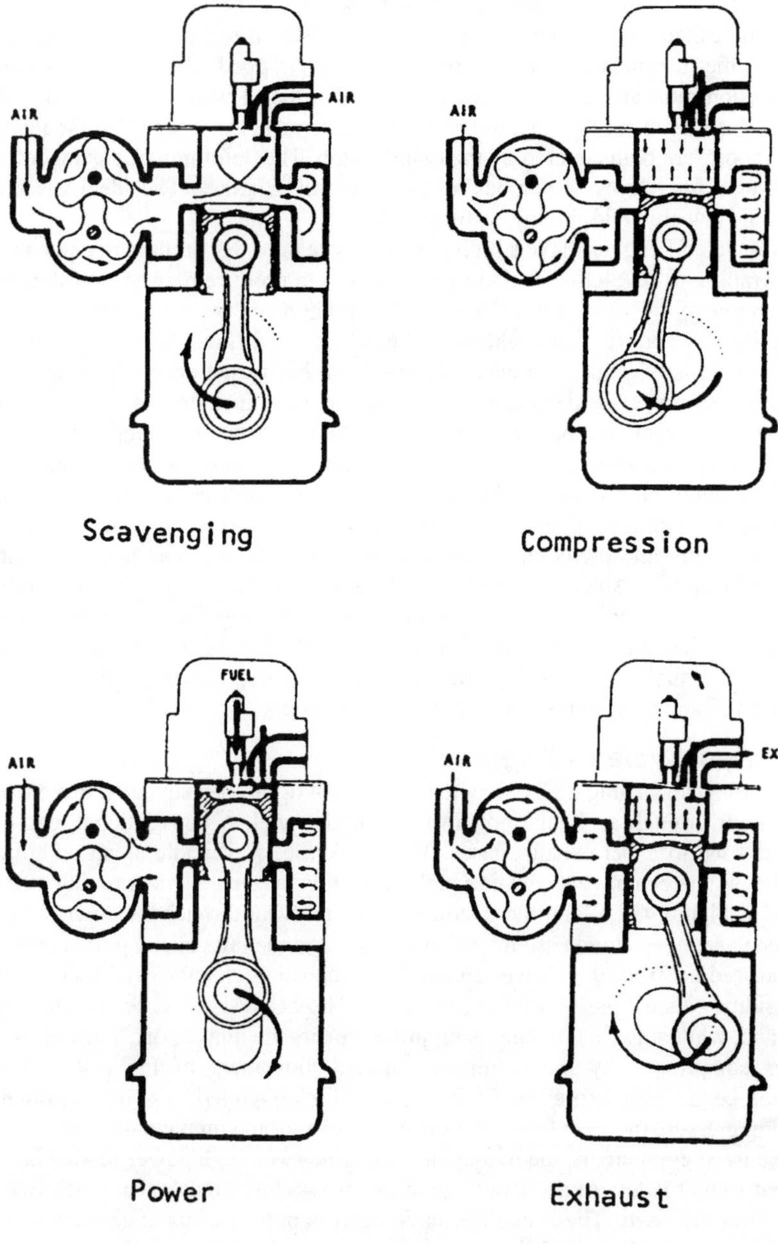

Figure 4.6. Two-cycle compression-ignition engine strokes.
(Printed by permission of copyright owner, Detroit Diesel Corporation, all rights reserved.)

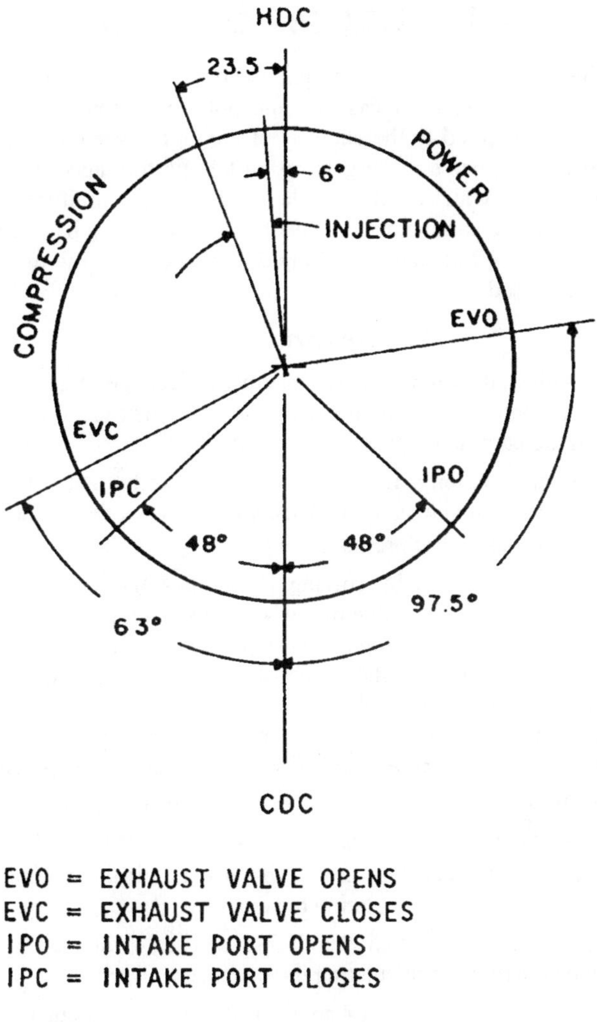

Figure 4.7. Typical timing of a two-cycle compression-ignition engine.

The dual cycle is not a good theoretical model for two-cycle CI engines. Also, as with their SI counterparts, there is some uncertainty as to what volume to choose for V_1 when calculating compression ratio and displacement. As shown in Figure 4.7, the piston can began compressing the trapped air when the exhaust valve fully closes, so V_1 could be taken as the volume at exhaust valve closure. Because of the blower, however, the trapped air at that Point is already partially pressurized.

4.6 Chapter Summary

In a brief history of engine development, the origin of some engine thermodynamic concepts was traced. The ideal gas law and polytropic compression/expansion law were reviewed and applied to the analysis of the Otto cycle and the dual cycle. The Otto cycle provides upper limit targets for certain performance parameters of actual four-cycle SI engines, while the dual cycle provides upper limit targets for four-cycle CI engine performance parameters. The dual cycle provided a basis for comparison of the indicated thermal efficiencies of SI and CI engines. Finally, two-cycle SI and CI engines were discussed.

Homework Problems

4.1 Starting with Equation 4.1, derive an equation for the mass density of air. Then use the equation to calculate the density of air when the ambient temperature is 30°C and the barometric pressure is 97 kPa.

4.2 Starting with Equation 4.1, derive an equation for the mass density of air. Then use the equation to calculate the density of air when the ambient temperature is 15°C and the barometric pressure is 99 kPa.

4.3 Derive a new equation by solving Equation 4.4A for n. Then use the new equation to calculate the value of n when the pressure increases from 100 kPa to 4000 kPa as a gas is compressed to one-sixteenth its initial volume. By inserting a pressure transducer in the combustion chamber and using an encoder to measure crank angle, researchers can obtain pressure volume data and use it to determine the value of n in actual engines. For a given r, the value of n is normally larger for compression and smaller for expansion processes.

4.4 Rework Problem 4.3, but calculate the value of n when the pressure decreases from 1266 kPa to 100 kPa as a gas is expanded to 7.5 times its initial volume.

4.5 Verify all the entries at Point 2 in Table 4.1, i.e., if the quantity is not a given, derive the equation for calculating the quantity.

4.6 Verify all the entries at Point 3 in Table 4.1, i.e., if the quantity is not a given, derive the equation for calculating the quantity.

4.7 Verify all the entries at Point 4 in Table 4.1, i.e., if the quantity is not a given, derive the equation for calculating the quantity.

4.8 For the Otto cycle, plot p_{cme}/p_1 versus r for $2 < r < 12$ if $T_3 = 9\ T_1$. Note that increased compression ratios put more stress on the engine. What value of r would you recommend to minimize such stress while gaining most of the improvement in p_{cme}?

4.9 Rework Problem 4.8, but let $T_3 = 10\ T_1$.

4.10 Plot e_{Otto} versus r for $2 < r < 12$ if $T_3 = 9\ T_1$. Note that simplifying assumptions underlying the Otto cycle are all violated in practice; thus, an actual spark-ignition engine cannot achieve the plotted efficiencies, but they can serve as an upper limit.

4.11 The following data were taken from a test of a four-cylinder SI engine running on gasoline: $N_e = 2250$ rpm, $\dot{m}_f = 10.8$ kg/hr, $H_g = 47{,}600$ kJ/kg, $D_e = 2.38$ L, $r = 7.4$ and $p_{ime} = 915$ kPa. Assume $p_1 = 100$ kPa.

(a) Calculate indicated power.

(b) Calculate fuel equivalent power.

(c) Calculate indicated thermal efficiency.

Now, assuming $T_3 = 9\ T_1$,

(d) Calculate theoretical Otto cycle efficiency.

(e) Calculate theoretical Otto cycle cycle mean effective pressure.

(f) Calculate theoretical Otto cycle cycle power.

(g) Compare the latter three quantities as upper limits to their corresponding measured quantities from the engine test.

4.12 The following data were taken from a test of a four-cylinder SI engine running on gasoline: $N_e = 2000$ rpm, $\dot{m}_f = 7.8$ kg/hr, $H_g = 47{,}600$ kJ/kg, $D_e = 2.59$ L, $r = 7.5$ and $p_{ime} = 705$ kPa. Assume $p_1 = 100$ kPa.

(a) Calculate indicated power.

(b) Calculate fuel equivalent power.

(c) Calculate indicated thermal efficiency.

Now, assuming $T_3 = 9\ T_1$,

(d) Calculate theoretical Otto cycle efficiency.

(e) Calculate theoretical Otto cycle cycle mean effective pressure.

(f) Calculate theoretical Otto cycle cycle power.

(g) Compare the latter three quantities as upper limits to their corresponding measured quantities from the engine test.

4.13 Plot p_{cme}/p_1 versus β_{dc} for $0.1 < \beta_{dc} < 0.7$ assuming $T_3 = 9\ T_1$. Plot two curves on the same graph, one for $r = 16$ and one for $r = 20$. Note that the simplifying assumptions underlying the dual cycle are violated in an actual engine and thus a compression-ignition engine cannot achieve the p_{cme} values plotted on your graph. The plots serve as an upper limit to performance.

4.14 Rework Problem 4.13, except plot curves for $r = 14$ and for $r = 18$.

4.15 Plot e_{dual} versus β_{dc} for $0.1 < \beta_{dc} < 0.7$ assuming $T_3 = 9\ T_1$. Plot two curves on the same graph, one for $r = 16$ and one for $r = 20$. Do the curves verify that increasing departures from constant-volume burning result in decreasing cycle efficiencies?

4.16 Rework Problem 4.15, but plot curves for $r = 14$ and $r = 18$.

References and Suggested Readings

Brain, Marshall. 2000. How car engines work. Available at: www.howstuffworks .com/engine.htm. Simulations and animations of engines and engine subsystems.

Gray, R.B. 1975. *The Agricultural Tractor: 1855-1950*. St. Joseph, MI: ASAE.

Rinschler, G.I., and R. Asmus. 1995. Power plant perspectives: Part I. *Automotive Engineering* 103(4): 37-42.

Sonntag, R.E., and G.J. Van Wylen. 1991. *Introduction to Thermodynamics,* 3rd Ed. New York, NY: John Wiley and Sons, Inc.

Tipler, P.A. 1991. *Physics for Scientists and Engineers,* 3rd ed. New York, NY: Worth Publishers.

CHAPTER 5

FUELS AND LUBRICANTS

5.1 Introduction

Through the refining process, crude oil is converted into a variety of petroleum products that are initially defined by their boiling ranges. The chemistry of ideal combustion of fuels provides insights into engine air-fuel ratios and an assessment of exhaust products. The importance of octane rating is shown through an analysis of combustion in SI engines, while an analysis of CI engine combustion shows the importance of cetane rating. Other important fuel properties are also discussed. Lubrication theory is presented. SAE viscosity rating and API service classification are presented as means of specifying lubricating oils for engines. A typical engine lubricating system is described.

5.2 Crude Oil Refining

Crude oil is found in rock formations that were floors of oceans thousands of years ago. Organic matter from the sea was trapped by the rocks and subjected to high pressure. Through progressive cracking of the molecules and elimination of oxygen, the organic matter eventually became petroleum. All petroleum is made up of approximately 86% carbon and 14% hydrogen. The *hydrocarbon molecules* in crude oil are found in a wide variety of sizes and are accompanied by dirt, water, sulfur, and other impurities. Thus, it is necessary to refine crude oil to produce fuels suitable for use in engines.

The molecular structures illustrated in Figure 5.1 include most of the molecular families found in refined fuels. Chemical formulas for these families are as follows:

Paraffin: C_nH_{2n+2} Diolefin: C_nH_{2n-2} Olefin: C_nH_{2n}

Naphthalene: C_nH_{2n} Aromatic: C_nH_{2n-6}

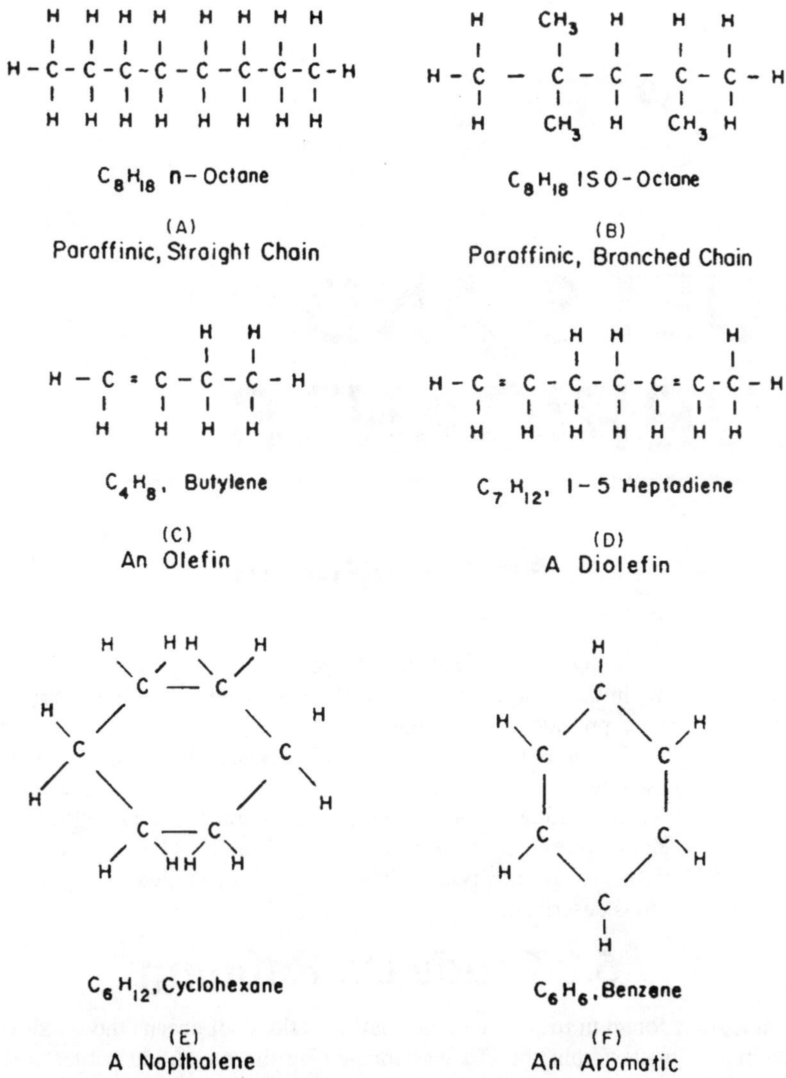

Figure 5.1. Molecular structures of some fuel hydrocarbon families.

Figure 5.1a shows normal octane as an example of a paraffinic molecule. Normal cetane has the same structure, except that it has 16 carbon molecules with associated hydrogen, rather than the eight carbon atoms shown in Figure 5.1a. When n is four or less in the formulas above, the molecules are small enough to form gases at normal temperatures and pressures. The boiling point of hydrocarbons increases with n—that is, with molecular weight—and this is the principle on which refineries work.

A flowchart for a typical refinery is shown in Figure 5.2. The physical principle upon which all refineries are based is that the boiling point of hydrocarbons varies

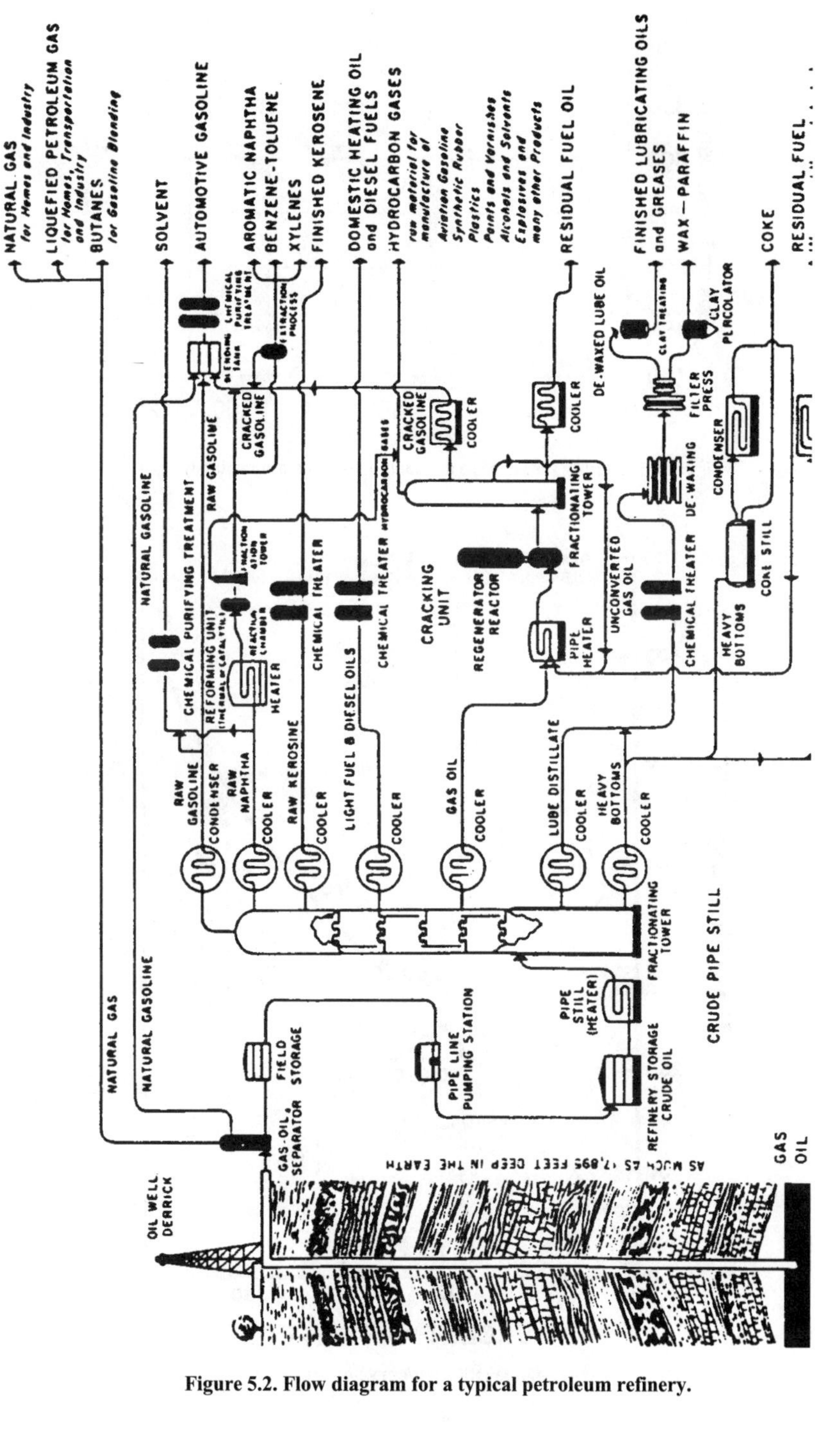

Figure 5.2. Flow diagram for a typical petroleum refinery.

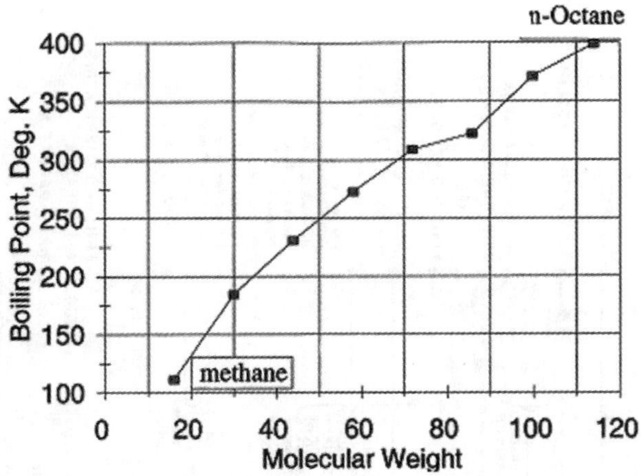

Figure 5.3. Boiling points of hydrocarbons versus molecular weight.

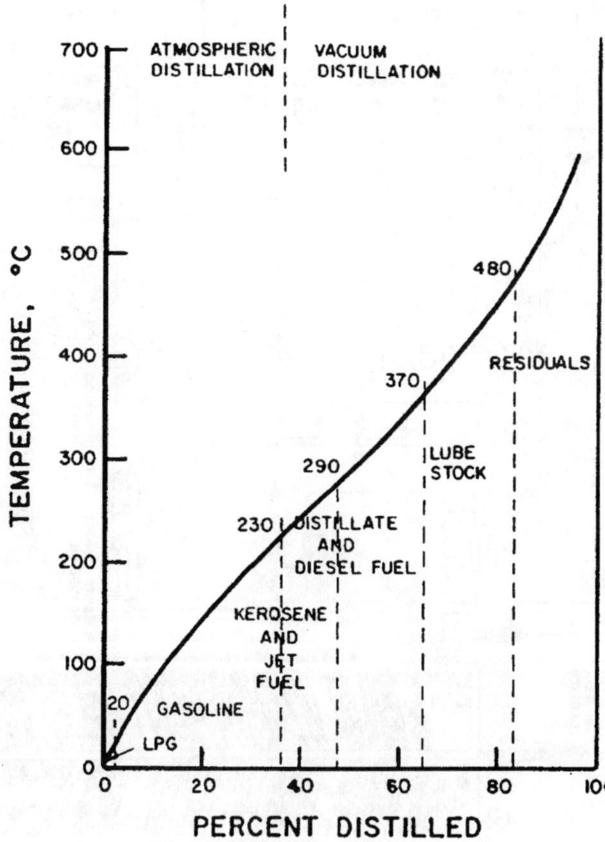

Figure 5.4. Distillation curve of crude oil.

almost proportionally with their molecular weight (Figure 5.3). The crude oil is heated beyond the boiling point of the heaviest hydrocarbon and fed into a fractionating tower. The hydrocarbons repeatedly re-condense and re-vaporize as they pass upward through the plates in the tower. In moving up through the tower, the hydrocarbons encounter progressively cooler temperatures, which permit only the progressively smaller molecules to re-vaporize and continue rising. Thus, heavier hydrocarbons can be withdrawn near the bottom of the tower, and lighter hydrocarbons can be withdrawn near the top.

A *distillation curve* shows the percentage of hydrocarbons that boil and distill at various temperatures. A distillation curve for one sample of crude oil is shown in Figure 5.4. For proper engine performance, fuels must have boiling points within the approximate ranges shown in Figure 5.4. That is, from 30°C to 230°C for gasoline and from 230°C to 370°C for diesel fuel. Typical crude oil yields about equal volumes of gasoline and diesel fuel, as Figure 5.4 shows. But, historically, the United States has needed much more gasoline than diesel fuel to provide fuel for automobiles. The *cracking unit* in Figure 5.2 uses catalysts with elevated temperatures and pressures to crack larger hydrocarbon molecules into smaller ones, shifting more of the refinery yield into the gasoline range.

5.3 Combustion and Exhaust Emissions

5.3.1 Ideal Combustion

Combustion is the chemical reaction of carbon and hydrogen in the fuel with oxygen in the air to form water and other exhaust products. The process liberates heat, which raises the cylinder pressure and drives the piston downward during the power stroke. The actual combustion process is a complex one, in which many intermediate species form and then disappear in subsequent reactions. However, insight can be gained by considering *ideal combustion*, in which all of the hydrogen in the fuel is converted into water and all of the carbon is converted into carbon dioxide. As a further simplification, air is assumed to be 21.3% oxygen and 76.9% nitrogen by weight, and all other trace elements are ignored. Air contains 21% oxygen and 79% nitrogen on a volume basis. Thus, in air, there are 3.76 molecules of nitrogen for every molecule of oxygen. The apparent molecular weight of air is 29.

Atomic and molecular weights of certain elements are needed in combustion chemistry, and acceptable values are given in Table 5.1. Actual petroleum fuels are mixtures of many hydrocarbons. However, the hydrocarbon molecules listed in Table 5.2 are commonly used to represent petroleum fuels when combustion processes are being illustrated. The alcohols listed in Table 5.2 are pure substances. Methyl soyate is an alternative fuel for compression-ignition engines. More specifically, it is the methyl ester of soybean oil and is formed by a chemical reaction involving soybean oil, methyl alcohol and a suitable catalyst.

The combustion of gasoline is illustrated as follows:

$$\begin{array}{ccccccccccc} C_8H_{18} & + & 12.5\,O_2 & + & 47\,N_2 & \rightarrow & 47\,N_2 & + & 8\,CO_2 & + & 9\,H_2O \\ 1(114) & & 12.5(32) & & 47(28) & & 47(28) & & 8(44) & & 9(18) \\ 1 & & 3.51 & & 11.54 & & 11.54 & & 3.05 & & 1.42 \end{array}$$

Table 5.1. Atomic and molecular weights of some elements.

Element	Symbol	Atomic Weight	Mole Formula	Mole Weight
Carbon	C	12	C	12
Hydrogen	H	1	H_2	2
Oxygen	O	16	O_2	32
Nitrogen	N	14	N_2	28

Table 5.2. Some representative fuel molecules.

Fuel	Molecule	Formula	Mole Weight
Natural gas	Methane	CH_4	16
LPG	Propane	C_3H_8	44
Gasoline	Octane	C_8H_{18}	114
Kerosene	Dodecane	$C_{12}H_{36}$	170
Diesel fuel	Cetane	$C_{16}H_{34}$	226
Methanol	Methyl alcohol	CH_4O	32
Ethanol	Ethyl alcohol	C_2H_6O	46
Butanol	Butyl alcohol	$C_4H_{10}O$	74
Biodiesel	Methyl soyate	$C_{19}H_{36}O_2$	296

For simplicity, one fuel molecule is used. Eight molecules of CO_2 must be created to use the carbon in the fuel, and nine molecules of water must be created to use the hydrogen. Then, 12.5 molecules of oxygen must be supplied to provide the oxygen in the H_2O and CO_2. Finally, each of the 12.5 molecules of oxygen is accompanied in the air by 3.76 molecules of nitrogen for a total of 47 molecules of N_2. The first row of numbers below the chemical equation shows the total mass of each reactant and product to be its molecular weight times the number of molecules present. The second row of numbers is the mass relative to the mass of fuel. For example, the mass of oxygen used is 3.51 ($12.5 \times 32/114$) times the mass of fuel used in the reaction. Notice that, although the dominant exhaust product is nitrogen, for each kilogram of fuel, 1.42 kilograms of water are produced. In a suitably warmed-up engine, the water would exit through the exhaust system as water vapor. Also, notice that the *stoichiometric* (chemically correct) air-to-fuel ratio is

$$A/F = (3.51 + 11.54) / 1 = 15.1$$

The air-fuel mixture in an engine is seldom perfectly uniform. That is, there may be leaner areas where the air-to-fuel ratio is lower. Gasoline engines are sometimes operated on a ratio richer than 15.1:1 to obtain more power. In other words, extra fuel is supplied to ensure the consumption of all of the oxygen in the combustion chamber. With *rich combustion*, there is not enough oxygen to convert all of the carbon to CO_2. Therefore, CO (carbon monoxide) appears in the exhaust. Even *lean combustion* is not uniform or sufficiently complete to eliminate all CO from the exhaust. Thus, engines must operate in a properly ventilated area to avoid carbon monoxide poisoning.

Calculations similar to those for gasoline show the *stoichiometric air-to-fuel ratio* for diesel fuel is 14.9:1. However, CI engines are operated on a much leaner-than-stoichiometric ratio to achieve more complete combustion of the fuel. Thus, some *free oxygen* will be found in the exhaust from diesel engines.

The *exhaust emissions* from engines will contain products in addition to those just described. Small quantities of fuel will escape combustion and will appear in the exhaust as gaseous, unburned hydrocarbons. At high temperatures, some of the nitrogen will react with oxygen to form nitrous oxide (NO) and nitrous dioxide (NO_2). The combination of NO and NO_2 is indicated as NO_x by emissions analysts. Also, some of the carbon in the fuel may appear as soot (free carbon) in the exhaust. The federal government has established limits as to the amount of CO, NO_x, and unburned hydrocarbons that may appear in the exhaust of automobile and truck engines. California has begun to regulate emissions from all types of engines. The federal government has begun to regulate exhaust emissions from large trucks and from other vehicles with CI engines.

Calculation of the stoichiometric air-to-fuel ratio and the theoretical exhaust emissions for ethyl alcohol are illustrated in Example Problem 5.1.

Example Problem 5.1

Calculate the stoichiometric air-to-fuel ratio and the exhaust products when ethyl alcohol (ethanol) is used as an engine fuel. Show the balanced combustion equation.

Solution

The formula and molecular weight for ethanol are given in Table 5.2. The exhaust products will include N_2, CO_2, and H_2O.

$$
\begin{array}{ccccccc}
C_2H_6O & + \ 3\,O_2 & + \ 11.28\,N_2 & \rightarrow & 11.28\,N_2 & + \ 2\,CO_2 & + \ 3\,H_2O \\
1(46) & 3(32) & 11.28(28) & & 11.28(28) & 2(44) & 3(18) \\
1 & 2.087 & 6.866 & & 6.866 & 1.913 & 1.174
\end{array}
$$

$$A/F = (2.087 + 6.866) / 1 = 8.95$$

Three water molecules are created to use the hydrogen in the fuel, and two CO_2 molecules are created to use the carbon. The H_2O and CO_2 require seven oxygen atoms; one of these is supplied by the fuel, and six atoms must be supplied from the air. Each oxygen molecule in the air is accomplished by 3.76 molecules of N_2. The stoichiometric air-to-fuel ratio is only 8.95:1. Ethanol thus burns so much richer than gasoline that a gasoline engine would have to be modified to run correctly on ethanol.

Electronic instruments used for analyzing exhaust gases typically work on a dry volume basis. The theoretical exhaust gases can be analyzed on a similar basis by counting exhaust molecules because, according to Avogadro's Law, each molecule occupies a definite volume. In the theoretical combustion of ethyl alcohol, for example, 11.28 molecules of N_2 and two molecules of CO_2 are produced. Thus, there are a total of 13.28 molecules, not counting those in the water. The theoretical composition of the dry exhaust gases on a volume basis is

$$11.28/13.28 \times 100 = 84.9\% \text{ nitrogen}$$

$$2/13.28 \times 100 = 15.1\% \text{ carbon dioxide}$$

The exhaust from most engines is primarily nitrogen.

5.3.2 General Combustion Equations

The above calculations were for stoichiometric combustion. General combustion equations are given below. They include the *equivalence ratio* (ϕ) that is a measure of mixture richness and is defined as

$$\phi = \frac{(F/A)_{actual}}{(F/A)_{stoich}} = \frac{(A/F)_{stoich}}{(A/F)_{actual}} \tag{5.1}$$

The fuel/air (F/A) is simply the inverse of the A/F ratio. The general combustion equation is

$$C_xH_yO_z + \frac{U}{\phi}O_2 + 3.76\frac{U}{\phi}N_2 = 3.76\frac{U}{\phi}N_2 + R\,CO_2 + V\,CO + W\,O_2 + \frac{y}{2}H_2O \tag{5.2}$$

where

x = number of carbon atoms in fuel molecule
y = number of hydrogen atoms in fuel molecule
z = number of oxygen atoms in fuel molecule
$U = x + y/4 - z/2$
$R = x$, when $\phi \le 1$
$R = x - 2U(1 - 1/\phi)$, when $\phi > 1$
$V = 0$, when $\phi \le 1$
$V = 2U(1 - 1/\phi)$, when $\phi > 1$
$W = U(1/\phi - 1)$, when $\phi < 1$
$W = 0$, when $\phi \ge 1$

The actual A/F ratio from the combustion is

$$A/F = \frac{137.3\,U}{\phi\,(12x + y + 16z)} \tag{5.3}$$

The theoretical concentrations of the *dry* exhaust gases on a *volume* basis are

concentration of $N_2 = 3.76U/(\phi\,T)$
concentration of $CO_2 = R/T$
concentration of $CO = V/T$
concentration of $O_2 = W/T$

where $T = 3.76U/\phi + R + V + W$.

5.3.3 Blended Fuels

Blended fuels are sometimes used in engines. For example, a blend containing 10% ethanol (by volume) in 90% gasoline is currently in widespread use for automobiles. The procedure given below is for determining an equivalent fuel molecule for a blended fuel. Subscripts p and s are used to indicate the primary and secondary fuels in the blend. The number of moles of secondary fuel per mole of primary fuel can be shown to be

$$r_s = \frac{\rho_s}{\rho_p}\frac{f_s}{f_p}\frac{m_p}{m_s} \tag{5.4}$$

where

ρ_p, ρ_s = mass densities of primary and secondary fuels

f_p, f_s = fraction of blended fuel volume occupied by primary and secondary fuel

m_p, m_s = mole weights of primary and secondary fuels

Then the composite fuel molecule is $C_{xc}H_{yc}O_{zc}$, where

$xc = r_s\, xs + xp$

$yc = r_s\, ys + yp$

$zc = r_s\, zs + zp$

and where xs and xp are the number of carbon atoms in the secondary and primary fuel molecules, etc. After the composite molecule is determined for the blended fuel, Equation 5.2 can be used for the combustion analysis of the blended fuel.

5.4 SI Combustion and Octane Rating

Octane rating is a measure of the knock resistance of gasoline. *Knock* is avoided in a spark-ignition engine when burning starts at the spark plug and a *flame front* sweeps smoothly across the combustion chamber to consume the fuel (Figure 5.5a). Knock occurs when the *end gases*–the gases ahead of the flame front–ignite spontaneously and generate a rapid, uncontrolled release of energy. The quick release causes a sharp rise in pressure and pressure oscillations, which may lead to an audible ping or knock. Under what conditions does knock occur or not occur? That question is answered next.

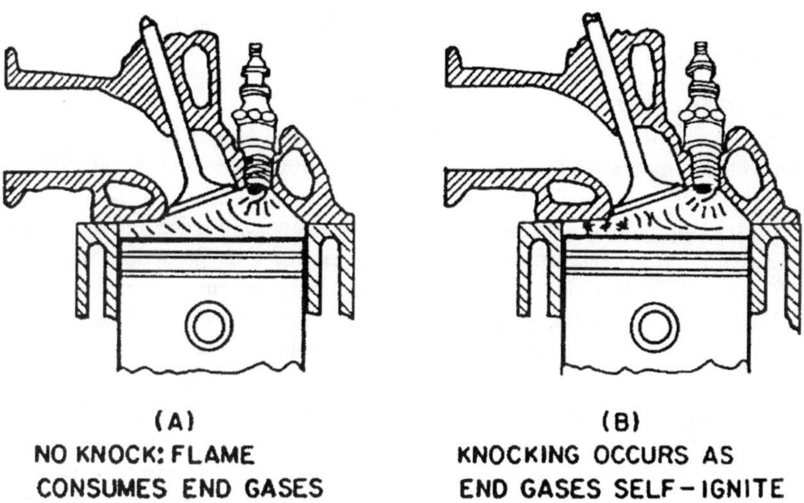

Figure 5.5. Smooth and knocking combustion in a spark-ignition engine.

5.4.1 Cause of Engine Knock

Burning behind the flame front typically releases enough energy to raise the temperature of the end gases above their *self-ignition temperature* (SIT). Fortunately, end gases do not immediately ignite upon reaching their SIT. Rather, there is an *ignition delay*, during which certain *pre-flame reactions* must occur. If the end gases have a sufficiently long ignition delay, the flame front will sweep through to consume the end gases before they can self-ignite, and knocking will not occur. Knocking is most likely to occur during acceleration from slow speeds. Slow engine speeds reduce turbulence in the combustion chamber and reduce the speed of the flame front. Thus, more time is available for the ignition delay to be completed. Also, pressure and temperature increase in the combustion chamber during acceleration, and these increases tend to reduce the length of the ignition delay. A variety of other conditions can promote knocking. It is more likely to occur with higher cylinder pressures and temperatures associated with high compression ratios. Lean mixtures also promote knocking, because the relative scarcity of fuel molecules retards the speed of the flame front. Sustained knock in an engine is audibly annoying and can cause burning of piston crowns and other engine damage.

5.4.2 Knock Reduction Techniques

Several techniques are available for reducing the likelihood of engine knock. For example, the wedge-shaped combustion chamber (the design shown in Figure 5.5) provides more surface area for cooling the end gases and lengthening the ignition delay. Gasolines with higher octane ratings also reduce the likelihood of knock. Hydrocarbons with branched-chain molecules (Figure 5.1b) or ring-structured molecules (Figures 5.1e and 5.1f) have considerably longer ignition delays than do straight-chain hydrocarbons (Figure 5.1a). Thus, gasolines with reduced percentages of straight-chain hydrocarbons are more knock-resistant and are assigned higher octane numbers.

5.4.3 Measurement of Octane Rating

When the anti-knock properties of gasoline first became an issue, a committee known at the Cooperative Fuels Research (CFR) Committee was established to discuss how to measure anti-knock properties. Members of the committee proposed using a special single-cylinder engine, similar to the one shown in Figure 5.6, except that a SI engine is used for octane measurements. The engine they proposed, which became known as a CFR engine, is driven by an electric motor to run at a fixed speed and has an adjustable compression ratio. The octane rating of a sample of gasoline is determined by comparing its performance to that obtained with blends of iso-octane and normal heptane. Iso-octane is extremely knock-resistant and has been assigned an octane rating of 100, while normal heptane is very knock-prone and has been assigned an octane rating of 0. To measure the octane rating of a sample of gasoline, the sample is burned in the CFR engine and the compression ratio is adjusted until the knock meter indicates that a standard level of knocking has been attained. Then, blends of the reference fuels are burned until a blend of iso-octane and normal heptane is found that gives the same knock as the gasoline. The octane rating of the sample is equal to the percentage of iso-octane in the blend of reference fuels.

Figure 5.6. CRF engine for measuring cetane rating.
(Courtesy of Wukesha Engine Division of Dressler Industries.)

5.4.4 Octane Rating Scales

The ASTM has developed two different methods for measuring octane ratings of gasoline. Both methods use the same CFR engine, but different operating conditions. The motor method is more severe and results in a lower octane rating than does the research method. The *Research Octane Number* (RON) is typically about eight numbers higher than the *Motor Octane Number* (MON) for a given gasoline sample. Many service stations now post an anti-knock index on their pumps. The *anti-knock index* is simply the numerical average of the RON and MON.

5.4.5 Octane Requirements

Octane requirements of an engine vary with altitude. The reduced air pressure at higher elevations reduces cylinder pressures and makes knocking less likely. The API has divided the country into a number of regions based on altitude. Gasolines sold in regions with higher altitudes typically have RONs that are several numbers lower than those sold for use in regions with lower altitudes. SAE Recommended Practice J312 (see References and Suggested Readings) provides details as to octane adjustments for various geographic regions in the United States.

5.5 CI Combustion and Cetane Rating

Octane number is not a good measure of the knock resistance of diesel fuel, because diesel combustion differs from spark-ignited combustion. There is no spark plug in a CI engine chamber. Rather, the fuel is injected and its temperature must be raised above the auto-ignition point by use of high compression ratios. Self-ignition temperatures and critical compression ratios for several fuels are shown in Figure 5.7.

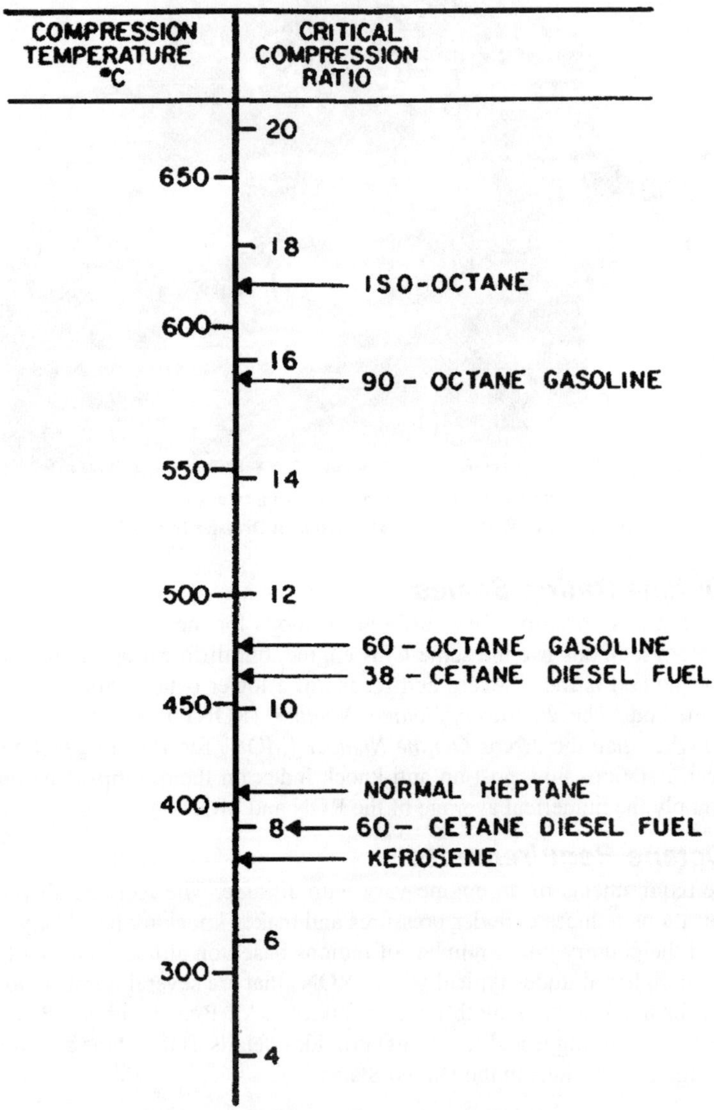

Figure 5.7. Critical compression ratios and temperatures.

Typically, fuel injection does not begin until the chamber temperature exceeds the SIT of the fuel, but the fuel does not ignite immediately upon injection. A two-part ignition delay—a physical and a chemical delay—must be completed before the fuel will self-ignite. During the physical delay, the fuel is atomized into droplets and begins to vaporize and mix with air in the chamber. Then, the chemical delay begins; this is a pre-flame reaction similar to the total ignition delay in a spark-ignition engine. For fuels of normal viscosity, the physical delay is much shorter than the chemical delay.

5.5.1 Combustion and Knock in Diesel Engines

In contrast to spark-ignition engines, knocking in a diesel becomes worse as the ignition delay is increased. The reason can be deduced from Figure 5.8, which is a plot of the rate of energy release from the fuel in a running CI engine. Typically, injection might begin at 20° before HDC. During the delay, fuel evaporates and mixes with air, and pre-flame reactions occur to prepare a quantity of combustible air-fuel mixture. When compression ignition occurs, all of the prepared mixture burns quite suddenly to produce the triangular-shaped peak identified in Figure 5.8 as *pre-mixed combustion.* To allow combustion to continue, fuel vapor must diffuse into the burned-out zones from one side and air must diffuse in from the other. The diffusion process slows the continuing combustion, which is identified in Figure 5.8 as *diffusion combustion.* Thus, diffusion combustion is smoother and quieter than pre-mixed combustion. As shown in Figure 5.8, the combination of pre-mixed combustion and diffusion combustion gives the total energy release.

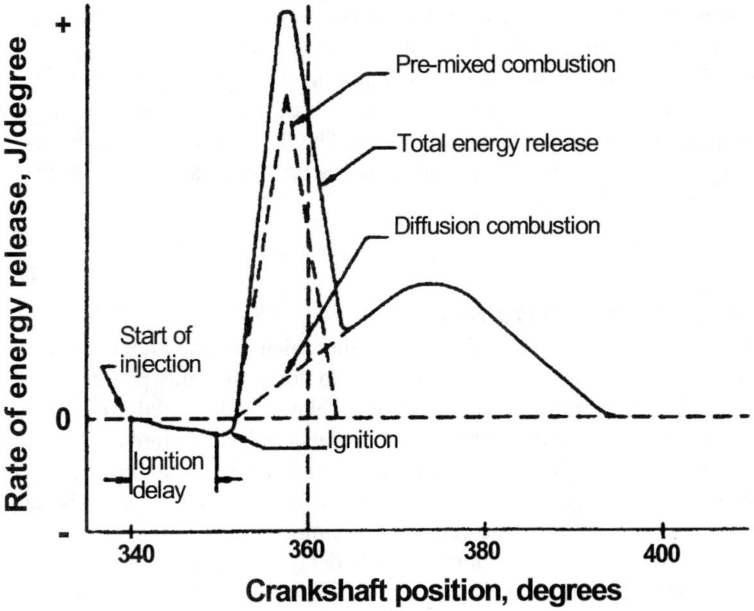

Figure 5.8. Energy release from compression-ignition fuels.

The ignition delay controls the proportioning between pre-mixed and diffusion combustion; the longer the ignition delay, the higher the percentage of the fuel that burns in pre-mixed combustion and the lower the percentage that burns in diffusion combustion. Since diesel knock comes from the rapid, pre-mixed combustion, it follows that a longer ignition delay gives more knocking and a shorter delay gives less knocking. Advancing the start of injection (SOI) increases the ignition delay because chamber pressures and temperatures are lower with earlier SOI. Increasing the torque load on an engine increases the chamber temperature and thus shortens the ignition delay.

5.5.2 Cetane Number of Fuels

The *cetane number* is an indicator of the anti-knock qualities of CI engine fuels. High cetane ratings are achieved by blending fuels with higher percentages of straight-chain hydrocarbons (Figure 5.1a), because these have shorter delays. It should be apparent that fuels with high cetane numbers will have low octane numbers, and the reverse is also true.

5.5.3 Measurement of Cetane Number

Cetane number is measured in the CFR engine shown in Figure 5.6. Originally, the cetane number of a sample of diesel fuel was determined by comparing its performance to a blend of cetane and alphamethylnapthalene. The latter fuel consists of ring-structured molecules and has been assigned a cetane rating of 0. Cetane consists of straight-chain molecules and has been assigned a cetane rating of 100. To measure the cetane rating of a sample of a CI fuel, the fuel is burned in the CFR engine and the compression ratio is adjusted until ignition occurs at HDC. Injection is always set to begin 13° before HDC. After the compression ratio is set, blends of the reference fuels are tried until a blend that has the same ignition delay as the test fuel is found. The cetane rating of the test fuel is equal to the percentage of cetane in the blend of reference fuels. The procedure was modified in 1962 by replacing alphamethylnapthalene with heptamethylnonane. The latter fuel has a cetane rating of 15. Thus, the following equation is used to calculate the cetane numbers from the reference blend:

$$\text{Cetane Number} = (\%\ \text{n-cetane}) + 0.15\ (\%\ \text{heptamethylnonane}) \qquad (5.5)$$

5.5.4 Effect of Cetane Rating

The cetane rating of a CI engine fuel affects more than engine knock. If the cetane rating is too low, the ignition delay may be so long that the piston is moving downward on the expansion stroke before the delay is completed, so starting the engine may be difficult or impossible. The appearance of white smoke, consisting of unburned fuel droplets, in the exhaust during starting is an indication of low cetane fuel. Use of excessively high cetane fuel also can cause difficulty. Then, the ignition delay may be too short to allow the adequate mixing of fuel with air and incomplete combustion may result. Also, a short ignition delay shifts more burning into the diffusion mode and diffusion burning is less efficient than pre-mixed burning. The

Table 5.3. Limiting requirements for diesel fuels.[a]

Property	ASTM Method [b]	Grade 1D [c]	Grade 2D [c]
Flash point, °C, min.	D93	38	52
Water % sediment, % of volume, max.	D1796	0.05	0.05
Distillation temp. °C, min.	D86	-	282
At 90% point, °C, max.		288	338
Kinematic viscosity, min [d]	D445	1.3	1.9
@40 °C, mm^2/s, max. [e]		2.4	4.1
Ash, % by mass, max.	D482	0.01	0.01
Sulfur, % by mass, max.	D2622	0.05[f]	0.05[f]
Copper strip corrosion	D130	No. 3	No. 3
Cetane number, min. [g]	D613	40[h]	40[h]

[a] To meet special operating conditions, modifications may be agreed upon between purchaser, seller and manufacturer.
[b] The test methods indicated are standard. Other methods may be acceptable.
[c] Grades 1D and 2D (but not low-sulfur 1D and 2D grades) are required to contain a sufficient amount of 1,4-dialkylaminoanthraquinone (blue dye) so its presence is visually apparent.
[d] When a cloud point less than –12°C is specified, the minimum viscosity shall be 1.7 mm^2/s and the minimum 90% recovered temperature shall be waived.
[e] 1 mm^2/s = 1 cSt (centistoke).
[f] Grades low-sulfur 1D and 2D shall contain a maximum of 0.05% sulfur by mass. ON all grades, other sulfur limits may apply in selected areas in the United States and in other countries.
[g] Where cetane number by Test Method D613 is not available, Method D4737 may be used as an approximation.
[h] Low ambient temperatures as well as engine operation at high altitudes may require use of fuels with higher cetane ratings.
[i] Grades 1D and 2D must have cloud point 6°C above tenth percentile minimum ambient temperature for region of use. Grades low-sulfur 1D and 2D must meet cloud-point requirement or have minimum cetane index of 40 or maximum aromatic content of 35%.

black smoke in the exhaust of a heavily loaded CI engine comes from diffusion burning. The SAE minimum for cetane rating of diesel fuels is 40, as given in Table 5.3. The cetane number of commercial diesel fuels seldom exceeds 50 and should not exceed 60.

5.6 Fuel Properties and Their Measurements

The internal combustion engine was invented more than one hundred years ago, and numerous improvements have been made since its invention. The development of fuels paralleled the development of the engine. Many standards concerning the required properties of engine fuels and tests for measuring those properties have been set. Most of the standards were developed through the cooperative efforts of the *American Society for Testing Materials* (ASTM), the *Society of Automotive Engineers* (SAE), and the *American Petroleum Institute* (API). Only the most important of the many standards will be discussed here. Some standards apply to only one type of fuel.

For instance, fuel viscosity is relevant only to CI engine fuels. Other standards, such as heating value, apply to all types of fuels.

5.6.1 Specific Gravity of Fuels

Specific gravity is a measure of the density of liquid fuels. It is the ratio of the density of the fuel at 15.6°C to the density of water at the same temperature. The density of water at 15.6° is 1 kg/L, so the specific gravity of a fuel is equal to its density in kg/L. Density of liquids decreases slightly with increasing temperatures. Therefore, densities must be measured at the standard temperature of 15.6°C or must be corrected to that temperature.

The API has devised a special scale for gravities. It is expressed in API degrees and is calculated as follows:

$$\text{API}^{\circ} = \frac{141.5}{\text{SG}} - 131.5 \tag{5.6}$$

where SG = specific gravity of fuel at 15.6°C.

The *gravity* of a fuel is measured with a *hydrometer*. The hydrometer in Figure 5.9 is calibrated in API degrees and is read at the liquid level. Some hydrometers include a thermometer and correction scale, which permit the measurement of gravity at any temperature and correction to 15.6°C.

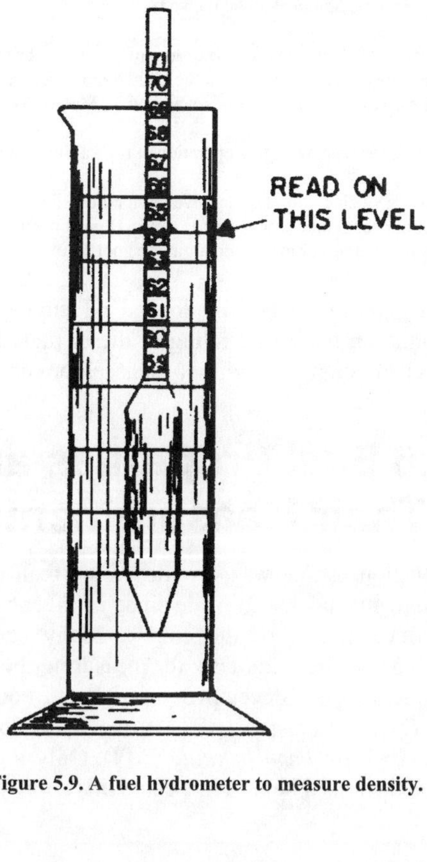

Figure 5.9. A fuel hydrometer to measure density.

5.6.2 Heating Value of Fuels

The purpose of fuels is to release energy for doing work. Thus, the heating value of fuels is an important measure of their worth. Heating values can be measured by burning the fuel in a bomb calorimeter. The combustion creates water, as previously discussed, and energy from the fuel is used to convert that water to vapor in the bomb. The heating value measured by the bomb is therefore called the *lower*, or *net*, *heating value* of the fuel. The *gross*, or *higher heating value* is found by adding to the net heating value the latent heat of vaporization of the water created in combustion. When engine efficiencies are calculated, as was done in Chapter 2, it is important to state whether the higher or lower heating value of the fuel is used in the calculation. Published *heats of combustion* (see Table 5.4) are usually higher heating values and are therefore often used to calculate engine efficiencies.

Equations have been developed for estimating the heating value of petroleum fuels from their API gravity. The equations are

$$Hg = 42{,}860 + 93\,(API - 10) \tag{5.7}$$

and

$$H_n = 0.7190\,H_g + 10{,}000 \tag{5.8}$$

where

H_g = gross or higher heating value, kJ/kg
H_n = net or lower heating value, kJ/kg
API = API gravity, degrees

The calculation of typical heats of combustion is illustrated in Example Problem 5.2.

Example Problem 5.2

Estimate the higher and lower heating values for the diesel fuel used in an OECD Nebraska Tractor Test in which the fuel specific gravity at 15.6°C was 0.8301.

Solution

By using Equation 5.6, the API gravity can be calculated as

$$API = (141.5 / 0.8301) - 131.5 = 39.0°$$

By using Equation 5.7, the estimated H_g can be calculated as

$$H_g = 42{,}860 + 93(39.0 - 10) = 45{,}557 \text{ kJ/kg}$$

By using Equation 5.8, the estimated H_n can be calculated as

$$H_n = 0.7190 \times 45{,}557 + 10{,}000 = 42{,}755 \text{ kJ/kg}$$

The H_g could be used in computing the thermal efficiency of the tractor used in the OECD Nebraska Test. The procedure was given in Chapter 2.

Table 5.4. Properties of selected fuels.

Fuel	API Gravity, Degrees	Density, kg/L	H_g, kJ/kg	Research Octane Number	Boiling Range, °C	Stoichiometric A/F Ratio
Butane	112	0.580	49,500	98	0	15.5
Propane	146	0.509	50,300	111	-44	15.7
Gasoline	61	0.735	47,600	93	30 to 230	15.2
D1 diesel	40	0.832	45,700	40[a]	160 to 260	15.0
D2 diesel	38	0.834	45,500	40[a]	200 to 370	
Methyl soyate		0.885	38,380	51[b]		12.5
Methanol		0.792	22,700	110	65	6.49
Ethanol		0.785	29,700	110	78	8.95
Butanol		0.805	36,100		118	11.2

[a] Minimum cetane rating for diesel fuel.

[b] Measured cetane rating for methyl soyate.

5.6.3 Fuel Volatility

Fuels must vaporize before they can burn. *Volatility* refers to the ability of fuels to vaporize: Fuels that vaporize easily at lower temperatures are more volatile than are fuels that require higher temperatures to vaporize. Reid vapor pressure and distillation curves are both indicators of fuel volatility.

Reid vapor pressure (RVP) is measured with a standardized vapor pressure bomb. Fuel is placed in the bomb, and the bomb is immersed in a water bath at 37.7 ± 0.1°C. With proper technique, a pressure gage on the bomb indicates the vapor pressure of the fuel. The RVP test is used on gasoline. Before 1990, winter gasolines typically had RVPs in the range of 60 to 80 kPa. RVPs of summer gasolines are typically 15 to 20 kPa lower to reduce the tendency for gasoline to vaporize in fuel lines and cause vapor lock of the fuel pump. Vapor lock is discussed later. With the passage of the 1990 Clean Air Act by the federal government, vapor pressure limits permitted in large cities were reduced. Currently, maximum RVPs are 56 kPa in certain large northern cities and 49 kPa in certain large southern cities.

Reid vapor pressure expresses volatility with a single number, but a *distillation curve* gives a more complete picture of fuel volatility. The apparatus shown in Figure 5.10 is used for distillation tests. A 100-ml sample of fuel is placed in the distilling flask and heated. A thermometer is inserted through the stopper to measure the temperature of vapors rising from the fuel. Vapors pass through a side tube in the neck of the flask and are condensed by passing through a metal tube immersed in an ice bath. Then, the condensed fuel is caught in a graduated cylinder. The thermometer is read when the first drop falls into the graduated cylinder, and this temperature is called the initial boiling point. Sample heating continues, and the thermometer is read as each one-tenth of the sample is distilled. The end point is the highest temperature read and is reached just as the last volatile portion of the fuel boils away. Nonvolatile residues may remain in the bottom of the flask; the recovery in the graduated cylinder is then less than 100%.

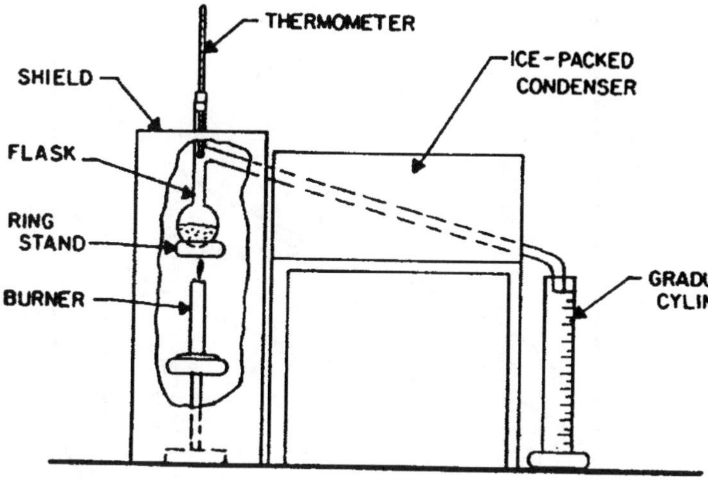

Figure 5.10. Fuel distillation apparatus.

Results of distillation tests are plotted as shown in Figure 5.11. The curves are especially important for gasoline, and three points on the distillation curve are of special interest. The points T_{10}, T_{50}, and T_{90} refer, respectively, to the temperatures on the curve at which 10, 50, and 90% of the fuel has been distilled. For the easy starting of a gasoline engine in winter conditions, the *T_{10} temperature* must be sufficiently low to allow enough fuel to evaporate to form a combustible mixture. The T_{50} point is associated with engine warm-up: a low *T_{50} temperature* will allow the engine to warm up and gain power quickly without stalling. The *T_{90} temperature* is associated with the crankcase dilution and fuel economy: if the T_{90} temperature is too high, the larger fuel molecules will condense on the cylinder liners and pass down into the lubricating oil in the crankcase instead of burning. Gasoline volatility is adjusted by petroleum refiners to suit the season and location (see SAE Recommended Practice J312 [SAE, 1999a] in the References and Suggested Readings). Gasoline sold for use in summer or at high elevations is made less volatile because evaporation is easier under such conditions and vapor lock must be avoided. *Vapor lock* is the formation of fuel vapor bubbles in the fuel line, which prevent the fuel pump from working. Conversely, gasoline must be made more volatile for use in winter conditions and/or at lower elevations, when vaporization is more difficult. Refiners adjust volatility through blending techniques–for example, more butane and lighter hydrocarbons are blended into gasoline to reduce the T_{10} temperature.

The volatility of diesel fuel is important, although it is not as critical as the volatility of gasoline. Diesel fuel must be volatile enough to ensure vaporization in the combustion chamber. But, if the fuel is too volatile, fuel droplets will evaporate too quickly to permit adequate spray penetration in the chamber. A low T_{10} temperature will permit easier starting of a diesel engine. A low T_{50} temperature helps to minimize diesel smoke and odor. A low T_{90} temperature helps to minimize crankcase dilution and improve fuel economy.

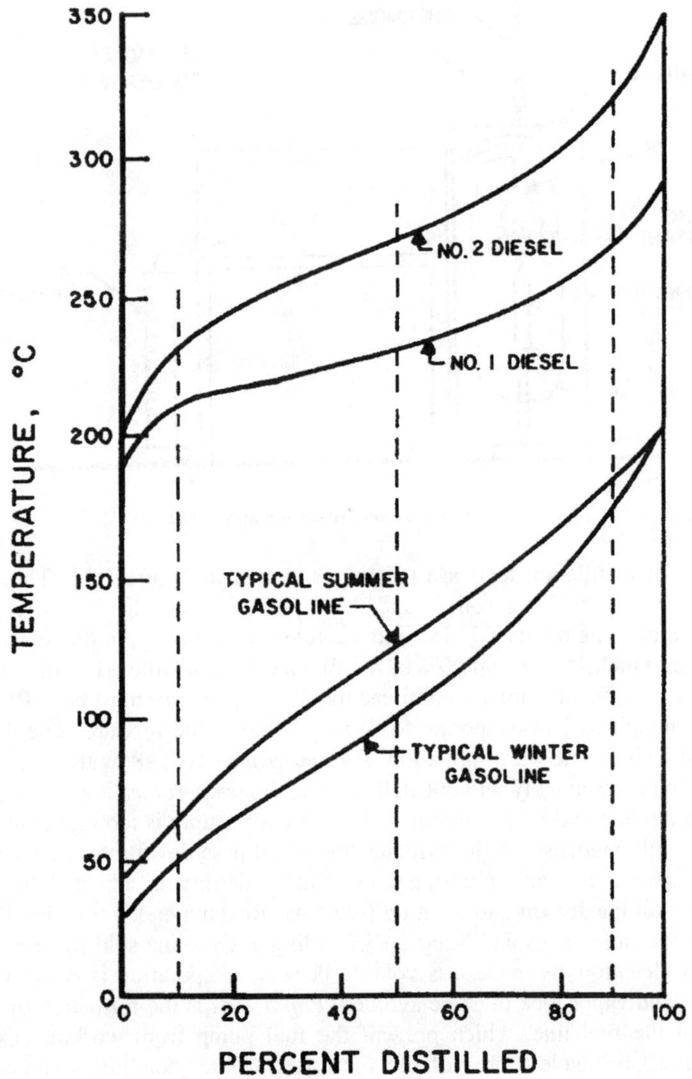

Figure 5.11. Distillation curves of fuels.

5.6.4 Flash Point of Fuels

The *flash point* varies with fuel volatility but is not related to engine performance. Rather, the flash point relates to safety precautions that must be taken when handling a fuel. The flash point is the lowest temperature to which a fuel must be heated to produce an ignitable vapor-air mixture above the liquid fuel when exposed to an open flame. At temperatures below the flash point, not enough fuel evaporates to form a combustible mixture. Insurance companies and governmental agencies classify fuels according to their flash points and use these points in setting minimum standards for the handling and storage of fuels. Gasolines have flash points well below the freezing point of water and can readily ignite in the presence of a spark or flame. Diesel fuels must have flash points exceeding the minimums listed in Table 5.3, unless governmental agencies specify higher minimums for the region in which the fuel is used.

5.6.5 Fuel Viscosity

Viscosity is a measure of the flow resistance of a liquid. For example, water has low viscosity, but honey and syrup have very high viscosity. More information about viscosity can be found in Section 5.10.2. Fuel viscosity is an important consideration when fuels are injected into combustion chambers by means of high-pressure injection systems, as in CI engines. The viscosity must be high enough to ensure proper lubrication of the injector pump. If viscosity is too low, the fuel will flow too easily and will not maintain a lubricating film between moving and stationary parts in the pump. If viscosity is too high, the injectors may not be able to atomize the fuel into small enough droplets to achieve good vaporization and combustion. Injection line pressure and fuel delivery rates also are affected by fuel viscosity. Viscosity limits for diesel fuel have been established by SAE; they are listed in Table 5.3.

5.6.6 Cloud and Pour Points

As a liquid is cooled, a temperature at which the larger fuel molecules begin to form crystals is reached. With continued cooling, more crystals form and agglomerate until the entire fuel mass begins to solidify. The temperature at which crystals began to appear is called the *cloud point*, and the *pour point* is the highest temperature at which the fuel ceases to flow. The cloud point typically occurs between 5 and 8°C above the pour point. Cloud and pour points become important for heavier fuels in the higher boiling ranges (Figure 5.4). Although the pourability of gasoline is not a problem, SAE provides guidelines for specifying pour points of diesel fuel (Table 5.3).

5.6.7 Fuel Impurities

Crude oil contains many impurities that could be harmful in the storage and use of petroleum fuels. Refining removes most but not all of the impurities. Therefore, fuel specifications must include the permissible concentration of each impurity.

Sulfur compounds are present in crude oil, but most of the sulfur is removed during refining. Sulfur is undesirable in fuels. *Sulfur oxides* formed during combustion can convert to acids and cause corrosion of engine parts and accelerated wear of rings and cylinder liners. Typically, sulfur concentrations in gasoline are much smaller than in diesel fuel. In the United States, the sulfur content of gasoline averages less than 0.03% by weight and nearly always is less than 0.015%. Historically, the SAE maximum for sulfur content of D1 or D2 diesel fuel has been 0.5% by weight, as listed in Table 5.3. More recently, low-sulfur grades of D1 and D2 diesel fuel have been developed to meet more stringent emissions requirements. Specifications of the new grades are similar to those of D1 and D2 diesel fuel listed in Table 5.3, except that the maximum allowable sulfur is 0.05%.

The potential corrosivity of sulfur is measured by immersing a copper strip in the fuel for three hours at 50°C and then comparing the resulting corrosion with that on standard corroded strips. Standard strips are numbered, with higher numbers indicating more corrosion. As Table 5.3 shows, the SAE requirements for diesel fuel indicate that corrosion shall not be worse than that on the No. 3 standard strip.

Ash consists of small solid particles and oil or water-soluble metallic compounds found in fuel. Ash is measured by burning a small quantity of the fuel until all the combustible material has been consumed. The unburnable residue is designated as ash and is reported as a percentage of the weight of the fuel sample burned. Ash is particularly harmful in diesel fuel. The abrasiveness of ash can contribute to wear of the close-fitting parts in the injection system and can contribute to plugging the fuel filter and injection nozzles. The SAE limits for ash in diesel fuel are 0.01% by weight, as indicated in Table 5.3.

Water and sediment can enter fuel during handling and storage. Water vapor can move into storage tanks through vented openings and condense into liquid when air temperatures drop. Rain may also leak into some tanks, and ground water may seep into leaky underground storage tanks. Sediment can consist of rust from metal tanks and/or small dirt particles that enter tanks with air or through careless handling. The presence of water in petroleum promotes the formation of slime and other organic sediment. Gasoline should be visually free of un-dissolved water and sediment; it should be clear and bright when observed at 21°C. The SAE limits for water and sediment in diesel fuel are 0.05% by volume, as listed in Table 5.3.

Gum is a viscous liquid or semisolid that forms in petroleum fuels containing unsaturated hydrocarbons. Thus, gum is more likely to form in gasoline that is made by cracking. Usually, the gum is soluble in gasoline, but leaves a sticky residue in carburetors and on intake valves when gasoline evaporates. Gasoline usually has a negligible gum content after refining, but gum may form during storage. High temperatures, the presence of oxygen, and passing time are all conducive to gum formation. The ASTM has developed a standard test for identifying existent gum in gasoline and an oxidation stability test as an indicator of the gum-forming tendency of gasoline. Most commercial gasoline contains antioxidant additives to retard gum formation, but extended storage of gasoline is not recommended. In the past, diesel fuel was more chemically stable and can be stored for much longer periods than gasoline. Increasing demand for distillate products has led to the use of cracked compounds in diesel fuel and, consequently, antioxidants are also beginning to be used in diesel fuels.

5.7 Fuel Additives

Many additives have been developed to improve the performance of petroleum fuels. Until 1970, nearly all automotive gasoline sold in the United States contained TEL (tetraethyl lead) to increase knock resistance and raise the octane number. Fuel refiners were able to use a wide variety of lower octane hydrocarbons in gasoline and then use lead additives to boost octane ratings to acceptable levels. However, lead poisons the catalysts in catalytic emission control systems, and, in 1970, petroleum

Table 5.5. Gasoline additives.

Additive	Type	Concentration, ppm	Function
Octane boosters	MTBE	15%	Extend chemical delay.
Deposit modifiers	Boron & phosphorous compounds		Alter chemistry of chamber deposits to reduce surface ignition and spark plug fouling.
Antioxidants	Amines	4 to 60	Inhibit gum formation.
Detergents	Alkyl amine phosphates	48	Prevent deposits in fuel system and intake manifold.
Metal deactivators	Amine derivatives	4	Destroy catalytic effect of copper on gum formation.
Rust inhibitors	Fatty acid amines	4 to 60	Inhibit rust formation from water and air.
Anti-icing additives	Alcohols and surfactants	50 ppm to 1%	Prevent ice formation and adhesion in fuel system.
Lubricants	Light mineral oils	0.1 to 0.5%	Lubricate valve guides and upper cylinder regions.

suppliers began to market unleaded gasoline. Various lead substitutes have been investigated, and *MTBE* (*methyl tertiary butyl ether*) has come into commercial use as an octane booster. The use of tetraethyl lead in gasoline has been discontinued. Other commonly used gasoline additives are listed in Table 5.5. Because pure petroleum is transparent, petroleum products usually also contain a dye for product identification.

Some gasoline additives (for example, antioxidants, metal deactivators, and rust inhibitors) are also used in diesel fuel. Ignition accelerators are used to raise the cetane rating of diesel fuel. Several nitrates are available to reduce the SIT of the fuel by serving as local ignition points. Reducing the SIT boosts the cetane rating of the fuel. Additives are available to reduce the pour point by preventing the agglomeration of fuel crystals. The development of diesel fuel additives is likely to increase due to the increasing demand for diesel fuel and the consequent use of a wider variety of hydrocarbon constituents to meet the demand.

5.8 Alternative Fuels

Certain fuels are becoming available as alternatives to conventional, petroleum-based gasoline and diesel fuel. Two factors have driven the search for alternative fuels. The first relates to concerns about the continuing availability of conventional fuels. At expected use rates, less than a century of world petroleum reserves remain, with the greatest remaining concentrations being in the Middle East and in the Commonwealth of Independent States (formerly the Union of Soviet Socialist Republics). The United States and other countries importing a growing proportion of their petroleum needs have a strong incentive to seek alternatives. A second factor promoting alternative fuels is the growing need to reduce environmental pollution from engine exhaust emissions. Some alternative fuels, especially those containing oxygen, are less polluting than conventional petroleum fuels.

Butane and propane, which are gases at normal temperatures and pressures, may be compressed and stored as liquids in heavy tanks as LPG (liquefied petroleum gas). LPG is burned in spark-ignition engines equipped with special fuel handling equipment for vaporizing the fuel before it enters the carburetor. LPG engines are clean burning. They are used in industrial vehicles and for stationary work but are no longer used in farm tractors.

Compressed natural gas (CNG) is being considered as a replacement fuel for vehicles because it is clean burning. It is a high-octane fuel (approximately 120 research octane number) and is best suited for use in spark-ignition engines. Its low energy density (about one quarter that of gasoline) is a major disadvantage, that is, a very large tank on the vehicle and/or frequent refilling would be required. Thus, CNG is best suited for use in urban vehicles near refueling stations.

Carbon monoxide and hydrogen can be obtained through pyrolysis of coal and then reacted together using the Fischer-Tropsch process to form liquid hydrocarbons. The process is used in South Africa, where coal is abundant but petroleum supplies are limited, to produce a majority of its gasoline and diesel fuel. The process is inefficient and used only where crude oil supplies are limited and coal is plentiful.

Ethanol derived from cornstarch has come into widespread use in the United States. The ethanol is blended with gasoline in a blend that contains 10% anhydrous ethanol by volume. The federal government provides financial incentives for use of such blends. The stoichiometric air-fuel ratio for the blend is 14.4:1, somewhat lower than for gasoline alone. Consequently, the blends can be burned in unmodified gasoline engines and help to reduce the CO emissions from the exhaust. In addition, the ethanol boosts the octane rating of the blend by approximately three octane numbers. Use of ethanol-gasoline blends is expected to increase as the United States works to reduce exhaust emissions in large cities that have air-quality problems. Other countries are using ethanol in even greater quantities; many automobiles in Brazil are designed to run on 100% ethanol.

Vegetable oils (soybean oil, sunflower oil, rapeseed oil and others) are being considered for use as a fuel for compression-ignition engines. The cetane rating of most vegetable oils is near 40 and the oils will ignite in an unmodified, compression-ignition engine. However, the vegetable oils are at least 10 times more viscous than typical petroleum diesel fuel. The high viscosity leads to poor spray atomization, incomplete combustion and rapid carbon buildup in diesel engines. Fortunately, the vegetable oils can be reacted with simple alcohols in the presence of a suitable catalyst to form esters. For example, reacting soybean oil with methanol forms methyl soyate, the methyl ester of soybean oil. The viscosity of methyl soyate is only a little higher than that of diesel fuel and the cetane rating of the ester is approximately 50. Ester fuels have been used experimentally in various countries and, in Europe, the methyl ester of rapeseed oil is being used commercially for farm tractors. An important advantage of the vegetable oils and the simple alcohols is that they are virtually free of sulfur.

5.9 Fuel Storage

There are local, state and federal regulations concerning the storage of vehicle fuels. The regulations address two principal concerns, minimizing the danger of fire and reducing the risk of environmental damage due to spills.

Governmental agencies classify fuels according to their flammability. Minimum flash point limits are specified for diesel fuel to reduce the danger of fire. Because of its very low flash point (typically -40°C), gasoline is classified into a category that requires more careful handling and storage procedures than diesel fuel. The governmental regulations typically relate to design of storage facilities and procedures for minimizing fire hazards. One usual requirement is the prohibition of smoking near fuel storage areas.

In the United States, federal regulations specify storage requirements to reduce environmental damage from spills. Secondary containment is required, so that the fuel will still be contained if the primary containment fails. In the case of underground tanks, double walls are used to provide secondary containment. In some cases, earthen berms are required to surround above-ground storage tanks for secondary containment.

5.10 Lubrication and Lubricants

Engines and vehicles contain numerous moving parts that require lubrication. Without lubrication, excessive friction would consume much power and would cause excessive wear to the moving parts. Lubrication reduces friction by interposing a film of oil between the parts sliding on other parts. In addition, a lubricant acts as a cushioning agent between mating parts, transports heat away from bearings, and serves as an effective seal to reduce blow by of combustion gases past piston rings. Finally, lubricating oil acts as a cleansing agent in an engine.

5.10.1 Lubrication Theory

An example of parts needing lubrication is shown in Figure 5.12; a section of a piston ring is shown sliding downward on a liner wall at velocity v. Gas forces behind the ring and ring tension combine to create a normal force, F_N, pressing the ring against the wall. A tangential force, F_T, appears to oppose the ring motion. If there is no lubricant, the two forces are related as

$$F_T = f\,F_T \tag{5.9}$$

where

F_T = tangential force

f = coefficient of dry friction

F_N = normal force

The tangential force arises from two primary sources. One source is adhesion due to attraction between molecules in one part to those in the mating part. Adhesion may be augmented by chemical reactions and tends to bind the two parts together. The other source is the mechanical interlocking of irregularities on the mating surfaces. As Figure 5.12b illustrates, surfaces that may appear smooth to the naked eye have

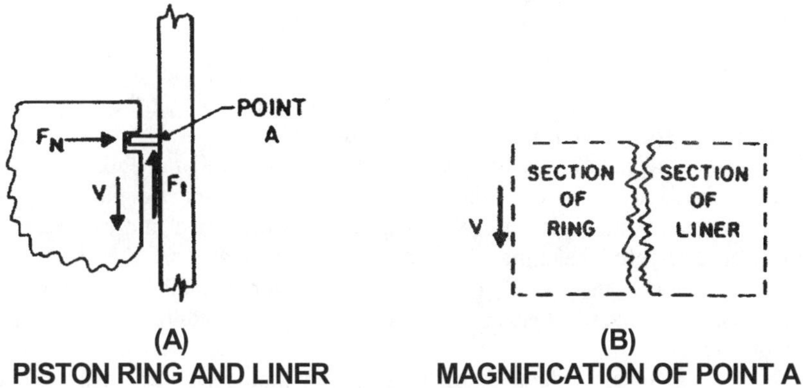

Figure 5.12. Illustration of lubricating theory.

irregularities when viewed with sufficient magnification. Mechanical friction is the shearing of these metal projections as the mating surfaces move relative to each other. The *coefficient of friction* in Equation 5.9 is a larger *static coefficient of friction* when sliding is being initiated and a smaller *dynamic coefficient of friction* when sliding is underway.

Interposing a film of lubricating oil between the mating surfaces can greatly reduce mechanical friction. If the film is thick enough to prevent contact of the irregularities (Figure 5.12b), the lubrication is referred to as *full film* or *hydrodynamic lubrication.* If the film is very thin (approximately the thickness of a single layer of oil molecules), the lubrication is referred to as *boundary lubrication*. *Mixed-film lubrication* occurs when the film thickness is intermediate between hydrodynamic and boundary lubrication. It should be apparent that hydrodynamic lubrication can virtually eliminate wear, because the mating parts are not in actual contact.

The thickness of the oil film between mating parts is affected by the relative speed of the parts. The normal force is continually tending to force the oil film thickness toward zero. However, as illustrated in Figure 5.13, relative motion tends to drag more

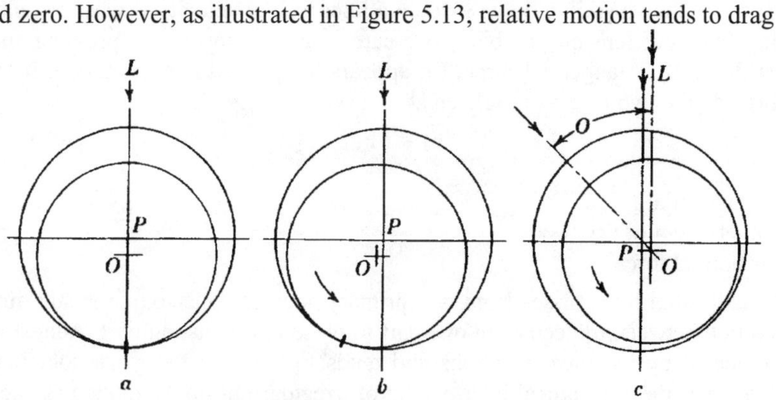

Figure 5.13. Action of journal bearings (a) at rest, (b) in mixed-film lubrication, and (c) in hydrodynamic lubrication.

oil into the space between the parts. All bearings in an engine are in boundary lubrication when the engine starts from rest, but hydrodynamic lubrication can be established at normal operating speeds. Piston rings are in hydrodynamic lubrication at mid-stroke of the piston, but move into mixed-film or boundary lubrication when the piston stops at each end of the stroke. Much of the wear in an engine or vehicle occurs during startup or when the engine and/or vehicle are coasting to a stop.

5.10.2 Oil Viscosity

Viscosity is the single most important property of a lubricating oil. The viscosity must be high enough to support an oil film between the mating parts, but excessive viscosity can cause unnecessary power consumption. Sir Isaac Newton deduced the following relationship for a fluid being shear-stressed between two plates (Figure 5.14):

$$F_T = \mu A \frac{v}{h} \tag{5.10}$$

where

F_T = tangential force, N
μ = dynamic viscosity, Pa·s
A = contact area, m^2
v = velocity, m/s
h = film thickness, m

By solving Equation 5.10 for μ, one can show that the *dynamic viscosity* is the ratio of shear stress to shear rate. A fluid is said to be Newtonian if, at constant temperature, shear stress is proportional to the shear rate.

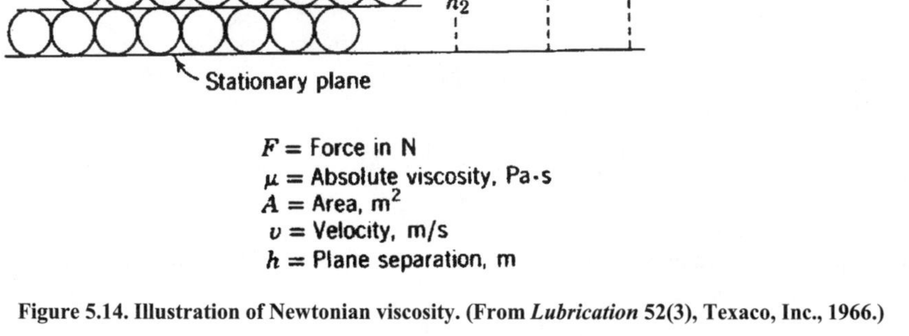

Figure 5.14. Illustration of Newtonian viscosity. (From *Lubrication* 52(3), Texaco, Inc., 1966.)

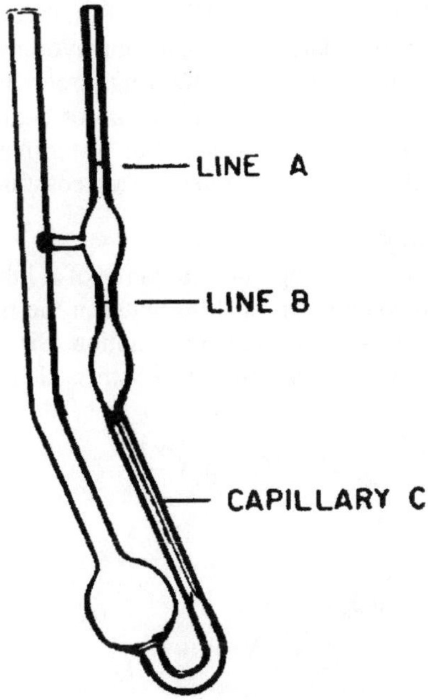

Figure 5.15. A Cannon-Fenske viscometer.
(Adapted from a drawing by Cannon-Fenske Instrument Co.)

There are a number of ways to measure fluid viscosity. One of the most popular methods is by use of the Cannon-Fenske viscometer illustrated in Figure 5.15. This orifice-type viscometer is an empirical instrument, in that the viscosity is not measured directly. Instead, the time in seconds is measured for a given quantity of oil to flow through the capillary tube in the viscometer. A calibration constant is then used to convert the time reading to a viscosity reading. The measured time period is also affected by the mass density of the oil. To eliminate the density effects, the viscosity is reported as *kinematic viscosity*, which is defined as:

$$\nu = \frac{\mu}{\rho} \tag{5.11}$$

where
ν = kinematic viscosity, mm^2/s
μ = dynamic viscosity, mPa·s
ρ = mass density, kg/L

Before the SI system was adopted, the centipoise (cP) was the popular measure of dynamic viscosity and the centistoke (cSt) was the popular measure of kinematic viscosity. To convert,

$$1\ cP = 1\ mPa{\cdot}s \quad \text{and} \quad 1\ cSt = 1\ mm^2/s$$

Still another type of viscometer is the Saybolt Universal Viscometer. It is also an empirical instrument in that the viscosity is reported as the number of seconds required for a given quantity of oil to flow through an orifice. The viscosity is reported in Saybolt Universal Seconds, or SUS. If $\nu > 50$ mm²/s, then ν (in SUS) $\approx 4.632\ \nu$ (in mm²/s). Nomographs are available for converting the non-linear relationship at smaller viscosities.

5.10.3 Classification of Oil by Viscosity

For user convenience, SAE has adopted a system for classifying lubricating oils by their viscosities (see Table 5.6). SAE viscosity numbers with a W (for winter) are classified by their behavior at low temperatures. Those without a W are classified by their viscosities at 100°C. The borderline pumping temperature indicates the lowest temperature at which the oil can be pumped by the lubricating system when the engine is being started in cold weather.

5.10.4 Viscosity Index

The viscosity of a motor oil is highly dependent upon its temperature, as indicated by the MacCoull equation,

$$\ln \ln (\nu + 0.7) = A + B \ln (T) \tag{5.12}$$

where

ν = kinematic viscosity of the oil, mm²/s
T = absolute temperature of the oil, K

Equation 5.12 is not valid as the temperature of an oil approaches its pour point, or at very low viscosities (approaching 1.0 mm²/s). Certain oil additives may also invalidate the equation. A and B are constants for a specific oil. If the viscosity of an oil is known at two temperatures, constants A and B can be determined. Table 5.7 lists viscosities at two temperatures for some typical SAE oils. Note that the limits in Table 5.6 permit some variation in viscosity between individual samples of a given SAE viscosity grade. Thus, the viscosities for any given SAE oil may vary somewhat from those in Table 5.7.

Table 5.6. SAE classification of motor oils.

SAE Viscosity Grade	Maximum Viscosity mPa·s	At Temperature, °C	Borderline Pumping Temperature, °C	Viscosity at 100 °C, mm²/s min.	Viscosity at 100 °C, mm²/s max.
0W	3250	-30	-35	3.8	
5W	3500	-25	-30	3.8	
10W	3500	-20	-25	4.1	
15W	3500	-15	-20	5.6	
20W	4500	-10	-15	5.6	
25W	6000	-5	-10	9.3	
20				5.6	< 9.3
30				9.3	< 12.5
40				12.5	< 16.3
50				16.3	< 21.9
60				21.9	< 26.4

Table 5.7. Kinematic viscosities of some typical oils.

SAE Viscosity Grade	Kinematic Viscosity at 40 °C, mm^2/s	Kinematic Viscosity at 100 °C, mm^2/s
10W	33	6.0
20W	50	6.8
20	60	9.0
30	110	10.0
40	180	16.0

The large change in oil viscosities with temperature is undesirable. For example, a given oil may be so highly viscous as to make engine cranking difficult in winter temperatures. Using an oil with a lower SAE number could alleviate the cold-cranking problem, but when the engine reaches a normal operating temperature, the viscosity might not be high enough to ensure hydrodynamic lubrication. The *viscosity index (VI)* was developed in 1928 to express viscosity change with temperature. Initially, a reference oil, in which the viscosity varied greatly with temperature, was assigned a VI of zero. Another reference oil, in which the viscosity had a much smaller variation with temperature, was assigned a VI of 100. The slope of the viscosity-temperature relationship of an oil sample was compared to those of the reference oils to determine the VI of the sample. Later, oils were developed with VIs above that of the 100-VI reference oil. In response, the ASTM developed viscosity index tables. The VI of any oil can be read from one of the tables if the viscosity of the oil is known at 40 and 100°C.

Multigrade oils are especially formulated to have a high viscosity index. For example, an SAE 10W-40 is a multigrade oil that meets the maximum viscosity limit of an SAE 10W at –20°C and the minimum viscosity limit of an SAE 40 oil at 100°C. Such an oil is formulated by adding a temperature-activated polymer to an SAE 10W oil. As the oil temperature rises above –20°C, for example, the polymer becomes increasingly active in linking smaller oil molecules together into larger units, thus reducing the high temperature "thin-out" of the oil and allowing it to meet the minimum viscosity limit of an SAE 40 oil at 100°C.

5.10.5 Service Classification of Motor Oils

The API has devised a system that allows engine manufacturers to specify oils to meet the working conditions of the engine. The *API service classification* involves a cooperative effort among SAE, ASTM, and API. Engine manufacturers, working through SAE, propose new oil service classifications to meet the needs of new engines. ASTM develops standard test criteria that the oils must meet to qualify for the new classifications. Usually, standard tests are conducted in special, single-cylinder engines, as described in the article by Muller and Papageorge in the list of Suggested Readings and References. Once a new classification is developed, it is publicized by API.

The API service classification consists of two letters. The first letter is either a C (for Commercial use or Compression-ignition engines) or an S (for Service station or Spark-ignition engines). The second letter denotes the severity of service, with higher letters indicating more severe service. Thus, an SG oil would be for use in SI engines with more severe operating conditions than for an SF oil. To increase market potential, oil suppliers attempt to formulate oils to meet more than one API service classification. For example, one might find an oil with an API service classification of SF-SG/CD-CE, i.e., an oil meeting four different API service classifications. The API service classification is an open-ended system, i.e., it allows higher service letters to be added as more severe operating conditions require higher classifications. Specific requirements of each API service classification can be found in SAE Standard J183, Engine oil performance and engine service classification (SAE, 1999b).

5.10.6 Oil Additives

Straight mineral oil would not be an adequate lubricant for modern engines. Increasing demands on engines in the form of higher pressures and temperatures, higher speeds and so forth have placed increasingly greater demands on the lubricating oil. Oil formulators have used additives in formulating oils for new API service classifications to help meet these increasing demands. Table 5.8 is a summary of additives used to improve the performance of lubricating oils. In addition, oils generally include an *oiliness* additive. Oiliness refers to the ability of an oil to adhere

Table 5.8. Lubricating oil additives. (Courtesy of Chevron Oil Co.)

Type	Typical Compounds	Reason for Use
Detergents, dispersants	Succinimides, neutral metallic sulfonates, phenates, phosphates, polymeric detergents, amine compounds	Keep sludge, carbon and other deposit precursors suspended in the oil.
Oxidation inhibitors	Zinc dialkyl dithiophosphates, compounds of nitrogen and sulfur, hindered phenols, Bis-phenols	Prevent or control oxidation of oil, formation of varnish, sludge, and corrosive compounds, slow the viscosity increase over time.
Alkaline compounds	Overbased metallic sulfonates and phenates	Neutralize acids, prevent corrosion.
Extreme pressure (EP) antiwear, friction modifiers	Zinc dialkyl dithiophosphates, tricresyl phosphates, organic phosphates, chlorine compounds	Form protective film on engine parts. Reduce wear and prevent galling and seizing.
Rust inhibitors	High-base additives, sulfonates, phosphates, organic acids or esters, amines	Prevent rust on metal surfaces by forming protective film or neutralizing acids.
Metal deactivators	Zinc dialkyl dithiophosphates, metal phenates, organic nitrogen compounds	Form film so that metal surfaces do not catalyze oil oxidation.
Viscosity index improvers	Polyisobutylene, methacrylate, acrylate polymers	Increase the viscosity index, allow easier cold starting.
Pour point depressants	Methacrylate polymers	Lower "freezing" point of oils, ensure free flow at low temperatures.
Antifoamants	Silicone polymers	Reduce foam in crankcase.

to metal surfaces and is related to the polarity of the molecules. Molecules are electrically neutral as a whole, but the positive and negative charges within the molecule may have an unsymmetrical arrangement. Typically, esters of vegetable oils can be added to engine oils to provide oiliness. Each ester molecule has a polar (electrically unsymmetrical) head to adhere to the metal surface and a hydrocarbon tail to provide boundary lubrication.

5.10.7 Engine Lubricating Systems

The lubrication system must continually supply clean, filtered oil to all moving parts in an engine while it is running. A typical lubrication system is illustrated in Figure 5.16. The oil level in the crankcase is maintained somewhat below the crankshaft, so that the moving crank throws do not dip into the oil. A screen-covered inlet near the bottom of the crankcase conveys oil to the *oil pump*, which is typically a positive-displacement gear pump. The pump delivers pressurized oil to the oil filter through passages drilled into the engine block. Either the pump or the filter is equipped with a pressure relief valve to regulate the oil pressure. Some earlier engines used *bypass filters*, in which only a portion of the oil from the pump was filtered and the remainder moved directly to the bearings, and other moving parts. Current practice is to use *full-flow filters* that filter all of the oil from the pump. A bypass valve must be provided to allow oil to bypass the filter if the filter becomes blocked. Oil filters can be of the surface or depth type. Accordion-pleated paper in *surface filters* trap contaminants on the surface as the oil flows through the paper. *Depth filters* are packed with cotton or other absorbent materials to trap contaminants as the oil flows through.

Oil leaving the filter enters a fore-and-aft drilled gallery in the block and flows via cross galleries to the main bearings. Drilled passages in the crankshaft convey oil to each connecting rod bearing from an adjoining main bearing. Cross galleries may also be provided to convey oil to the camshaft bearings. One or more vertical galleries convey oil to the rocker arm shaft in the engine head. The hollow rocker arm shaft conveys oil to each of the rocker arms, from where it exits to oil the valve stems. The rocker arm area is enclosed by a valve cover to prevent escape of the oil. Oil returns from the valve cover area to the crankcase via vertical passages that are provided. Typically, there are no galleries to deliver pressurized oil to lubricate the cylinder walls. Instead, oil escaping from the connecting rod bearings creates an oil mist that coats the cylinder walls. Oil control rings on the pistons wipe off most of the oil on each down stroke of the piston and oil passes through holes in the ring grooves to oil the piston pins before falling back into the crankcase. In some heavy-duty engines, drilled passages in the connecting rod convey oil to nozzles, which spray oil on the underside of the piston to provide cooling.

If the crankcase has insufficient capacity to keep the oil from overheating, an *oil cooler* may be provided between the oil pump and the oil filter. Typically, the cooler is a heat exchanger in which the oil transfers heat to the engine coolant.

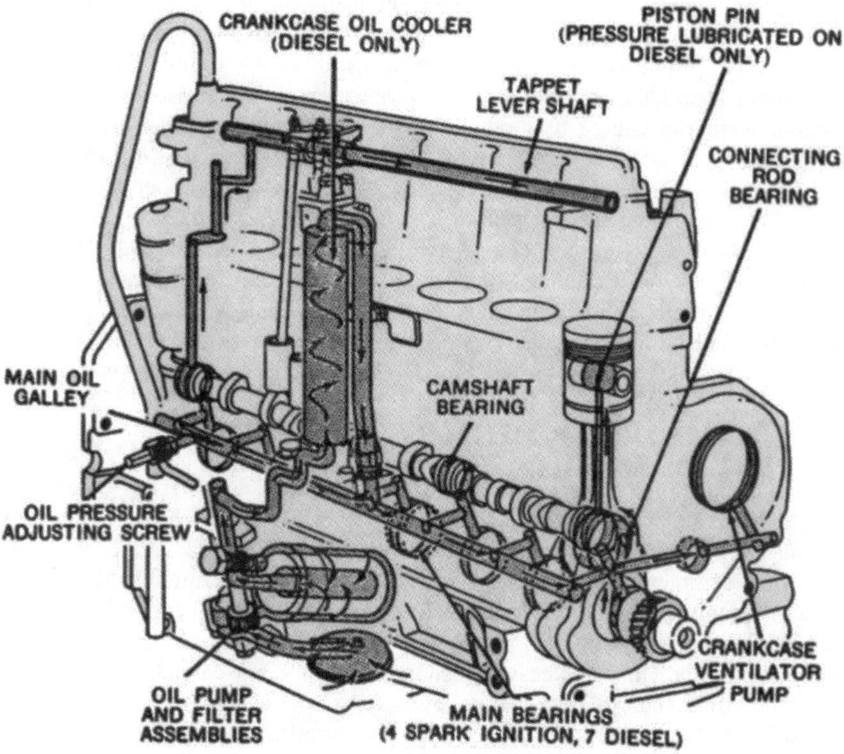

Figure 5.16. A pressure-feed and splash lubrication system. (Courtesy of Deere and Co.)

Two indicators are normally provided for monitoring the performance of an engine lubricating system. A *dipstick* extends from a special opening in the block to some depth into the crankcase. Marks on the dipstick indicate the proper oil level. A *pressure sensor* is provided in one of the oil galleries to measure the oil pressure and transmit a signal to an indicating device. This device may provide a readout of the actual pressure, or it may simply be a warning light. In the latter case, the pressure sensor is a normally closed switch that opens when sufficient pressure is present in the oil gallery. The warning light illuminates when the ignition switch is turned on, but is turned off when the oil pressure rises to a sufficient level.

5.10.8 Oil Selection and Maintenance

An engine manufacturer normally uses two criteria for specifying the oil to be used in the engine. The manufacturer specifies the API service classification of the oil that must be used to protect the engine. Also, the manufacturer may specify SAE viscosity grades that should be used for various ambient temperatures. Oil formulators provide the API service classification and the SAE viscosity grade on each oil container to guide the engine user in selecting an appropriate oil for the application.

Oil testing laboratories have become widely available for assisting engine users in engine maintenance. Typically, an engine user collects a small sample of the used oil each time the engine oil is changed and sends it to the testing laboratory. The laboratory measures the concentrations of various metals in the used oil. Laboratory personnel warn the engine user when any of these concentrations exceed certain norms, which indicate abnormal wear in the engine. The identity of metals that wear off in an engine can be correlated to specific parts, e.g., high copper concentrations could indicate abnormal bearing wear or high chromium concentrations could indicate excessive ring wear. High concentrations of silicon could indicate entry of foreign material into the crankcase. As indicated in Table 5.8, alkaline additives in oil provide buffering to neutralize acid compounds that are formed during combustion. The *total base number* (TBN) indicates the level of buffering still remaining in the oil. Testing laboratories thus measure and include the TBN in their reports to engine users.

5.11 Chapter Summary

Through the refining process, crude oil is converted into a variety of petroleum products that are distinguished primarily by their boiling ranges. For example, diesel fuel boils over a higher temperature range than gasoline. While actual fuel combustion chemistry is complex, many insights can be gained by considering ideal combustion, in which all hydrogen in the fuel is used to form water, while all carbon in the fuel is used to form carbon dioxide.

Gasolines in the United States are blended to suit the season and geographic area. The octane rating and volatility can be lower for gasolines used at higher elevations, and gasoline volatility must be increased for cold weather starting. The preferred fuel for CI engines is No. 2 diesel during most of the year. In winter months, to void fuel filter plugging, No. 1 diesel may be used as a fuel alone or in blends with No. 2 diesel. New, low-sulfur grades of diesel fuel have been developed to meet current and future exhaust emissions requirements. The SAE limiting requirements for diesel fuels were presented. Several fuels were discussed as alternatives to gasoline or diesel fuel. Requirements for safe storage of fuels were discussed.

A review of lubrication theory led to a discussion of boundary, mixed-film and hydrodynamic modes of lubrication. Oiliness is an important oil property in boundary lubrication, while viscosity is the most important oil property in hydrodynamic lubrication. The SAE viscosity classification system and the API service classification system were discussed as the means for engine manufacturers to specify the lubricating oils to be used in their engines. A typical engine lubricating system was described. Finally, we discussed how an engine user can make use of a commercial oil-testing laboratory to monitor the condition of the engine.

Homework Problems

5.1 Assuming $\phi = 1$,

(a) Calculate the air-fuel ratio for the combustion of propane.

(b) Calculate the mass of water produced per unit mass of fuel.

(c) Calculate the theoretical exhaust emissions on a dry volume basis.

5.2 Rework Problem 5.1, but let $\phi = 0.7$ and use diesel fuel for the fuel.

5.3 Rework Problem 5.1, but let $\phi = 1.1$ and use gasoline for the fuel.

5.4 Rework Problem 5.1, but let $\phi = 1.0$ and use methanol for the fuel.

5.5 Rework Problem 5.1, but let $\phi = 0.7$ and use methyl soyate for the fuel.

5.6 Plot the theoretical exhaust gases from the combustion of gasoline versus ϕ as ϕ varies from 0.8 to 1.2.

5.7 Rework Problem 5.6, but use ethanol for the fuel.

5.8 Rework Problem 5.6, but use butanol for the fuel.

5.9 Calculate the equivalent molecule for a volume blend of 10% ethanol and 90% gasoline. Then rework Problem 5.1 using the blend as the fuel. Assume $\phi = 1.0$.

5.10 Rework Problem 5.9, except use a volume blend of 20% ethanol in 80% gasoline (this is a popular blend in Brazil). Assume $\phi = 1.0$.

5.11 Using the specific gravity of the fuel reported in Figure 3.4,

(a) Calculate the H_g for the fuel used in the test.

(b) Calculate the H_n for the fuel used in the test.

(c) On the basis of the H_g, calculate the fuel equivalent power during the PTO test at rated engine speed.

5.12 Calculate the H_n for each of the fuels in Table 5.4 for which the API degrees are given.

5.13 A certain CI engine has a 96 mm bore and the thickness of the top ring is 3 mm. At mid-stroke, the piston is moving downward at a speed of 11 m/s. Gas forces behind the ring combine with ring tension to hold the ring against the cylinder wall with an interfacial pressure of 600 kPa.

(a) If there is no lubricant and the coefficient of dry friction is 0.42, calculate the total vertical force resisting ring motion.

(b) Also calculate the power consumed by friction on the ring.

5.14 Rework Problem 5.13, except assume an oil film 0.1 mm thick has been interposed between the ring and cylinder wall to put the ring in full hydrodynamic lubrication. Also, assume the dynamic viscosity of the SAE 30 oil is 10 mPa·s.

5.15 Rework Problem 5.14, except assume that cold temperatures have increased the oil viscosity to 4000 mPa·s and the film thickness to 0.3 mm.

5.16 Assuming that an oil has a viscosity ν_1 at temperature T_1 and viscosity ν_2 at temperature T_2,

(a) Derive equations to determine values for A and B to fit Equation 5.12 to the oil.

(b) Solve Equation 5.12 explicitly for ν. (*Hint:* Write Equation 5.12 twice, once for each temperature, then solve the system of two equations for A and B.)

5.17 Using the equations developed in Problem 5.16, prepare a spreadsheet for calculating the kinematic viscosity of each of the oils in Table 5.7 at temperatures ranging from –20°C to +120°C. For each oil, plot ln ln(ν) versus T. The plot for each oil should be nearly a straight line.

5.18 Assuming the specific gravity of oil is 0.9, modify the spreadsheet of Problem 5.17 to calculate the dynamic viscosity of the oil over the range of temperatures given in Problem 5.17.

5.19 A 3 mm thick piston ring in a bore of 96 mm is moving downward at a speed of 10 m/s. The oil film thickness between the ring and wall is 0.12 mm. Using the dynamic viscosities calculated in Problem 5.18, calculate and plot the power absorbed by ring friction versus temperature for temperatures ranging from –20 to +70°C. To make the graph more readable, omit any friction power levels greater than 10 kW.

References and Suggested Readings

Alexander, D.L. 1992. The viscosity of lubricants. *Lubrication* 78(3): 1-16.

Bolt, K., and B.R. Hall.1977. *Significance of Tests for Petroleum Products.* ASTM STP 7C. Philadelphia, PA: ASTM.

Contreras, L. 1996. Scientific oil check. *Resource* 3(12): 13-14.

Muller, R., and S. Papageorge. 1988. Engine oil performance tests in the United States. *Lubrication* 74(1): 1-16.

Owen, K., and T. Coley. 1990. *Automotive Fuels Handbook.* Warrendale, PA: SAE.

SAE. 1992a. Automotive gasolines. SAE Recommended Practice J312. *SAE Handbook,* vol. 3. Warrendale, PA: SAE.

SAE. 1992b. Engine oil performance and engine service classification. SAE Standard J183. *SAE Handbook,* vol. 3. Warrendale, PA: SAE.

SAE. 1992c. Engine oil viscosity classification. SAE Standard J300. *SAE Handbook,* vol. 3. Warrendale, PA: SAE.

CHAPTER 6

ENGINE DESIGN

6.1 Introduction

A number of design problems must be solved to achieve a successful engine design. Some general considerations include choice of the number of cylinders, bore and stroke to achieve the desired power output. The load factor must be considered in choosing the speed at which to rate the engine. Certain events must be timed to the crankshaft rotation to achieve successful operation. Reciprocating and torsional imbalances must be considered to avoid excessive vibration. A flywheel must be designed to smooth the torque impulses generated via the pistons, connecting rods and crankshaft. A harmonic balancer may be needed to protect the crankshaft from premature failure. All of these issues are considered in Chapter 6.

6.2 General Considerations

The first step in designing a new engine is to choose the desired rated power output and rated speed. The *load factor* must be considered in choosing the speed. It is defined as

$$\text{Load Factor} = \frac{\text{Average power output}}{\text{Maximum power output}} \tag{6.1}$$

Engines in tractors, heavy-duty trucks and other working vehicles have a much higher load factor than automotive engines. An automotive engine typically uses only a small fraction of its maximum available power while cruising at typical speeds on a highway, but may need to accelerate the engine to very high speed and use maximum power for brief periods. For example, an automobile engine might produce its maximum power for a brief period while accelerating to 5000 rev/min when the automobile is passing another vehicle on a two-lane road, but might have a normal load factor of 0.3 while cruising at 2000 rev/min. Conversely, a tractor engine might run at 2000 rev/min with a load factor close to one for long periods of time while pulling an implement in a field. Consequently, it is customary to rate an automotive engine at a much higher speed than a tractor engine.

After the rated speed and power output are chosen, Equation 2.6 can be used to calculate the *engine displacement* that would be required to produce the power output at an acceptable p_{bme} level. Experience with other engines has shown that p_{bme} levels of 700 to 900 kPa are reasonable. After the engine displacement is chosen, the designer must decide on the number of engine cylinders needed to achieve the required displacement with a reasonable piston size. Then the displacement per cylinder can be calculated. Choice of a bore-stroke ratio will then allow calculation of the bore and stroke of the engine.

Choice of the *bore-stroke ratio* of an engine involves a design compromise. Smaller bore-stroke ratios permit higher compression ratios and the consequent higher combustion efficiencies. For a given compression ratio and cylinder displacement, however, higher bore-stroke ratios have higher surface-volume ratios and thus more heat loss through the combustion chamber walls. Engines with shorter strokes have less piston travel and thus less friction. Use of higher bore-stroke ratios permits use of larger valves, thus allowing greater air throughput and higher power output. Typically, bore-stroke ratios are chosen within the range from 0.84 to 0.96, but ratios as low as 0.79 and as high as 1.3 have been used.

The above design procedure was presented as a set of step-by-step calculations. In practice, the procedure would be iterative and would involve judgment at each step. For example, if the answer from using Equation 2.6 was a displacement of 2.97 liters, the designer might round the displacement to 3.0 liters. Similarly, the chosen bore-stroke ratio might be modified to achieve round-off in the engine bore.

6.3 Engine Timing, Firing Orders, and Intervals

6.3.1 Firing Orders and Intervals

An engine will not run unless the following events are timed to the rotation of the crankshaft:

1. Opening and closing of the valves,

2a. Firing of the spark plug in SI engines, or

2b. Start of fuel injection in diesel engines.

Also, in a multi-cylinder engine, the cylinders must fire in the proper order. Thus, conventions have been established regarding the numbering of cylinders. Generally, cylinders are numbered from the belt-pulley (front) end of the engine toward the flywheel (rear) of the engine. If the engine is of the in-line type, the cylinders are merely numbered in the order in which their connecting rods are attached to the crankshaft. On V-engines, there are two numbering conventions. Cylinders can still be numbered in the order in which their connecting rods are connected to the crankshaft, in which case the numbering alternates between cylinder banks. Alternatively, some manufacturers number all cylinders, front to back, in the right bank and then number all of them in the left bank.

The *standard direction of rotation* of the engine crankshaft is clockwise, as viewed from the front of the engine. Before electric starters, engines were hand-cranked for starting and, for right-handed persons, turning the crank clockwise is much easier than turning it counterclockwise.

Increasing the number of cylinders in an engine produces smoother power flow by reducing the average *firing interval*. The average firing interval can be calculated by

$$AFI = \frac{180\,cy}{n} \tag{6.2}$$

where

AFI = average firing interval in crankshaft degrees
cy = 2 for two-cycle engines or 4 for four-cycle engines
n = number of cylinders in the engine

The firing order (FO) is simply the order in which the cylinders fire. In a two-cylinder, four-cycle engine, the firing order is 1 - 2, and the average firing interval is 360°. Because of the crankshaft arrangement, however, the firing intervals are not uniform (Figure 6.1a). There are two power strokes in one crankshaft revolution, followed by another revolution in which there are no power strokes. The student can determine AFIs for engines with three or more cylinders, all of which have uniform firing intervals. If the front half of the crankshaft in Figure 6.1c was used as the crankshaft of a three-cylinder engine with a firing order of 1 - 2 - 3, the firing intervals would be non-uniform. Instead, crank throws 2 and 3 are reversed from those shown in Figure 6.1c, so that piston 3 reaches HDC before piston 2. The reader can verify that the firing intervals of such a three-cylinder engine are uniform at 240°.

More than one firing order is possible for some multi-cylinder engines. A *firing order tree* is used to display all possible firing orders of an engine. The six-cylinder engine will be used to illustrate the preparation of a firing order tree. The tree begins with the firing of cylinder No. 1. The end view of the crankshaft in Figure 6.1c shows that either cylinder 2 or 5 would be in position to fire next. Thus, two branches are started on the tree. After either 2 or 5 fires, either 3 or 4 must fire, giving two choices for each of the first two branches for a total of 4 branches. Next, either 1 or 6 is in position to fire, but 1 has already fired. Thus, cylinder 6 is the only choice and the tree branches cease multiplying. The rest of the tree is completed easily. Either 2 or 5 are in position for fire next but, in the branch 1 - 2 - 3 - 6, for example, 2 has already fired and the next element in the branch must be cylinder 5. By similar logic, the last element in this branch must be cylinder 4. The full tree is shown below.

```
            3 - 6 - 5 - 4
       2 <
      /     4 - 6 - 5 - 3
   1
      \     3 - 6 - 2 - 4
       5 <
            4 - 6 - 2 - 3
```

(A)

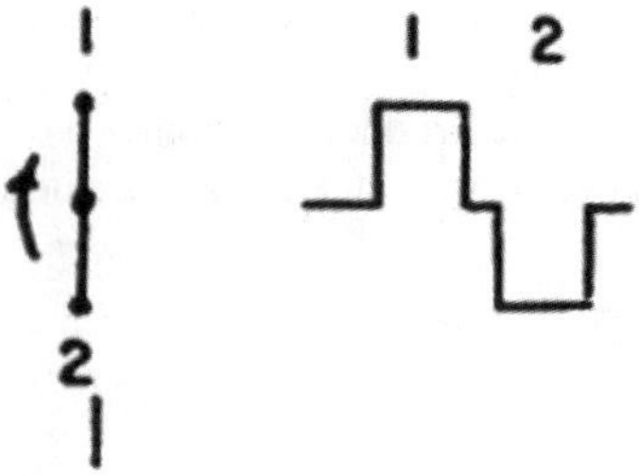

Crankshaft Angle, deg.	Cylinder No. 1	Cylinder No. 2
0–180	Power	Compression
180–360	Exhaust	Power
360–540	Intake	Exhaust
540–720	Compression	Intake

(B)

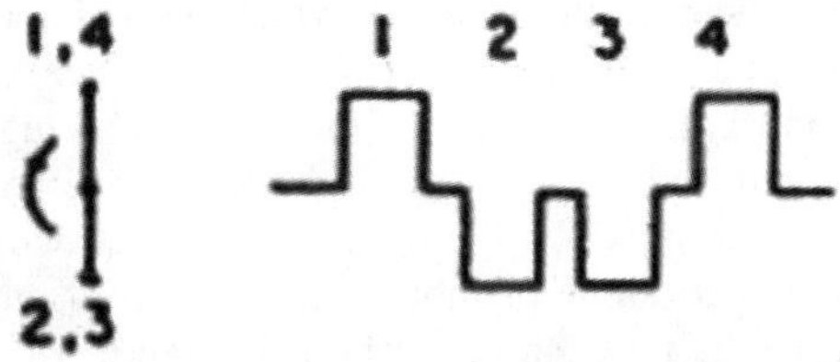

Crankshaft Angle, deg.	Cylinder No. 1	Cylinder No. 3	Cylinder No. 4	Cylinder No. 2
0–180	Power	Compression	Intake	Exhaust
180–360	Exhaust	Power	Compression	Intake
360–540	Intake	Exhaust	Power	Compression
540–720	Compression	Intake	Exhaust	Power

(C)

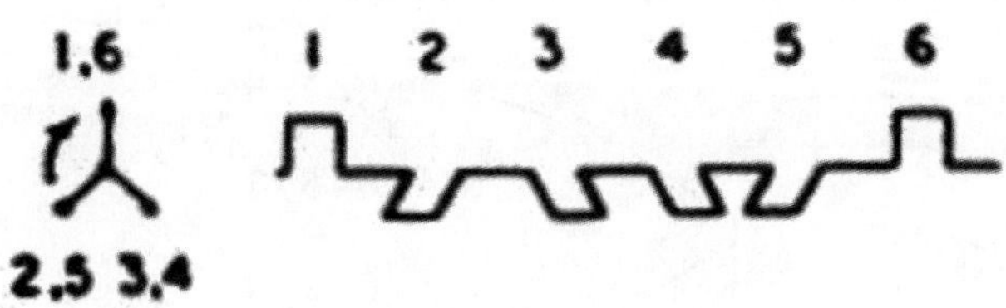

Figure 6.1. Firing order sequences for several engines (continued on next page).

(C) (diagram on previous page)

Crankshaft Angle, deg.	Cylinder No.					
	1	5	3	6	2	4
0						
60						
120	Power	Compression		Intake	Exhaust	
180			Compression			Exhaust
240		Power			Intake	
300	Exhaust			Compression		
360			Power			Intake
420		Exhaust			Compression	
480	Intake			Power		
540			Exhaust			Compression
600		Intake			Power	
660	Compression			Exhaust		
720			Intake			Power

(D)

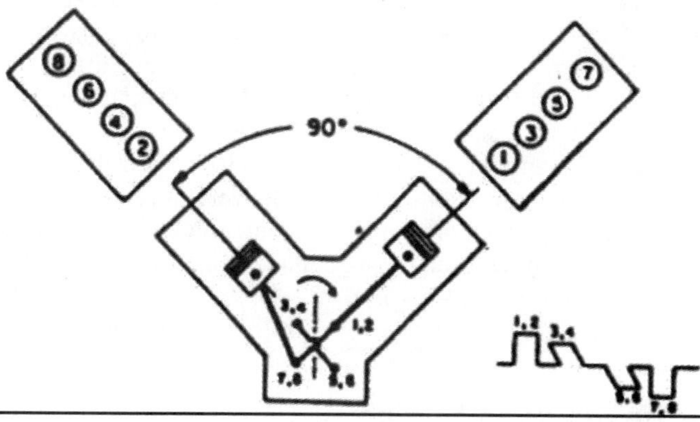

Crank-shaft Angle, deg	Cylinder No.							
	1	2	3	4	5	6	7	8
0								
90	Power			Intake		Exhaust	Compres-sion	
180		Exhaust	Intake		Compres-sion			Power
270	Exhaust			Compres-sion		Intake	Power	
360		Intake	Compres-sion		Power			Exhaust
450	Intake			Power		Compres-sion	Exhaust	
540		Compres-sion	Power		Exhaust			Intake
630	Compres-sion			Exhaust		Power	Intake	
720		Power	Exhaust		Intake			Compres-sion

Figure 6.1. Firing order sequences for several engines (continued from previous page).

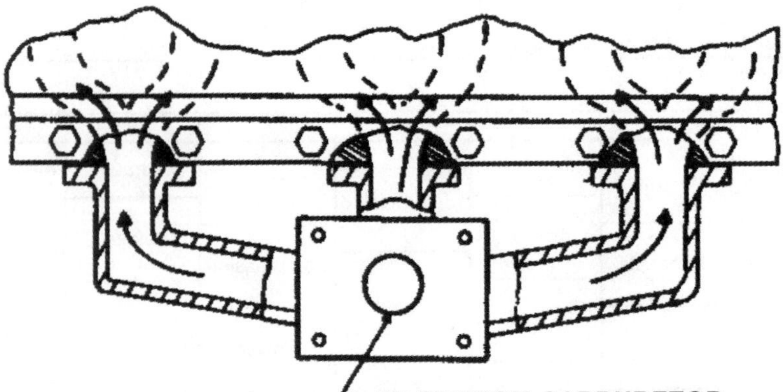

Figure 6.2. Intake manifold for an in-line, six-cylinder engine.

The need to balance air delivery to the various cylinders influences the choice of which firing order is actually used. A typical intake manifold for a six-cylinder spark-ignition engine is shown in Figure 6.2. The design is often called the Siamese-twin design because each branch serves two cylinders. Note that, if either of the FO branches beginning 1 - 2 were selected, there would be two consecutive air surges to the front branch of the manifold and then oscillation between the center and rear branches for remaining air surges. For uniform air delivery, it is better to eliminate the two consecutive surges in a single branch by selecting a FO beginning with 1 - 5. By far the most common firing order used for six-cylinder engines is

1 - 5 - 3 - 6 - 2 - 4

With this introduction, the student can work out firing orders and firing order trees for any engine.

6.3.2 Valve Timing in an Engine

Within the combustible range of air-fuel ratios in an engine, the required air delivery rate is much larger than the required fuel delivery rate. It is always easy to deliver enough fuel to combust the air, and thus it is the air-handling capacity that limits the power output of an engine. Engine designers have several methods for maximizing the air pumping capacity of engines. Valve diameters are made as large as possible within the constraints of the allowable diameter of the combustion chamber. Sometimes four valves (two intake and two exhaust) are used within each cylinder to achieve greater flow area. Valve lift is increased to the maximum extent consistent with smooth operation of the valves. Also, the timing of the valves is arranged to maximize airflow through the engine.

In a four-stroke-cycle engine, two revolutions of the crankshaft are required per engine cycle. A *valve-timing spiral* (Figure 6.3) provides a convenient way of displaying the full two revolutions of the crankshaft needed to show all valve events. Note that intake valve opening (ivo, when the valve just starts to open) occurs before head dead center (bHDC) to have the intake valve fully open when the piston starts the intake stroke. Inertia causes the air to continue moving into the cylinder even after the piston begins moving toward HDC; thus, intake valve closure (ivc, when the valve becomes fully closed) occurs well into the compression stroke. Due to diminishing gas pressure and unfavorable crank geometry, there is little torque production during the latter part of the power stroke. Thus, exhaust valve opening (evo, when the valve just starts to open) occurs well before crank dead center (CDC). The exhaust valve is kept open throughout the exhaust stroke, so exhaust valve closure (evc, when the valve is fully closed) occurs after HDC. Note that there is *valve overlap*, i.e., the intake and exhaust valves are open simultaneously for a small part of the cycle. Some of the exhaust gas can move into the intake ports because of the overlap, resulting in dilution of the incoming charge in a SI engine. A richer mixture is needed at low idle to overcome this dilution.

The optimum valve timing is a function of engine speed. The valve events are closest to the respective ends of the piston strokes at low engine speeds and move

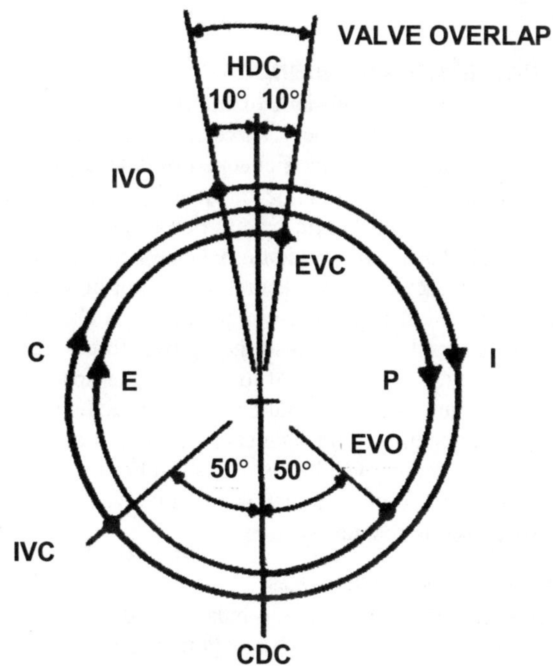

C = COMPRESSION
E = EXHAUST
P = POWER
I = INTAKE
IVO = INTAKE VALVE OPENS
IVC = INTAKE VALVE CLOSES
EVC = EXHAUST VALVE CLOSES
IVO = EXHAUST VALVE OPENS

Figure 6.3. Typical valve-timing spiral for an engine.

farther away from these dead centers as the engine speed increases. At present, it is impractical to change the valve timing while the engine is running, so the timing is compromised to provide maximum airflow over the normal range of engine speeds. However, engine manufacturers are working on electro-hydraulically actuated valves whose timing could be changed as a function of engine speed.

6.4 Engine Balance

There are several sources of potential unbalance in an engine. These include:

- Rotating masses
- Reciprocating masses
- Rotational speed fluctuations resulting from torque pulses
- Crankshaft twist resulting from torque pulses

In the following sections, we will discuss methods that are used to deal with these unbalances.

6.4.1 Rotating Mass Unbalance

The crank throws on the crankshaft of multi-cylinder engines (Figure 6.4) are a possible source of rotating unbalance. There are two types of crankshaft balancing, *static* and *dynamic*. Static balancing is checked by laying the crankshaft on a pair of knife edges. The crankshaft is in static balance if it can be placed in any angular orientation and not roll on the knife edges. For example, the crankshaft of a two-cylinder engine could have inherent static balance because the two throws are 180° apart. However, the two-cylinder crankshaft would not be in dynamic balance. When the two-cylinder crankshaft is spun at a high speed, an oscillating couple appears in the side view of the crankshaft because the inertial forces generated by each crank throw are separated by a distance equal to the cylinder spacing. Dynamic balance is achieved by use of counterweights installed opposite each crank throw, as can be seen in Figure 6.4. Engine manufacturers spin each crankshaft on a balancing machine and remove mass from the counterweights as needed to bring the crankshaft into dynamic balance. Analysis of reciprocating unbalance involves the study of piston-crank kinematics, as discussed in the next sections.

6.4.2 Piston-Crank Dynamics

Piston-crank kinematics affect engine balance and torque production. The piston movement as a function of crankshaft rotation is given by an equation similar to Equation 3.1, i.e.,

$$\frac{S}{R} = [1 - \cos\theta] + \frac{L}{R}[1 - \sqrt{1 - (\frac{R}{L}\sin\theta)^2}] \tag{6.3}$$

where

S = piston displacement from HDC, m
R = crank throw radius, m
θ = crankshaft position, radians, measured from HDC
L = connecting rod length, m

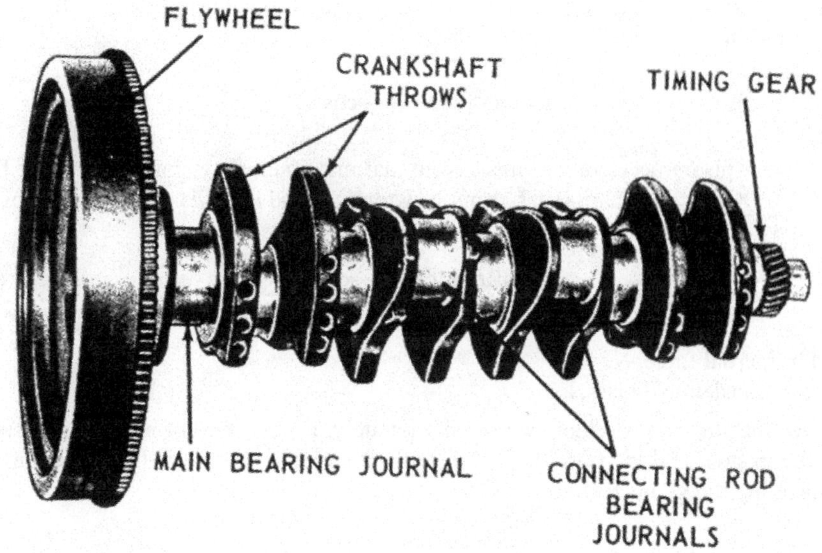

Figure 6.4. A crankshaft with counterweights and a flywheel.

Through the binomial series expansion, Equation 6.3 can be simplified to the following approximate equation:

$$\frac{S}{R} = 1 - \cos\theta + \frac{R}{2L}\sin^2\theta \tag{6.4}$$

By differentiating Equation 6.4, the approximate piston velocity is

$$v = R\omega[\sin\theta + \frac{R}{2L}\sin 2\theta] \tag{6.5}$$

where

$v = dS/dt$ = piston velocity, m/s

$\omega = d\theta/dt$ = crankshaft speed, rad/s

Note that ω is used instead of N to avoid the need for conversion factors in the piston-crank dynamics equations. However, $\omega = (2\pi/60)$ N.

By setting the derivative of Equation 6.5 equal to zero, the maximum piston speed can be shown to occur when the crankshaft is at

$$\theta = \arccos[\frac{L}{4R}(-1 + \sqrt{1 + \frac{8R^2}{L^2}})] \tag{6.6}$$

Typically, the L/R ratio is close to three and, from Equation 6.6, the maximum piston speed occurs when the crankshaft is 1.27 to 1.31 radians (73° to 75°) from

HDC. Differentiating Equation 6.5 gives the following equation for piston acceleration:

$$a = R\omega^2[\cos\theta + \frac{R}{L}\cos 2\theta] \quad (6.7)$$

where a = piston acceleration, m/s^2. Note that maximum acceleration occurs at HDC and at CDC. From Equation 6.7, using Newton's second law, the piston inertia force is given by

$$F = mR\omega^2[\cos\theta + \frac{R}{L}\cos 2\theta] \quad (6.8)$$

where

F = inertial force, N

m = translational mass, kg

The connecting rod is seldom analyzed separately. Instead, part of its mass is assigned to the piston and part to the crankshaft pin (see Figure 6.5 for illustration of connecting rod geometry) as

$$m_{c1} = \frac{b}{L}m_c \quad (6.9)$$

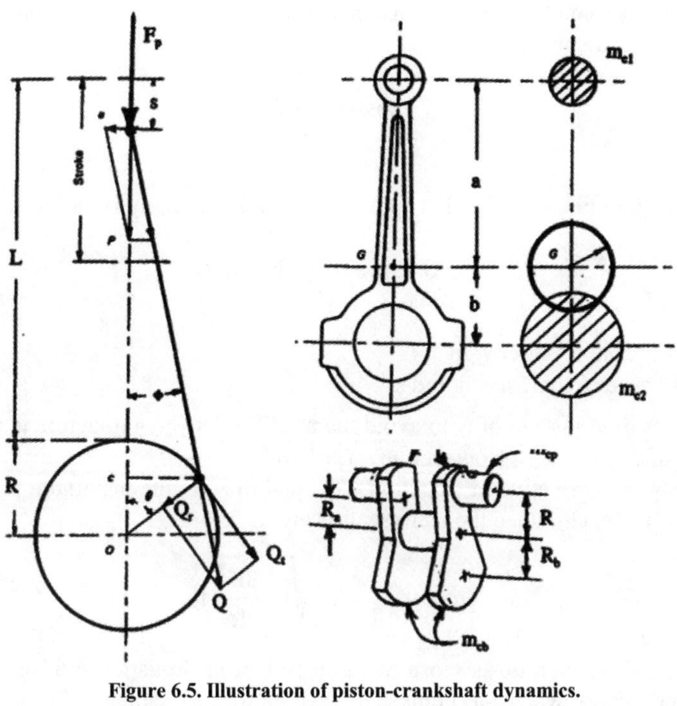

Figure 6.5. Illustration of piston-crankshaft dynamics.

and

$$m_{c2} = \frac{L-b}{L} m_c \tag{6.10}$$

where

m_c = mass of connecting rod, kg
m_{c1} = translational mass assigned to piston, kg
m_{c2} = rotational mass assigned to crankpin, kg
b = distance from crankpin center to center of gravity of connecting rod, m

The appropriate mass to use in Equation 6.8 is the mass of the piston and piston pin, plus mass m_{c1}.

6.4.3 Reciprocating Unbalance—Single-Cylinder Engines

In a single-cylinder engine, part of the vibration arising from the reciprocating piston can be offset by increasing the mass of the crankshaft counterweights beyond that needed to balance the crankshaft alone. The total oscillating force in the x-direction (the direction of piston movement, usually vertical) is

$$F_x = (m_p + m_{c1})R\omega^2(\cos\theta + \frac{R}{L}\cos 2\theta) + (m_e + m_{c2})R\omega^2 \cos\theta \tag{6.11}$$

where

m_p = mass of piston and piston pin, kg
m_e = equivalent mass at the crankpin, kg

The equivalent mass at the crankpin includes the crankpin and counterweights, and can be calculated using

$$m_e = \frac{m_{cp}R + 2m_{ca}R_a - 2m_{cb}R_b}{R} \tag{6.12}$$

where

m_e = equivalent mass at crankpin, kg
m_{cp} = mass of crankpin, kg
m_{ca} = material mass supporting crankpin, kg
m_{cb} = counterweight mass opposite crankpin, kg
R = crankpin radius = half of piston stroke, m
R_a = radius to center of gravity of m_{ca}, m
R_b = radius to center of mass mcb, m

See Figure 6.5 for identification of the masses and radii in Equation 6.12. The first term in Equation 6.11 arises from the reciprocating masses, while the second term is the x-direction component of the rotating mass. In addition, the rotating mass produces a force F_y perpendicular to the crankshaft centerline and to F_x, i.e.,

$$F_y = (m_e + m_{c2})R\omega^2 \sin\theta \tag{6.13}$$

From Equation 6.11, note that the oscillating force, F_x has two frequency components, one at crankshaft speed (called the *primary shaking force*) and one at twice crankshaft speed (called the *secondary shaking force*). From Equation 6.12, by using large counterweights, the equivalent mass, m_e, could be made sufficiently negative to completely cancel out the primary shaking force, but at the cost of generating a large lateral oscillating force, F_y. As a compromise in single-cylinder engines, the crankshaft counterweights are made large enough to cancel out about half of the primary shaking force, F_x.

6.4.4 Reciprocating Unbalance—Multi-Cylinder Engines

In multi-cylinder engines, adding the forces, F_x, of all of the cylinders (being sure to provide the proper phase angles between the crank angles of the various cylinders) shows that some of the forces cancel out while others do not. Table 6.1 summarizes the primary and secondary shaking (inertia) forces of various engines. Note that the two-cylinder and four-cylinder engines each have a secondary inertia force. In four-cylinder engines that are bolted directly to the vehicle frame, as in a farm tractor, a *Lanchester balancer* is used to cancel out the secondary shaking force to protect the operator from the vibration. The balancer consists of two counterweights that rotate at twice crankshaft speed (see Figure 6.6). The counterweights are arranged to provide zero net force in the y-direction and, if properly timed to the crankshaft, will completely cancel the secondary inertia force. The counterweights are properly timed when the masses are farthest from the crankshaft when any two pistons are at HDC. Four-cylinder automotive engines do not require a Lanchester balancer; the rubber engine mounts absorb the vibration and protect the operator. The two-cylinder engine also has a secondary shaking force with half the amplitude of that of the four-cylinder engine. Although a Lanchester-type balancer could be used on the two-cylinder engine, the benefit has not justified the cost and secondary balancers are generally not used on two-cylinder engines.

Table 6.1. Amplitudes of shaking forces and couples.

No. Cylinders	Primary Force, N	Secondary Force, N	Primary Couple, N·m	Secondary Couple, N·m
1	K[a]	KR/L	Zero	Zero
2	Zero	2KR/L	Ka	Zero
3	Zero	Zero	$\sqrt{3}$ Ka	$\sqrt{3}$ KaR/L
4	Zero	4KR/L	Zero	Zero
6	Zero	Zero	Zero	Zero

[a] Half of this force can be counterbalanced.
All forces and couples are in a vertical plane.
Standard crankshafts are assumed, as shown in Figure 6.1.
$K = mR\omega^2$
ω = crankshaft rotational speed, rad/sec
R = crank-throw radius, m
L = connecting rod length, m
a = spacing between cylinders, m

Figure 6.6. A Lanchester balancer.

6.4.5 Inertia Couples

Table 6.1 also shows inertia couples or moments. These are calculated from the inertia forces by taking moments about the center of the crankshaft as viewed along the y-axis. Thus, the couples tend to make the engine rock (pitch) about the y-axis. The primary inertial couple on the two-cylinder engine and the secondary inertia couple on three-cylinder engines are generally left unbalanced. However, most three-cylinder engines have provision for balancing about half of the primary inertial couple through use of a counterweighted front pulley. By proper phasing of the counterweighted pulley to the crankshaft, an inertia couple can be set up to oppose the inherent inertia couple shown in Table 6.1. Unfortunately, the counterweighted pulley sets up a new inertia couple around the z (vertical) axis, i.e., a yawing couple. Thus, only about half of the rocking couple can be balanced without creating an excessive yawing couple.

6.4.6 Instantaneous Torque and Flywheels

Piston-crank kinematics can also be used in calculating the instantaneous torque produced by each cylinder. The instantaneous torque is force, Q_t, (see Figure 6.5) multiplied by the crank radius, R. The end result of the analysis is summarized as

$$\frac{T}{F_p R} = [1 + \frac{\cos\theta}{\sqrt{\frac{L^2}{R^2} - \sin^2\theta}}]\sin\theta \qquad (6.14)$$

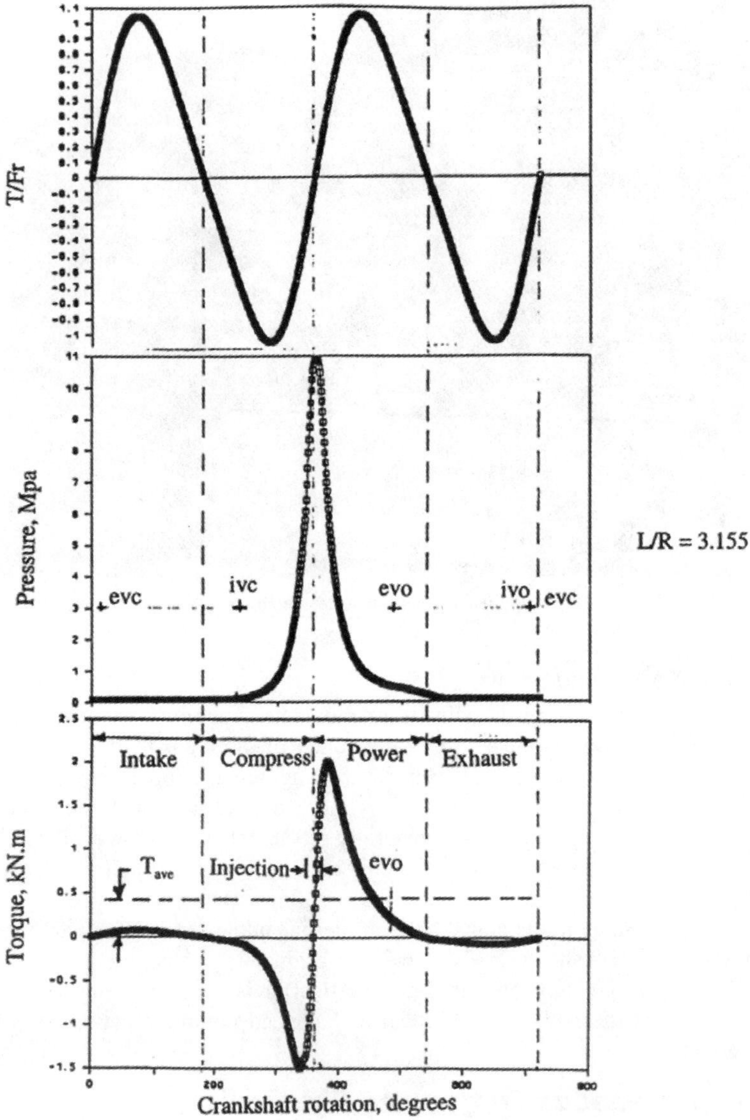

Figure 6.7. Instantaneous torque from an IH5288 engine.

where

T = instantaneous torque, N·m

F_p = instantaneous force on piston, N

R = crank-throw radius, m

The force on the piston at any time is the piston top area multiplied by the instantaneous cylinder pressure. The radial force, Q_r, is also of interest, since it provides loading on the main bearings. It is

$$\frac{Q_r}{F_p} = \cos\theta - \frac{\frac{R}{L}\sin^2\theta}{\sqrt{1-\frac{R^2}{L^2}\sin^2\theta}} \tag{6.15}$$

Neither the instantaneous torque nor the radial bearing loads can be calculated until the cylinder pressure is known as a function of crank angle. Figure 6.7 shows instantaneous torque for a specific CI engine. Inserting the engine L/R ratio into Equation 6.14 produced the diagram T/F_pR shown at the top of Figure 6.7. The middle graph shows instantaneous cylinder pressures as measured with a transducer, and the bottom graph shows instantaneous torques. The piston strokes and valve events are also shown on the diagram. Note that the exhaust valve was opened well before CDC, to permit the early start of the exhaust process with very little loss of instantaneous torque.

6.4.7 Flywheel Design

Note that the instantaneous torque in Figure 6.7 is less than the average torque, T_{ave}, during most of the cycle. The average torque output is equal to the average torque imposed by the load on the engine; when T_{ave} is less than the load torque, the engine will stall unless a flywheel is used to supply part of the torque demand. The flywheel accelerates and stores kinetic energy when the instantaneous torque is greater than T_{ave}; it decelerates and gives up kinetic energy when T_{ave} is greater than the instantaneous torque. The flywheel design consists of providing sufficient inertia such that cyclic speed changes are within desired limits.

In a multi-cylinder engine, the contributions of the cylinders combine to produce an instantaneous torque curve that is less peaked than the one in Figure 6.7. Figure 6.8 shows typical combined instantaneous torque curves for a four-cylinder and a six-cylinder engine. Each of the shaded areas illustrated for the six-cylinder engine are where the instantaneous torque exceeds the mean load torque; on a torque versus crank-angle plot, the area of each shaded region is equal to the kinetic energy that can be stored by the flywheel for release when the mean torque exceeds the instantaneous engine torque. It can be shown that the required mass moment of inertia of the flywheel is

$$I_f = \frac{900\,\Delta E\,k}{\pi^2 N_e^2} \tag{6.16}$$

where

I_f = required mass moment of inertia, kg m^2
N_e = average engine speed, rev/min
ΔE = kinetic energy transfer (J), = λW
 where λ = ratio taken from Table 6.2
 W = indicated work per revolution, J/rev
k = 50/p = coefficient of speed variation
 where p = percent by which the speed is permitted to vary from N

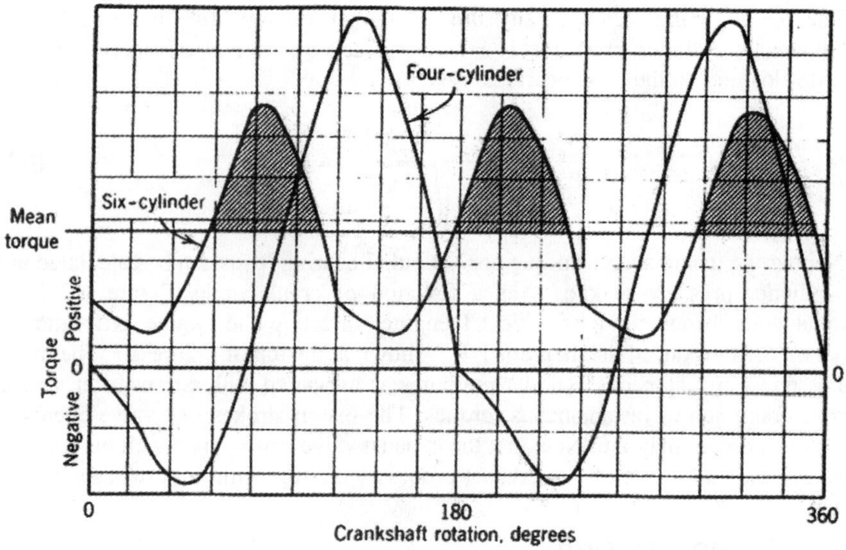

Figure 6.8. Total instantaneous torque curves for four- and six-cylinder engines.

Table 6.2. Approximate values of flywheel constants for standard crankshaft arrangements (Figure 6.1) and four-stroke cycle operation.

Engine Type	$\lambda = \Delta E/W$
One-cylinder	2.40
Two-cylinder	1.50
Three-cylinder	0.70
Four-cylinder	0.30
Six-cylinder	0.10

The indicated work per cycle can be estimated by

$$W = \frac{60{,}000\, P_b}{N_e e_m} \tag{6.17}$$

where

P_b = brake power, kW

e_m = mechanical efficiency, decimal (estimated at 0.8 when P_b is full load)

The inertia calculated using Equation 6.16 is the total inertia required to control the speed variation through each engine cycle. The clutch may contribute up to 40% of the required inertia, leaving 60% to be provided by the flywheel itself.

6.4.8 Crankshaft Twist and Vibration Dampeners

The torsional flexibility of the crankshaft, coupled with the inertia of the flywheel, forms a torsional spring-mass system. The natural frequency of oscillation of this system depends on the crankshaft torsional stiffness and on the flywheel inertia. The longer the crankshaft and the heavier the flywheel, the lower is the natural frequency of the crankshaft-flywheel system. The instantaneous torque fluctuations excite oscillating twisting of the crankshaft. Each cylinder generates an excitation cycle once each two revolutions of the crankshaft and, at normal speeds, this fundamental frequency is typically far below the natural frequency of the crankshaft-flywheel system. However, as shown in Figure 6.7, the instantaneous torque curve is not sinusoidal. Through a Fourier series expansion, the instantaneous torque curve can be represented as a sine wave at the fundamental frequency plus other sine waves whose frequencies are multiples of the fundamental frequency, i.e., as a fundamental sine wave plus higher *harmonics*. Some of these higher harmonics can approach the system natural frequency, causing the oscillations in crankshaft twist to become large enough to cause crankshaft failure. The problem is most likely to occur in long crankshafts, for example, in an in-line, six-cylinder engine. To correct the problem, a small flywheel is flexibly mounted to the front pulley of the engine. Figure 6.9 illustrates

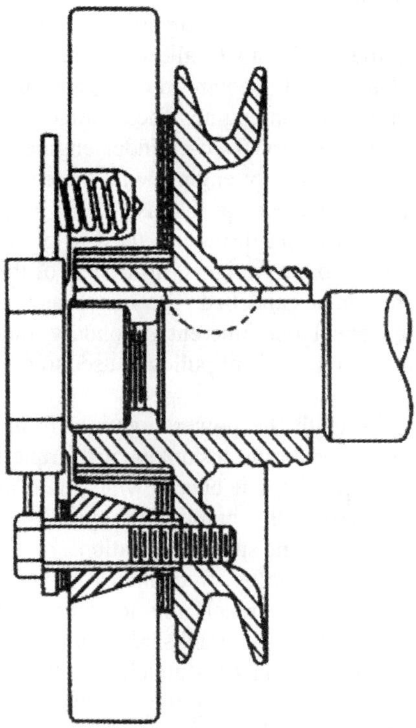

Figure 6.9. A vibration dampener.

two types of flexible connections. One is a spring-loaded friction disk between the small flywheel and the pulley. The other consists of rubber bushings around the bolts that hold the small flywheel to the pulley. Of course, only one type is used in each installation. In attempting to transmit the oscillations from the pulley to the small flywheel, the flexible connection tends to convert the vibrational energy to heat, which is dissipated by convection and radiation. In absorbing the vibrational energy, the vibration damper thus prevents crankshaft failure. The modification to the front pulley is called either a *vibration dampener* or a *harmonic balancer*. Typically, a vibration dampener is used on in-line, six-cylinder engines, but not on four-cylinder engines.

6.5 Chapter Summary

By choosing the desired brake power output and rated speed of a new engine and assuming a reasonable value of brake mean effective pressure, the engine designer can calculate the required engine displacement, and then choose the number of cylinders needed to achieve the displacement with a reasonable piston size. Selection of the bore-stroke ratio then determines the engine bore and stroke. In choosing the power output and rated speed, the designer must consider the engine load factor. Automotive engines have very low load factors and, by being rated at a very high speed, can achieve a much higher power output from a given displacement than can heavier duty engines with high load factors.

The valves and either the injection timing in a CI engine or the spark in an SI engine must be timed to the crankshaft to allow the engine to run. For a four-cycle engine, valve timing is conveniently shown on a valve timing spiral. Valve timing is selected to maximize airflow into an engine. The average firing interval of an engine decreases with numbers of cylinders. Two-cylinder engines have nonuniform firing intervals while intervals are uniform for engines with three or more cylinders.

A number of engine unbalance problems were presented in the chapter. Counterweights opposite each crank throw are used to put a crankshaft in dynamic balance. By use of oversized counterweights, up to half of the primary shaking force in a single-cylinder engine can be cancelled. The Lanchester balancer is used on large, four-cylinder engines to cancel out inherent secondary shaking forces. On three-cylinder engines, a counterweighted front pulley is used to cancel out about half of an inherent pitching couple.

A flywheel is used to smooth the power impulses from engines. The flywheel accelerates and stores energy when the instantaneous torque is greater than the load torque, then decelerates and gives up the energy when the instantaneous torque is less than the load torque. Increasing the number of cylinders decreases the required flywheel size needed to control cyclic speed fluctuations.

In engines with long crankshafts, frequencies of higher harmonics of the instantaneous torque curve can approach the torsional natural frequency of the crankshaft-flywheel system, resulting in crankshaft failure. To prevent such failures, a vibration dampener or harmonic balancer is attached to the front pulley. The flexible connection between the balancer and the pulley drains off vibrational energy, thus preventing crankshaft failure. Typically, vibration dampeners are used on in-line, six-cylinder engines but are not used on engines with four or less cylinders.

Homework Problems

6.1 (a) Prepare a firing order tree for a four-cylinder engine with a standard crankshaft.

(b) Calculate the average firing interval for this engine if it is a four-stroke cycle engine.

6.2 Rework Problem 6.1, except the engine is a three-cylinder engine.

6.3 Draw a complete valve timing spiral for an engine with the following valve events: ivo, 12° bHDC; ivc, 48° aCDC; evo, 50° bCDC; evc, 10° aHDC. Label all strokes and all valve events.

6.4 Draw a complete valve timing spiral for an engine with the following valve events: ivo, 8° bHDC; ivc, 55° aCDC; evo, 47° bCDC; evc, 12° aHDC. Label all strokes and all valve events.

6.5 (a) Assuming L/R = 3.0, plot S/R for one complete revolution of the crankshaft using Equation 6.3.

(b) On the same graph, plot S/R using Equation 6.4.

(c) Comment on the extent to which the approximate equation (6.4) is accurate in calculating piston movement.

6.6 Rework Problem 6.5, but use an L/R ratio of 3.2.

6.7 (a) Assuming L/R = 3.0 and using Equation 6.5, plot normalized piston velocity, $V/R\omega$, for one complete revolution of the crankshaft.

(b) On the same graph, mark the crank angle at which maximum piston velocity occurs as indicated by Equation 6.6.

(c) Does the crank angle of (b) agree with the maximum velocity plot of (a)?

6.8 Rework Problem 6.7, but use L/R = 2.95.

6.9 Beginning with Equation 6.8, verify all entries for the one-cylinder engine in Table 6.1.

6.10 Beginning with Equation 6.8, verify all entries for the two-cylinder engine in Table 6.1.

6.11 Beginning with Equation 6.8, verify all entries for the three-cylinder engine in Table 6.1.

6.12 Beginning with Equation 6.8, verify all entries for the four-cylinder engine in Table 6.1.

6.13 Using Equation 6.14, plot T/F_pR for one complete revolution of the crankshaft. Plot three curves on the same graph, one each for L/R = 2.8, 3.0, and 3.2. How much influence does the L/R ratio have on the curves?

6.14 Rework problem 6.13, but plot curves for L/R = 2.9, 3.1, and 3.3.

6.15 Using Equation 6.15, plot Q/F_p for one complete revolution of the crankshaft. Plot three curves on the same graph, one each for L/R = 2.8, 3.0, and 3.2. How much influence does the L/R ratio have on the curves? At what crank angle would the absolute bearing load be greatest? (Note that the cylinder pressure at HDC is not the same as at CDC).

6.16 Rework problem 6.15, except plot curves L/R = 2.9, 3.1, and 3.3.

6.17 Derive Equation 6.16. (*Hint:* Let ΔE = the change in rotational kinetic energy between the states of maximum and minimum cyclic speeds.)

6.18 You are to evaluate possible rated speeds for a four-cylinder, four-cycle engine with displacement of 4.8 L.

(a) Assuming the brake mean effective pressure is 800 kPa, calculate and plot the brake power output for speeds ranging from 2000 to 2400 rev/min.

(b) On another graph, plot the required inertia of the flywheel-clutch assembly if the cyclic speed is to vary no more than 0.5% from the mean crankshaft speed.

6.19 Rework Problem 6.15, but use a six-cylinder engine with a displacement of 7.2 L.

6.20 Rework Problem 6.15, but use a two-cylinder engine with a displacement of 2.4 L.

6.21 Rework Problem 6.15, but use a three-cylinder engine with a displacement of 3.6 L.

References and Suggested Readings

Bremer, R. 1979. A practical treatise on engine crankshaft torsional vibration control. SAE Paper No. SP-445-750763. Warrendale, PA: SAE.

Erickson, M.A. 1971. Fatigue testing and crankshaft development. *Transactions of the ASAE* 14(1): 1-2, 6.

John Deere Service Training. 1991. *Fundamentals of Service–Engines*. Moline, IL: John Deere Corp.

SAE. 1992a. Maximum allowable rotational speed for internal combustion engine flywheels. SAE Standard J1456. *SAE Handbook,* vol. 3. Warrendale, PA: SAE.

SAE. 1992b. Flywheel spin test procedure. SAE Standard J1240. *SAE Handbook,* vol. 3. Warrendale, PA: SAE.

SAE. 1992c. Engine flywheel housings. SAE Standard J617c. *SAE Handbook,* vol. 4. Warrendale, PA: SAE.

SAE. 1992d. Flywheel dimensions for truck and bus applications. SAE Standard J1857. *SAE Handbook,* vol. 3. Warrendale, PA: SAE.

SAE. 1992e. Flywheels for single-plate spring-loaded clutches. SAE Standard J618. *SAE Handbook,* vol. 3. Warrendale, PA: SAE.

SAE. 1992f. Flywheels for two-plate spring-loaded clutches. SAE Standard J619. *SAE Handbook,* vol. 3. Warrendale, PA: SAE.

CHAPTER 7

COMPRESSION-IGNITION ENGINES

7.1 Introduction

Rudolph Diesel intended his engine to burn the coal dust that was a useless byproduct of manufacturing in the 1880s. Although coal dust was unsuccessful as a fuel, Diesel was able to develop a compression-ignition engine that ran successfully on oil. Early CI engines were large and slow and used in stationary applications or to propel ships. The heavy distillate petroleum that was used as fuel was forced into the engines by means of compressed air.

Further development of the CI engine awaited the development of improved fuel-injection systems in the 1920s. *Robert Bosch* began mass producing fuel-injection systems in 1927. The development of injector pumps was paralleled by the development of smaller CI engines suitable for mobile applications. CI engines began appearing in crawlers and in some large wheel tractors in the 1930s. Large-scale use of CI engines began in the 1960s and today, almost all new tractors, excavators, dozers, haul trucks, etc., have CI engines. These engines are also used in large highway trucks and city busses. Some automobiles and pickup trucks also are equipped with CI engines.

In this chapter, you will learn the major differences that distinguish CI engines from SI engines. A large part deals with the injection system because it is primarily this system that controls the operation of a diesel engine. Mechanical injection systems are discussed, because they are still in widespread use. Electronic injection systems will also be discussed, because they are also in commercial use and will eventually supplant mechanical systems. Finally, the smoke chart is presented because it presents insights into CI engine combustion.

7.2 Comparison of CI and SI Engines

The CI engine cycle can be carried out in either two or four strokes of the piston. The four-cycle CI engine is more widely used and four-cycle operation will be assumed unless two-cycle CI engines are specifically mentioned. CI and SI engine have many similarities, including their external appearance. However, differences are present in their construction, made necessary by fundamental variations in operation, as summarized in Table 7.1.

Unlike the SI engine, the air and fuel are not united in a CI engine until fuel is injected into the combustion chamber. The fuel injected into a CI engine (typically starting at about 20° bHDC) has very little time to mix with air. Therefore, the mixture in the combustion chamber is heterogeneous, i.e., it is very rich within the fuel spray plumes and very lean outside the plumes. In contrast, the air-fuel mixture of a SI engine has much more time to mix and is nearly homogeneous by the time of ignition in the combustion chamber. The fuel-air equivalence ratio of the homogeneous mixture in a SI engine must remain close enough to unity to be combustible. Because the mixture in a CI engine is heterogeneous, small pockets of combustible mixture can form in the combustion chamber and the overall fuel-air equivalence ratio can be far less than unity. Because the air in a CI engine is not throttled, the equivalence ratio is very low when the engine is running at high idle. Although the equivalence ratio increases as additional load is placed on a CI engine, it is necessary to limit the equivalence ratio to about 0.7 at full load to avoid excessive exhaust temperatures and heavy exhaust smoke. The lower overall equivalence ratio and greater expansion ratio of a CI engine allows it to run with a lower exhaust temperature than a SI engine.

The compression ratio of a CI engine must be high enough to cause *auto-ignition* of the *air-fuel mixture*. The compression ratio of a SI engine must be low enough to prevent auto-ignition. Therefore, the compression ratio of the CI engine is much higher than that of a SI engine. The high compression ratio increases the stress on a CI engine and thus it must be constructed more robustly than a SI engine.

A spark ignites the mixture in a SI engine, and a flame front sweeps smoothly across the combustion chamber. Initial combustion in a CI engine is rough and uncontrolled because the mixture may ignite spontaneously at more than one place in the combustion chamber. Because their mixtures are heterogeneous, CI engines must

Table 7.1 Comparison of SI and CI engines.

Factor	SI Engine	CI Engine
Intake/compression	Air and fuel	Air only
Speed control	Throttled air-fuel mixture	Air unthrottled, fuel control only
Mixture uniformity	Nearly homogeneous	Very heterogeneous
Equivalence ratio	0.85 to 1.25	0 to 0.7
Exhaust temperature	Higher	Lower
Compression ratio range	7 to 10	14 to 20
Required strength	Lower	Higher
Relative efficiency at full load	1.0	1.4
Relative $P_b / (D_e N_e)$	1.0	$0.7 \times 1.4 = 0.98$

be supplied with excess air to reduce the richness of the fuel-rich regions in the chamber. The power output of an engine is limited by its air-handling ability, which tends to limit the power output of CI engines. This power deficit is offset to a large degree by the higher thermal efficiency of a CI engine. Thus, a CI engine can produce about the same power output as a SI engine of equal displacement running at the same speed.

CI engines are more efficient than are SI engines at both full load and part load. The excess air supplied to the CI engine and its higher compression ratio both help to increase the indicated thermal efficiency. Also, as discussed in Chapters 4 and 5, the indicated thermal efficiency of a diesel engine improves at part load because fuel injection is ended sooner. The SI engine has no corresponding improvement in indicated thermal efficiency at part load. Figure 7.1 is a comparison of the brake thermal efficiencies of two engines from the same manufacturer. The brake thermal efficiencies for the SI (gasoline) and CI (diesel) engines at full load are, respectively, 24.9% and 32.0%. Thus, this CI engine provides 28.5% better fuel economy at full load. The brake thermal efficiency of the CI engine declines at part load due to declining mechanical efficiency, but the brake thermal efficiency of the SI engine declines faster. At 30% load, for example, the CI engine provides 73% better fuel economy than does the SI engine.

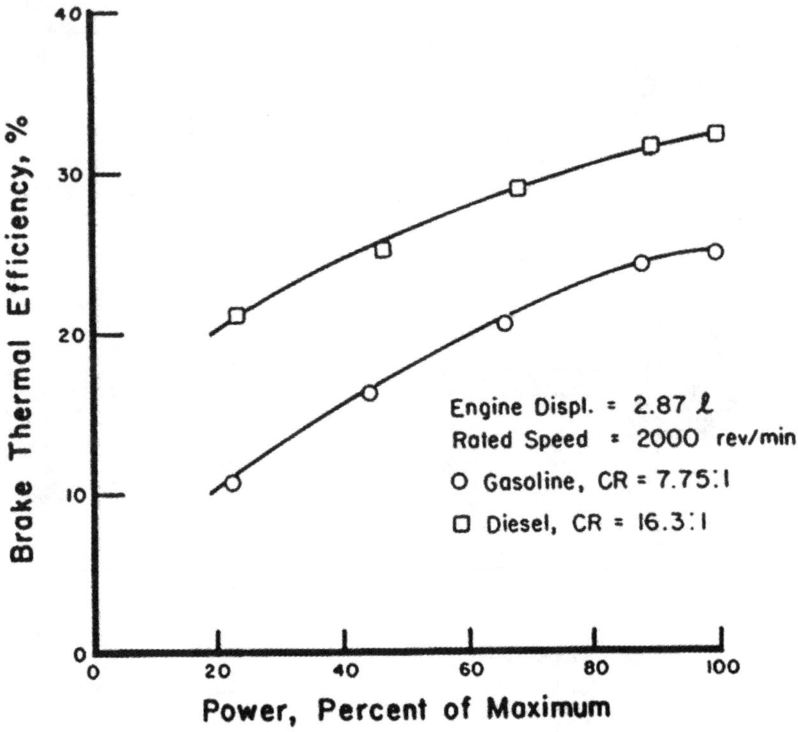

Figure 7.1. Typical brake thermal efficiencies of spark-ignition and compression-ignition engines.

7.3 Types of CI Combustion Chambers

Manufacturers have designed a number of different types of combustion chambers for CI engines. The principal types are illustrated in Figure 7.2. Figures 7.2b, c, and d are variations of *indirect injection* (IDI) CI engines. Figure 7.2a illustrates a *direct injection* (DI) CI engine. In all IDI designs, an extra chamber for starting the combustion process is provided, and combustion is continued in the main chamber. Ignition delay is shorter with the IDI designs; only a small amount of fuel needs to

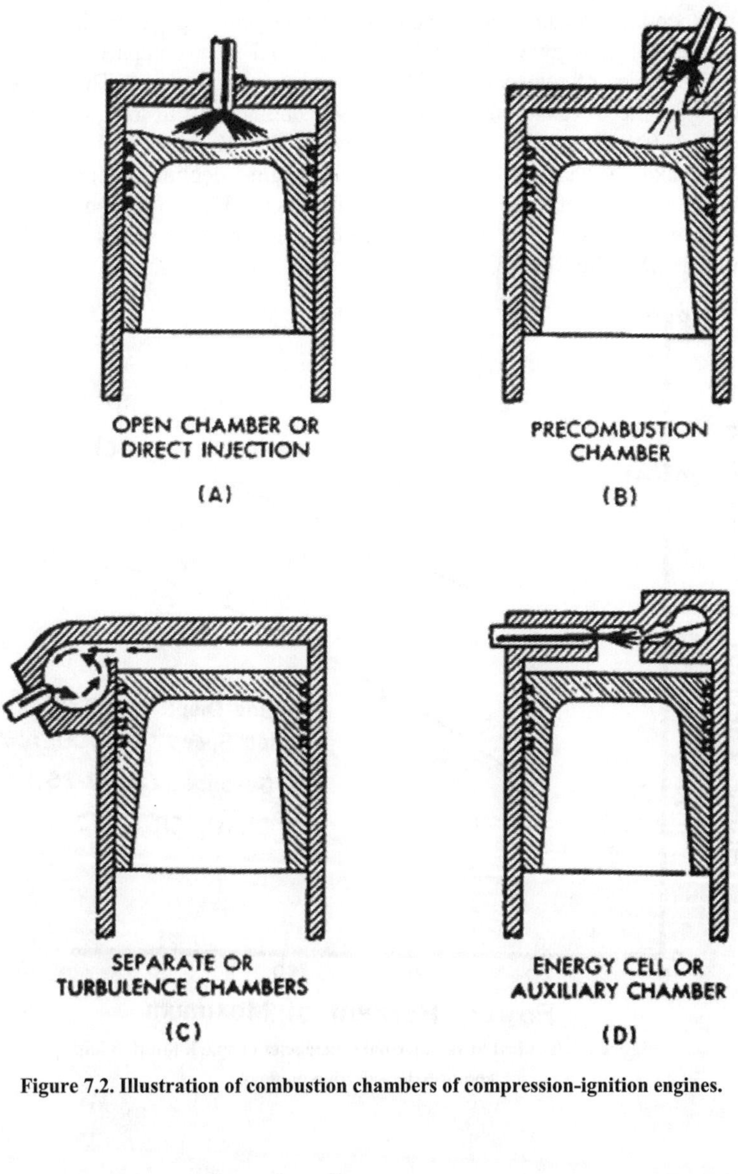

Figure 7.2. Illustration of combustion chambers of compression-ignition engines.

accumulate in the small extra chamber before a combustible mixture is formed. Thus, IDI combustion-ignition engines are quieter than DI combustion-ignition engines and/or can be run on fuel with a lower cetane number. Also, IDI engines can use fuels with higher viscosity because the fuel does not need to be atomized as finely as in a DI engine. Finally, because of the shorter ignition delay and reduced pre-mixed combustion, there is less stress on IDI engines so they can be lighter in weight. Because of their lighter weight and quieter operation, IDI engines have been used in automotive applications instead of the heavier DI engines. Advantages of DI engines are discussed later.

7.3.1 Precombustion Chambers

A *precombustion chamber* (PC) is illustrated in Figure 7.2b. The PC contains about 20% to 30% of the total clearance volume of the cylinder and 70% to 80% of the clearance volume is in the main chamber. The PC may be insulated from the remainder of the head by airspace, and therefore kept very hot (600 to 750°C). Mild turbulence is created in the PC as air flows in during the compression stroke. Fuel is injected into the hot PC and ignites. As the flaming fuel rushes into the main chamber through the restricted throat, considerable turbulence is generated to mix the unburned fuel vapors with the air in the main chamber.

7.3.2 Swirl Chambers

The extra chamber in Figure 7.2c is called either a *swirl chamber* or a *turbulence chamber*. It contains 50% to 90% of the total clearance volume and is connected to the main chamber by a tangential throat. A counterclockwise flow is established in the swirl chamber when air enters during the compression stroke. Fuel is injected into the hot swirl chamber and ignites. Since the flow must reverse in direction as the burning fuel and unburned vapors rush into the main chamber, considerable turbulent mixing occurs. CI engines with swirl chambers are clean burning and are able to achieve low levels of smoke, NO_x, and unburned hydrocarbons in the exhaust. Typically, emissions of CO are low in all CI engines.

7.3.3 Energy Cells

The auxiliary chamber in Figure 7.2d is called either an *air cell* or an *energy cell*. It contains 5% to 15% of the total clearance volume and is connected to the main chamber through a very restricted orifice. The injector nozzle is located directly across the chamber and directs a narrow spray plume toward the auxiliary chamber. Ignition occurs in the main chamber and the pressure rise from combustion forces about 60% of the unburned fuel into the auxiliary chamber. The fuel ignites in the hot cell and creates a strong turbulence as it reenters the main chamber and meets the remainder of the fuel being sprayed by the nozzle. Combustion lasts longer in the engine in Figure 7.2d. Thus, although peak pressures in the combustion chamber are lower, the indicated thermal efficiency is also lower.

7.3.4 Direct Injection

Beginning about 1960, there has been a continuing trend toward the use of *direct injection* (DI), CI engines in large trucks, busses, construction equipment and farm tractors so that now DI engines are used almost exclusively in these vehicles. The combustion chambers of DI engines have less surface area per unit of displacement volume than do the combined combustion and pre-combustion chambers of IDI engines. Also, compared to an IDI engine, more of the combustion of a DI diesel occurs in the pre-mixed mode. Therefore, the DI engines offer significantly better fuel economy (typically 8% to 10%) than do IDI engines.

The air must swirl in the combustion chamber of a DI engine if adequate mixing of fuel and air is to be achieved. A multi-hole nozzle sprays the fuel out in several distinct plumes, as shown in Figure 7.3. The swirling air moves perpendicular to the spray to achieve mixing. The intake valve and port are designed to start the air swirling when it enters the combustion chamber during the intake stroke. Swirl continues during the compression stroke, and it intensifies as air is forced into a cup on the top of the piston as the piston approaches HDC. The amount of swirl is a design compromise. Increased swirl improves mixing of fuel with air, but also tends to narrow the boundary layer and increase heat loss through the walls of the combustion chamber. The swirl speed increases with engine speed. Thus, DI engines are classified according to their swirl ratio. *Swirl ratio* is defined as

$$\text{swirl ratio} = \frac{\text{swirl speed}}{\text{engine speed}} \tag{7.1}$$

where

swirl speed = rotational speed of air in chamber at start of injection, rpm
engine speed = rotation speed of crankshaft, rpm

The concept of air swirl is illustrated in Example Problem 7.1.

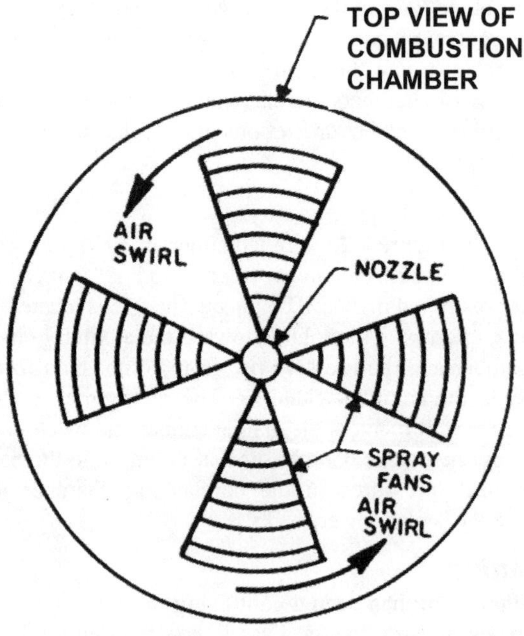

Figure 7.3. Illustration of air swirl in a direct-injection engine.

Example Problem 7.1

The injector nozzles in a certain DI engine each have four holes. If the engine runs at 2400 rev/min and the ignition delay is 18°, what is the theoretical swirl ratio that is needed to fan one spray plume halfway to the next plume during the ignition delay? (The ignition delay is the period between fuel injection and fuel ignition.)

Solution

The spray plumes from a four-hole nozzle are 90° apart. Theoretically, then, the air must move 45° during the delay period. The time available for this movement is

$$\text{time} = 18° \times 60 \text{ s/min}/(2400 \text{ rev/min} \times 360°/\text{rev}) = 0.00125 \text{ s}$$

Therefore, the theoretical air rotational speed must be

$$\text{speed} = 45° \times 60\text{s/min}/(0.00125 \text{ s} \times 360°/\text{rev}) = 6000 \text{ rpm.}$$

The required swirl ratio is

$$\text{swirl ratio} = 6000/2400 = 2.5$$

The actual swirl ratio must be higher than the theoretical ratio because if the air is to drag the spray droplets to a new location, the air must move faster than the spray droplets.

The DI engines used in large trucks, busses, construction equipment and farm tractors traditionally used injector nozzles with four holes and were medium-swirl engines. Medium swirl includes swirl ratios of four to eight at the time of injection. A recent trend has been to use five-hole nozzles to help reduce exhaust emissions. Very large DI engines can use even lower swirl because their high fuel-delivery rates permit the use of injector nozzles with six or more holes. High-swirl engines with single-hole nozzles have been designed, but these are less popular because of their higher heat loss.

When the piston of a DI engine approaches HDC, the swirling air that had been rotating above the piston is forced into the piston cup; this radial movement into the cup is referred to as *squish*. The air squishing in from all sides meets in the center of the cup. Often, the cup is designed with a peak in the center to cause the air to curl back toward the edges of the cup, thus setting up turbulent eddies in radial planes. The cup with the peak in the center is referred to as the Mexican hat design. The effect of swirl and squish combine to increase turbulence in the combustion chamber and thus reduce the heterogeneity of the air-fuel mixture.

7.4 Conventional Fuel Systems for CI Engines

Figure 7.4 illustrates the complete fuel system that is conventionally used in CI engines. The fuel tank is the same as those in SI engines. Fuel is moved from the tank by a transfer pump. The transfer pump is omitted in some systems and the diesel fuel is transferred by a charge pump located in the same housing as the injector pump.

Fuel filters on CI engines are more critical than those on SI engines because dirt or other small particles can ruin the expensive injector pump. Usually, two fuel filters are

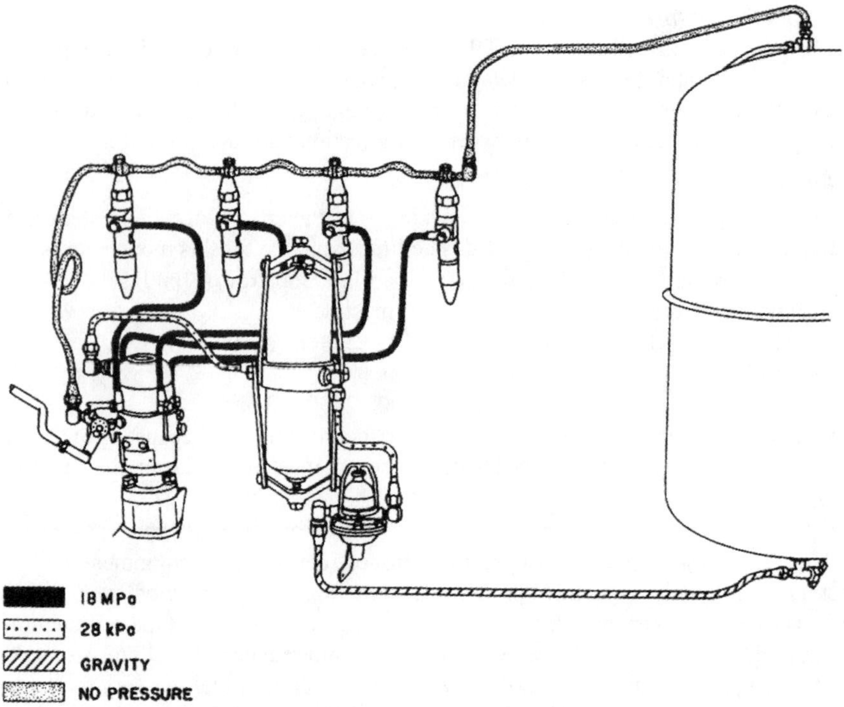

Figure 7.4. A complete conventional fuel system for a compression-ignition engine. (Adapted from a drawing courtesy of Deere and Co.)

used on CI engines. The fuel first passes through a primary filter, which removes the coarser particles, and then passes through a secondary filter, where finer particles are removed. It usually is necessary to change the primary filter more often than the secondary one because the primary filter collects more foreign material.

The filtered fuel flows to the *injector pump*. At the proper times, the injector pump sends bursts of high-pressure fuel to the *injector nozzles* through *injector lines*. The injector lines must be of heavy wall construction to resist the high pressure. All the injector lines on an engine are made the same length to ensure uniform injection timing in all cylinders.

A portion of the fuel going into each injector nozzle leaks out of the top end of the nozzle. The leakage fuel provides for lubrication of the injector needle and for cooling the injector. A low-pressure *return line* is provided to return the hot leakage fuel to the tank for cooling. The injector pump in Figure 7.4 has a case drain to allow air and fuel vapor to escape from the pump. Discharge from the case drain is returned to the tank through the same line used to convey injector leakage. Fuel injection nozzles of widely varying design are used on CI engines. Fuel injectors are discussed later.

Injector pumps have difficult tasks to accomplish. They must meter an amount of fuel that is appropriate for the engine speed and load to each cylinder. The fuel must be metered equally to all cylinders at the proper time and at the correct rate. The injection line pressure must rise sharply to a level high enough to produce a fine spray and must fall sharply enough to cause an abrupt end to each injection. Injector pumps of widely varying design are in use in diesel engines. The common rail system was one of the first systems used for injection; a variation of it is still being used on some modern diesel engines. However, jerk pumps are in much wider use. A jerk pump uses one or more plungers to deliver bursts of high-pressure fuel to the injectors. *In-line jerk pumps* have a separate pump for each cylinder in the engine. *Distributing pumps* have only one pump, but they have a distributor to direct the pump discharge to the proper injector. Jerk pumps are discussed in more detail later.

7.4.1 In-Line Injector Pumps

In-line injector pumps have one individual pump plunger for each cylinder, but the individual plungers are grouped together in a common housing. Details of one individual plunger are shown in Figure 7.5. A reservoir of fuel surrounds the plunger. The fuel can flow in through ports A and A' when the plunger is at the bottom of its

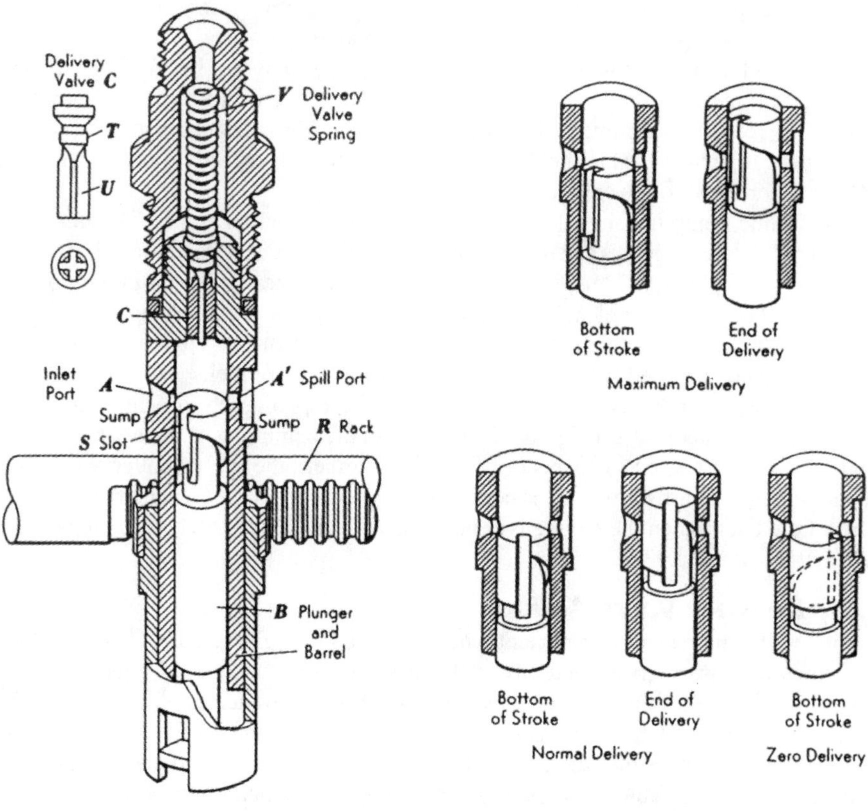

Figure 7.5. A plunger from an in-line injection pump. (Courtesy of United Technologies Diesel Systems, Inc.)

stroke. The plunger is driven by a cam (not shown). When the plunger begins to rise, fuel flows back into the reservoir until the top of the plunger covers ports A and A'. As the plunger continues to rise, the fuel trapped in the pump is forced out through the delivery valve at the top and flows through an injector line to the injector nozzle. Notice that the lower part of the plunger is smaller in diameter and the reduced section is bound at top by a helical edge called a *scroll*. When the scroll reaches spill port A', fuel above the plunger can escape through a slot in the plunger, flow into the reduced section, and then flow out of spill port A' to the reservoir. Thus, the delivery valve closes, and injection ceases when spill port A' is uncovered by the scroll. By a slight modification in design, some pumps can work with only one port and port A is omitted.

The amount of fuel delivered per injection can be controlled by rotating the plunger. For example, rotating the plunger clockwise (as viewed from the top) would cause spill port A' to be uncovered later in the stroke and therefore would increase the size of the injection. A gear on each plunger meshes with a *rack*. Moving the rack forward or backward rotates all plungers in unison and thus controls the size of fuel injections. A cam is located below each pump, and the cam lobes are arranged to provide the proper timing between cylinders. The entire pump is timed to the crankshaft through a set of gears at the front of the engine.

A flyweight governor within the pump controls fuel delivery by moving the rack. Governor principles were discussed in Chapter 2. Engines with in-line pumps can be stopped by rotating the plungers (Figure 7.5) until the slot is aligned with inlet port A. Since fuel can then escape to the sump at all times, the delivery valves never open, and no fuel is delivered to the engine.

The in-line pump has the advantage of starting the injection at a fixed point in the cycle. Fuel delivery always starts when the top of the plunger covers ports A and A' in the pump in Figure 7.5. Thus, injection can be timed to start at the optimum point in the cycle. The rate of fuel delivery is determined by the shape of the cams in the injector pump. The end of delivery varies with the load on the engine, i.e., delivery ends later, when the load is heavy and the engine needs more fuel.

Some in-line pumps have a *retard notch* at the top of each plunger. The notch aligns with the inlet port A (Figure 7.5) and delays the beginning of injection when the engine is being started. The piston thus rises farther, the air is hotter when fuel injection begins, and starting is easier. The scroll can be shaped to uncover the *spill port* a little later, when the engine is being started. Thus, excess fuel can be injected to help start the engine.

7.4.2 Delivery Valve Action

Fuel will compress to some extent when subjected to high pressure. If the injection line pressure were permitted to fall back to zero between injections, the compressibility of the fuel would cause the line pressure to increase too slowly at the start of injection and fall too slowly at the end of injection. As a result, large fuel droplets would dribble into the combustion chamber at the start and end of injection. The dribbled fuel would burn poorly, and engine performance would suffer. The purpose of the *delivery valve* in Figure 7.5 is to prevent fuel dribble. When the scroll uncovers the spill port at the end of injection, the spring forces the delivery valve downward. Initial downward movement closes the delivery valve, and continued movement creates a clearance space where the top of the delivery valve was previously located. Since fuel rushes back from the injection line to fill this clearance space, the line pressure falls abruptly. However, the total movement of the delivery valve is limited in order to trap some residual pressure in the injection line. Therefore, the trapped fuel cannot fully expand, pressure can rise sharply at the beginning of the next injection, and fuel dribble can be avoided. In-line pumps have a delivery valve above each plunger.

Equation 7.2 is useful in the analysis of the need for delivery valves. Pressure changes in the injection lines result in volume changes as

$$\frac{\Delta V}{V_t} = \frac{\Delta p}{\beta_e} \tag{7.2}$$

where

V_t = volume of fuel trapped in injector line, mL,
ΔV = change in volume, mL,
Δp = change in pressure, MPa,
β_e = effective bulk modulus, MPa.

The effective bulk modulus has three components and can be calculated as

$$\frac{1}{\beta_e} = \frac{1}{\beta_f} + \frac{1}{\beta_c} + \frac{1}{\beta_g} \tag{7.3}$$

where

β_f = bulk modulus of the fuel ≈ 1860 MPa
β_c = E / 2.5 = 208,205 MPa for steel lines
β_a = 1.4 p for entrained air; p = peak injection pressure = 20 MPa max

Typical values making up the effective bulk modulus are

$$1/\beta_e = 0.538 \times 10^{-3} + 0.479 \times 10^{-5} + 0.357 \times 10^{-1}$$

The last term dominates if air or vapor is present and causes such a large)V that the injection system will not deliver fuel. It is thus necessary to bleed the air and/or vapor from the injection lines before the engine will run. If there is no air or vapor in the lines, then the bulk modulus is dominated by fuel shrinkage under pressure; the steel lines appear 100 times stiffer than the fuel.

7.4.3 Distributing Injector Pumps

It is expensive to manufacture a separate pump for each cylinder in a diesel engine. Use of distributing injector pumps reduces costs by providing only one pump for the entire engine. A special distributor connects the pump to each injector line at the proper time.

A cutaway view of a distributing pump is shown in Figure 7.6. Fuel enters through a port at the right end and flows to a vane-type charge pump. Since the flow rate and pressure from a vane pump increase with speed and fuel viscosity, a pressure regulator valve is provided. The regulator valve opens farther as pressure increases and releases part of the flow from the charge pump back to the inlet.

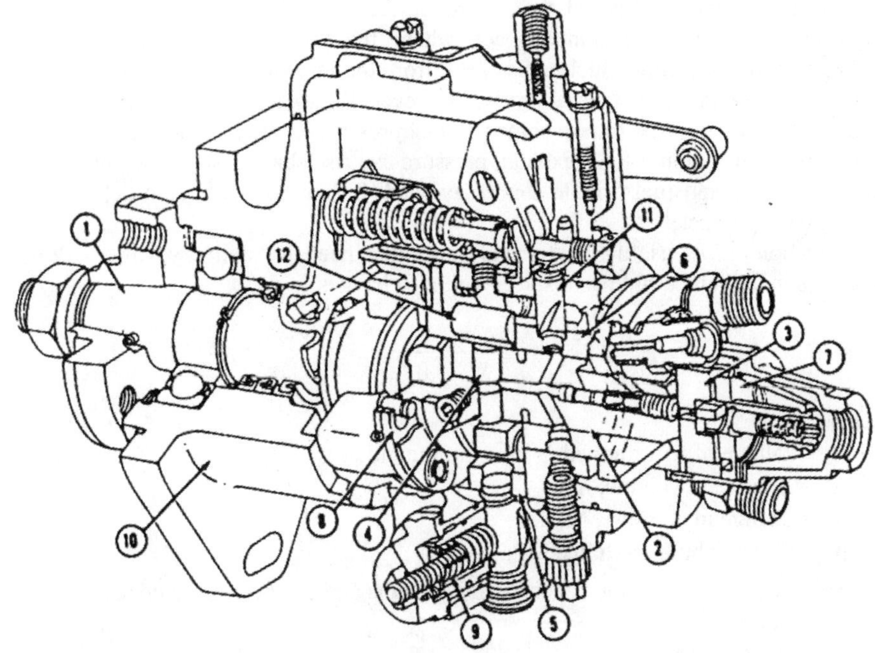

1—Drive shaft
2—Distributor rotor
3—Transfer pump blades
4—Pumping plungers
5—Cam ring
6—Hydraulic head
7—Regulator assembly
8—Governor
9—Automatic advance
10—Housing
11—Metering valve
12—Rollers

Figure 7.6. A cutaway view of a distributing injection pump. (Courtesy of Stanadyne Automotive Corp.)

Fuel from the charge pump flows through internal passages to the *rotary metering valve*, shown near the top of Figure 7.6. A flyweight governor (near the left end of Figure 7.6 works through a linkage to operate a double-sided metering valve. The front side of the metering valve is gradually opened as the engine slows down under increasing load.

Fuel passing the metering valve flows to an annular passage surrounding the central rotor of the pump. Holes in the side of the passage align with radial holes in the rotor (see Figure 7.6) when the main injector pump is ready to receive fuel. Fuel flows into the main pump and forces its rollers apart. An end view of the main pump is shown in Figure 7.7. An internal cam surrounds the pumping plungers, but fuel can force the rollers apart when the rollers are aligned with the low part of the cam. The rollers may or may not contact the cam when the pump is filling. That is, the metering valve allows only enough fuel into the pump to satisfy the load on the engine.

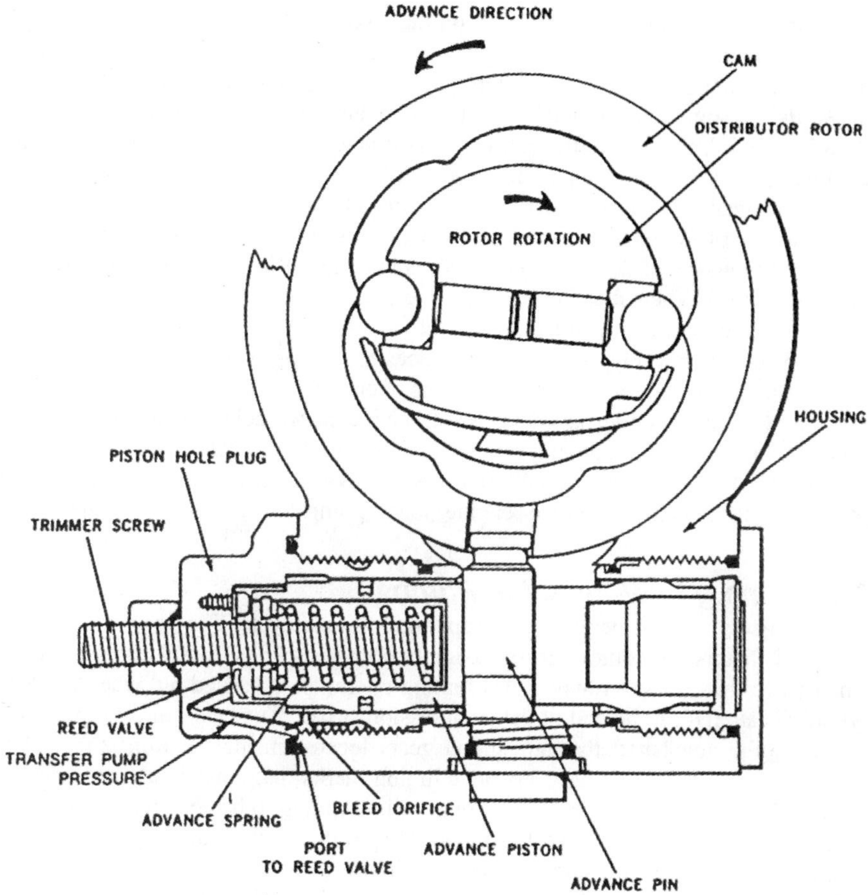

Figure 7.7. End view of the cam and advance mechanism in a distributing injection pump. (Courtesy of Stanadyne Automotive Corp.)

Discharge from the main pump begins when the rollers reach a point where the cam begins to force them inward. At this time, since the radial inlet passages in the rotor are no longer aligned with openings in the annual passage, fuel cannot flow back to the metering valve. Instead, fuel is forced to flow through the center of the rotor and through a delivery valve. After passing through the delivery valve, fuel flows into a radial outlet passage in the distributor rotor. The rotor is timed to have the outlet passage aligned with a passage to one of the injector lines when fuel is being discharged. The delivery valve in the rotor serves the same purpose as those in in-line pumps. That is, it prevents fuel dribble. In some pumps, a delivery valve is placed in each line connector.

If the cam in Figure 7.7 was fixed in position, injections would always end at the same point in the cycle. The end of injections would occur at the point shown in

Figure 7.7, i.e., when the rollers are on the high point of the cam. The start of injection would then vary according to the load on the engine. With a light load, the metering valve would not allow the plungers to move very far apart; therefore, the rollers would contact the cam later, and injection would be retarded. An *automatic advance mechanism* is sometimes provided to offset the inherent light-load retard.

In Figure 7.7, notice that the entire cam is free to rotate to any position dictated by the advance pin. Fuel is pumped into a small piston on the left and can rotate the cam to a more advanced position when fuel pressure is high. When engine load is light and the front side of the metering valve in Figure 7.6 is closing to reduce fuel delivery to the main pump, the back side of the valve is opening to increase fuel pressure in the advance piston. When fuel pressure declines, the spring in the piston expands and rotates the cam to retard the injection timing.

The mechanism in Figure 7.7 also advances the timing when engine speed is high. Recall that pressure from the charge pump increases with pump speed. Therefore, when the engine and pump speeds increase, the increased fuel pressure compresses the spring in the advance piston and rotates the cam to advance the timing. The advance mechanism also influences engine torque reserve. As the cam rotates toward the retard position with heavily increased engine load, pump fill time and torque reserve increases.

7.4.4 Timing Fuel-Injection Pumps

Fuel injections must begin at the proper point in the cycle if engine performance is to be satisfactory. In contrast to spark-ignition engines, most diesel engines must be timed when they are not running. The timing operation is carried out when the No. 1 cylinder is at HDC at the end of the compression stroke. Typically, injector pumps are driven by the crankshaft through timing gears located inside the front cover of the engine. Generally, the timing gears have timing marks that must be aligned when the pump is in time. However, the timing gears allow timing to be changed in increments of one or more teeth. For finer adjustments, the pump usually is mounted to the engine by means of a flange with slotted holes (see Figure 7.6). Thus, the entire pump can be rotated slightly (relative to the engine) to advance or retard the timing. An inspection window with a removable cover usually is provided for viewing timing marks on the pump shaft. For proper timing, the marks on the timing gears must be aligned as specified by the manufacturer

7.4.5 Conventional Fuel-Injection Nozzles

The two principal types of conventional fuel-injection nozzles are the *single-hole* type and the *multi-hole* type (Figure 7.8). Single-hole nozzles nearly always have a *pintle* at the lower end of the plunger. The pintle blocks the orifice when the nozzle is not injecting and partially blocks it when the needle is lifting to open the valve or falling to close it. The partial blocking helps to prevent a weak spray from entering the combustion chamber. Single-hole nozzles are used in IDI diesel engines. DI diesel engines require the use of multi-hole nozzles

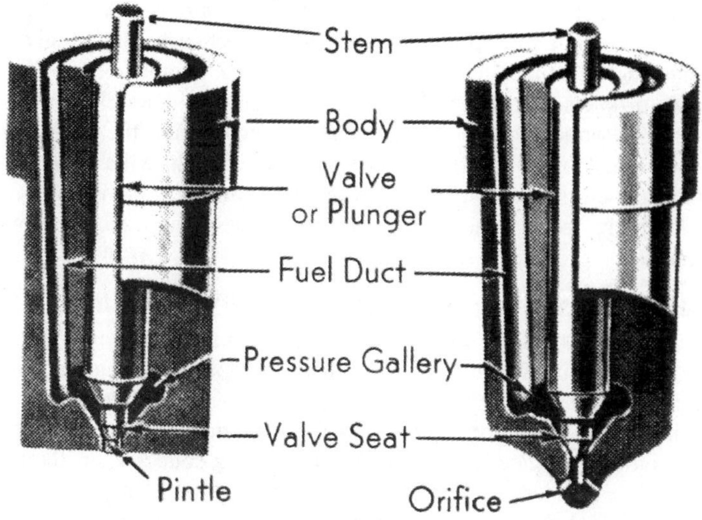

Figure 7.8. Types of fuel injection nozzles for CI engines. (Courtesy of United Technologies Diesel Systems, Inc.)

A cutaway view of a multi-hole nozzle is shown in Figure 7.9. A *pintle-type nozzle* would be similar to the one in Figure 7.9, except for the details of the nozzle tip. A spring-loaded needle valve blocks flow to the orifices when fuel is not being injected. When the pump delivers a surge of fuel, fuel pressure acts on the lower part of the needle to force the needle back against spring pressure. Fuel is injected until the pressure is insufficient to hold back the spring; then, the needle closes to stop the injection. An adjusting screw is provided for setting the pressure at which the needle valve opens. Another adjustment may be provided for controlling the maximum needle lift. Both adjustments should be set in accordance with the manufacturer's specifications. A portion of the fuel leaks past the valve guide to provide lubrication of the guide. The escaped fuel passes through the threads of the opening pressure adjustment and into the fuel return line, as previously discussed in Section 7.3.

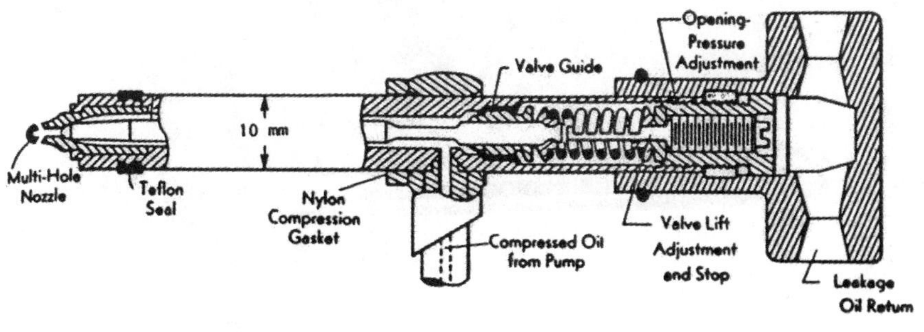

Figure 7.9. Cutaway view of a conventional injection nozzle. (Adapted from a drawing courtesy of Stanadyne Automotive Corp.)

Maintenance of correct orifice size and opening pressure is especially important for multi-hole nozzles. The fuel spray from the nozzle must seek out and mix with the air in the combustion chamber, while avoiding excessive impingement on the chamber walls. Spray droplets that are too large will penetrate to the chamber wall, while droplets that are too small will not penetrate far enough into the air in the chamber. Opening pressures for multi-hole nozzles are typically 20 MPa or higher. Typical orifice diameters are in the range from 150 to 320 μm. On some injector nozzles, the entire tip can be replaced if the orifices become excessively large or become blocked with carbon.

7.4.6 Testing Conventional Fuel-Injection Nozzles

Nozzle testers are available for testing the performance of injection nozzles (Figure 7.10). The tester includes a fuel reservoir, hand pump, pressure gage, and valves. It can be used to measure the pressure at which the needle valve pops open and to observe whether a satisfactory spray pattern is produced. A properly functioning nozzle also should produce a characteristic chattering sound when the nozzle pops open. The tester can be used to verify that the nozzle tip does not leak when the nozzle pressure is slightly below the opening pressure. Finally, the tester can be used to determine whether fuel leaks past the valve guide at the proper rate. Because of the emphasis on proper opening of the nozzle, the tester is often called a *pop tester*.

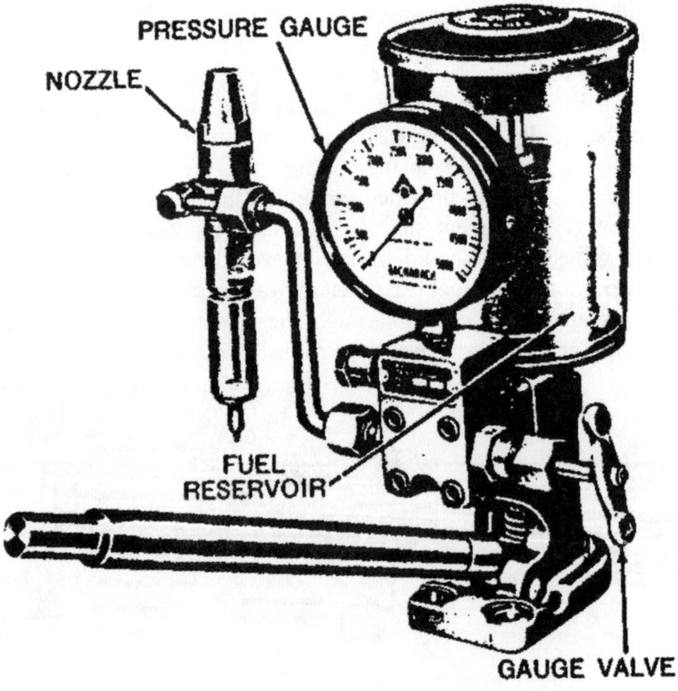

Figure 7.10. A nozzle pop tester.

7.4.7 Bleeding Conventional Injection Systems

As discussed in Section 7.4.2, the presence of air or fuel vapor in the high-pressure lines of a diesel engine will prevent the engine from running. Both air and fuel vapor are far more compressible than is liquid fuel. Thus, air or vapor bubbles in the high-pressure injection lines will compress when the pump delivers a surge of fuel, and pressure will not rise enough to lift the needle in the injector nozzle. The bubbles will expand when the delivery valve closes and thus be ready to prevent the next injection. Air or fuel vapor bubbles can form when the filters are changed, when the high-pressure lines are opened or removed, or when the engine runs out of fuel. Any air or vapor in the low-pressure portions of the system should be bled from the system before the high-pressure lines are bled. Usually, a plug can be removed from the top of each filter to allow air to escape when the engine is being cranked. Some injector pumps have one or more bleed plugs for bleeding air or vapor from the pump housing. All of the bleed plugs must be retightened when the air has been removed.

Air or fuel vapor is removed by *bleeding the injection system*. Fuel is forced into the high-pressure lines by cranking the engine or, on some engines, by means of a hand pump. The collars joining the high-pressure lines to the injectors are loosened enough to allow air and vapor to escape through the connections while fuel is being forced into the lines. The collars are retightened when all of the air and vapor have escaped.

7.5 Mechanically Controlled Unit Injectors

Jerk pumps generate large transient pressures in the injector lines. Closing the needle valve in an injector causes a pressure wave to travel back and forth in the injection line. Under some conditions, the reflected pressure wave can be strong enough to lift the needle and cause a secondary injection. Secondary injections are bad because the fuel is injected too late to be used properly by the cylinder. Thus, there would be benefits if injection lines could be eliminated.

The injection system shown in Figure 7.11 eliminates the high-pressure injection lines. An individual pump and injector are combined into the same unit, and the device is therefore called a *unit injector*. The engine camshaft has an extra cam lobe for each cylinder to operate the unit injector. As with in-line pumps, the spill port and scroll metering principle is used to control the size of injections, and a rack and gear is provided to rotate the plunger. Injection ceases when the scroll uncovers the spill port. The pump discharges directly into a passageway to the bottom of the needle. When pressure is sufficiently high, the needle lifts against spring pressure and opens to release fuel to the spray orifices. The unit injector principle eliminates the injection line but brings new problems. A more complicated camshaft and extra rocker arms are needed to operate the unit injectors. Also, a linkage is needed to permit the governor to operate all of the control racks simultaneously.

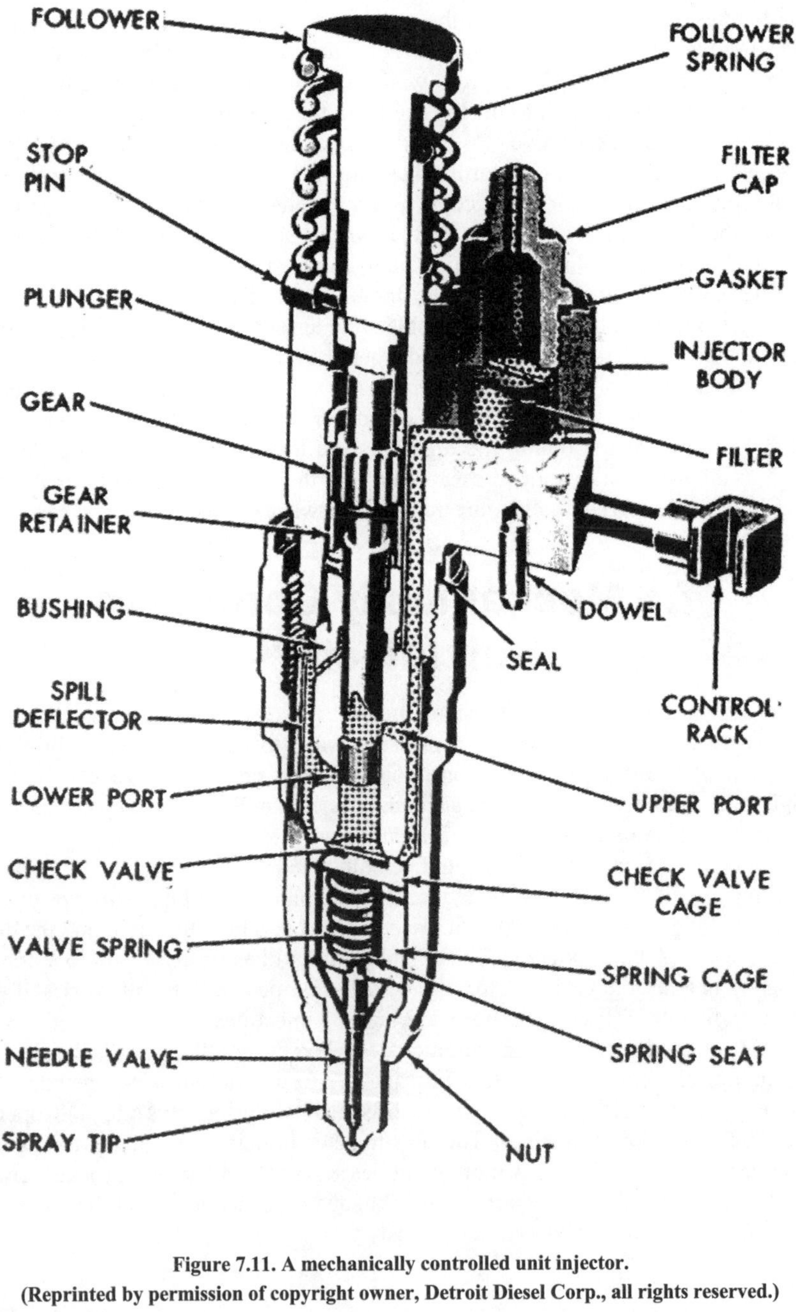

Figure 7.11. A mechanically controlled unit injector.
(Reprinted by permission of copyright owner, Detroit Diesel Corp., all rights reserved.)

7.6 Electronic Fuel Injection Systems

The development of fast, compact, inexpensive microprocessors has made it possible to electronically control individual fuel injections to a CI engine. Both the quantity and the timing of individual injections can be controlled by the microprocessor. There are two types of electronic fuel injection systems for CI engines. The initial systems, which are still in use today, are operated by the camshaft, similar to the unit injectors discussed in Section 7.5, but timing of injections is controlled electronically. More recent systems require neither mechanical actuation nor mechanical control.

7.6.1 Electronically Controlled Unit Injectors

An electronically controlled unit injector is shown in Figure 7.12. As with the mechanical type, the pumping plunger in this unit injector is driven by a lobe on the

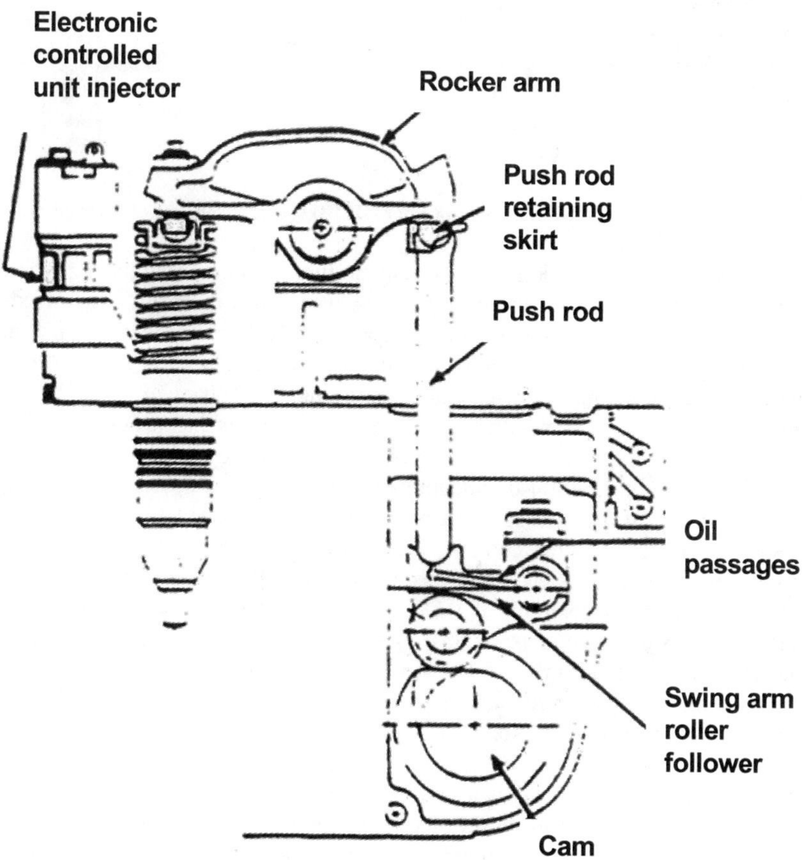

Figure 7.12. An electronically controlled unit injector. (Courtesy of Caterpillar, Inc.)

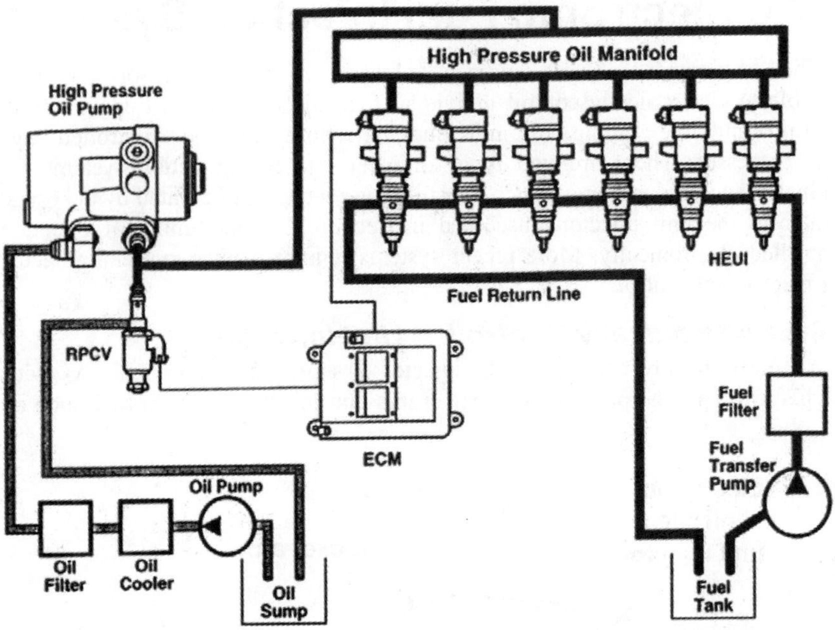

Figure 7.13. A hydraulic-electronic unit injector (HEUI) system. (Courtesy of Caterpillar, Inc.)

engine camshaft. The pumping plunger delivers fuel via an internal passage to a multi-hole nozzle at the bottom of the injector. However, scroll metering is not used in the electronically controlled unit injector. Instead, a special solenoid valve on the spill port is computer controlled. There are several advantages to computer control of injections. In conventional systems, injection timing is not changed while the engine is running; the governor controls the size of the fuel injections, as previously discussed. A computer-controlled injector can be programmed to use multiple inputs to control both the timing and duration of injections to meet designer objectives. These inputs include position of the speed control pedal, engine speed, crankshaft position, turbocharger boost, engine coolant temperature, and possibly others. Also, the computer program can reside on an electronically programmable, read-only memory (EPROM). By simply changing the EPROM chip, the engine characteristics can be changed. For example, one program might optimize the engine for fuel economy, while another might optimize it for low emissions. Another advantage is that the expense of machining a scroll on the pumping plungers is eliminated. Because the spill port is controlled by a solenoid valve, the pumping plunger can be a simple cylinder.

There are some disadvantages to the electronically controlled unit injector system. The first is the added cost of an on-board computer, plus the sensors that are needed for the multiple inputs to the computer. Also, the computer is typically mounted on the engine and must be protected from the high surface temperature of the engine.

Typically, the computer is mounted on a cooling plate attached to the engine; by circulating fuel through the cooling plate, the temperature of the computer mounting surface can be kept low enough to prevent damage to the computer.

7.6.2 Hydraulic-Electronic Unit Injection Systems

A *hydraulic-electronic unit injection* (HEUI) system is schematically illustrated in Figure 7.13. In this system, a high-pressure, variable-displacement pump delivers engine oil to an oil manifold. This pump maintains manifold pressure in the range from 4 to 23 MPa. Figure 7.14 shows a cross section of one of the fuel injectors. A fuel transfer pump delivers fuel to the injectors at approximately 200 kPa pressure. When the computer in the electronic control module (ECM) sends an energizing signal to the control valve in a unit injector (Figure 7.14), high-pressure oil from the

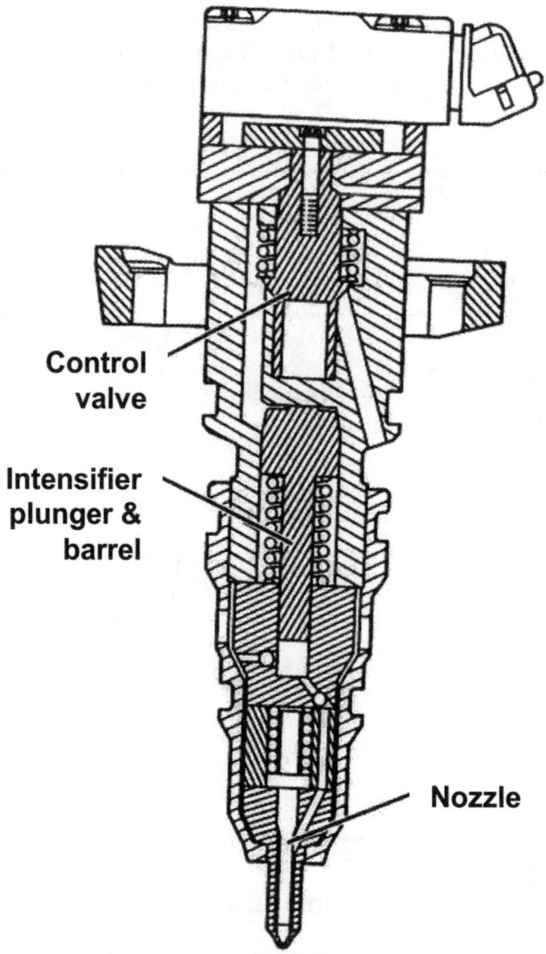

Figure 7.14. Cross-sectional view of a HEUI injector. (Courtesy of Caterpillar, Inc.)

manifold is able to reach the intensifier plunger, causing it to move downward to deliver fuel to the injector. The high fuel pressure causes the injector needle to lift and fuel enters the combustion chamber through the multiple holes in the injector tip. When the control valve is de-energized, spent oil is ejected from the unit injector to the valve cover area. Falling fuel pressure allows the injector needle to reseat, thus ending the injection. A small amount of fuel from each injector is returned to the fuel tank via a return line.

The HEUI system has the advantages mentioned in Section 7.6.1 for electronically controlled unit injectors. In addition, the HEUI system has a more desirable injection-rate pattern, as illustrated in Figure 7.15. The actuating current is initially high to achieve rapid opening of the control valve (Figure 7.14), then is lowered for more rapid closure when the ECM turns off the valve. Consequently, the valve (poppet) lift achieves almost a square pattern, as does the fuel injection rate. In contrast, the injection rate from a conventional system is more triangular and requires a longer duration to inject a given quantity of fuel. The shorter injection times of the HEUI system allow the engine to achieve more efficient and complete combustion of the fuel.

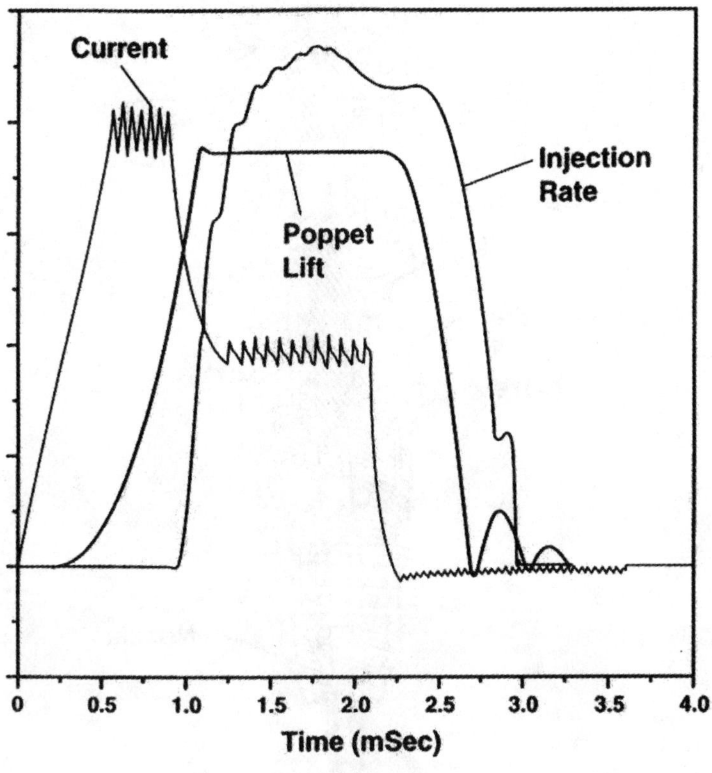

Figure 7.15. Waveforms for a HEUI system. (Courtesy of Caterpillar, Inc.)

For cold-weather starting of an engine, the high oil viscosity poses a special challenge for the HEUI system. The high viscosity tends to delay start of injection and reduce injection size for a given input signal to the control valve. To eliminate these disadvantages, an oil temperature sensor is used as an input to the ECM. The ECM program uses the temperature data to raise the actuating oil pressure and adjust the timing and duration of injection to offset the inherent injection delay and reduced injection size. The HEUI is able to function well at temperatures as low as –40°C.

7.6.3 Common-Rail, Electronic-Injection Systems

A common-rail, electronic-injection (CREI) system is illustrated in Figure 7.16; a cross-sectional view of a nozzle is shown in Figure 7.17. In the CREI system, a variable-displacement pump delivers fuel at very high pressure to a manifold or common rail. The manifold pressure must be much higher than the peak gas pressures in the combustion chamber and thus manifold pressures up to 140 MPa are used. The internal volume of the manifold must be sufficient to damp the pressure oscillations created by the high-pressure pump and the injection process. An electronic-control unit (ECU) regulates pump delivery to maintain constant pressure in the manifold; typically, pressure deviations are less than 5 MPa from the mean pressure.

When an injection is to begin, the ECU provides an electrical input to open the solenoid valve within the injector, thus allowing high-pressure fuel from the manifold to begin spraying fuel through the multi-hole nozzle tip. Injection ceases when the ECU de-energizes and closes the solenoid valve. As with other electronic injection

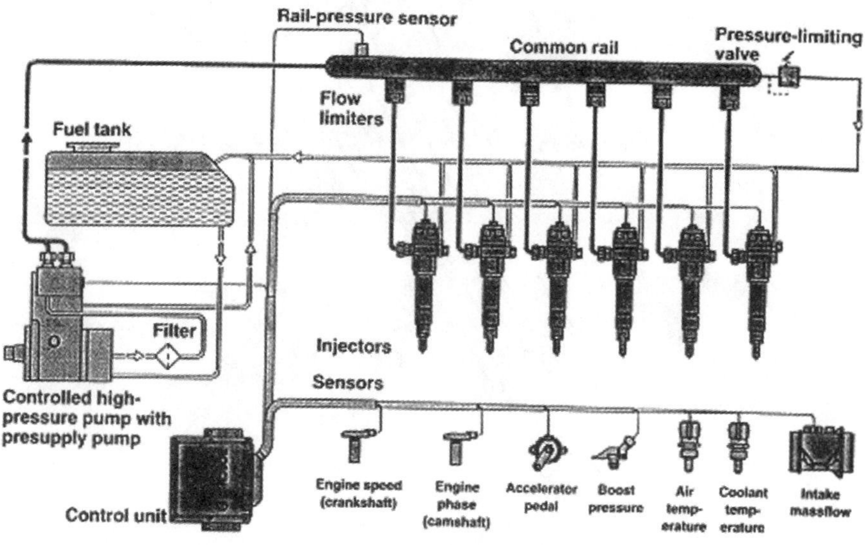

Figure 7.16. A common-rail electronic injection (CREI) system. (Courtesy of Robert Bosch Corp.)

systems, the ECU uses inputs from a variety of sensors to determine the state of the engine at any time. Also, as with other electronic systems, the ECU program can be easily modified to change the performance characteristics of the engine. Waveforms from a CREI system are shown in Figure 7.18. Note the desirable shape of the injection-rate waveform, i.e., it is nearly a square pulse. Figure 7.18 shows another advantage of the CREI system; it is able to provide a small pilot injection before each main injection. The pilot injection is able to initiate the ignition delay period, thus decreasing the amount of pre-mixed combustion and reducing diesel knock. The pilot injections also help to reduce NO_x emissions.

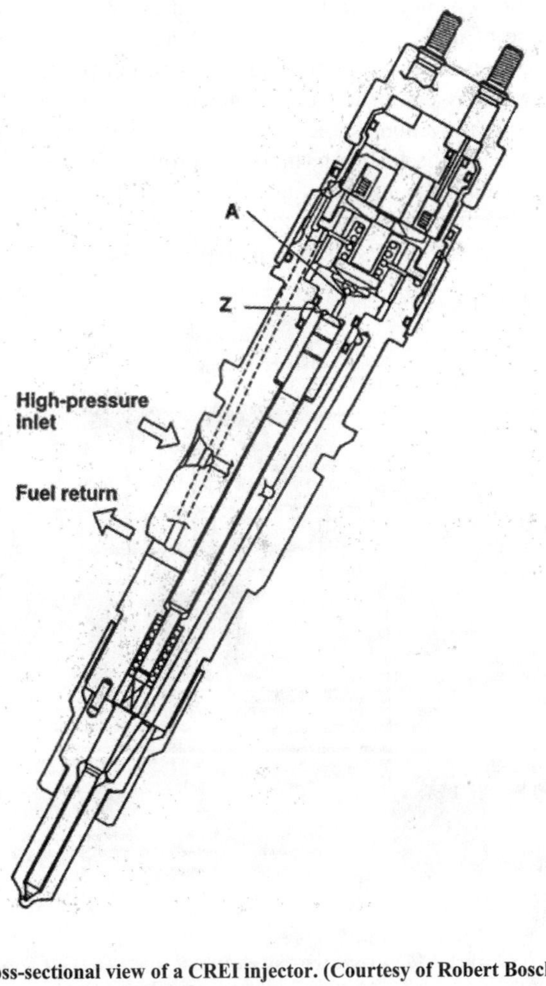

Figure 7.17. A cross-sectional view of a CREI injector. (Courtesy of Robert Bosch Corp.)

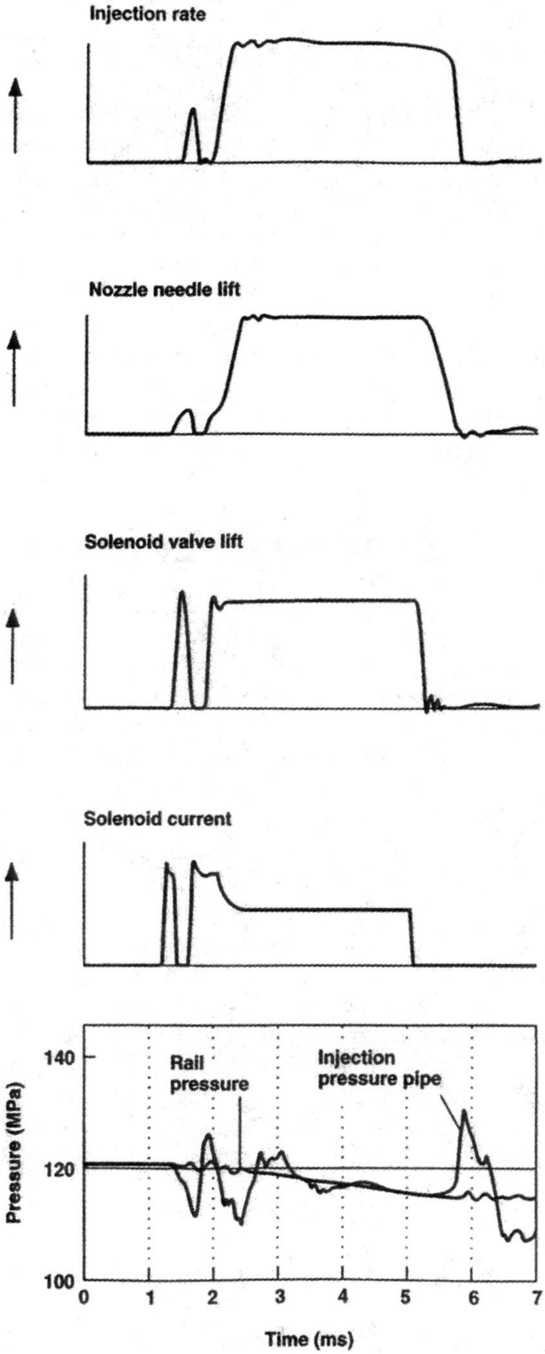

Figure 7.18. Waveforms for a CREI system. (Courtesy of Robert Bosch Corp.)

7.7 Diesel Combustion and Smoke

Uniform mixing of air and fuel is not fully achieved in a CI engine, and smoke may therefore appear in the exhaust. Generally, smoke is classified as cold or hot (see Figure 7.19).

Cold smoke is white in color and consists of a fog of unburned fuel particles. It is most likely to appear when a cold engine is being started. Low air temperatures, light loads, low cetane fuels and high air-fuel equivalence ratios also are conducive to white-smoke formation.

Hot smoke is black or gray in color and consists of carbon particles (soot). Hot smoke is a product of incomplete combustion and occurs when the air supply is insufficient. Thus, hot smoke is most likely to occur under heavy engine load when the air-fuel equivalence ratio is lowest. Hot smoke often is negligible when the engine is idling, but it increases in intensity as more load is placed on the engine and greater quantities of fuel are consumed.

7.8 Starting CI Engines

The temperature in the combustion chamber of a CI engine must be raised above the SIT (self-ignition temperature) of the fuel if the engine is to start and run. Usually, an electric starter is used to crank the engine. On some larger CI engines, batteries may be connected in series to provide 24 V for starting. However, 12 V starters are more commonly used.

The electric starter usually can start a CI engine when air temperatures are 5 to 10°C or above. Starting aids may be required when air temperatures are lower. *Electric glow plugs* are typically used for cold weather starting of IDI engines. A glow plug is placed in each precombustion chamber. A switch is used to electrically energize the glow plug and heat the small prechamber prior to cranking the engine. However, glow plugs usually are not used with DI, CI engines. Instead, *ether* is sprayed into the combustion chamber while the engine is being cranked. Because ether has an extremely low SIT it can start an engine in cold weather. After the engine starts, combustion quickly generates enough heat to keep the engine running on its regular fuel. Usually, a remote control is provided to allow control of the ether spray from the operator's station. Some diesel engines have a heater installed in the intake manifold to heat the incoming air. An ether starting aid should never be used when a heater is used in the intake manifold. Block heaters are also used to improve starting. They can provide some warming of the combustion chamber and also allow the starter to crank faster in cold weather.

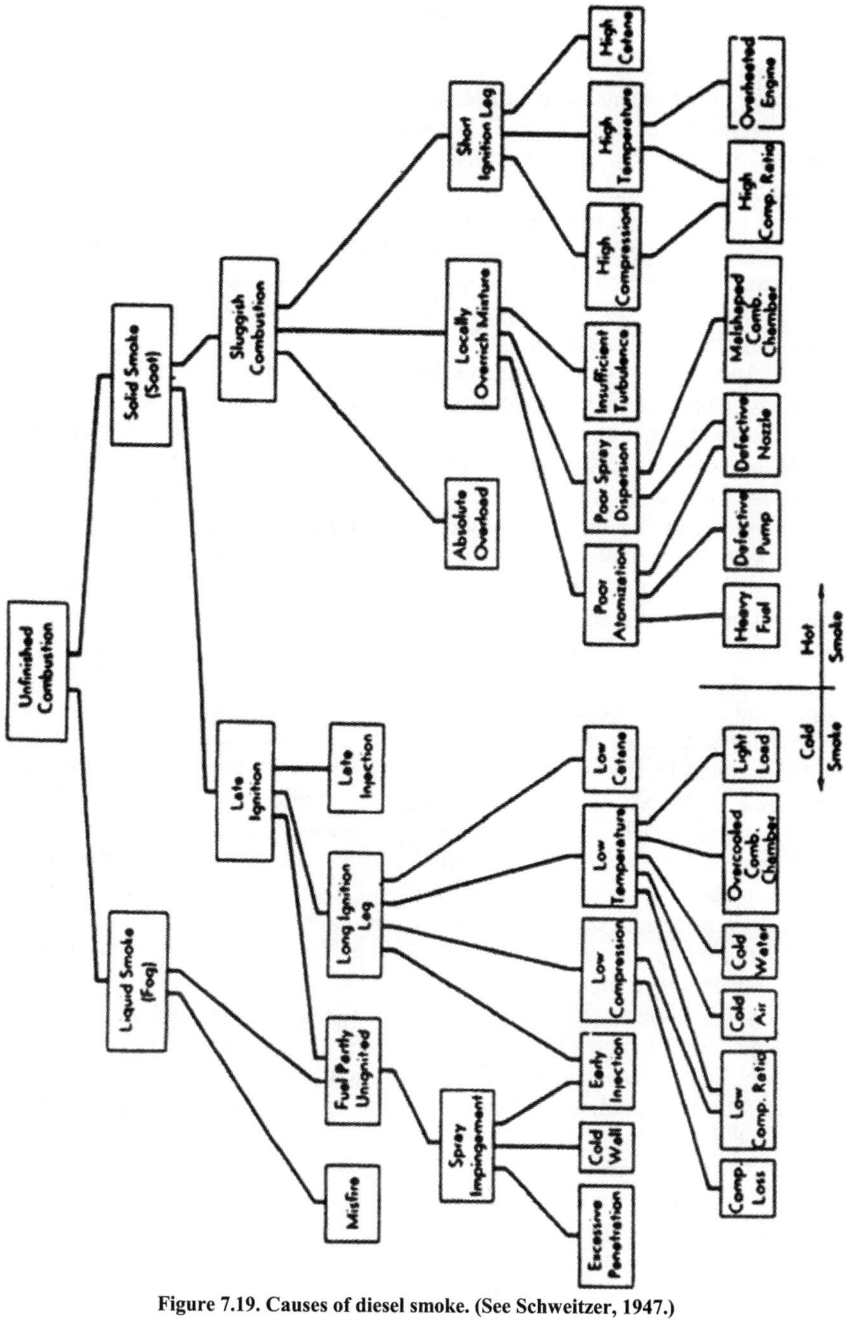

Figure 7.19. Causes of diesel smoke. (See Schweitzer, 1947.)

7.9 Chapter Summary

Although CI and SI engines have many similarities, they have fundamental differences in their means of fuel metering, speed control, and ignition, which require differences in engine construction. Because of these differences, CI engines run leaner, have better full- and part-load fuel economy, and must be of stronger construction than SI engines.

Several types of IDI chambers have been developed for starting the combustion process with CI engines. IDI engines are quieter and can run on fuels of higher viscosity and lower cetane number than DI engines. Direct injection yields greater fuel economy, however, and DI, CI engines are used in most construction equipment, farm tractors, large trucks, and busses.

The fuel-injection system dominates the operation of diesel engines. Conventional systems include a fuel tank, transfer pump, filters, injection pump, and injectors. Lines of heavy walled construction convey high-pressure bursts of fuel from the injector pump to the injectors. Low-pressure return lines convey injector leakage fuel back to the tank. Injector pumps must meter fuel equally to all cylinders at the proper time and at a rate appropriate for the engine speed and torque. The injection pressure must rise sharply, open the injectors, and create a fine spray; then, it must fall sharply to cause an abrupt end to each injection. The two most common types of injector pumps are in-line pumps and distributing pumps.

Injection nozzles include a spring-loaded needle valve, which is forced open by each burst of fuel from the injector pump. The spring closes the needle valve and helps trap residual pressure in the fuel line at the end of each injection. An adjusting screw is provided for setting the nozzle's opening pressure. Some fuel leaks past the needle guide to provide lubrication and is conveyed back to the tank through low-pressure return lines. There are two principal types of injection nozzles: single-hole, pintle nozzles are used in IDI diesels; multi-hole nozzles are used in DI diesels.

The presence of air or fuel vapor bubbles in the high-pressure line of a CI engine will prevent the engine from running or producing full power. Compression of the bubbles prevents the nozzle from opening when line pressure increases. The bubbles can enter the fuel system during engine disassembly or when the engine runs out of fuel. In either case, bubbles must be bled from each line through a loosened connection to the nozzle while fuel is being forced into the other end of the line.

Unit injectors combine a pump plunger and fuel injector into a single unit, thereby eliminating the problems that are possible with high-pressure lines. Scroll metering is used, as with in-line injector pumps. A lobe on the engine camshaft causes the plungers to reciprocate.

Several types of electronic fuel-injection systems have been developed for CI engines. Each type has an ECU (electronic control unit) or ECM (electronic control module) to control the quantity and timing of each injection. A variety of sensors are provided to sense the engine condition and the speed request of the operator. The program in the ECU or ECM uses these inputs to optimize the engine performance for specific requirements. For example, the ECU or ECM could be programmed to provide maximum power, minimum fuel consumption, minimum emissions, or to meet other specific goals.

The earliest electronic injection systems for CI engines used cam-driven unit injectors; the opening and closing of the spill port was accomplished by a solenoid valve controlled by the ECM or ECU. The HUEI system eliminates the need for the cam drive. Instead, the plungers are driven by high-pressure oil supplied via a manifold by a variable-displacement hydraulic pump. In the CREI system, the fuel itself is highly pressurized in a common rail and fed to each injector nozzle through a solenoid valve. Injection occurs when the pressurized fuel lifts the spring-loaded needle in the injector. A special advantage of the CREI system is that a small pilot injector can precede the main injection, thus initiating the ignition delay period. The resulting reduction in pre-mixed combustion reduces diesel knock and NO_x emissions.

Non-uniform mixing of fuel and air results in smoke appearing in the exhaust of CI engines. Cold smoke consists of a fog of unburned fuel, is white in color, and is most likely to appear when a cold engine is being started. Hot smoke is black or gray, consists of carbon particles, and is most likely to appear when the fuel-air equivalence ratio is too high.

A CI engine will start only if the combustion air is compressed to a temperature above the SIT of the fuel. An electric starter usually can start a diesel engine when the temperature is above 5 to 10°C, but starting aids must be used with lower temperatures. Electric glow plugs are used in the prechambers of IDI engines and preheat the chamber before the engine is cranked. Ether, a volatile fuel with a very low SIT, is sprayed into the intake manifold to aid the starting of DI engines .

Homework Problems

7.1 Develop a spreadsheet to synthesize curves similar to those in Figure 7.1.

(a) First, using theory from Chapter 2, derive the equation: $e_m = (1+P_f/P_b)^{-1}$.

(b) Next, use the equation to calculate P_f/P_b at maximum power if $e_m = 0.8$ at maximum power. Then, calculate e_m versus P_b/P_{bmax} for $0.2 < P_b/P_{bmax} < 1.0$. Assume these e_m values apply to both the SI and CI engines.

(c) Next, use Equation 4.26 to estimate the indicated thermal efficiency of the CI engine as a function of load. Replace k with n = 1.3; let $T_3/T_1 = 9$ and r = 16. Also assume the following relationship for β: $\beta_{dc} = 0.0222 + 0.778\ P_b/P_{bmax}$.

(d) Assume the indicated thermal efficiency of the SI engine remains constant at 0.72 times that of the CI engine at maximum power. Calculate $e_{bt} = e_m \times e_{it}$ for the CI engine and

(e) Likewise, calculate $e_{bt} = e_m \times e_{it}$ for the SI engine.

(f) On the same graph, plot e_{bt} versus P_b/P_{bmax}.

(g) Finally, add to the spreadsheet a column showing the percentage by which the brake thermal efficiency of the CI engine exceeds that of the SI engine at each load. Note that this problem is for illustration only; Equation 4.26 does not give precise predictions of indicated thermal efficiency, nor is the equation relating β_{dc} to P_b/P_{bmax} an experimentally verified relationship.

7.2 Rework Problem 7.1, but use a compression ratio of 14.

7.3 Rework Problem 7.1, but use a compression ratio of 18.

7.4 A six-cylinder DI, CI engine consumes 16.4 kg/h of fuel while running at its rated speed of 2300 rev/min. The fuel density is 0.834 kg/L. The engine displacement is 5.392 L and the compression ratio is 16.

(a) Calculate the clearance volume of each cylinder.

(b) Calculate the average volume of each fuel injection.

(c) Calculate the volume of individual fuel injections as a percentage of the clearance volume.

(d) Assuming mixing of fuel and air must occur more than 20° of crankshaft rotation, calculate the time (in ms) available for mixing. The point of this problem is to illustrate the impossibility of achieving uniform mixing of fuel and air in a CI, DI engine.

7.5 Rework Problem 7.4, except use a compression ratio of 17.

7.6 Rework Example Problem 7.1, except use a five-hole nozzle.

7.7 Rework Example Problem 7.1, except let the engine speed be 2000 rev/min.

7.8 Rework Example Problem 7.1, except use a three-hole nozzle.

7.9 A 3-liter, naturally aspirated CI engine is running on No. 2 diesel fuel with a stoichiometric A/F ratio of 15. The air temperature is 20°C and the barometric pressure is 98 kPa. The p_{bme} is 700 kPa and the BSFC is 0.25 kg/kWh at rated speed and power. The high idle speed is 2.540 rpm and the rated speed is 2400 rpm.

(a) Calculate the ambient air density.

(b) Assuming the cylinders are able to fill to 85% of capacity with ambient air, calculate the air consumption of the engine in kg/hr when the engine is running at a high idle speed.

(c) As (b), but when the engine is running at rated speed.

(d) Calculate the fuel consumption in kg/h at rated speed.

(e) Calculate the equivalence ratio at rated speed.

(f) Assuming that the fuel consumption rate falls to 2.6 kg/h at high idle, calculate the equivalence ratio at high idle. Note that this wide change in equivalence ratio of the CI engine is in sharp contrast to that of the SI engine, in which the equivalence ratio stays relatively close to unity.

7.10 Rework Problem 7.9, but let the p_{bme} = 750 kPa and the BSFC = 0.20 kg/kW·h at rated conditions.

7.11 (a) Using Equation 7.3, calculate the effective bulk modulus if the peak injection pressure is doubled to 40 MPa.

(b) Would such a doubling of the peak pressure allow an injection system to function properly without bleeding the injection lines?

7.12 (a) Using Equation 7.3, calculate the effective bulk modulus if the lines are bled, but the high-pressure lines are made of aluminum instead of steel. Assume the modulus of elasticity of aluminum is 35% that of steel.

(b) Based on the effective bulk modulus, would aluminum or steel be the better material for constructing the high-pressure injection lines?

7.13 Assume a conventional injection pump is driven by a 30-tooth gear. If the pump mounting flange is not slotted, calculate the smallest change in timing of start of injection (in terms of crankshaft degrees) that can be made by changing the timing by one gear tooth.

7.14 Rework Problem 7.13, but let the pump be driven by a 40-tooth gear.

7.15 (a) Assuming the manifold in a CREI system is constructed of steel and no air or vapor is present in the manifold, calculate the percentage change in the volume of the fuel in the manifold as the manifold was initially charged from zero to 140 MPa.

(b) Calculate the percentage change due to a 5 MPa pressure deviation from the mean pressure in the manifold.

(c) If the injection pressure was 100 MPa at the nozzle tip when the solenoid valve opened in the injector, by what percentage would the fuel volume expand en route from the manifold?

7.16 Rework Problem 7.15, but assume the injection pressure is 80 MPa.

7.17 (a) Assume the compression ratio of a CI engine is 16 and $n = 1.33$ during the compression stroke. Calculate the lowest theoretical ambient temperature at which the engine would start on No. 2 diesel fuel (SIT of 387°C).

(b) Rework part a, but use diethyl ether (SIT of 304°C).

7.18 Rework Problem 7.17, but use a compression ratio of 14.

7.19 Rework Problem 7.17, but use a compression ratio of 18.

7.20 Rework Problem 7.17, but use a compression ratio of 20.

References and Suggested Readings

Brezonick, M. 1998. Lucas Varity's new common rail system. *Diesel Progress* (30 October): 30, 32-34.

SAE. 1997. Decoupling pressure generation and injection. *Off-Highway Engineering* (August): 51-56.

Glassey, S.F., A.R. Stockner, and M.A. Finn. 1993. HEUI–A new direction for diesel engine fuel systems. SAE Paper No. 930270. Warrendale, PA: SAE.

Goering, C.E. 1992. *Engine and Tractor Power,* 3rd ed. St. Joseph, MI: ASAE.

Obert, E.F. 1973. *Internal Combustion Engines and Air Pollution.* New York, NY: Harper and Row, Publishers.

Otten, R.L., W. Von Wrisberg, D.E. Rieser, and R.L. Miller. 1993. Enhanced series of mid-size agricultural tractors. ASAE Paper No. 931571. St. Joseph, MI: ASAE.

Schweitzer, P.H. 1947. Must diesels smoke? *SAE Transactions* 1(3): 476.

CHAPTER 8

INTAKE AND EXHAUST SYSTEMS

8.1 Introduction

The intake system of an engine includes an *air cleaner* for removing dust and dirt from the incoming air and an intake manifold. Depending upon the engine, the incoming air may also pass through a carburetor or throttle-body fuel injector and/or through a turbocharger compressor. The exhaust system includes an *exhaust manifold* and *muffler*, but the exhaust may also pass through the turbine of a turbocharger. A turbocharger can be used to boost the air consumption of an engine, thereby allowing greater fuel consumption and higher power output. Procedures are presented for calculating the air consumption of an engine and for fitting a turbocharger to an engine. When turbocharger boost is very high, an intercooler may be required to cool the air leaving the compressor. Air-to-water and air-to-air intercoolers are discussed. Some manufacturers have used turbocharging and intercooling to produce a family of engines from one basic engine block and head. An example of this technique is presented.

8.2 Air Cleaners and Intake Manifolds

Dust particles in the air must be prevented from entering an engine to prevent rapid wear. For many years, oil-bath air cleaners were used. The entering air was directed through a cup of oil in the bottom of the cleaner, and then exited upward to carry oil into a wire mesh maze. The oil in the maze trapped dirt and dust particles and, when dripping back downward, carried the dust and dirt into the cup. Periodically, the cup had to be removed, emptied, and refilled with clean oil. The oil-bath air cleaners have been replaced by dry-type air cleaners (see Figure 8.1). The entering air is directed into a swirling motion outside of a paper filter. The heavier dust particles are separated by centrifugal action and fall into a dust cup at the bottom. The air flows through the filter to a central pipe that exits to the intake manifold of the engine. The dry-type air cleaners are able to remove more than 99% of the dust in the air while providing an initial pressure restriction of less than 0.5 Pa. The filter gradually becomes restricted with dust. A service indicator (see Figure 8.2) is often provided to warn the operator when the air cleaner needs servicing. The indicator is simply a vacuum gage that measures the vacuum in the exit pipe of the air cleaner. Sometimes, the filter can be cleaned by blowing low-pressure air from the inside toward the outside of the filter. More often, the used filter is simply replaced with a new one.

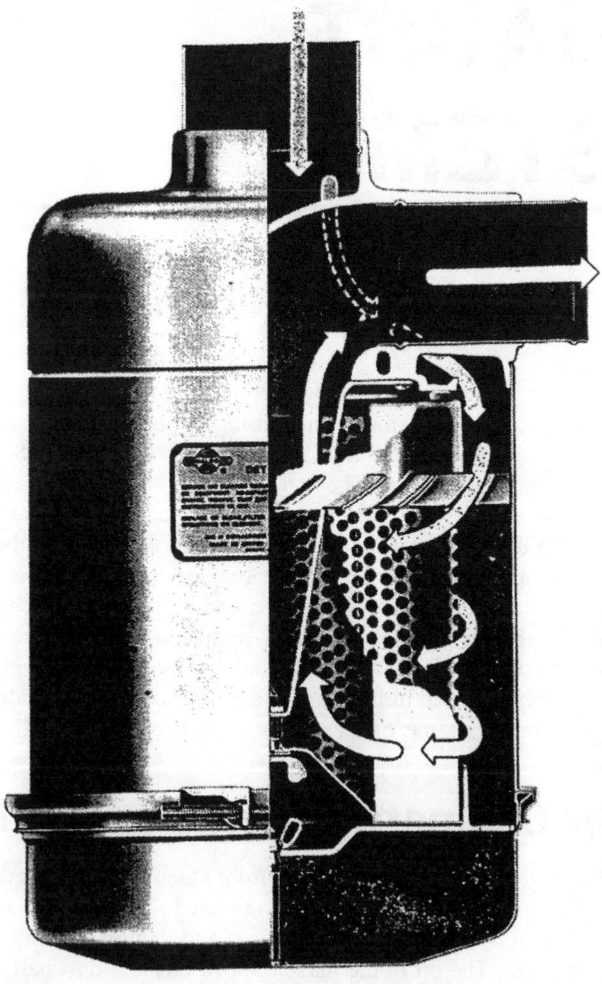

Figure 8.1. A dry-type air cleaner. (Courtesy of Donaldson Co., Inc.)

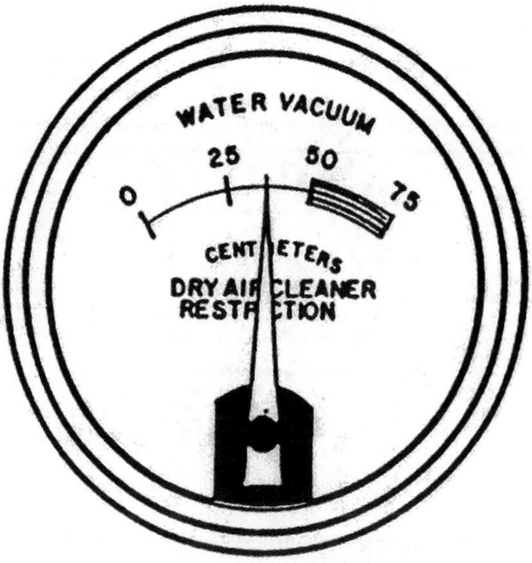

Figure 8.2. An air cleaner service indicator.

In SI engines, air exiting the air cleaner may be piped to a carburetor or throttle-body injector before entering the intake manifold. If an SI engine uses *port injection*, the air from the air cleaner is piped directly to the intake manifold. In *naturally aspirated* (NA) CI engines, air from the air cleaner is piped directly to the intake manifold. In *turbocharged* engines, air leaving the air cleaner passes through a compressor before entering the intake manifold.

In some SI engines, the intake manifold receives an air-fuel mixture from a carburetor or throttle-body fuel injector and distributes the mixture to the cylinders. The intake manifold is subdivided into branches that serve individual cylinders or pairs of cylinders, as illustrated in Figure 6.2. Some of the fuel that is not fully vaporized moves in a film along the manifold walls; the inner surface of the manifold must be smooth to minimize the film thickness. Also, as was discussed in Section 6.3.1, there are pressure oscillations in the manifold that are associated with the opening and closing of the valves. The frequency of the pressure oscillations is usually much higher than the frequency of opening and closing of the intake valves. A *tuned manifold* is one in which the manifold branches are long enough the lower the frequency of the oscillations to match the valve frequency. Tuned manifolds can increase the air delivery to the engine, but the long branch lengths cause the manifold to be bulky.

In SI engines equipped with port injection, a fuel injector is positioned to inject fuel into each intake port of the engine; the injected fuel enters the combustion chamber when the intake valve opens. Although the intake manifold handles only air, the manifold is still subdivided into branches. Manifold tuning is still possible, as was discussed in the previous paragraph.

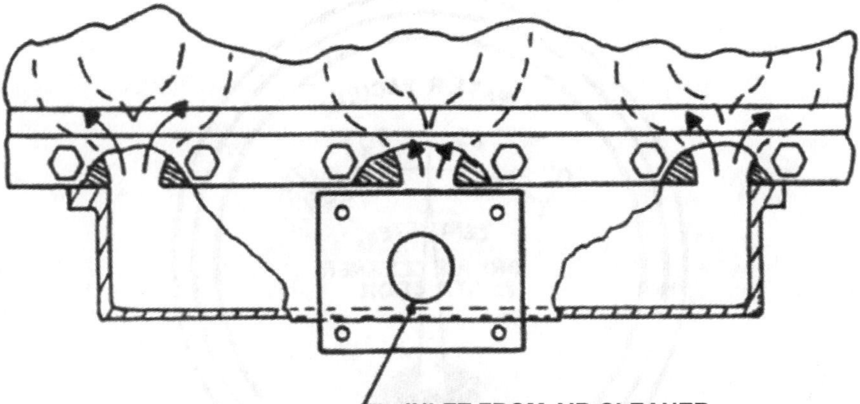

Figure 8.3. An intake manifold for a CI engine.

In CI engines, the fuel is injected directly into the combustion chambers and the intake manifold transmits only air. The intake manifold on CI engines is typically a box-like enclosure on the side of the engine (see Figure 8.3), i.e., it is not subdivided into branches. The enclosure covers the intake valve ports on the engine head and has a circular opening for connection to the line from the air cleaner or turbocharger.

8.3 Exhaust Manifolds, Mufflers, and Spark Arrestors

Exhaust manifolds for SI or CI engines are subdivided, i.e., are similar in appearance to the intake manifold shown in Figure 6.2. In NA engines, gases exiting the exhaust manifold are piped directly to the vehicle muffler. On turbocharged engines, the exhaust gases pass through a turbine before being piped to the muffler. Two types of mufflers are shown in Figure 8.4. The outer chamber of the straight-through muffler is about three times larger in diameter than the central pipe and the intermediate space may be packed with a heat-resistant, sound-absorbing material. The baffles in the reverse-flow muffler cause the exhaust gases to reverse direction several times in passing through the muffler. Both types of muffler cause a reduction in the velocity of the exhaust gases and an increase in pressure, i.e., kinetic energy in the gases is converted into potential energy. Slowing the exhaust gases helps to reduce their sound level. The large surface area of the mufflers helps to dissipate heat, thus cooling the exhaust gases. When an off-road vehicle works in an environment containing dried plants or other combustible material, the muffler serves the important function of eliminating sparks before they reach the exit of the tail pipe. A well-designed muffler will produce as little back pressure on the exhaust as possible, since the pressure drop across the muffler is subtracted from the brake mean effective pressure of the engine.

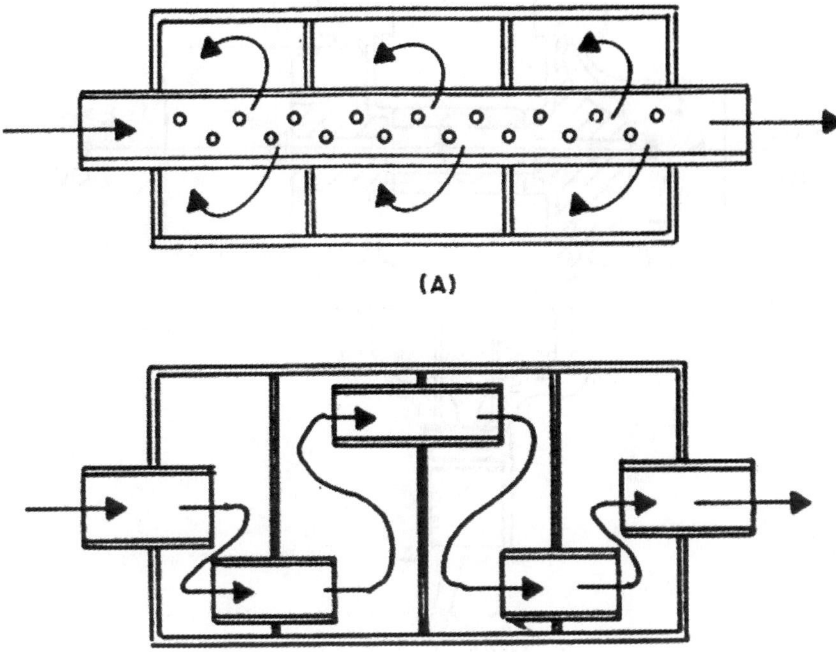

(A)

(B)

Figure 8.4. Cutaway views of (a) straight-through and (b) reverse-flow mufflers.

8.4 Supercharging an Engine

The objective of supercharging an engine is to increase airflow, which allows greater fuel consumption and greater power output from an engine of given displacement. Superchargers may be mechanically driven or be driven by the engine exhaust; the latter are called turbochargers. A turbocharger consists of an exhaust-driven turbine directly connected to a compressor wheel. The spinning compressor wheel receives air from the air cleaner, compresses it, and delivers it to the engine intake manifold. The first mechanically driven supercharger was used on an engine in 1905. The first turbocharger was patented in 1915. However, turbochargers were seldom used until the 1940s and these were very bulky. Following the burgeoning gain in popularity of CI engines in the 1960s, light-weight, high-speed turbochargers came into widespread use on CI engines in farm tractors and other heavy-duty vehicles. A turbocharger increases the air-delivery ratio of an engine, as discussed in Section 8.4.1.

The concept of a turbocharger is illustrated in Figure 8.5. Points in Figure 8.5 are numbered for identifying pressures and temperatures. Point 1 is between the air cleaner and the compressor inlet. Point 2 is between the compressor outlet and the intake manifold. Point 3 is in the exhaust manifold. Point 4 is between the turbine

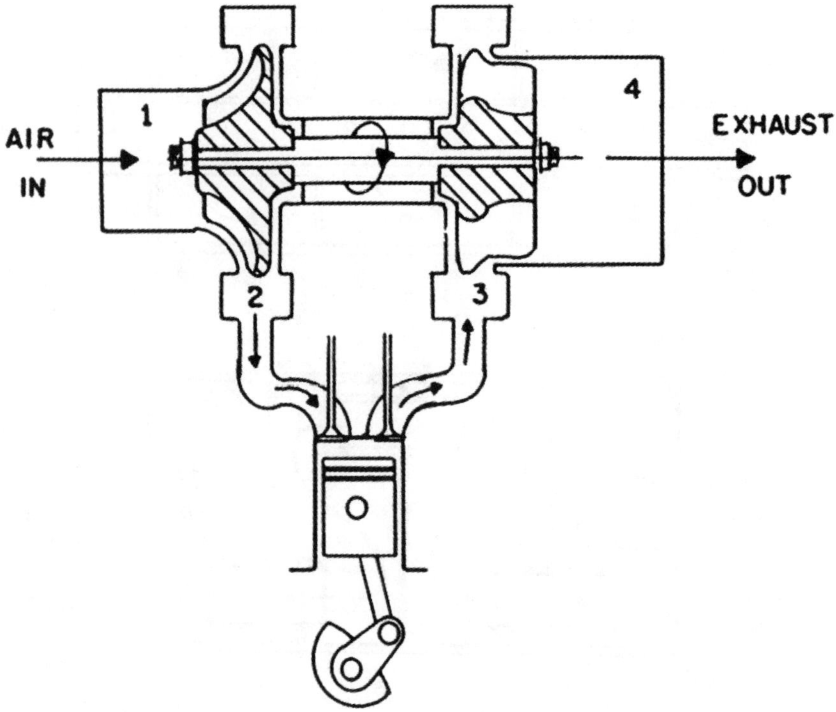

Figure 8.5. Illustration of the concept of a turbocharger.

outlet and the muffler. Air from the air cleaner enters the compressor at temperature T_1 and pressure p_1 and exits at temperature T_2 and pressure p_2. Absolute temperatures and pressures must be used in all turbocharger calculations. If the engine is equipped with an *intercooler*, the temperature of the air leaving the intercooler is designated as T_{2i}. The exhaust gas enters the turbine at condition T_3, p_3 and exits at condition T_4, p_4. The increase in pressure across the compressor is referred to as ***boost.***

8.4.1 Air Consumption and Air-Delivery Ratio

The theoretical air consumption rate of a four-cycle engine is given by the following equation:

$$\dot{m}_{at} = 0.03\,D_e N_e \rho_a \tag{8.1}$$

where

$\dot{m}_{at}$ = theoretical air consumption rate, kg/h

D_e = engine displacement, L

N_e = engine speed, rpm

ρ_a = density of air entering compressor, kg/m^3

The factor 0.03 is a units constant and also accounts for the two crankshaft revolutions per combustion cycle in a four-cycle engine. The entering air density can be calculated using Equation 3.3, with R = 8.314/29.

The air-delivery ratio is the ratio of the measured over the theoretical air consumption of an engine, i.e.,

$$e_v = \frac{\dot{m}_a}{\dot{m}_{at}} \tag{8.2}$$

where

e_v = air-delivery ratio

$\dot{m}_a$ = actual air consumption of the engine, kg/h

For naturally-aspirated (NA) engines, e_v is called the volumetric efficiency of the engine because it is a measure of the efficiency of the combustion chambers in filling with air during the intake stroke. The same term is sometimes used for turbocharged engines, but the correct term is air-delivery ratio. The condition, $e_v = 1$, would be achieved if each combustion chamber filled completely with air at ambient pressure during each intake stroke, i.e., each combustion chamber would be filled with air whose density was equal to that of the air in the environment around the engine. In a NA engine, pressure losses during the intake process cause the air density in the cylinders to be less than that of the ambient air, and the volumetric efficiency is less than one. For a NA engine running at rated load and speed, the volumetric efficiency is typically about 0.85.

A turbocharger is able to increase the density of the air entering the combustion chambers to well above that of the ambient air. By using the ideal gas law (Equation 3.3) to compare the air density leaving the turbocharger compressor to that of the ambient air, the following approximate equation for air-delivery ratio can be derived:

$$e_v = \left(\frac{p_2}{p_1}\right)\left(\frac{T_1}{T_2}\right) \tag{8.3}$$

where

p_1, p_2 = absolute pressure of air entering and leaving the compressor, kPa

T_1, T_2 = absolute temperature of air entering and leaving the compressor, °K

Equation 8.3 provides a good estimate of the air-delivery ratio when a turbocharged CI engine is running at rated load and speed, but is inaccurate when an engine is lightly loaded.

8.4.2 Turbochargers

The performance characteristics of a turbocharger can be shown using two maps, one for the compressor and one for the turbine. Typical examples are shown in Figures 8.6 and 8.7. On a compressor map, the compressor pressure ratio is on the y-axis and the compressor airflow is on the x-axis. Two sets of contours are plotted on the map. The oval-shaped contours are lines of equal compressor efficiency. More will be said about compressor efficiency in the next section, but typical values range from 0.5 to 0.8. The other set of contours are lines of equal compressor speed. Note that operating speeds of compressors are measured in tens of thousands of revolutions per minute, i.e., compressors run very fast. The surge line on Figure 8.6 marks the region where the constant-speed contours begin to slope downward with increasing airflow. It can be shown that operation of a compressor to the left of the surge line is very unstable and results in air surges between the compressor and the intake manifold.

There are several forms of turbine maps, one of which is shown in Figure 8.7. Here, the turbine flow rate is on the y-axis, while the turbine speed is on the x-axis. Two sets of contour lines are plotted. One is for contours of constant turbine efficiency while the other is for contours of constant pressure ratio across the turbine. Referring to Figure 8.7, the turbine pressure ratio is defined as $\kappa_{pt} = p_3 / p_4$. The turbine efficiency will be defined later.

The flow rates through the compressor and turbine are related through the fuel-air ratio of the engine. On a mass basis, the exhaust flow rate out of the engine must be equal to the sum of the air and fuel flow rates into the engine. The turbine flow rate is thus related to the air or compressor flow rate as

$$\frac{\dot{m}_t}{\dot{m}_c} = 1 + FA \tag{8.4}$$

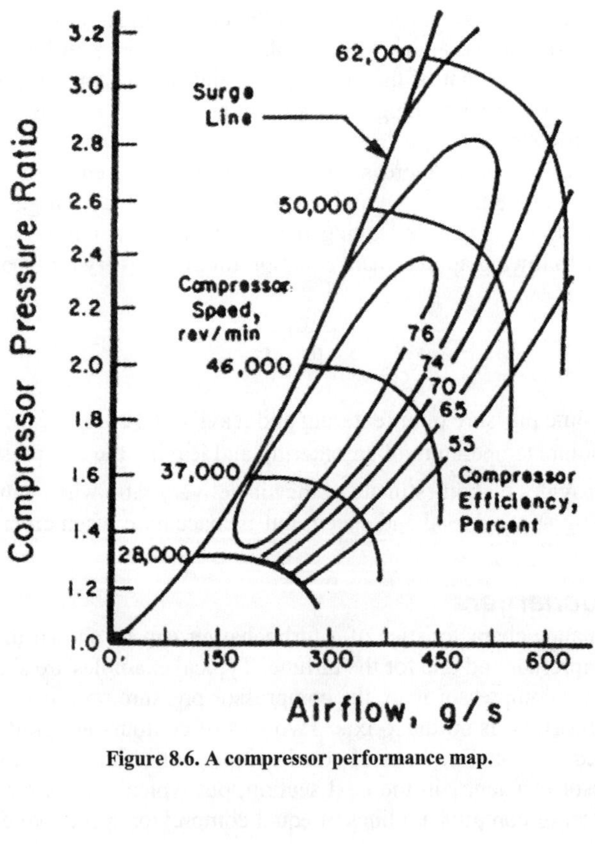

Figure 8.6. A compressor performance map.

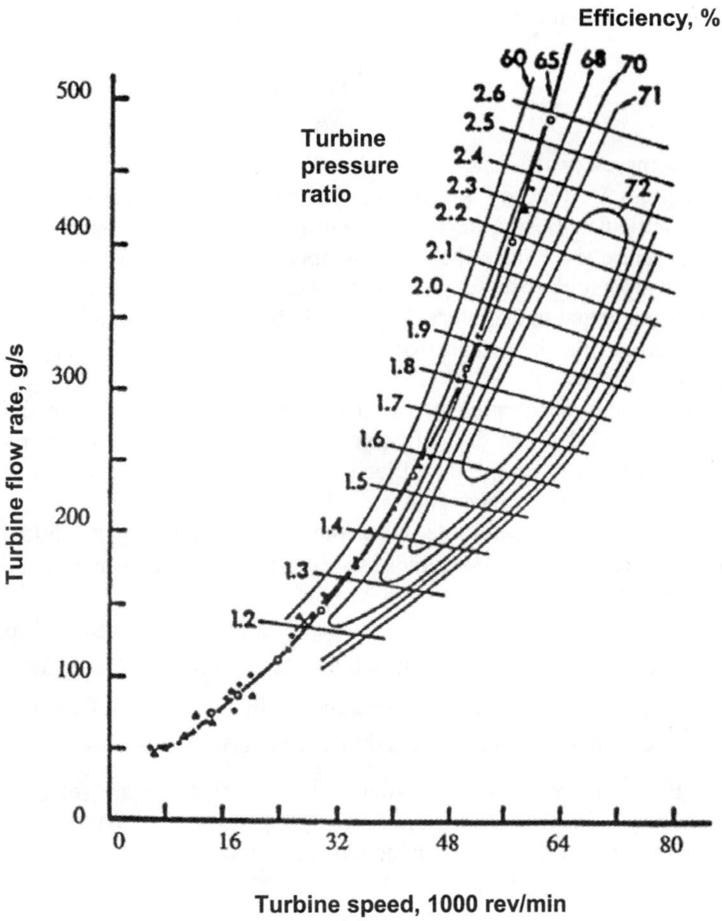

Figure 8.7. A turbine performance map.

where

$\dot{m}_t$ = mass flow rate through the turbine

$\dot{m}_c$ = mass flow rate through the compressor

FA = fuel-air ratio, i.e., the inverse of the air-fuel ratio

8.4.3 Selecting a Turbocharger for an Engine

In addition to Equations 8.1 through 8.4 above, two additional equations are needed in selecting a turbocharger. The first is for the pressure ratio across the compressor, κ_{pc}, as

$$\kappa_{pc} = \frac{p_2}{p_1} = 1 + \frac{\text{boost}}{p_1} \tag{8.5}$$

where boost = pressure increase across the compressor.

The second equation is for the temperature ratio across the compressor,

$$\frac{T_2}{T_1} = 1 + \frac{\kappa_{pc}^{0.286} - 1}{e_c} \tag{8.6}$$

where e_c = compressor efficiency, decimal.

The compressor efficiency is defined as the theoretical temperature rise across the compressor (calculated using the pressure ratio and the ideal gas law) divided by the actual temperature rise. Since the actual temperature rise is always greater than the theoretical temperature rise, e_c is always less than 1.0. In contrast, the turbine efficiency is the actual temperature drop across the turbine divided by the theoretical temperature drop. Like the compressor efficiency, the turbine efficiency is also always less than 1.0.

The purpose of turbocharging is to increase the power output of an engine, and thus the first step is to select the desired power output that would be appropriate for the engine at the desired rated speed and load. The following procedure may be used in selecting a turbocharger for an engine.

1. Select the desired, achievable power output, P_b; use Equation 2.6 to verify that the chosen power level does not require an excessive p_{bme}. Realistically, $p_{bme} \leq 1250$ kPa is achievable.

2. Calculate $\dot{m}_f = P_b \times BSFC$, using an achievable value for BSFC. Typically, for a well-designed engine, it is possible to achieve $0.2 < BSFC < 0.25$ kg/kW h.

3. Calculate $\dot{m}_a = \dot{m}_f \times (A/F)$, using the desired A/F ratio of the turbocharged engine. For a CI engine running on diesel fuel, typically $25 < (A/F) < 32$.

4. Select the compressor and the point on the compressor map (see Figure 8.6 for an example map) at which the compressor will operate at rated load and speed of the engine. Equations 8.1 through 8.6 can be reworked into

$$\frac{\kappa_{pc} e_c}{e_c + \kappa_{pc}^{0.286} - 1} = \frac{\dot{m}_a}{0.03 D_e N_e \rho_a} \tag{8.7}$$

Note that all terms on the right side of Equation 8.7 are known. By assuming a value for e_c, Equation 8.7 can be iteratively solved to determine κ_{pc}. Then the map for the selected compressor can be entered at the calculated values of m_a and κ_{pc} to find the corresponding value of e_c on the map. If the map value of e_c differs from the value used in Equation 8.7, a new value of κ_{pc} can be calculated using the map value of e_c. After a few iterations, values of e_c, κ_{pc} and $\dot{m}_a$ will be found that are compatible with the compressor map and with Equation 8.7. Then the compressor selection is completed.

5. Select the turbine and the operating point on the turbine map. Speed, flow and power constraints must be met in matching the turbine to the compressor. The turbine and compressor must rotate at the same speed, the turbine flow must equal the compressor flow times (1 + FA), and the turbine must supply enough power to drive

the compressor while overcoming bearing friction. The power available from the turbine (P_t) and the power needed to drive the compressor (P_c) are related through the mechanical efficiency of the turbocharger, i.e.,

$$e_m = \frac{P_c}{P_t} = \frac{\dot{m}_c C_{pc}(T_2 - T_1)}{\dot{m}_t C_{pt}(T_3 - T_4)} \quad (8.8)$$

where

P_c = power needed to drive the compressor
P_t = power available from the turbine

Equation 8.8 can be reworked into characteristic-value equations that incorporate the speed, flow and power constraints:

$$\tau_{available} = (1 + FA)e_c e_t e_m \frac{T_3}{T_1} \quad (8.9A)$$

and

$$\tau_{required} = \left(\frac{C_{pc}}{C_{pt}}\right)\left(\frac{\kappa_{pc}^{0.286} - 1}{1 - p_{rt}^{u}}\right) \quad (8.9B)$$

where

τ_{avail} = characteristic value available
$\tau_{required}$ = characteristic value required
$u = -(k' - 1)/k'$
e_t = turbine efficiency, decimal
e_m = turbocharger mechanical efficiency, decimal
C_{pc} = constant-pressure specific heat of ambient air, kJ/kg·K
C_{pt} = constant-pressure specific heat of heated air, kJ/kg·K

The available characteristic value depends upon the FA ratio, the turbocharger efficiencies, and the temperature ratio across the engine, as shown in Equation 8.9A. By entering a turbine map (Figure 8.7) at coordinates of engine speed and theoretical mass flow rate, values of pressure ratio and turbine efficiency can be found. The value for turbine efficiency is typically about 0.98. The fuel-air ratio is simply the inverse of the air-fuel ratio that was chosen at the start of the turbocharger fitting process. The temperature ratio, T_3/T_1, can be estimated from engine performance data. Then the available characteristic value can be calculated using Equation 8.9A. Next, Equation 8.9B can be used to find the required characteristic value. Note that values for C_{pc}, C_{pt} and k′ can be taken from handbooks. C_{pt} and k′ are temperature dependent and must be selected appropriately for the exhaust temperature, T_3. The available characteristic value from Equation 8.9A must be as large or larger than the required characteristic value obtained from Equation 8.9B, or the turbine will not drive the compressor fast enough to provide the desired boost.

The engine must be able to supply the pressure, p_3 and temperature, T_3, which are calculated in Step 5 above. Some experimentation will be necessary to determine these values for the engine in question. As a guideline, the temperature rise, $(T_3 - T_2)$ across the engine is a function of the fuel-air ratio and is typically between 480 and 580°C at rated load and speed. Experiments in which the exhaust pipe is subjected to varying restrictions can provide data on how T_3 varies as p_3 varies. Engine manufacturers have developed computer programs to simplify the fitting of turbochargers to engines.

Figure 8.8 is a plot of Equation 8.9B with typical values as follows: $C_{pc} = 1.107$ kJ/kg·K, $C_{pt} = 1.189$ kJ/kg·K, and $k' = 1.325$. When typical values are inserted for the quantities in Equation 8.9B, typically $1 < \tau < 2$. Also, typically, $1.4 < \kappa_{pc} < 3$, and $1.4 < \kappa_{pt} < 3$. Finally, for good engine operation, it is desirable that $p_2 > p_3$ and thus $\kappa_{pc} > \kappa_{rt}$. The cross-hatched area on Figure 8.8 is the area in which all of the above conditions are met. Note that a large characteristic value is desirable to achieve the greatest boost with the least back pressure on the engine exhaust. Maximizing the quantities on the right side of Equation 8.9A provides the maximum value of the characteristic value.

The procedure for fitting a turbocharger is illustrated in Example Problem 8.1.

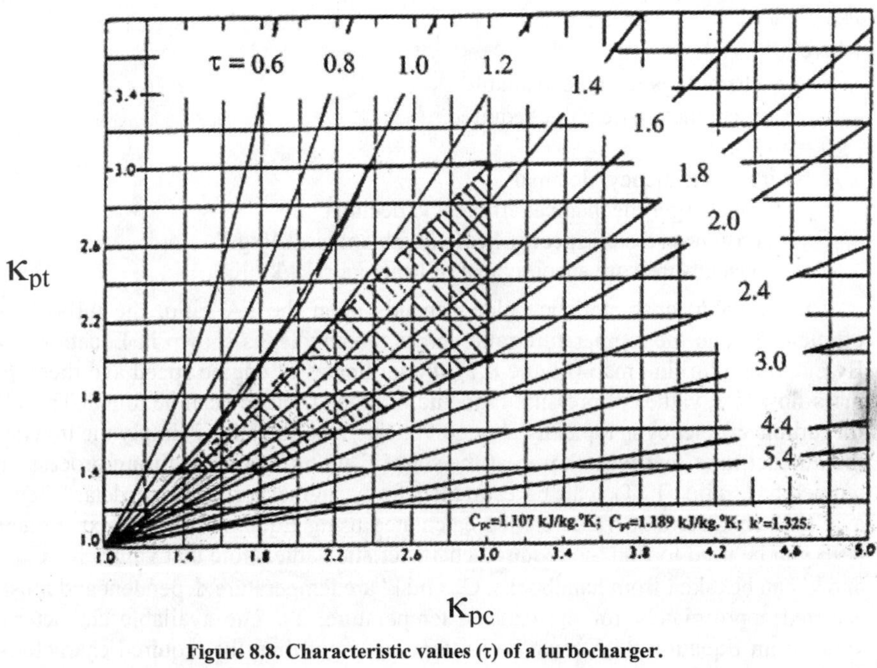

Figure 8.8. Characteristic values (τ) of a turbocharger.

Example Problem 8.1

A turbocharger is to be fitted to a 10-liter diesel engine, which is to run at 2400 rpm rated speed and provide a brake power output of 180 kW. The density of the ambient air is 1.16 kg/m^3, the desired A/F ratio of the turbocharged engine is 30:1, and it is estimated that the engine can achieve a BSFC of 0.25 kg/kW h.

Solution

Follow the numbered steps in the procedure above.

1. $P_b = 180$ kW (given).
2. $\dot{m}_f = 180 \text{ kW} \times 0.25 \text{ kg/kW h} = 45.0$ kg/h fuel consumption
3. $\dot{m}_a = 45.0 \times 30 = 1350$ kg/h (375 g/s)air consumption.
4. $e_v = \dot{m}_a / (0.03\, D_e\, N_e\, \rho_a) = 1350 / (0.03 \times 10 \times 2400 \times 1.16) = 1.62$.
5. Next, using Equation 8.7 with the right side = 1.62 and an initial guess of $e_c = 0.7$, find $\kappa_{pc} = 2.2$. Then, enter the compressor map (Figure 8.6) at $m_a = 375$ g/s and $\kappa_{pc} = 2.2$ to find $e_c = 0.77$. Repeat the solution of Equation 8.7 with $e_c = 0.77$ and find $\kappa_{pc} = 2.11$. Reentering Figure 8.2, we find that $m_a = 375$ g/s, $\kappa_{pc} = 2.11$ and $e_c = 0.77$ are compatible with the map; therefore, the compressor design is completed. The compressor map shows that the compressor would run at approximately 47,000 rev/min. From Equation 8.5, assuming $p_1 = 100$ kPa, the boost would be

$$P_2 = \kappa_{pc} \times p_1 = 2.11 \times 100 = 211 \text{ kPa}$$

$$\text{Boost} = p_2 - p_1 = 211 - 100 = 111 \text{ kPa}$$

If a solution could not be found on the compressor map of Figure 8.6, another compressor would have to be used. Points to the left of the surge line on the compressor map are unstable and cannot be used.

6. Next, the turbine must be selected. The turbine speed is $N_t = N_c = 47{,}000$ rev/min. The compressor flow is $\dot{m}_t = (1 + FA)\ \dot{m}_a$. Since A/F = 30, FA = 0.033. Therefore, = $\dot{m}_t = 1.033 \times 1359 = 1688$ kg/h (469 g/s). Entering the turbine map (Figure 8.7) with these coordinates, we find that $e_t = 0.67$ and $\kappa_{pt} = 2.46$, approximately. Next we use Equation 8.9A to calculate the available characteristic value. We will assume $e_m = 0.98$ and $T_3/T_1 = 3$. Then

$$\tau_{available} = 1.033 \times 0.77 \times 0.67 \times 0.98 \times 3 = 1.57$$

Now we use Equation 8.9B with typical values of C_{pc}, C_{pt} and k' to calculate the required characteristic value

$$\tau_{required} = (1.107/1.189) \times (2.11^{0.286} - 1)/(1 - 2.46^{-0.245}) = 1.12$$

The required characteristic value is well below the available characteristic value, so the turbine will easily drive the compressor; in fact, it will drive it faster than 47,000 rev/min, thus increasing the boost until an equilibrium point is found. In this case, the turbine selection is poor. Another turbine should be selected that will run near 47,000 rev/min while operating at a κ_{rt} nearer to 2.46 when 469 g/s of flow are moving through the turbine.

Using Equation 8.6, we can find the temperature in the intake manifold of the engine of Example Problem 8.1 as

$$T_2/T_1 = 1 + (2.11^{0.286} - 1) / 0.77 = 1.31$$

Then, assuming T_1 = 293°K (20°C), the temperature in the intake manifold would be 384°K (111°C). This is a high temperature, i.e., well above the boiling point of water. If the temperature in the intake manifold could be reduced, the engine life could be increased and the volumetric efficiency could be further improved. An intercooler can be used to reduce the temperature of the air in the intake manifold.

8.4.4 Turbocharger Operation

The operating point on the compressor map varies with the load and speed of the engine. Figure 8.9 illustrates the changes in operating point for a specific engine. The compressor speed and efficiency contours were omitted from Figure 8.9 for clarity. Note that pressure ratio across the compressor is less than unity at Point A, which corresponds to engine operation at low idle. At low idle, the turbine cannot extract enough exhaust energy to drive the compressor and the turbocharger may actually stall. As the engine speed increases while the engine moves toward high idle at Point B, the turbocharger is still ineffective, but airflow increases due to the increased pumping capacity of the engine. The engine is losing speed between Points B and C as increasing load is placed on the engine. However, the turbocharger becomes increasingly effective as the engine moves toward governor's maximum at Point C; thus the compressor pressure ratio rises sharply and airflow increases. As the engine enters the load-control region in moving toward Point D, fuel delivery begins to fall, the turbine is able to extract less energy and the compressor pressure ratio falls. The decrease in turbocharger effectiveness and the fall in engine speed combine to produce a marked reduction in airflow as the engine moves into load control. Ideally, if the compressor is well matched to the engine, the curve between Points B and C on Figure 8.9 will lie within the maximum compressor efficiency contour of the compressor map (see Figure 8.6).

Because a turbocharger spins at very high speeds, lubrication of the turbocharger bearings is critical. An oil line is connected from the engine oil gallery to the turbocharger bearings and a return line returns the oil to the crankcase for cooling. By starting an engine and then allowing it to run at low idle for a short time, an operator can ensure that oil pressure builds and lubricates the turbocharger bearings before the turbocharger begins to spin at high speed.

One of the disadvantages of turbochargers is *turbocharger lag*. When fuel delivery is suddenly increased to accelerate a CI engine, there is a time lag until the turbine can extract enough additional energy to boost air consumption to match the increase in fuel delivery. The fuel-air equivalence ratio rises sharply during turbo lag, resulting in

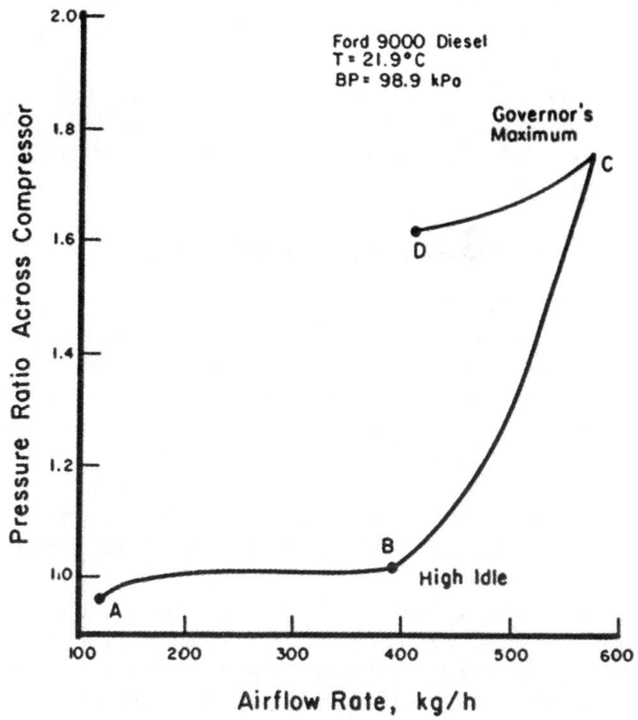

Figure 8.9. Intake manifold pressure and flow variations in a turbocharged CI engine.

incomplete combustion and a transient increase in black smoke in the exhaust. To combat turbocharger lag, at least one organization has developed and tested a prototype of an *electroturbocharger,* which incorporates an electric motor. When a fueling increase is detected, the motor quickly boosts the compressor speed until the turbine can extract enough exhaust energy to allow the turbocharger to supply the needed air. At very high speeds, the motor becomes a generator and feeds current back into the vehicle electrical system. During acceleration from low idle, the electroturbocharger is able to provide boost pressure more than double that of a conventional turbocharger.

8.4.5 Altitude Compensation

As barometric pressure declines at higher elevations, the ambient air density declines in accordance with the ideal gas law. Consequently, the ability of NA engines to pump air decreases at higher elevations and fuel delivery must be reduced accordingly to prevent over-rich fuel-air mixtures. An engine is thus able to produce less power at higher elevations, i.e., it must be *de-rated for altitude.* When a turbocharger is installed on an engine (see Figure 8.5), the pressure at the turbine outlet decreases at higher altitudes causing the turbine and compressor to spin faster. The faster turning compressor is able to offset the reduction in pressure at the compressor inlet, resulting in a stabilization of the intake manifold pressure. In brief, the turbocharger is able to compensate for changes in altitude and de-rating is no longer necessary. In mountainous regions, vehicle owners may desire such altitude compensation but have no desire to increase the fueling rate for more power, as is normally done when a turbocharger is added to an NA engine. If the device of Figure 8.5 is installed on an NA engine but fueling is not increased, the device is called an altitude compensator instead of a turbocharger.

8.5 Intercoolers

In Example Problem 8.1, it was shown that a turbocharger compressor can greatly elevate the temperature of the air passing through it. Such high air temperatures reduce the air-delivery ratio and can be harmful to the engine. The air temperature can be reduced by use of an intercooler, so called because the device is located intermediately between the compressor and the engine. Alternatively, the device has also been called an aftercooler because it is located after the compressor. We will use the more common term, intercooler. Two types of intercoolers are available, the air-to-water intercooler and the air-to-air intercooler. With an air-to-water intercooler, engine coolant is passed through a finned tube in the intake manifold. The hot air from the compressor passes across the finned tube, transferring heat to the coolant, which in turn rejects the heat through the engine radiator. Figure 8.10 illustrates typical changes in temperatures of the intake manifold air and the coolant in the intercooler. Note that, in an intercooler of finite length, the air temperature can never fall to the water temperature. Typically, there is a 15 to 20°C difference between the temperatures of the air and water exiting the intercooler. Because the coolant passes through the vehicle radiator before entering the water pump and intercooler, there is a 10 to 15°C difference between the coolant thermostat setting and the temperature of the air leaving the intercooler. Note in Figure 8.10 that the change in water temperature is much smaller than the change in air temperature. Temperature changes through an air-to-water or air-to-air intercooler are related through the following equation:

$$\frac{\Delta T_{ai}}{\Delta T_s} = \frac{\dot{m}_s C_{ps}}{\dot{m}_{ai} C_{ai}} \tag{8.10}$$

where

ΔT_{ai} = temperature drop of the engine air across the intercooler
ΔT_s = temperature rise of the secondary fluid (air or water) across the intercooler
$\dot{m}_s$ = mass flow rate of secondary fluid through the intercooler, kg/h
$\dot{m}_{ai}$ = mass flow rate of engine air through the intercooler, kg/h
C_{ps} = specific heat at constant pressure of secondary fluid in the intercooler
C_{ai} = specific heat at constant pressure of the engine air in the intercooler

The specific heat of water is 4.2 times greater than that of air and the mass flow rate of water is typically much higher than the airflow rate. Thus, as Equation 8.10 shows, the water needs very little temperature increase to cool the air passing through the intercooler.

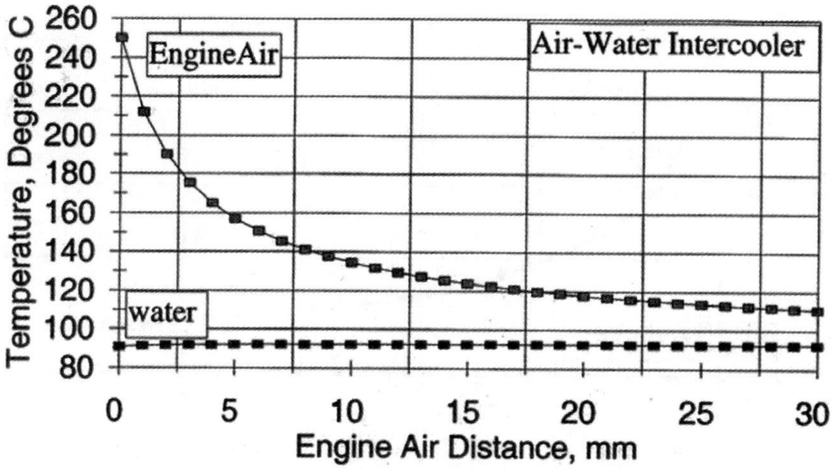

Figure 8.10. Temperature variations in an air-to-water intercooler.

An air-to-air intercooler has an appearance similar to a vehicle radiator and is often positioned just ahead of the radiator. Air leaving the turbocharger compressor is piped to the intercooler, where it passes through finned tubes before being piped back to the intake manifold. A fan is used to pull ambient air across the finned tubes to absorb heat. Temperature variations in an air-to-air intercooler are illustrated in Figure 8.11. The air from the compressor and the secondary (ambient) air both have the same specific heat. However, the mass flow rate of the secondary air is far greater than that of the air from the compressor. Thus, the secondary air has a much smaller temperature change than does the air from the compressor. The secondary air temperature, and thus the final temperature of the intercooled air, varies with the temperature of the vehicle environment. However, from a comparison of Figures 8.10 and 8.11, an air-to-air intercooler can nearly always cool the engine air to an appreciably lower temperature than can an air-to-water intercooler. Thus, despite the bulkiness of air-to-air intercoolers and the need for additional piping to move the air to and from them, air-to-air intercoolers are gaining in popularity over air-to-water intercoolers.

Use of an intercooler can provide a gain in the air-delivery ratio of a turbocharged engine. The ideal gas law can be used to calculate the gain in the air-delivery ratio. The approximate air-delivery ratio for a turbocharged, intercooled engine operating at rated load and speed is

$$e_v = (\kappa_{pc} \frac{T_1}{T_2})(\frac{T_2}{T_{2i}}) \tag{8.11}$$

where T_{2i} = temperature of the engine air leaving the intercooler, °K.

Comparison of Equations 8.3 and 8.11 shows that the ratio, T_2/T_{2i}, is the intercooler factor. Thus, Equation 8.3 is for turbocharged engines without intercoolers. There is no pressure ratio in the intercooler factor because intercooling is a constant-pressure

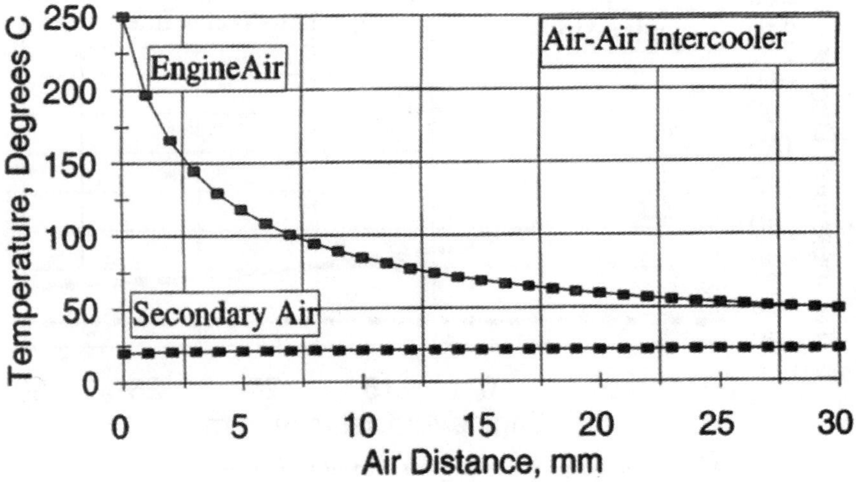

Figure 8.11. Temperature variations in an air-to-air intercooler.

process. Typically, the turbocharger makes a much bigger contribution to an increased air-delivery ratio than does the intercooler. The more important benefit of the intercooler is to improve engine life by lowering the air temperature.

8.6 Turbocharging and Engine Families

Some engine manufacturers have used turbocharging and intercooling to reduce the costs of manufacturing a family of engines and of maintaining an inventory of replacement parts for the engines. The procedure is illustrated in Example Problem 8.2.

Example Problem 8.2

An engine manufacturer wants to provide four engine models from one basic design of a six-cylinder, 7.636 L CI engine that is to run at a rated speed of 2200 rpm. The desired power levels of the four engines are to increase in approximately 15 kW increments from a base level of 75 kW for the NA version of the engine. How might turbocharging and intercooling be used to accomplish this objective? The intercooler is to use the engine coolant as the secondary fluid, and the thermostat setting for the engine coolant is 94°C. Tests have shown that the NA engine can achieve BSFC of 0.282 kg/kW h.

Solution

The assumed atmospheric conditions will be 300°K and 100 kPa. Thus, the ambient air density is

$$\rho_a = (29/8.314) \times (100/300) = 1.16 \text{ kg/m}^3$$

Assuming $e_v = 0.85$ for the base NA engine, the air consumption is

$$\dot{m}_a = 0.03 \times 7.636 \times 2200 \times 1.16 \times 0.85 = 497 \text{ kg/h}$$

The fuel consumption of the 75 kW, NA engine would be

$$\dot{m}_f = P_b \times BSFC = 75 \times 0.282 = 21.2 \text{ kg/h}$$

The air-fuel ratio for the NA engine is

$$A/F = 497/21.2 = 23.4$$

Turbochargers will be added for the increased airflow needed to support increased fuel flow and increased power. For the 120 kW engine, a boost of 91 kPa is selected based on needed airflow. Then, from Equation 8.5, $\kappa_{pc} = 1.91$. A reasonable value for compressor efficiency is $e_c = 0.67$ and, from Equation 8.6,

$$T_2/T_1 = 1 + (1.91^{0.286} - 1)/0.67 = 1.303$$

Since T_1 = 300°K (27°C), note that the temperature of the air leaving the compressor would be 1.303 × 300 = 391°K (118°C), which is very hot. Thus, an intercooler will be added; it can reduce the air temperature to within approximately 10°C of the thermostat setting, i.e., to 104°C (377°K). From Equation 10.10, the estimated volumetric efficiency of the 120 kW, turbocharged, intercooled (TC, IC) engine would be

$$e_v = (1.91/1.303) \times (391/377) = 1.52$$

Then, using the new value for e_v, the calculated air consumption of the TC, IC engine would be 890 kg/h. The decreased equivalence ratio provided by the turbocharger will decrease the BSFC. Experience with similar installations suggests an improvement of 11% is realistic, giving a BSFC of 0.251 kg/kW h. Then the approximate fuel delivery rate would be

$$\dot{m}_f = P_b \times BSFC = 120 \times 0.251 = 30.1 \text{ kg/h}$$

After the engine was constructed, the actual fueling rate could be adjusted to produce exactly the 120 kW of brake power desired. The A/F ratio of the TC, IC engine would be

$$A/F = 885/30.1 = 29.6$$

The proposed increase in fueling rate, from 21.2 kg/h for the 75 kW, NA version to 30.1 kg/hr for the 120 kW, TC, IC engine, would be acceptable; the TC, IC engine would run leaner than the NA version. Mixture richness is defined by the equivalence ratio, i.e.,

$$\phi = FA_{Actual}/FA_{Stoichiometric} = AF_{Stochiometric}/AF_{Actual}$$

The equivalence ratio is less than one for a lean mixture, equal to one for a chemically correct mixture and greater than one for a rich mixture.

The intermediate sized engines (90 and 105 kW) could be achieved by using appropriate fueling rates; the rates would not exceed 90 × 0.282 = 25.4 kg/h and 105 × 0.282 = 29.6 kg/h for the 90 and 105 kW engines, respectively. These fueling rates, being less than the 30.1 kg/h used for the 120 kW engine, would reduce the turbocharger boost below 91 kPa and the temperature of the air leaving the compressor (T_2) would fall below the 118°C calculated for the TC, IC engine. If T_2 fell close to 104°C or below, not enough temperature differential above the coolant temperature would exist to permit cooling of the air. Thus, intercooling would probably be impractical except for the 120 kW engine.

Table 8.1. Summary of an engine family.

Engine	BSFC, kg/kW h	Brake Power, kW	Fuel Flow, kg/h	Boost, kPa	T_2, °C	Air-Delivery Ratio	Airflow, kg/h	Equivalence Ratio
NA	0.282	75	21.2	0	27	0.85	497	0.637
TC	0.272	90	24.5	69	99	1.36	794	0.467
TC	0.261	105	27.4	88	115	1.45	848	0.482
TC & IC	0.251	120	30.1	91	118	1.52	890	0.503

Stoichiometric A/F = 14.9, $e_c = 0.67$
NA = naturally aspirated
TC = turbocharged
TC & IC = turbocharged and intercooled

Table 8.1 is a summary of the solution to Example Problem 8.2. Note that the 90 kW engine has the lowest equivalence ratio; addition of the turbocharger has boosted the air delivery appreciably for this engine, but the accompanying increase in fuel delivery is the least of the turbocharged engines. In effect, some of the allowable fuel increases have been held in reserve to achieve three different brake power levels for the turbocharged engines.

To summarize Example Problem 8.2, a family of four engines ranging from 75 to 120 kW in power was obtained from one basic engine. The family included a 75 kW NA engine, 90 and 105 kW TC engines and a 120 kW TC, IC engine. The advantage is that all four engines could share a common inventory of parts except that a larger injector pump and fuel injectors might have to be used on the largest engine. The three smaller engines would be over-designed, since they would have to have the same strength as the 120 kW engine. That disadvantage would be more than offset by the greatly reduced parts inventory that was achieved. Many engine manufacturers adopt the strategy of Example Problem 8.2.

8.7 Chapter Summary

An air cleaner protects an engine by removing dust and dirt from the incoming air. A service indicator may be provided to warn the operator when dust accumulation on the air filter is producing excessive pressure loss. A muffler is normally provided to reduce the sound level of exhaust gases. The muffler can also serve as a spark arrestor to prevent ignition of combustible material in the vehicle environment.

The objective of turbocharging an engine is to increase airflow, which allows greater fuel delivery and higher power output from an engine of given displacement. A turbocharger consists of an exhaust driven turbine coupled to a compressor. Boost is the increase in pressure produced by the compressor. The increased pressures are accompanied by increased temperatures and, if the boost is high enough, use of an intercooler may be beneficial to reduce the temperature. Air-to-water and air-to-air intercoolers are available. The latter are more bulky and expensive, but are becoming more popular because they can provide a greater reduction in the temperature of the intake manifold air.

In fitting a turbocharger to an engine, the desired power level is selected, from which the necessary fuel and air-delivery rates can be calculated. The compressor and compressor operating point can be selected. The required turbine speed and exhaust flow rate can then be calculated and the turbine can be selected. It is desirable for the turbocharger to have a high characteristic value to allow maximum boost with minimum exhaust back pressure. If the air leaving the compressor is at too high a temperature, an intercooler is used to cool the air. Intercooling is a constant-pressure process.

A turbocharger is ineffective until load is applied to the engine; then there is enough energy in the exhaust stream to drive the turbine. The turbocharger is most effective at governor's maximum; both boost and air delivery fall as the engine moves into the load-controlled range. The turbocharger is able to compensate for altitude changes because, as the inlet air pressure falls, the pressure in the exhaust pipe also falls and allows the turbine to spin faster. Thus, the turbocharger can accommodate reductions in ambient pressure with little change in pressure in the intake manifold.

Engine manufacturers use turbocharging and intercooling to produce a family of engines from one basic block and head. The smallest (lowest-power) engine in the family is naturally aspirated. Adding a turbocharger allows increased fuel delivery and increased power output. Adding an intercooler allows even greater fueling and power output. The technique has the disadvantage that all engines in the family must be built to handle the maximum power in the family line. This disadvantage is more than offset by the reduction in design, manufacturing and inventory costs that are achieved by the technique.

Homework Problems

8.1 Using Equations 8.3, 8.5 and 8.6, plot e_v versus boost for boost values ranging from 50 to 250 kPa. Assume atmospheric pressure is 100 kPa and $e_c = 0.65$. Note that boost selection is a compromise; increased boost provides greater airflow as the graph shows, but also increases mechanical stress on the engine. (This is a good spreadsheet problem.)

8.2 Rework Problem 8.1, but let $e_c = 0.55$.

8.3 Rework Problem 8.1, but let $e_c = 0.75$.

8.4 Using Equation 8.7, plot air-delivery ratio versus compressor-pressure ratio for pressure ratios from 1.0 to 2.0. Plot two curves, one for compressor efficiency equal to 0.6 and one for 0.7. (Note that the right side of Equation 8.7 is equal to the air-delivery ratio of the engine.)

8.5 Rework Problem 8.4, except use compressor efficiencies of 0.65 and 0.75.

8.6 Rework Example Problem 8.1, except use an 8.0 L engine running at 2400 rpm to provide a power output of 140 kW. Also, assume that the turbocharged engine has an A/F ratio of 32 and can achieve a BSFC of 0.24 kg/kW h.

8.7 Rework Example Problem 8.1, except use an 8.5 L engine running at 2400 rpm to provide a power output of 145 kW. Also, assume that the turbocharged engine has an A/F ratio of 35 and can achieve a BSFC of 0.22 kg/kW h.

8.8 Derive Equation 8.6, beginning with the assumption that the compressor efficiency is equal to the theoretical temperature rise across the compressor divided by the actual temperature rise. (*Hint*: Factor T_1 out of the numerator and denominator, then use the ideal gas law and the polytropic compression law to find a pressure ratio to replace the temperature ratio in the numerator.)

8.9 Using data from the power take-off test at rated engine speed, for the engine of Figure 3.4,

(a) Calculate the fuel consumption in kg/h.

(b) Calculate the theoretical air consumption in kg/h.

(c) Calculate the estimate air-delivery ratio using the measured boost and assuming the compressor efficiency is 0.65.

(d) Calculate the actual air consumption in kg/h.

(e) Calculate the actual A/F ratio.

(f) Calculate the fuel-air equivalence ratio assuming the stoichiometric A/F ratio for the diesel fuel used in the test is 15.

8.10 Rework Problem 8.9, but assume the compressor efficiency is 0.55.

8.11 Rework Problem 8.9, but assume the compressor efficiency is 0.75.

8.12 Note that the engine of Problem 8.9 was turbocharged but not intercooled.

(a) Calculate the minimum boost that would be needed for an air-to-water intercooler to be feasible.

(b) Calculate the air-delivery ratio of the engine without the intercooler.

(c) Calculate the air-delivery ratio of the engine with the intercooler.

8.13 Measurements on a certain turbocharged CI engine showed the following relationship for T_3:

$$T_3 = -141.3 + 0.267\, N_{HI} + 889\, \phi$$

where

T_3 = exhaust temperature, °K

N_{HI} = engine high idle speed, rpm

ϕ = equivalence ratio

Using Equation 8.9A, plot the available characteristic value versus the fuel-air ratio for fuel-air ratios ranging from 0.02 to 0.05. Assume the turbocharger has a mechanical efficiency of 0.98, a compressor efficiency of 0.7, and a turbine efficiency of 0.7. Also, the air temperature at the compressor inlet is 300°K and the high idle speed is 2200 rpm.

8.14 Rework Problem 8.13, but use compressor and turbine efficiencies equal 0.65 and a high idle speed of 2000 rpm.

8.15 Rework Problem 8.13, but use compressor and turbine efficiencies equal 0.60 and a high speed of 2100 rpm.

8.16 Modify Example Problem 8.2 and Table 8.1 by adding a 135 kW engine to the top of the line. Assume the latter engine has an air-to-water intercooler that brings the air temperature down to 100°C. Assuming the compressor efficiency is 0.65, the BSFC is 0.231 kg/kW h, and the equivalence ratio is 0.50,

(a) Calculate all other entries in the new line of the table.

(b) Would the brake mean effective pressure of the latter engine be allowable or excessive?

8.17 Rework Problem 8.16 except let the top-of-the-line engine be a 150kW engine.

8.18 Rework Problem 8.16, except let the top-of-the-line engine be a 180 kW engine.

References and Suggested Readings

Buchi, A.J. 1953. Exhaust turbocharging of internal combustion engines. *Journal of the Franklin Institute* (July).

Harr, T.B., and D.L. Hoag. 1974. Predicting relative sound radiation from muffler shells. *Transactions of the ASAE* 17(6): 1048-1052.

McCunn, J.C., R.R. Yoerger, and J.A. Weber. 1961. Pulsating pressures in an internal combustion engine induction system. *Agricultural Engineering* 42(9): 490-493.

SAE. 1979. Turbochargers and Turbocharged Engines. SAE Special Publication SP-442. Warrendale, PA: SAE.

SAE. 1994a. Air cleaner elements. SAE Standard J1141. *SAE Handbook,* vol. 3. Warrendale, PA: SAE.

SAE. 1994b. Air cleaner test code. SAE Standard J726. *SAE Handbook,* vol. 3. Warrendale, PA: SAE.

SAE. 1994c. Turbocharger nomenclature and terminology. SAE Standard J922. *SAE Handbook,* vol. 3. Warrendale, PA: SAE.

SAE. 1994d. Turbocharger gas stand test code. SAE Standard J1826. *SAE Handbook,* vol. 3. Warrendale, PA: SAE.

SAE. 1994e. Engine charge air cooler nomenclature. SAE Standard J1148. *SAE Handbook,* vol. 3. Warrendale, PA: SAE.

Sekar, R.R. 1982. A primer on air charge cooling. *Automotive Engineering* 90(5): 28-35.

Wollenweber, W.E. 1998. New system looks at dealing with turbo lag. *Diesel Progress* (August): 22, 24.

References and Suggested Readings

CHAPTER 9

COOLING SYSTEMS

9.1 Introduction

The temperature of the combustion gases in an internal combustion engine exceeds the melting point of the piston and cylinder walls. Thus, a cooling system is necessary to prevent failure of the engine. Excessive engine temperatures can cause many problems, including head cracking, ring sticking, and piston seizure. The cooling system must be able to function in an environment of high ambient temperatures, dirt and chaff, and vibration. Engines may be *air cooled* or *liquid cooled.* Among larger engines, liquid cooling is more common and most of this chapter will deal with liquid cooling. The liquid used in cooling is a blend of water and antifreeze. In this chapter, we will also use the term, water cooling, even though half of the coolant volume may be liquid other than water.

9.2 Heat Loads

The heat to be dissipated through the cooling system can be generated from a number of sources, which are summarized as

$$\dot{E}_T = \dot{E}_{CC} + \dot{E}_{IC} + \dot{E}_{OC} + \dot{E}_{AC} \tag{9.1}$$

where

$\dot{E}_T$ = total heat load, kW
$\dot{E}_{CC}$ = heat rejected by combustion chambers, kW
$\dot{E}_{IC}$ = heat rejected by intercooler, kW
$\dot{E}_{OC}$ = heat rejected by oil cooler, kW
$\dot{E}_{AC}$ = heat rejected by air conditioner, kW

For a water-cooled engine with an air-to-water intercooler and oil-to-water oil cooler, the first three terms to the right of the equal sign in Equation 9.1 can be grouped into $\dot{E}_e$, the heat rejected by the engine. Typically, $\dot{E}_e$ ranges from 60 to 94% of the maximum engine brake power. A typical heat load imposed by a cab air conditioner is 7 kW at an ambient air temperature of 32°C and 60% relative humidity. If the engine has an air-to-air intercooler, its heat is not rejected through the engine radiator, but is instead rejected through a separate air-to-air heat exchanger. If the engine is water cooled, the latter heat exchanger can be placed directly ahead of the radiator.

The cooling system must reject heat at a sufficient rate to keep critical engine parts at safe temperatures. The temperature on the inside surface of the cylinder walls must be kept below 220°C to prevent deterioration of the oil film. The heat can be rejected via air cooling or water cooling.

9.3 Water Cooling

The capacity of a water cooling system can be limited by the engine water flow to the radiator, the heat-exchanging capacity of the radiator, or the airflow through the radiator. Thus, all three of these elements must have sufficient capacity to handle the heat load to prevent overheating of the engine. The engine designer normally provides heat rejection and water flow rates from the engine at a specified speed and torque. The most conservative approach is to specify heat load and water flow rates at the engine maximum power point. It is a common practice for the engine designer to select a supplier for the radiator and fan. The supplier is a cooling specialist and is responsible for choosing the correct sizes for these elements.

9.3.1 Water Flow

The water flow from the engine must be sufficient to carry the engine-rejected heat to the radiator. The heat-carrying capacity of the cooling water is calculated by

$$\dot{E}_{RW} = Q_W \rho_W C_W \Delta T_W / 60 \qquad (9.2)$$

where

$\dot{E}_{RW}$ = heat transport rate, kW

Q_W = coolant flow rate, L/min

ρ_W = coolant density = 1.0 kg/L

C_W = specific heat of coolant, kJ/kg ·°C

ΔT_W = temperature drop as water moves through radiator, °C

As shown in Section 9.3.4, the specific heat of the coolant is a function of the coolant formulation and temperature. Equation 9.2 can be used to estimate the coolant flow needed to handle the heat transport from the engine for a specified temperature drop across the radiator. Typically, a temperature drop of 3 to 8°C can be achieved in the coolant temperature. Typical coolant flow rates are 1.3 to 2.5 L/min per kW of brake power.

The water pump must be large enough to handle the required coolant flow rate. Typically, a centrifugal pump is used to pump the coolant and its discharge rate will vary with the amount of back pressure on the pump. Normally, the pump is arranged to discharge into the engine *water jacket* surrounding the cylinders in the engine block. The coolant tends to rise as it is heated and moves through holes in the head gasket to the engine head, where it can cool the valves. Coolant collected at the front of the head is then routed to the top of the radiator for cooling and exits the bottom of the radiator to return to the water pump inlet. Cooling in the radiator creates a natural flow downward in the radiator. The coolant's natural circulation thus assists the *water pump* in circulating coolant in the cooling system. The pump must deliver the required flow volume against the flow resistance imposed by the engine water jacket, the thermostat, the radiator and the hoses connecting the radiator to the engine. The lowest coolant pressure occurs at the pump inlet and must be kept close to or above atmospheric pressure to prevent vaporization and *cavitation* in the pump. Minimizing the flow restriction in the radiator and pressurizing the cooling system help to maintain higher pressures at the pump inlet. The engine manufacturer normally specifies to the radiator supplier the required conditions at the pump inlet to avoid cavitation.

9.3.2 Radiator Design

The *radiator* (Figure 9.1) must be capable of transferring the required rate of heat from the cooling water to the air while withstanding a harsh environment. The water velocity in the radiator must be below the saturation range, which is defined as that velocity beyond which there is little or no increase in heat transfer. A water flow rate of 1.4 kg/min per radiator tube is generally below the saturation range. Thus, knowing the total water flow rate from Equation 9.2, the required number of radiator tubes can be calculated. The cross section of each tube is rectangular, with approximately a 4:1 depth-to-width ratio. The long axis of each tube is parallel to the airflow to promote heat transfer with minimal airflow restriction. Tube spacing across the width of the radiator ranges from 1.0 cm for automotive vehicles to 1.6 cm for heavy-duty vehicles. To reduce radiator width for a given number of tubes, the tubes may be arranged in two to six rows. The rows are usually staggered, i.e., tubes in the second row are positioned halfway between those in the first row, etc. If the engine is to operate in an environment with many particulates, however, the rows may be placed in line to reduce plugging tendency. The radiator tubes are equipped with fins to enhance heat transfer to the air. Heat transfer increases with the number of fins per centimeter, but fin spacing must be wide enough to prevent plugging with dirt or chaff. Fin density is normally 3.5 fins/cm. However, for harvesting machines with a high chaff environment, the fin density should be reduced to 1.5 fins/cm. The *radiator frontal area* must be large enough to allow transfer of the heat from the water to the air. A conventional rule of thumb is to allow 20 to 30 cm^2 of frontal area per kilowatt of brake power. The *cooling fan* power increases substantially with air pressure restriction across the radiator. The restriction can be minimized by use of a thin radiator, i.e., with fewer rows of tubes across the depth of the radiator. The required number of tubes to accommodate the water flow can be achieved by increasing the width of the radiator.

The heat transfer capacity of the radiator is enhanced by allowing a large difference, ΔT_r, between the water and air temperatures. Figure 9.2 illustrates

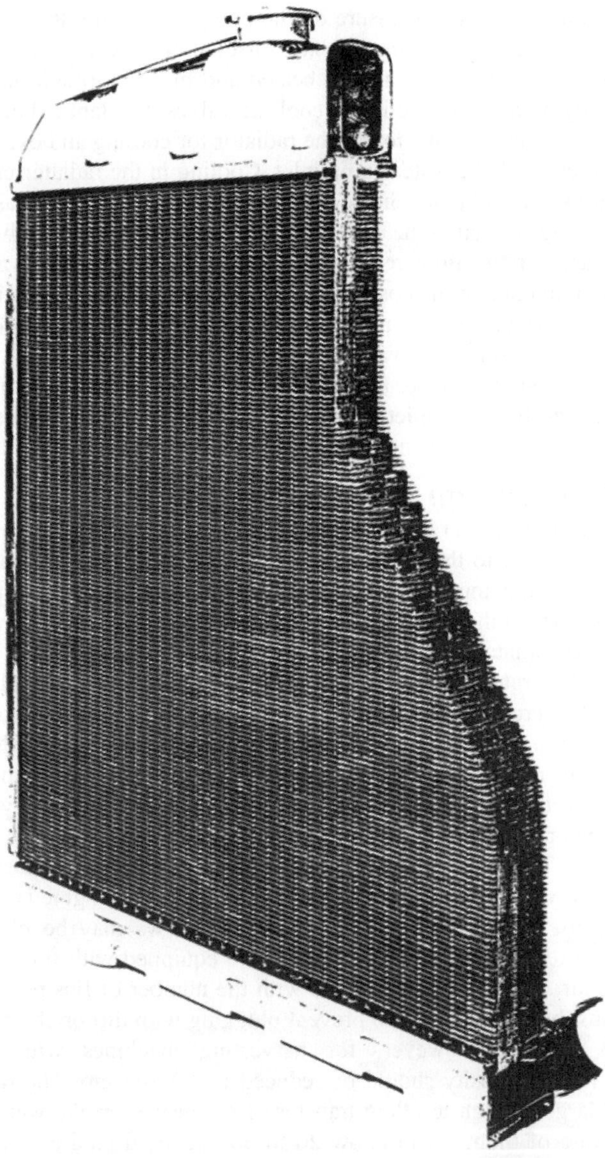

Figure 9.1. A cutaway view of a radiator. (Courtesy of Young Radiator Co.)

reasonable assumptions for design purposes. The liquid temperature entering the radiator is 98°C. The ambient air temperature is shown as 40°C, i.e., a hot day is assumed for a conservative design. If the drop in coolant temperature, ΔT_w, is 8°C and the rise in air temperature, ΔT_a, is 15°C, then the heat transfer difference, ΔT_r, will be

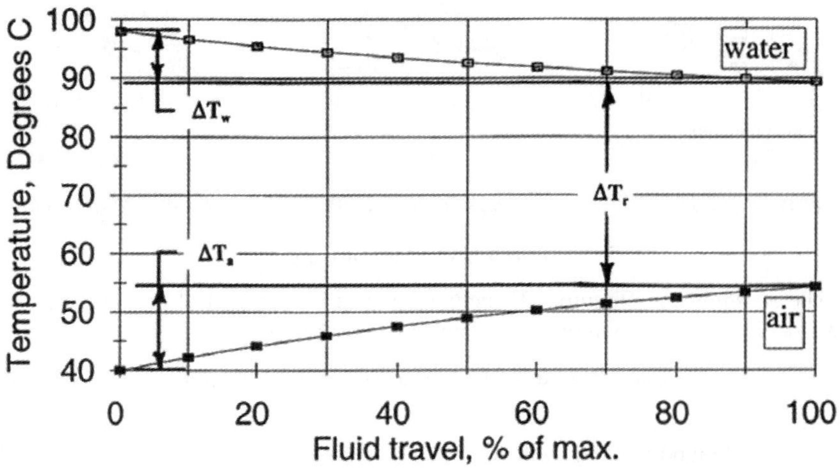

Figure 9.2. An example of temperatures in a radiator.

35°C. Some radiator manufacturers base temperature differences on inlet conditions, while others base them on average temperatures in the radiator. The engine designer must have a clear understanding as to which system is used by the radiator manufacturer. Engine manufacturers normally specify the maximum coolant temperature based on the type of operation and on warranty goals. For applications allowing higher temperatures, a radiator cap with increased relief pressure must be indicated to prevent boiling the coolant at the elevated temperatures.

The tops of the radiator tubes terminate inside a top tank, while the bottoms terminate in a bottom tank. Ordinarily, the tanks are soldered to the radiator core, but the tanks may be bolted to the core in heavy-duty radiators. The radiator hose connections should be arranged to cause equal flow distribution across all radiator tubes. Thus, if the upper hose connection is on the right side of the radiator, the bottom hose connection should be on the left, or vice versa. The top tank must be able to accommodate approximately a 5% expansion in water volume as the water heats from cold to warm engine conditions. The radiator filler neck should extend down into the tank far enough to allow the 5% expansion. A small bleed hole at the top of the neck will prevent air from being trapped in the top of the tank when the radiator is being filled. Air or gas can become entrained in the cooling water by combustion chamber leaks, by leaks on the suction side of the water pump, or by turbulence in the top tank of the radiator. Top-tank volume large enough to allow 200 changes of its coolant per minute will keep flow velocities low enough to allow continuous escape of the air from the water. Also, a horizontal de-aeration baffle may be placed in the top tank to reduce turbulence and allow air to escape as the water settles into the radiator tubes. If inadequate space is available in the top tank, a remotely located surge tank can be used to obtain de-aeration. *De-aeration* is important because entrained air or gas lowers the cooling capacity of the liquid coolant.

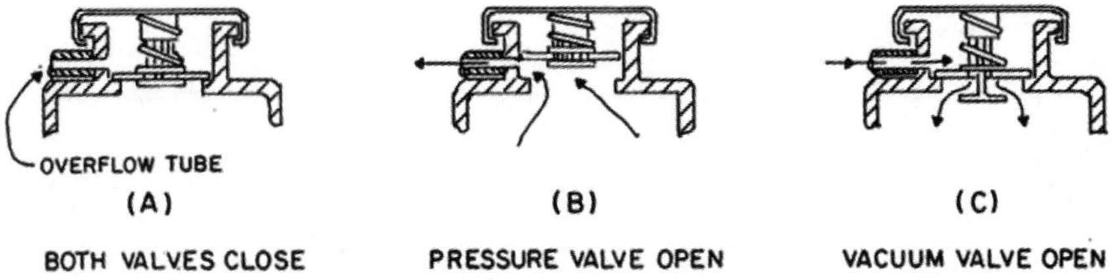

Figure 9.3. A typical radiator cap.

The *radiator cap* (Figure 9.3) includes two valves to control pressure within the radiator. The boiling point of the coolant can be estimated using

$$T_{BP} = 100 - 0.053\,\text{Conc} + 0.00367\,\text{Conc}^2 + \Delta p/4 \tag{9.3}$$

where

T_{BP} = boiling point of coolant, °C
Conc = concentration of ethylene glycol, %
Δp = radiator pressurization, kPa

As the equation shows, the boiling point of water is 100°C at atmospheric pressure, but increases with antifreeze concentration and also increases approximately 1°C for every 4 kPa pressure increase. A radiator may be pressurized as much as 100 kPa but 50 kPa *overpressure* is more common. The pressure valve in the cap must open and allow coolant to escape when the pressure reaches the maximum safe level for which the radiator is designed. When the engine cools and the pressure falls, a vacuum valve in the cap opens to relieve the vacuum and prevent implosion of the radiator. An overflow tank (Figure 9.4) can be used to collect escaping coolant when the pressure valve opens and to return the coolant to the radiator when the vacuum valve opens.

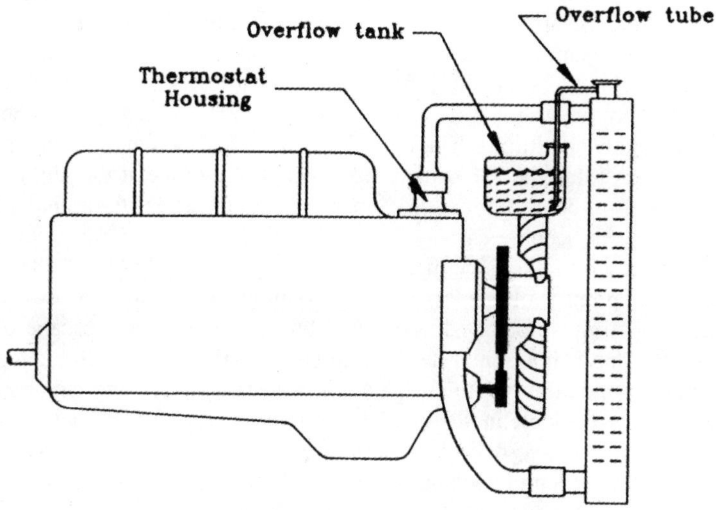

Figure 9.4. A radiator overflow tank.

There are many soldered joints in radiators, and they may not be able to maintain integrity when subjected to the vibration and jolting that can occur on many vehicles. Thus, radiators should be isolated from the vehicle frame by means of flexible radiator mounts.

9.3.3 Airflow

The airflow through the radiator must be sufficient to carry away the heat transferred from the liquid coolant. The following equation can be used to calculate the heat-carrying capacity of the air:

$$\dot{E}_{Ra} = Q_a\, \rho_a\, C_{pa}\, \Delta T_a \tag{9.4}$$

where

$\dot{E}_{Ra}$ = heat transport rate, kW
Q_a = airflow rate, m^3/s
ρ_a = density of air, kg/m^3
C_{pa} = specific heat of air = 1.0 kJ/ kg ·°C
ΔT_a = temperature rise as air moves through radiator, °C

The air density may be calculated by using an equation derived from the ideal gas law, i.e.,

$$\rho_a = \frac{3.49 p_a}{T_a} \tag{9.5}$$

where

p_a = absolute pressure of air passing through radiator, kPa
T_a = absolute temperature of air passing through radiator, °K

A reasonable choice for air temperature rise, ΔT_a, is 15 to 20°C. For a selected air temperature rise, Equation 9.4 can be used to calculate the *airflow rate*, Q_a, needed to transport the heat from the radiator.

The next step is to select a fan to move the air. Figure 9.5 illustrates performance curves for a specific fan, i.e., a nine-bladed fan of 700 mm diameter with a 30° blade angle. The lower family of curves relates the *fan static pressure* to airflow at various fan speeds. The resistance curve shows static pressure versus airflow for a specific radiator installation. The upper set of curves show fan power requirements at various speeds. The designer can follow the resistance curve to find the fan speed needed to provide the required airflow. Then the designer uses the upper set of curves to find the *fan power* requirement at the selected speed and airflow. Figure 9.5 is for a specific fan, which might or might not be ideal for the required airflow needs of the installation being designed. The designer would have a set of such curves for fans of different capacities and would choose the fan best suited to the specific installation. Alternatively, fan performance equations could be used to calculate performance of geometrically similar fans, as explained later in this section.

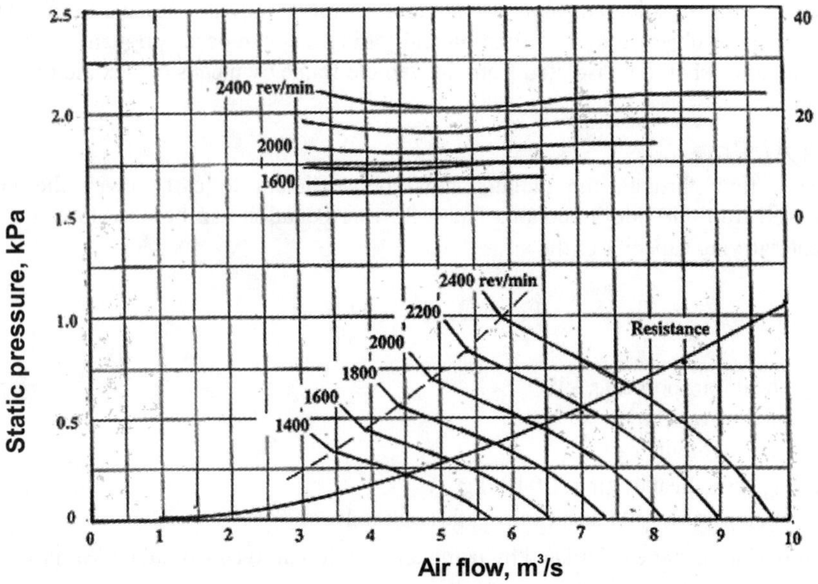

Figure 9.5. Performance curves for a nine-blade, 700-mm fan with 30° blade angle. (Data courtesy of Crowley Co., Inc.)

The curves of Figure 9.5 are for the fan performance at standard atmospheric conditions at sea level (p_a = 101.2 kPa and T_a = 20°C). The standard air pressure at other elevations can be calculated by

$$p_{atm} = 101.1 - 0.01116\,h \tag{9.6}$$

where

p_{atm} = atmospheric pressure, kPa
h = altitude, m

The fan power can be corrected to the new conditions by

$$P_{fo} = \frac{294.1 p_o}{101.3\,T_o} P_{fc} \tag{9.7}$$

where

P_{fo} = fan power observed at non-standard conditions, kW
p_o = observed absolute pressure, kPa
P_{fc} = fan power from performance curve at standard conditions, kW
T_o = observed air temperature, K

Equations 9.6 and 9.7 show that the observed fan power decreases at higher air temperatures and/or higher elevations. As Equation 9.5 shows, however, the air density also decreases under such conditions, thus decreasing the heat-carrying capacity of the air (see Equation 9.4). If the vehicle were to be used consistently at higher elevations, it would be necessary to increase the fan capacity accordingly.

Equations 9.6 and 9.7 also show the observed fan power increases as the ambient temperature falls below the standard 20°C. It is prudent to calculate the correction factor and observed fan power for the lowest temperature at which the vehicle is likely to be used.

The curves on Figure 9.5 are for a given fan diameter and for six constant speeds. Fan performance for other diameters and speeds can be calculated by using *fan performance equations*. The first fan equation adjusts flow, i.e.,

$$\frac{Q_{fn}}{Q_{fc}} = \frac{N_{fn}}{N_{fc}}\left(\frac{D_{fn}}{D_{fc}}\right)^3 \tag{9.8}$$

where

Q_{fn}, N_{fn}, and D_{fn} = fan airflow, speed and diameter of the new point to be calculated

Q_{fc}, N_{fc}, and D_{fc} = fan airflow, speed and diameter of current point on the curves

Note that additional subscripts, c, refer to current points on the curves of Figure 9.5. Additional subscripts, n, refer to new performance points to be calculated using the fan equations.

The second fan equation adjusts static pressure, i.e.,

$$\frac{\Delta p_{fn}}{\Delta p_{fc}} = \left(\frac{N_{fn}}{N_{fc}}\right)^2 \left(\frac{D_{fn}}{D_{fc}}\right)^2 \tag{9.9}$$

where Δp_{fc} and Δp_{fn} = static pressures at a current point on the curves and at a new point to be calculated.

The third fan equation adjusts the fan power, i.e.,

$$\frac{P_{fn}}{P_{fc}} = \left(\frac{N_{fn}}{N_{fc}}\right)^3 \left(\frac{D_{fn}}{D_{fc}}\right)^5 \tag{9.10}$$

where P_{fc} and P_{fn} = fan power at a current point on the curves and at a new point to be calculated.

An example will illustrate the use of the fan equations. In Figure 9.5, the fan diameter is 700 mm. The curve for 1800 rpm passes through the point where the flow is 6.5 m^3/s and where static pressure is 0.24 kPa; the corresponding fan power is 10 kw. Thus, D_{fc} = 700 mm, N_{fc} = 1800 rpm, Q_{fc} = 6.5 m^3/s, Δp_{fc} = 0.24 kPa and P_{fc} = 10 kW. If the fan diameter was kept at 700 mm, but the speed was increased to 1875 rpm, then D_{fn} = 700 mm, N_{fn} = 1875 rpm and application of the fan equations would give Q_{fn} = 6.77 m^3/s, Δp_{fn} = 0.26 kPa and P_{fn} = 11.3 kW. The fan equations could also be used to calculate the performance of a fan of a different diameter if the fan was geometrically similar, i.e., had the same number of blades, same blade angle, etc.

If the *fan efficiency* was known, the fan power could be calculated by using the equation

$$P_f = \frac{Q_a \Delta p_a}{e_f} \tag{9.11}$$

where

Q_a = airflow rate through fan, m^3/s

Δp_a = static pressure rise across fan, kPa

P_f = fan power, kW

e_f = fan efficiency, decimal

The torque required to drive the fan can be calculated by

$$T_f = \frac{60{,}000\, P_f}{2\pi\, N_f} \tag{9.12}$$

where

T_f = fan torque, N·m

N_f = fan speed, rpm

Some fans are equipped with a modulating drive that allows them to turn more slowly when less cooling is required. The total power required to drive the fan is

$$P_{ft} = P_f + \frac{2\pi\, T_f (N_{fd} - N_f)}{60{,}000} + P_{dl} \tag{9.13}$$

where

P_{ft} = total power required to drive fan, kW

N_{fd} = input speed to the modulating drive unit, rpm

P_{dl} = power loss in fan belt drive, kW

The viscosity of the special liquid in a *modulating drive unit* increases with increasing temperature, thereby causing the fan speed, N_f, to approach the input speed, N_{fd}, and move more air. Conversely, the fan runs slower at cooler temperatures, thus reducing fan power. A belt drive is typically used to allow the fan shaft to be positioned higher than the engine crankshaft and to rotate faster than the engine crankshaft. The power loss in the belt drive, P_{dl}, is typically small i.e., 1 or 2% of the power transmitted by the belt.

A transition *shroud* is needed between the rectangular (or square) frontal area of the radiator and the circular area traced by the fan tips. Ideally, the fan circular area will provide a reasonably close match to the radiator frontal area. The fan to radiator clearance should be 10 cm or more. The clearance between the fan tips and the shroud is very important. In an off-road vehicle, some relative movement between the fan shroud and fan can occur. To avoid interference, the fan-to-shroud clearance should be at least 0.6 cm, but fan efficiency will suffer if the clearance is more than 1.25 cm. The depth of penetration of the fan into the shroud affects fan performance. Fan suppliers may provide performance curves for various depths of penetration, e.g., 30%, 50%, or 70% penetration into the shroud. Some manufacturers use suction fans, which pull air through the radiator; others use blower fans, which push air through the radiator. A blower fan helps to prevent accumulation of trash on the radiator front. Typically a suction fan will have greater penetration into the shroud than a blower fan. A cowling or grille between the radiator and the ambient air may restrict airflow into the radiator and an engine cover may restrict the flow of air leaving the fan. Since such restrictions increase the static pressure on the fan and decrease the airflow, care must be taken to avoid excessive restriction. The resistance curve of Figure 9.5 must include the effect of all such restrictions on the fan. Care must be taken to prevent re-circulation of warm air from the engine compartment back through the radiator, since the higher temperature of re-circulated air reduces the heat transfer rate from the liquid coolant to the air. Re-circulation is more likely to be a problem with blower fans if the radiator is ahead of the engine and vehicle movement tends to carry the exhausted air back toward the fan inlet.

The mean velocity of air movement through the radiator can be estimated by

$$v_a = \frac{Q_a}{A_{FR}} \tag{9.14}$$

where

v_a = mean *air velocity*, m/s,

A_{FR} = frontal area of radiator, m^2

Typically, mean air velocities range from 5 to 7.5 m/s.

9.3.4 Coolants

Its high specific heat and ready availability makes water a natural choice for a liquid coolant, but water alone has several limitations. Vehicles are often parked outdoors at ambient temperatures below the 0°C freezing point of water. The resulting expansion when water freezes could break the radiator or engine block. Also, water could cause severe rusting of metal parts in the engine. Thus, other liquids have been used with water to provide a suitable engine coolant.

Ethylene glycol ($C_2H_6O_2$) is among the *antifreeze* compounds that have been adopted for engine cooling systems. As shown in Figure 9.6, adding ethylene glycol reduces the freezing point of the solution until the glycol reaches a concentration of $68\%_v$. Typically, a solution of 50% water and 50% *ethylene glycol* is used. The ethylene glycol also raises the boiling point of the coolant, as indicated by Equation 9.3. Several additives are used with the ethylene glycol to improve its properties. Chemical inhibitors containing an alkaline base help to neutralize any acids that may enter the cooling system from exhaust gas leakage and thereby protect metal surfaces. Before such additives were used, *cavitation* sometimes caused pitting on the outer walls of the combustion chambers. Foam suppressors help to reduce surface tension and bubble formation. Rust inhibitors form a chemical film over metal parts to retard rust formation. Water softening agents help to minimize scale formation on the outer walls of the cylinders; scale formation retards heat transfer through the cylinder walls. A dye is normally added to the antifreeze to give it a distinctive color. Engine manufacturers often specify coolant formulations to be used in their engines. Note in Figure 9.6, the specific gravity of the coolant varies with the glycol concentration and thus with the freezing point of the coolant. This property is used in designing hydrometers to determine the freezing point of the coolant in an engine cooling system.

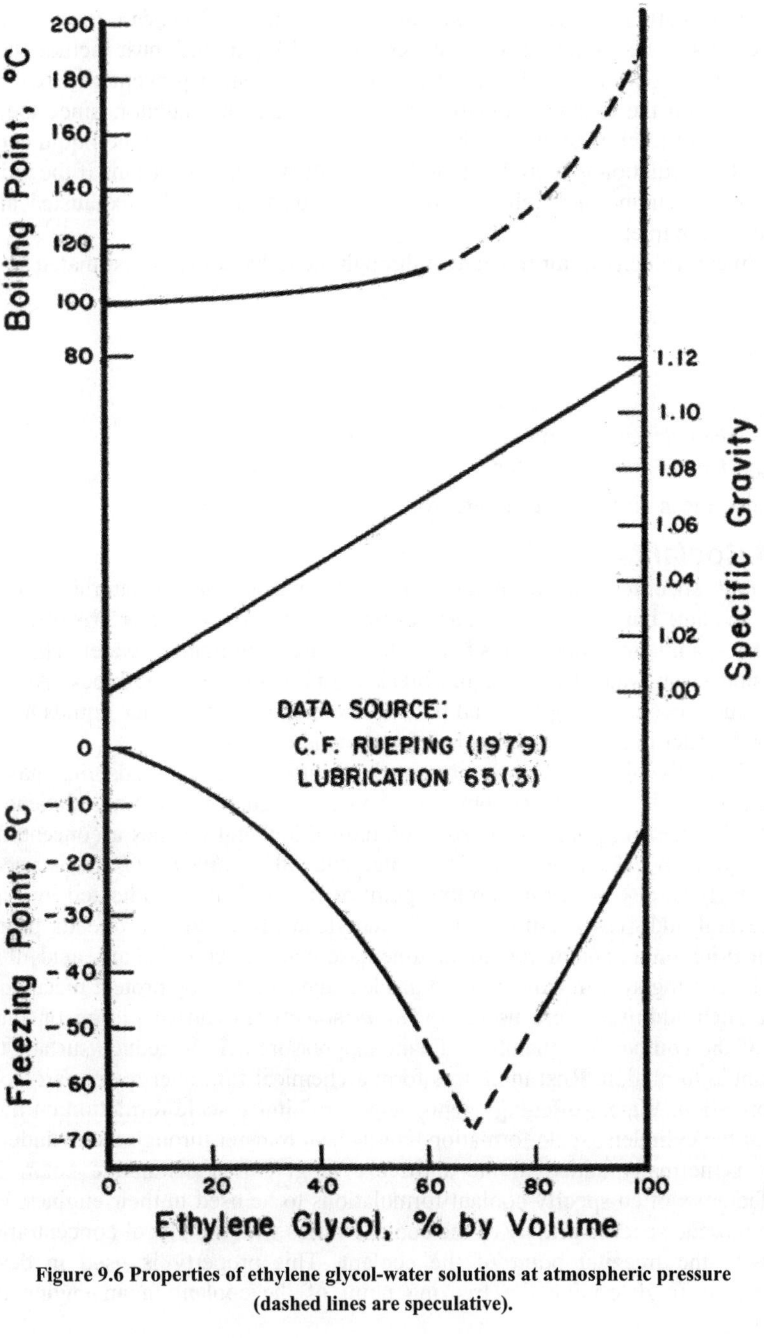

Figure 9.6 Properties of ethylene glycol-water solutions at atmospheric pressure (dashed lines are speculative).

The *specific heat* is a function of the coolant formulation and temperature. The following equations can be used to calculate the specific heat of the coolant. For water alone, the specific heat is

$$C_W = 4.127 + 0.0002892\, T_W \tag{9.15}$$

For a 50:50 blend of ethylene glycol in water (a typical concentration), the specific heat is

$$C_W = 3.409 + 0.003897\, T_W \tag{9.16}$$

where

C_W = specific heat, kJ/kg·°C

T_W = average temperature of coolant in radiator core, °C

The inclusion of ethylene glycol has the detrimental effect of lowering the specific heat of the coolant and must be considered in calculating the required coolant flow rate. However, this disadvantage is far outweighed by the advantages of including ethylene glycol.

9.3.5 System Considerations

As previously noted, excessive temperatures can cause great harm to engine parts. The liquid cooling system must be designed to prevent excessive temperatures under full engine loading while the vehicle is operating with the highest ambient temperatures it will ever encounter. The maximum cooling capacity is not needed when ambient temperatures are more moderate and/or when engine loading is less than the maximum. Under such conditions, the engine would run too cool, thereby decreasing fuel-use efficiency and increasing exhaust emissions, unless the cooling capacity was reduced. The function of the engine *thermostat* is to control the coolant temperature by causing some or all of the coolant to bypass the radiator when maximum cooling capacity is not needed.

The thermostat is mounted in a housing on the engine at the point where the coolant leaves the engine head. The sensing element must be immersed into the water departing from the engine, as shown in Figure 9.7. Two types of engine thermostats are illustrated in Figure 9.7. The bellows type (Figure 9.7a) contains ether in a short, sealed tube. The elasticity of the bellows holds the thermostat closed when the ether is cold. The ether expands as it absorbs heat when the coolant temperature rises and the expansion causes the thermostat to open. In Figure 9.7b, a bimetal strip uncoils as it heats, causing the thermostat to open. When either type of thermostat is closed, the coolant pump should still circulate coolant through the engine water jacket to maintain a more uniform engine temperature and prevent hot spots. A *bypass* is provided to allow coolant to exit from the head and enter the pump inlet when the thermostat is closed. Some bypasses are very small in diameter and remain open when the thermostat opens. The flow capacity of the bypass is negligible compared to the flow capacity to the radiator when the thermostat is open. In other designs, the bypass has a larger capacity, but an extension on the thermostat moves to block the bypass when the thermostat opens the passage to the radiator.

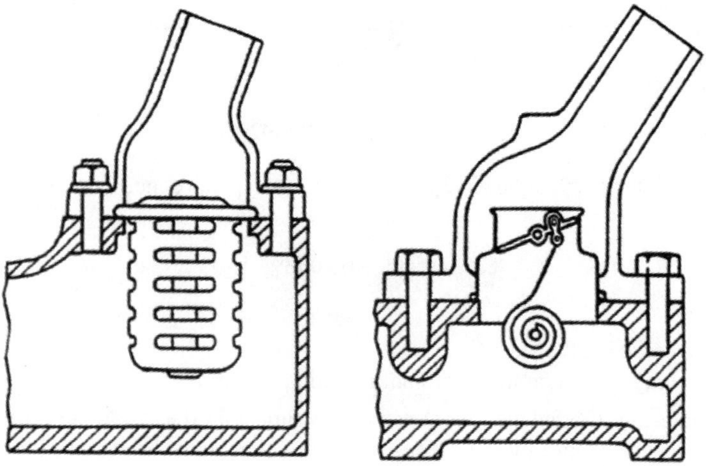

Figure 9.7. Engine thermostats.

A variety of thermostats are available to provide a range of engine control temperatures. The engine designer normally specifies the temperature at which the engine is to be controlled. A given thermostat operates over a temperature range, i.e., it begins to open at some temperature and becomes fully open at a somewhat higher temperature. A thermostat may fail in service; for example, it may stick in the open or closed position. Sticking in the open position will cause the engine to warm up too slowly, while sticking in the closed position will cause the engine to overheat.

Entrapped air or gas is detrimental to the operation of a liquid cooling system. Provisions for de-aeration were previously discussed, but it is also important to minimize or eliminate air entry into the cooling system. The cooling system should be designed to prevent trapping of air when the cooling system is being filled with liquid coolant. The liquid coolant should have a direct, low-restriction path from the radiator fill neck to the bottom of the engine water jacket. Also, high spots in the cooling jacket that are not vented to the radiator top tank will trap air and should be avoided, otherwise they will prevent complete filling of the cooling system with liquid.

9.4 Air Cooling

Air cooling offers several advantages and some disadvantages compared to liquid cooling. Air cooling eliminates the need for a water pump, radiator, thermostat, water hoses, and the coolant, including antifreeze. All coolant leakage problems are also eliminated. Because the temperature of the combustion chamber walls is higher in an air-cooled engine, emissions of unburned hydrocarbons and carbon monoxide are generally reduced. Also, the heat from an air-cooled engine is rejected at a higher temperature then from a liquid-cooled engine; the resulting larger ΔT between the heat source and the ambient air permits less airflow than is needed to remove heat from the radiator of a liquid-cooled engine. However, the cooling fins required around the combustion chambers of an air-cooled engine do not allow the cylinders to be positioned as close together as in a liquid-cooled engine. Engine temperature is not controlled as closely in an air-cooled engine, particularly at low engine speeds. Finally, the engine coolant may be used for such purposes as intercooling the engine air or cooling the engine oil on a liquid-cooled engine but is not available for such purposes on an air-cooled engine. Although liquid cooling is more popular, some air-cooled engines have found a niche in the engine market.

The heat loads discussed in Section 9.1 apply to either liquid-cooled or air-cooled engines. Since there is no liquid coolant in the air-cooled engine, an intercooler would have to be an air-to-air intercooler and an oil cooler would have to be of the oil-to-air type. The cooling system must be capable of handling the total heat load under the most unfavorable environmental conditions to be encountered by the vehicle.

An air-cooled engine is illustrated in Figure 9.8. A fan and shrouding are used to force air to move across the engine to remove heat. Equation 9.4 can be used to calculate the airflow rate the fan must provide to remove the heat from the air-cooled engine. If the ambient temperature was 40°C, an air temperature rise of approximately 22°C would still leave a substantial temperature difference between the engine surface and the exiting air. Under those conditions, which would promote heat transfer from the engine to the air, Equation 9.4 can be used to develop a rule of thumb. The rule is that approximately 0.04 m^3/s of airflow should be provided per kW of brake power.

If air already heated by the engine is allowed to re-circulate back to the fan inlet, the cooling capacity of the air will be greatly reduced. Therefore, the shrouding provided to direct the air should be designed to minimize air recirculation.

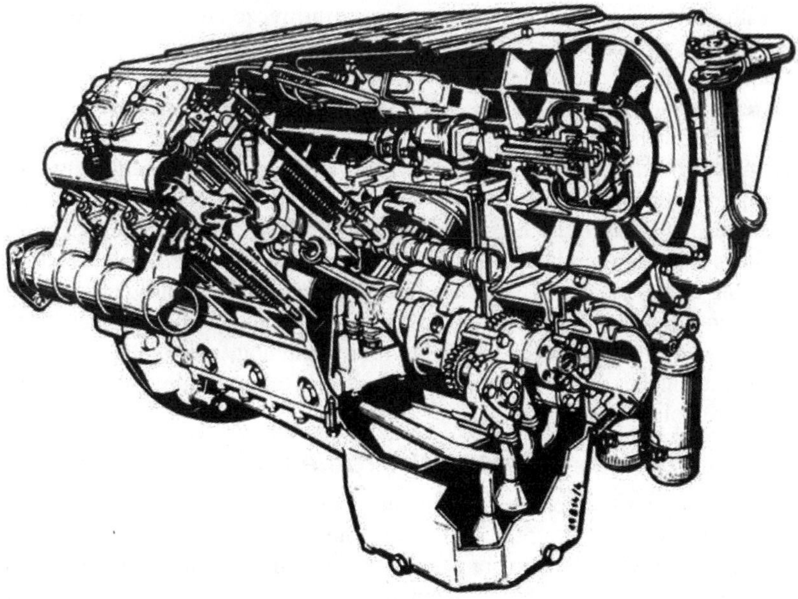

Figure 9.8. An air-cooled engine.

9.5 Chapter Summary

Engines must be liquid cooled or air cooled to prevent severe damage to engine parts. In either case, the heat to be rejected by the cooling system includes heat from the engine and from engine or vehicle accessories. An equation was presented for estimating the heat load to be handled by the cooling system.

In a liquid-cooled engine, the rejected heat must be carried to the radiator by the liquid coolant, transferred to the air by the radiator, and removed from the vehicle vicinity by the air. The heat transfer capacity of each of these elements must be equal to the heat rejection from the engine. Procedures were given to calculate the required flow rates of liquid coolant and air and to choose a radiator of adequate size. Suggestions were given for limiting air entry into the cooling system and for de-aerating the coolant because the presence of entrapped air or gas reduces the cooling capacity of the coolant.

A procedure was presented to choose a fan of adequate size to remove heat from the radiator and to calculate the power requirement of the fan. The fan curves were for performance at standard atmospheric conditions, but a procedure was presented for correcting the power requirement to non-standard conditions. A modulating fan includes a drive element to automatically reduce fan speed and save power when less cooling is needed. Fan-to-shroud clearances affect fan efficiency, and appropriate clearances were given.

Ethylene glycol or other antifreeze compounds are blended with water to provide a coolant with better properties than water alone. Antifreeze additives were discussed and properties of various water-antifreeze blends were given.

The liquid cooling system must have sufficient capacity to remove heat under the most extreme conditions to be encountered by the vehicle. Because the vehicle frequently encounters less severe conditions, the cooling system often has excess capacity. A thermostat is used to prevent overcooling and allow the engine to operate at an optimum temperature. The thermostat closes and restricts or prevents coolant flow to the radiator until the engine reaches the desired operating temperature. A bypass allows some coolant to circulate in the engine when the thermostat is closed.

Air cooling an engine eliminates the numerous parts involved in liquid cooling, but engine temperature is not controlled as closely as with liquid cooling, especially at low engine speeds. Although liquid cooling is more common, air-cooled engines have found a market niche. A rule of thumb was given for estimating the airflow rate needed for air cooling.

Homework Problems

9.1 You are to plan the cooling system specifications for a 120 kW, liquid-cooled engine. The coolant is a 50:50 blend of water and ethylene glycol. Assume that the heat rejected to the coolant is equal to 91% of the brake power and that the coolant cools 8°C and the air temperature rises 15°C in passing through the radiator. Assume the coolant enters the top of the radiator at a temperature of

98°C. Assume the radiator has 1.6 cm tube spacing and three rows of tubes. Design for ambient air temperature at 40°C and 100 kPa barometric pressure.

Calculate (a) the specific heat of the coolant, (b) the required coolant flow rate, (c) the required number of tubes in the radiator, (d) the radiator width, (e) the radiator frontal area assuming a height 15% greater than the width, and (f) the required airflow rate through the radiator.

(g) Select the required fan speed of the fan whose performance is shown in Figure 9.5, assuming the fan operates along the restriction curve shown in the figure. Use Figure 9.5 to determine (h) the static pressure rise across the fan and (i) the fan power.

Finally, calculate (j) the fan efficiency and (k) the mean velocity of airflow through the radiator. Note that you may have to adjust the answer of part c; the number of radiator tubes must be an integer number and the middle row has one less tube than the front and rear rows.

9.2 Rework Problem 9.1, but assume the coolant temperature falls 5°C and the air temperature rises 13 °C.

9.3 Rework Problem 9.1, but assume the brake power is 130 kW.

9.4 Rework Problem 9.1, but assume the brake power is 130 kW, the coolant temperature falls 5°C, and the air temperature rises 18°C.

9.5 (a) Plot the specific heat of water versus the water temperature for temperatures ranging from 80 to 120°C. Add a second curve on the same graph, but use a 50:50 blend of water and ethylene glycol as the coolant.

(b) What change in the design of the cooling system would be needed to compensate for the reduced specific heat of the coolant containing antifreeze?

9.6 Assume a fan is driven with a constant 2450 rpm input speed to a modulating drive unit. For each of the points on Figure 9.5 at which the resistance curve crosses a fan speed curve, determine (a) the fan power, (b) the fan torque, (c) the power absorbed in the modulating unit, and (d) the total power required to drive the fan assuming the fan belt drive is 98% efficient. Plot three curves, i.e., fan power, power absorbed in the modulating unit, and total power, all versus fan speed. Note: This is a good spreadsheet exercise.

9.7 Rework Problem 9.6, but assume the input speed to the modulating unit is 2500 rpm.

9.8 Assume a fan is operating at 2000 rpm on the resistance curve of Figure 9.5.

(a) Assuming the air temperature remains at 20°C, calculate the fan power at 250 m increments in elevation from 0 to 1500 m elevation.

(b) At each elevation, also calculate the mass flow rate of air delivered by the fan.

(c) Plot the fan power versus elevation and mass flow rate versus elevation.

(d) What change in cooling system design would be needed to allow the engine to operate at the higher elevations?

9.9 Rework Problem 9.8, but assume the fan is operating at 2400 rpm.

9.10 Digitize the 2400 rpm fan curves on Figure 9.5, i.e., (a) take data points along the static pressure versus flow curve and (b) the fan power versus flow curve. Enter the digitized curve data into a spreadsheet. Calculate (c) a static pressure flow curve and (d) fan power versus flow curve for a speed of 2500 rpm. Use the spreadsheet to plot (e) the two static pressure versus flow curves and (f) the two fan power versus flow curves.

9.11 Rework Problem 9.10 but change the fan speed to 2300 rpm.

9.12 Rework Problem 9.10 but, instead of changing the fan speed to 2500 rpm, change the fan diameter to 750 mm.

9.13 Rework Problem 9.10 but, instead of changing the fan speed to 2500 rpm, change the fan diameter to 650 mm

9.14 Rework Problem 9.10, but change the fan diameter to 650 mm and the fan speed to 2500 rpm.

9.15 (a) Calculate the airflow rate needed to cool a 120 kW, air-cooled engine.

(b) Compare your answer with the airflow needed for the engine of Problem 9.1.

9.16 (a) Calculate the airflow rate needed to cool a 130 kW, air-cooled engine. (b) Compare your answer with the airflow needed for the engine of Problem 9.3.

References and Suggested Readings

Hochkonig, M., and M. Rauser. 1992. Cooling system layout for high performance cars. In SAE Special Publication SP-918, *New Engine Design and Cooling Systems*, 163-183. Warrendale, PA: SAE.

Macherle, J. 1972. *Air-Cooled Automotive Engines,* 2nd ed. New York, NY: John Wiley and Sons, Inc.

Oban, H.C., and H.J. Slothower. 1965. Cooling systems for self-propelled combines. *Agricultural Engineering* 46(9): 499.

SAE. 1992a. Method for determining power consumption of engine cooling fan-drive systems. SAE Standard J1342. *SAE Handbook,* vol. 3. Warrendale, PA: SAE.

SAE. 1992b. Glossary of engine cooling system terms. SAE Standard J1004. *SAE Handbook,* vol. 3. Warrendale, PA: SAE.

Sprick, W.L., and T.H. Becker. 1985. The application and installation of diesel engines in agricultural equipment. ASAE Distinguished Lecture Series No. 11. St. Joseph, MI: ASAE.

CHAPTER 10

ELECTRICAL AND ELECTRONIC SYSTEMS

10.1 Introduction

Electrical and electronic systems have evolved to become an essential element of modern off-road vehicles. The earliest successful agricultural tractors, the 1892 Froelich (USDA, 1954), for example, incorporated a battery and electric ignition system. A modern off-road vehicle generally incorporates an electrical system with power generation, storage, and distribution. In addition, it may have dozens of electronic computer-based controllers integrated into its systems. A functional understanding of electrical and electronic systems and an ability to apply basic design information with regard to these systems is necessary for engineers working with off-road vehicles. Functional descriptions and basic design information is provided here for both electrical and electronic systems.

10.2 Electrical Systems

10.2.1 Functional Elements of Electrical Systems

Functions that are generally provided by the electrical system of a conventional off-road vehicle include

- Engine starting
- Ignition (in the case of a spark-ignition engine)
- Lighting, both work lighting and safety lighting
- Sensing, display, and control
- Ventilation
- Accessory functions including windscreen wiping, entertainment systems, radio systems, etc.

To provide these functions, the electrical system must incorporate power generation, power storage, power distribution elements and generally some type of power control elements. An example system incorporating these components is shown schematically in Figure 10.1. The charging system (power generation) includes an alternator and voltage regulator. The power storage system is the battery. The power distribution system is composed of both the current distribution system and the current return system. Both systems are required to form a circuit and both should be considered together in examining the design of power distribution. Overload protection in the distribution system commonly includes a fuse panel with common conductors from the battery and individual fuses with conductors to each of the loads. The fuse panel circuits may be supplied with either switched or unswitched power from the battery.

A common conductor is often used for the return system even though individual conductors are used. In some cases, the vehicle frame may be used as a return conductor, for example, with the engine starter motor. Loads supplied by the electrical system are shown generically in Figure 10.1. The loads could be engine starting, ignition, lighting, sensing devices, ventilation, entertainment systems, etc. The switches (the power control elements in Figure 10.1) could be manually operated or might be activated by some component of the vehicle. For example an engine low oil pressure switch might switch on a load, e.g., a low oil pressure indicator lamp in a dashboard. Similarly, the power control element might be an electronic control unit (ECU), an embedded computer-based controller, which could be regulating current to set the position of a solenoid controlled valve. In the latter case, power would be distributed to the ECU. Within the ECU, power is generally distributed independently to power the computer and to the power control element of the ECU.

Connectors (not shown in Figure 10.1) are a fundamental element of the distribution system. They are required to facilitate assembly of the vehicle, to allow vehicle service, and are in some cases part of the functionality of the vehicle; for

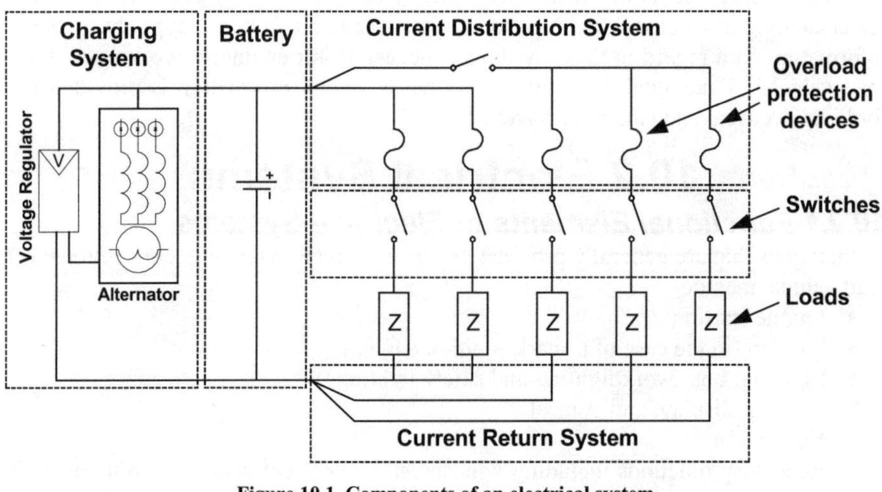

Figure 10.1. Components of an electrical system.

example, a hitch connector on an agricultural tractor allows power to be connected to implements. Connectors must be part of the electrical system design because their effects on the physical, electrical, and thermal characteristics of the system may be important.

Batteries

Batteries are the primary electrical energy storage elements in off-road vehicles. During an operating cycle of the vehicle between start-up and shut-down, the battery undergoes a discharge and charge cycle. In most designs, the battery supplies high current to an electrical starting motor for starting. During vehicle operation, the on-board alternator provides current to recharge the battery so that when the vehicle is shut down, the battery will be charged sufficiently to allow restarting. During vehicle operation, the generated current not needed for vehicle electrical loads is available for recharging the battery. Adequate charge storage must be provided in the battery to start the vehicle. Adequate alternator current capacity must be provided to power vehicle electrical loads and recharge the battery within the vehicle operating cycle.

Battery characteristics are critical in determining the capacity necessary for a particular vehicle. Figure 10.2 depicts a cell in a lead-acid battery. This type of battery is the primary type used in off-highway vehicles, although gel cell and other technologies continue to be improved and may eventually see application (Denton, 1997).

During discharge, current flow at a potential difference is due to the chemical reactions that occur in the battery (Figure 10.2). A single cell in the battery develops a potential difference of slightly more than 2 V at full charge. If a load is applied across the battery terminals, the reactions depicted in the diagram occur. At the positive plate, lead peroxide (PbO_2) combines with hydrogen that is ionized in the electrolyte of the battery in a reaction that consumes electrons and reduces the plate to lead, releases water, and causes the electrode to be more positive. At the same time, lead is

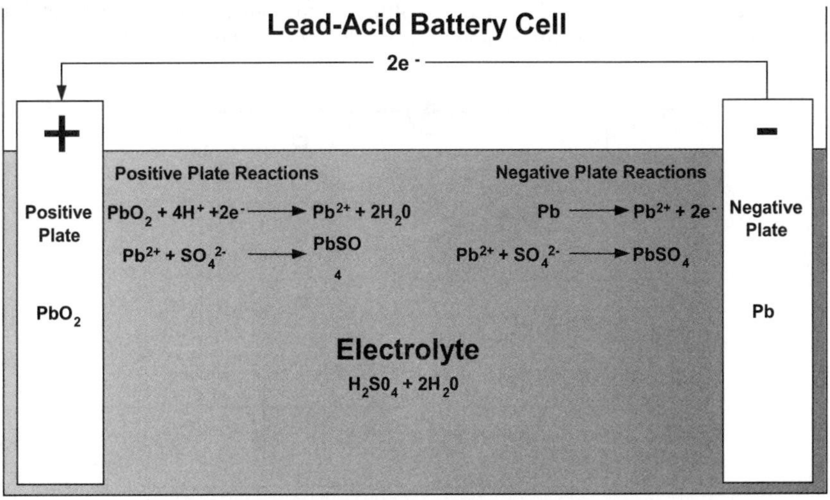

Figure 10.2. Diagram of a cell in a lead-acid storage battery during discharge.

combined with sulfate to form lead sulfate at the positive plate. At the negative plate, lead gains electrons making the plate more negative. The lead also combines with sulfate at the negative plate resulting in lead sulfate. Water released at the positive plate and both hydrogen and sulfate ions used by the reactions cause dilution of the acid in the electrolyte. Charging the battery reverses the reactions at both plates and increases the acid concentration in the electrolyte.

Fully charged, at a charging voltage of 2.3 to 2.45 V per cell, lead sulfate is no longer available for conversion into lead peroxide or lead, and further charging causes hydrolysis of the water in the electrolyte, releasing hydrogen at the positive plate and oxygen at the negative plate. These gasses are released from the battery posing a flammability hazard and decreasing the amount of water in the electrolyte. Allowing the battery to discharge completely or to a low-acid concentration results in high lead sulfate concentrations at the electrodes. If the battery is allowed to stand for a length of time in this state, large lead sulfide crystals form at the plates and are very difficult to convert back to lead and lead peroxide by charging. This results in a "sulfated" battery that is difficult or may be impossible to charge. Also, the freezing point of the electrolyte rises from approximately -60°C at full charge to as high as -3°C at full discharge (Bauer, 1999) making a fully discharged battery more susceptible to freezing.

Voltage levels in a lead-acid battery are a function of the acid concentration, temperature, and battery load. The internal resistance, R_i, (Figure 10.3) is a function of temperature and state of charge of the battery. Individual cells in the battery will produce a voltage dependent on the state of charge and cell temperature. Nominal states of charge for various electrolyte specific gravities and cell voltages are given in Table 10.1. The charge levels are based on the battery supplying a 12 V system and current can be supplied by the battery at voltages lower than that shown. Starting would be an example of this situation. Note that the cell voltage is not very sensitive to the state of charge. Electrolyte specific gravity is therefore used as an indicator of the state of charge. A device with a load resistor, R_L, and ammeter as shown in Figure 10.3 can be used to test the battery and indicate the charge state of the battery.

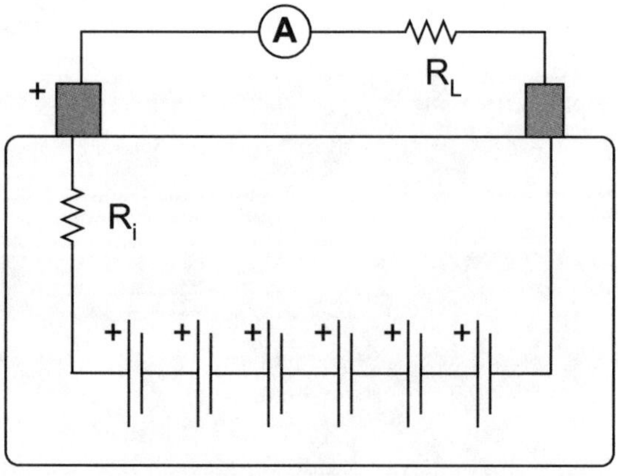

Figure 10.3. Circuit model for a 12 V lead-acid battery.

Table 10.1. Theoretical cell voltages and electrolyte density as a function of charge level at 27°C (Denton, 1997).

Electrolyte Specific Gravity	Cell Voltage, V	Charge, %
1.28	2.12	100
1.24	2.08	75
1.20	2.04	50
1.16	2.00	25
1.12	1.96	0

A typical 12 V lead-acid battery will be constructed of six cells connected internally in series. The battery case will typically be made of an ABS plastic. The battery plates will be made of a lead calcium alloy to minimize gassing, strengthen the plates, and minimize self-discharge. A porous material, commonly a glass-fiber mat or polyethylene film, will be used to separate the plates and to provide strength and vibration resistance in the battery.

Most modern batteries are designed to be low maintenance and incorporate lead-calcium alloys in the battery plates. Such batteries produce minimal gassing (and associated water loss) during charging at charge voltages below 2.3 V per cell. Careful maintenance of charging voltage allows the battery to be sealed and water loss to be insignificant throughout the life of the battery. Earlier designs typically incorporated lead antimony alloys and required regular addition of distilled water to compensate for water loss. The model shown in Figure 10.3 is not completely accurate in depicting all of the resistances in a battery but can be used to specify internal resistance. Some resistance exists in the battery in parallel with the cells. As a result of the parallel resistance, lead-acid batteries tend to self-discharge if left disconnected. Modern low-maintenance batteries using lead-calcium alloys plates have very low parallel resistance and may remain charged while disconnected for periods of a year or longer.

Many performance criteria are used to characterize batteries (SAE, 1994; SAE, 1984). Battery capacity is typically rated in terms of *cold cranking current* and *reserve capacity* (see SAE J537). Cold cranking current is the current that can be delivered by the battery for 30 seconds at the rating temperatures (-18°C, -29°C) and provide an ending voltage of at least 1.2 V per cell. Reserve capacity is the time that a fully charged battery can deliver 25 A current in a 27°C ambient environment and provide an ending cell voltage of at least 1.75 V per cell.

Sizing the Battery

The starter imposes the most significant battery load. Other loads generally have longer durations, draw lower current, occur while the engine and alternator are running and can generally be met by the alternator without significant demand from the battery. The battery must primarily meet the starting load and be able to be recharged during the operating cycle to meet the next starting load. The alternator must have the capacity to meet the permanent, long, and short duration loads within the vehicle while also providing adequate current to recharge the battery during the operating cycle. Battery sizing then becomes primarily one of meeting the starting demand. Energy efficiency of the battery can be used to estimate charging requirements. Energy efficiency of the battery, e_b, is the ratio of discharging energy to charging energy and is typically 75% (Denton, 1997).

$$e_b = \frac{E_d}{E_c} = \frac{P_d t_d}{P_c t_c} \qquad 10.1$$

where

e_b = energy efficiency of the battery
E_d, E_c = discharging energy, charging energy
P_d, P_c = discharging power, charging power
t_d, t_c = discharging time, charging time

Charging Systems

The charging system, shown in the complete electrical system in Figure 10.1, normally consists of an engine-driven alternator and a voltage regulator. The alternator generally consists of a set of stationary coils, a stator, arranged in a toroid within which a rotating coil, or rotor, is spun. Figure 10.4 is a cutaway side view showing the rotor and stator of a claw-pole alternator. The rotor consists of a shaft with two sets of claw poles surrounding a coil. Voltage is applied to the rotating coil through the brush and slip ring assembly, causing current to flow and the rotating coil and claw poles to be magnetized. The claw poles are marked with N and S in the figure, indicating the magnetic field associated with the poles. The poles rotate past coils within the stator inducing currents in the stator coils. The magnitude of the voltage applied to the rotor coil controls the strength of the magnetism in the claw poles and the magnitude of the current induced in the stator coils. Though many other physical arrangements are used, the claw-pole design is a common one.

The current induced in the stator coils of the alternator is an alternating current. The alternating current must be rectified to DC current to charge the battery and supply vehicle electrical loads. Figure 10.5 shows a schematic of a diode-bridge with a sinusoidal input current similar to that generated by one pair of stator coils of an alternator. When the diode-bridge is driven with a positive current from the stator,

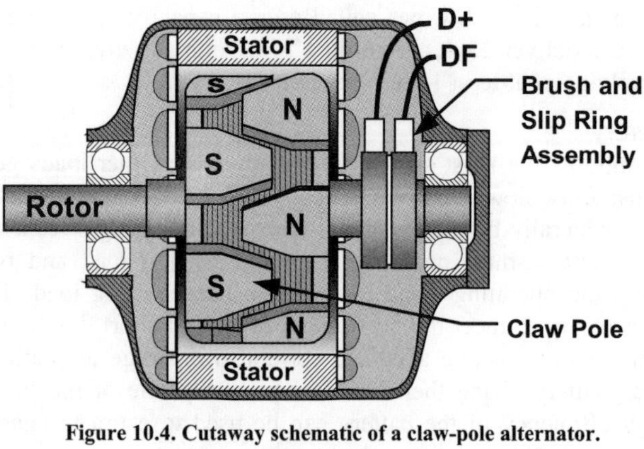

Figure 10.4. Cutaway schematic of a claw-pole alternator.

current flows downward from the top of the bridge (terminal "c") through the diode to the bridge terminal marked "a" and is blocked from flowing downward from the top through the diode to the bridge terminal marked "b." This results in the positive side of the stator being connected to the positive side of the load. Current returns from the load through the bottom side of the bridge through the bridge terminal marked "b," through the diode to terminal "d," to the negative side of the stator. When the bridge is driven with a negative current from the stator, current flows upward from the bottom of the bridge (terminal "d") through the diode to the bridge terminal marked "a" and is connected to the positive side of the load. Current is blocked from flowing upward from the bottom (terminal "d") through the diode to the bridge terminal marked "b." Current returns from the load to the terminal marked "b" to the top side of the bridge (terminal "c") and to the top of the stator. The bridge switches the connections of the stator to the load so that the positive end of the stator is always connected to the positive side of the load. The output signal has a DC component plus a large ripple.

Stator windings have a coil for each magnetic pole in the rotor. The example arrangement in Figure 10.6 has six poles (three pole pairs). Typical alternators have 12 to 16 poles. The frequency of the sinusoidal signal from the alternator can be calculated using Equation 10.2

$$f = \frac{Np}{2 \cdot 60} \qquad 10.2$$

where

N = rotational speed of the rotor, rpm

p = number of poles in the alternator

Most alternators use star-connected three-phase windings. Each of the three sets of stator windings (Figure 10.6) is spaced at 120° and has the same number of poles as on the rotor. The resulting alternator output when connected as shown in Figure 10.7 has overlapping sinusoidal current waveforms and, when rectified, has less ripple than a single-phase alternator. The result is a more efficient alternator.

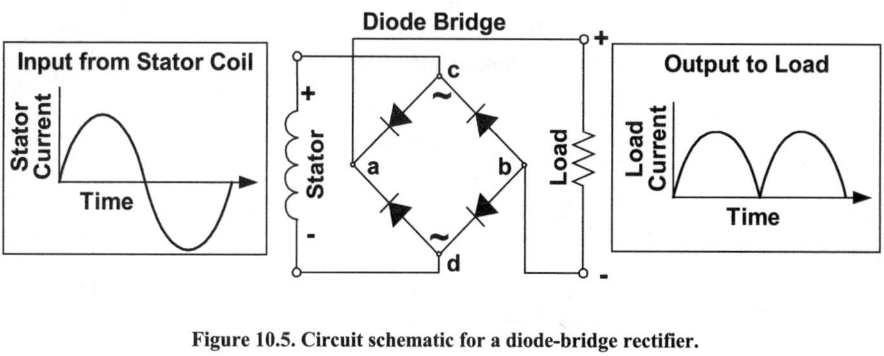

Figure 10.5. Circuit schematic for a diode-bridge rectifier.

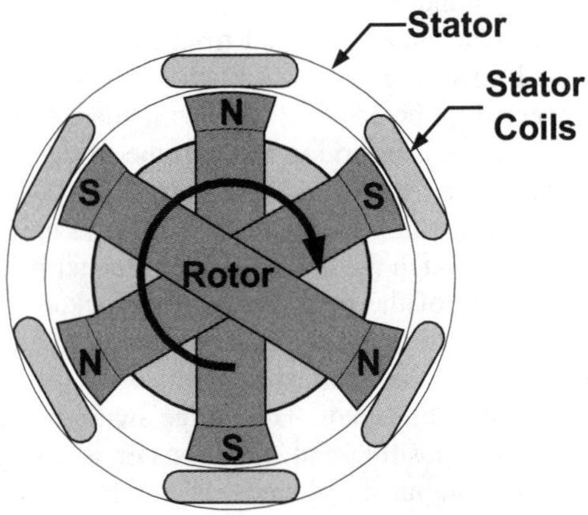

Figure 10.6. Arrangement of a six-pole single-phase alternator.

Current generated by the alternator is rectified in a bridge arranged to handle currents from the three-phase windings (Figure 10.8). The current generated across any two of the phase windings is applied to a four-diode rectifier as shown earlier. The three-phase arrangement requires six main diodes in the bridge to handle all three phases (labeled "Power Diodes" in Figure 10.8). Three additional "exciter" diodes are used to provide current for excitation of the rotor. The exciter diodes form the upper half of a six-diode bridge along with the three main negative diodes that form the lower half of the bridge. When the ignition switch is engaged, current from the battery may be used to provide current to energize the rotor. This current flows from the positive terminal on the battery through the charge indicator lamp, to the "D+" terminal, through the rotor, to the "DF" terminal, through transistor T3 to "D-," and back to the battery. T3 conducts initially due to low base current to T1 as explained later. As current and then magnetism builds in the rotor, current is generated in the stator and is rectified by both the main diodes providing potential between B+ and B- terminals and is also generated between the D+ and D- terminals. When the potential at the D+ terminal has increased to the level in the battery, current no longer flows from the battery through the charge indicator lamp, extinguishing the lamp and indicating that the alternator is generating sufficient current to initiate charging of the battery.

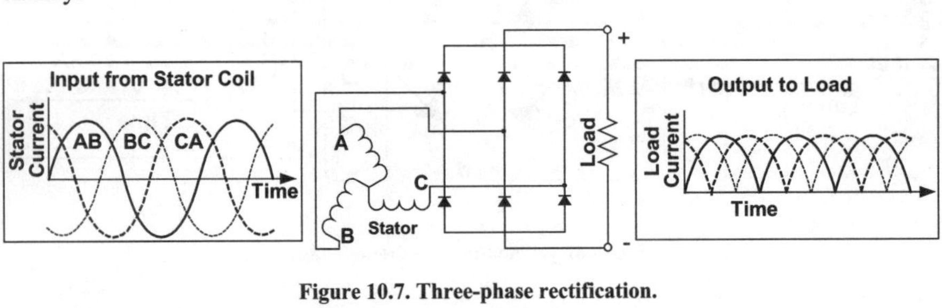

Figure 10.7. Three-phase rectification.

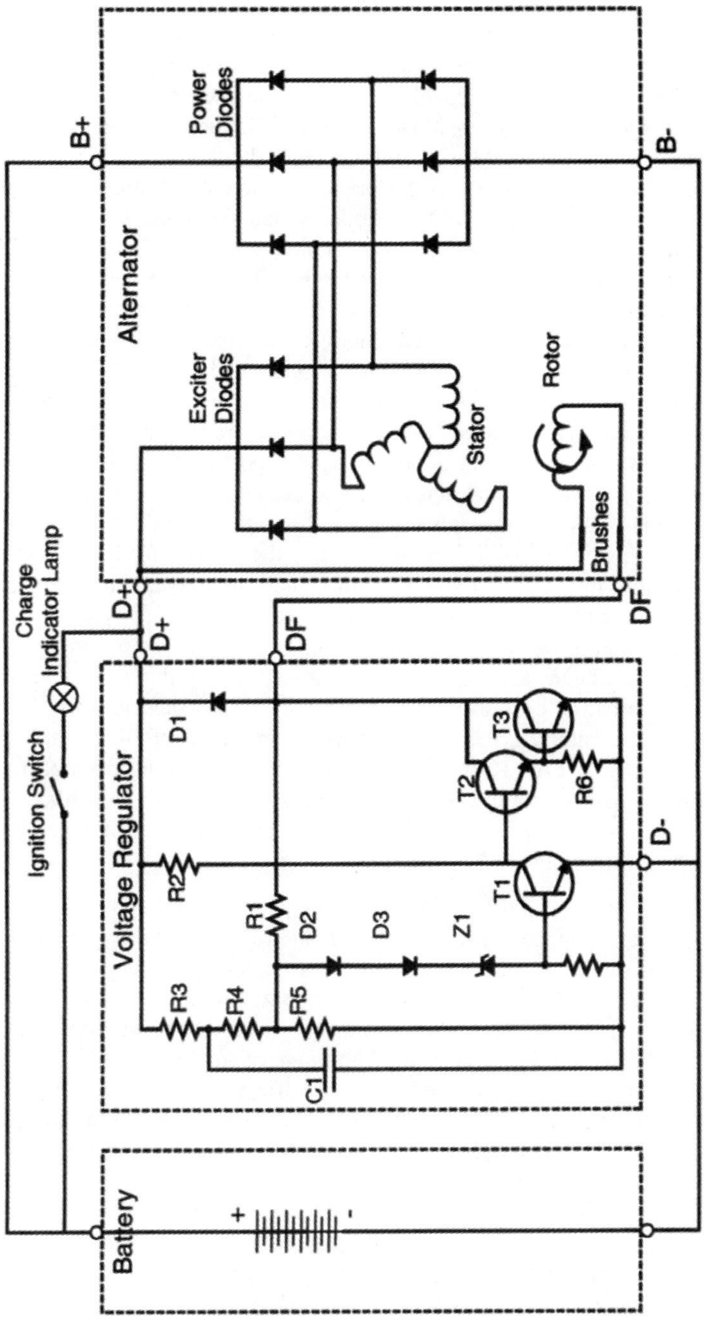

Figure 10.8. Circuit schematic of an electronic regulator and alternator charging circuit.

The solid-state voltage regulator shown in Figure 10.8 is similar to those used on most modern off-highway vehicles. The regulator senses voltage between the D+ and D- terminals. If voltage is too high, the regulator disables excitation current through the rotor. If not, excitation is enabled. Voltage between the D+ and D- terminals (Figure 10.8) is applied to the resistors R3, R4, and R5, and at the same time a portion of that voltage is applied to D2, D3, and the zener diode Z1. The zener diode acts to block current from flowing through this branch of the circuit until the voltage level is above the breakdown voltage of this diode. When the voltage level is below the breakdown level of the zener diode, current is prevented from flowing into T1 causing the collector-to-emitter path to be non-conductive. This action results in a very small current through R2, and consequently a high base-to-emitter voltage on T2, enabling T2, which in turn enables T3 to be conductive. Current from D+ then flows through the rotor and returns through DF, through T3 to D-, and returns to the battery. When the voltage level at D+ climbs to the level that the voltage drop across the zener diode is above the breakdown voltage of diode, it conducts, raising the potential at the base of T1, causing it to conduct. This action causes the potential at the base of T2 to drop, preventing current from flowing across T2 and causing T3 to be non-conductive. The path for current from D+ through the rotor, back to D- and to the battery is then blocked. During operation of the regulator, when the load on the alternator is low and the speed is high, the voltage regulator will switch the excitation current on and off continuously to regulate the voltage.

The voltage output from the regulator needs to be matched to battery conditions. Lower charging voltages are required when battery temperatures are high. The zener diode or an alternative portion of the circuit may be selected to vary the regulated voltage with temperature. For a 14 V alternator, an output of 14.5 V at 25°C with a temperature coefficient of -10 mv/°C for temperatures deviating from 25°C is typical (Bauer, 1999). SAE recommends that in construction and industrial machinery, temperature compensation be applied only when the battery and regulator are in the same ambient environment (SAE, 1987).

Alternator Sizing

Alternator performance curves (shown in Figure 10.9 for a hypothetical vehicle) may be used in selecting an alternator for a particular application. Standardized test procedures are available to allow alternator manufacturers to provide comparable test information (SAE, 1999). The full load current versus speed curve, for example, is produced by the standard test methods. The range of load expected for the alternator must be determined to allow selection of a specific alternator. In addition, the speed range of the alternator must be set by the design of the coupling of the alternator to the engine. The alternator speed is generally a set ratio to engine speed through a belt or gear drive.

The alternator must provide the necessary current to supply the expected loads on the vehicle while also supplying adequate current to recharge the battery during the operating cycle. The alternator must also provide the power necessary for all long-term or permanently connected loads while the vehicle is at idle. Long-term or permanently connected loads include electric fuel pumps, headlamps, work lighting, instrument panel lamps, ignition systems, electronic control systems, and charging the battery. Bauer (1999) recommends the alternator should be able to provide 130% of the long-term or permanently connected loads at the engine idle speed for passenger cars. Denton (1997) makes a similar recommendation that the alternator be sized to accommodate 150% of the long-term or permanently connected loads at idle. The alternator should also be able to provide the power to meet the sum of short-duration and long-term or permanently connected loads at the "normal" engine operating speed (Bauer, 1999). The "normal" engine speed would be the average speed over the operating cycle. The short-term loads occur for small parts of the operating cycle, and some duty cycle factor is normally used to multiply the continuous power use of the load to estimate the average use during the operating cycle.

Table 10.2 shows a hypothetical calculation of alternator size. Charging load was estimated assuming a 2kW starter operating at 15 seconds and a charging efficiency of 75%, and a vehicle operating cycle of 60 minutes. At idle, using the recommendations above, the alternator should be able to provide 1.5 × 362 W or approximately 39 amperes at 14 volts. At the normal engine operating speed, the alternator in this example should be able to deliver 421 W or 30 amperes. This example is very limited in the number of loads identified, but illustrates a method that can be used to size the alternator.

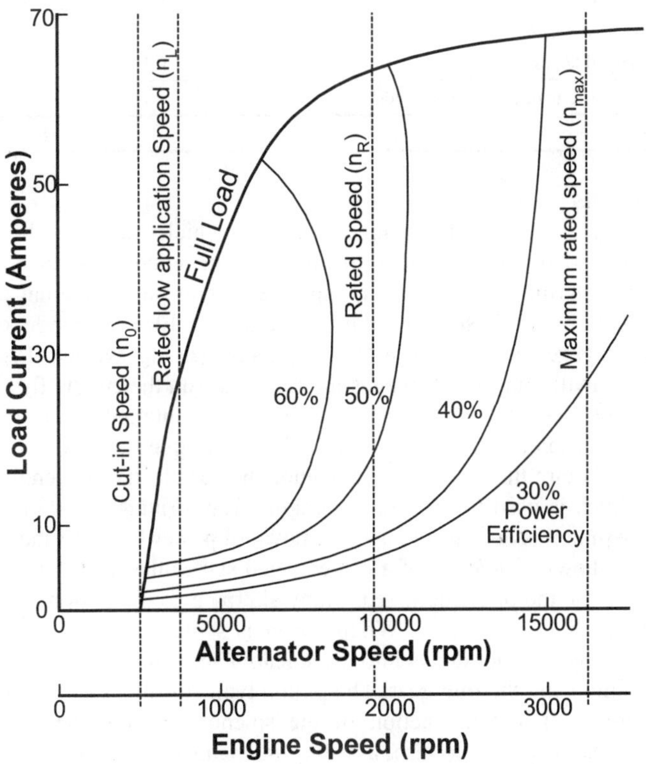

Figure 10.9. A typical performance curve for an alternator.

Table 10.2. Loads for a worst-case operating cycle.

Source			Power, W
Fuel pump			75
Electronic control units (ECUs)			
Engine controller			15
Transmission controller			25
Armrest controller			15
Hitch controller			25
Dashboard controller			15
Work lights			120
Radio			10
Tail lamps			10
Flashing beacon lamp			40
Charging load			12
Total long-term or permanently connected loads			362
Short-term or intermittent loads	Power, W	Use factor	
Ventilation blower	80	0.50	40
Windscreen wipers	60	0.25	15
Turn signal lamps	40	0.10	4
Total short-term or intermittent loads			59
Total			421

Starting Systems

Electric starters are a critical component in most vehicles and must be selected to allow the engine to be started under the most trying conditions the vehicle may encounter. Some starting systems use compressed air or small internal combustion starting engines, however those will not be covered here. Electric starters consist of a DC motor, an engagement system, and an optional gearing system. Typically, the starter pinion (a small spur gear) engages the ring gear (on the engine flywheel) upon the application of electrical power. Some starters have internal planetary reduction gears between the motor and pinion to allow higher torque designs. The DC motor drives the pinion to turn the engine at a high enough speed to start the engine. Then an over-running clutch is used to prevent the engine from driving the starting motor at high speed. The pinion is disengaged upon removal of power from the starter.

Figure 10.10 shows a cutaway of a pre-engaged starter that is designed to engage the pinion before the motor is energized. Many alternate configurations are available, but the major functions are shown. When power is switched to the solenoid pull-in winding, the solenoid plunger draws into the solenoid and pulls the solenoid shift lever to engage the pinion and the ring gear. The pinion typically slides over a helical spiral on the armature shaft and the action of the solenoid driving the pinion toward engagement causes the pinion to turn, assisting in engagement. If the pinion does not engage, the spring behind the pinion is compressed. When the solenoid is fully closed,

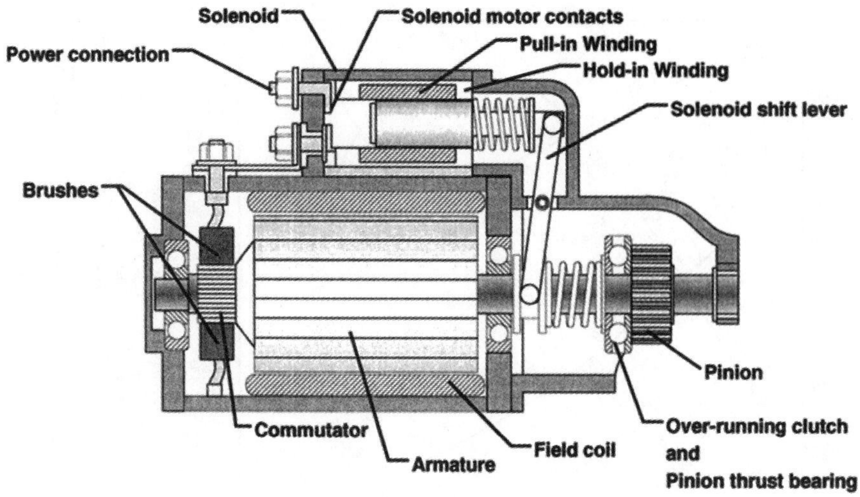

Figure 10.10. Cutaway schematic of a pre-engaged starter.

the pull-in winding is de-energized and the hold-in winding is powered, holding the solenoid in place with much less current than is drawn by the pull-in winding. Some solenoids may have a single winding. The solenoid also connects power to the starter motor when it closes, starting the pinion to turn. The pinion is engaged and the flywheel of the engine is driven at a speed that starts the engine. The pinion-to-flywheel gear ratio may be between 10:1 to 15:1. When the engine starts, the over-running clutch allows the pinion to turn at a high speed without turning the starter motor at that speed. Upon removal of power from the solenoid, the spring on the solenoid retracts the pinion from the flywheel, disengaging the pinion.

The starter motor is a conventional DC motor design, with power being used to generate a magnetic field in the field coils and also in the armature. The current to the armature is supplied from one of the brushes through the commutator, the armature, and the other brush to the frame of the starter. The brushes connect a single coil of many that make up the armature. The particular coil that is connected is dependent on the angular position of the armature, which places a particular commutator contact pair in contact with the brush. The particular armature coil selected by a given position is, through design, at the best advantage through magnetic repulsion and attraction to turn the armature. As the armature turns, the commutator turns repeatedly, connecting the most optimally aligned armature winding. The starter may include one or more pairs of field windings and an appropriate number of armature windings to optimize its performance.

Starters may have the field windings in series with the armature (series-wound), in parallel with the armature (shunt-wound), or may be designed to take advantage of both the series and parallel characteristics (compound winding). Series-wound motors have very high locked-rotor or low-speed starting torque. Figure 10.11 shows starter performance characteristics for a series-wound starter. Shunt-wound motors provide a

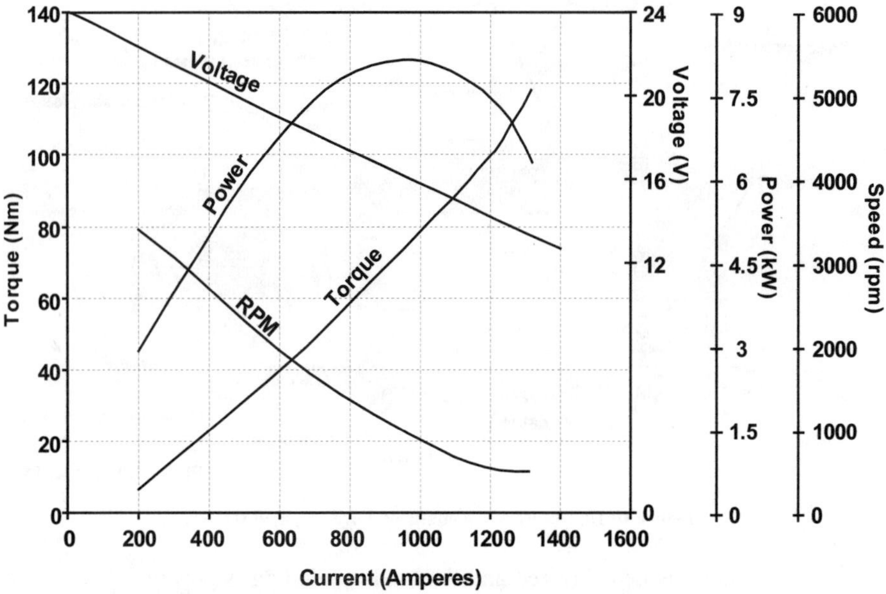

Figure 10.11. Performance characteristics for a heavy-duty starter.

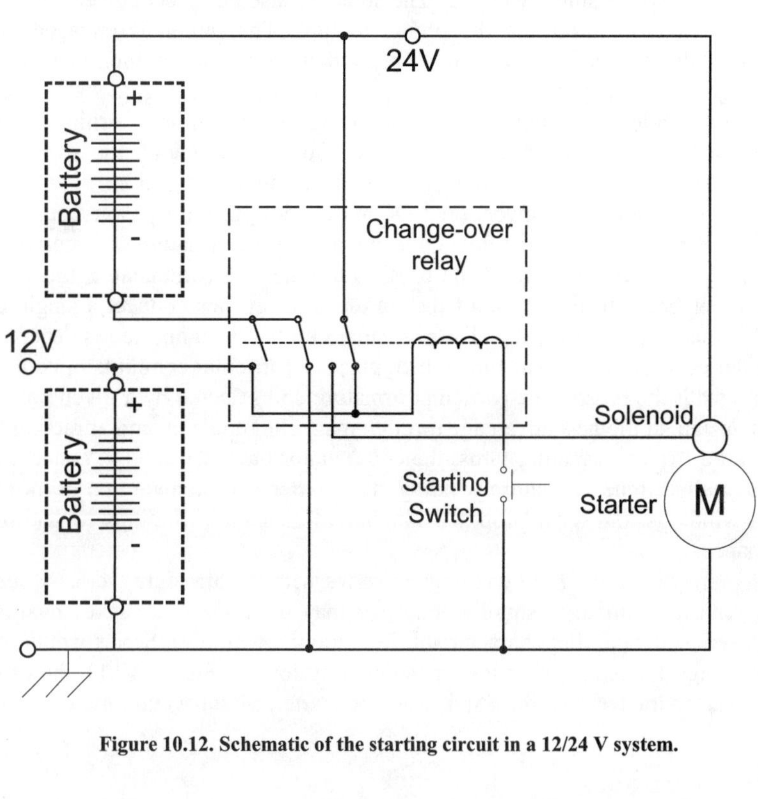

Figure 10.12. Schematic of the starting circuit in a 12/24 V system.

very constant speed over a large torque range and compound windings are sometimes employed in heavy-duty starters to limit the maximum speed of the motor and to allow a low-engagement torque (Denton, 1997). Speed-based switching is used to switch the shunt winding in and out.

Many off-road vehicles use 24 V starting systems. Two 12 V batteries are used in these systems and are normally connected in parallel producing a 12 V system voltage as shown in Figure 10.12. When the vehicle is started, a change-over relay is used to switch the batteries into a serial configuration during cranking and to return the system to a parallel connection after cranking.

Figure 10.13 shows hypothetical engine and starter torque curves for -20°C ambient temperature. These torque curves may be determined through testing or obtained from manufacturers' performance specifications and can be used to aid in selecting the proper starter for an engine under the conditions for which the curves were determined. The engine torque curve is primarily a function of compression resistance and viscous friction of the lubricant, which varies with temperature. The intersection of the starter torque curve and the engine cranking torque curve will provide an estimate of the starter operating speed. This speed must be in the 80-200 rpm range for diesel engines without a starting aid and 60-140 rpm with a starting aid (Bauer, 1999).

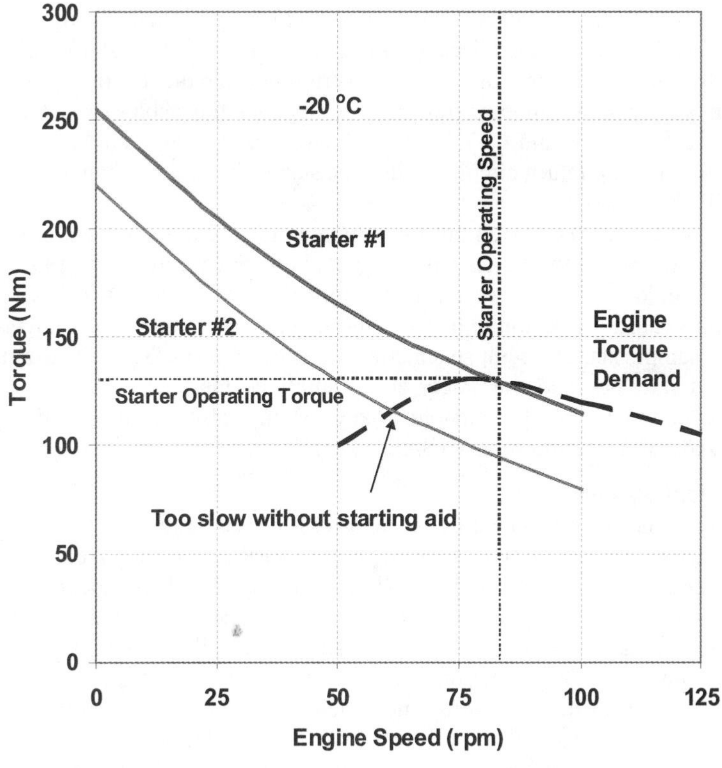

Figure 10.13. Starter and engine torque and speed curves.

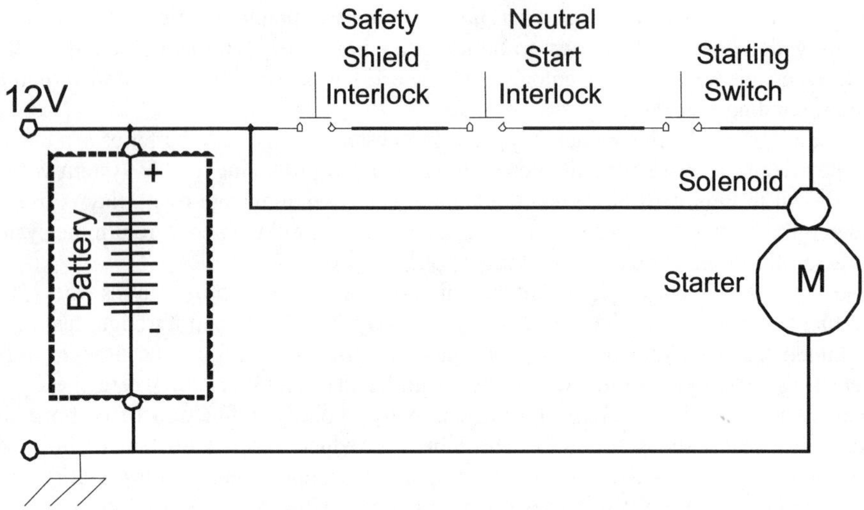

Figure 10.14. Starter circuit with safety interlocks.

Stalled torque has been estimated at three to four times the minimum steady-state cranking torque (Denton, 1997). This rule of thumb allows stalled torque and peak current demand to be estimated during starting. Performance of the starter may also be significantly influenced by the power distribution circuit to the starter. SAE recommends that the maximum voltage drop through starting cables should not exceed 0.1 V for 12 V systems and 0.17 V for 24 V systems in heavy-duty vehicles (SAE, 1996a). Design of the return circuit through the starter frame to the battery should also be carefully considered.

For safety reasons, the design must discourage or prevent inadvertent or deliberate electrical connection at the starter motor solenoid or starter relay (SAE, 1997). Such a connection could cause the vehicle to be started while bypassing safety interlock systems, thus causing a person to be run over or some other catastrophe. Figure 10.14 shows an example starter circuit schematic with safety interlocks. In the example, the neutral start interlock and the safety shield interlock must be closed to allow the starting switch to energize the solenoid. A key switch is also recommended to prevent unauthorized starting of the vehicle (SAE, 1985).

Fuel Control Systems

Electronic control systems are being employed in modern diesel engines to control fuel rate. These systems employ some type of electrically manipulated fuel injector described in detail in Chapter 7, sections 7.6.1 through 7.6.3. An electronic control unit (ECU) is used to manipulate both the timing of injection and the amount of fuel injected during each injection cycle. This ECU is typically known as the "engine" ECU or "engine" ECM (electronic control module). A hydraulically actuated, electronically controlled direct-injection system is illustrated in Figure 10.15. The ECU in this system manipulates the fuel flow through the injector by varying the opening of a solenoid valve controlling high-pressure oil. The high-pressure oil actuates a plunger that acts to release fuel and to intensify the pressure of the fuel. The ECU employs a pulse width modulation technique (PWM) to drive the solenoid valve. PWM is described in more detail in section 10.3.1. Other types of electronically controlled direct-injection systems use PWM modulation to manipulate fuel injection pressure directly during the injection cycle.

The ECU in a fuel control system computes fueling rate and injection timing based on a set of sensed inputs. The relationship between the sensed inputs and the calculated fueling rate is commonly referred to as a "map," but in practice, the effects of some inputs are determined through equations while other effects may be retrieved from a pre-programmed table or map. The computational performance required of the CPU in a direct-injection control system is high as the fueling rate is calculated and varied during each injection cycle. The performance of current microcontrollers limits the complexity of the equations that may be used in the fueling calculations.

The example in Figure 10.15 shows typical input variables used to compute fueling rate and injection timing. Information used by the fuel control system may be obtained through sensors connected directly to the ECU or may be obtained through a communications network attached to the ECU (described further in section 10.3). Examples of engine control parameters can be found in J1939-71.

Lighting

Lighting is one of the primary functions provided by the electrical system. Three major purposes are met by the lighting system:

- Safety lighting functions
- Work lighting
- Instrument lighting

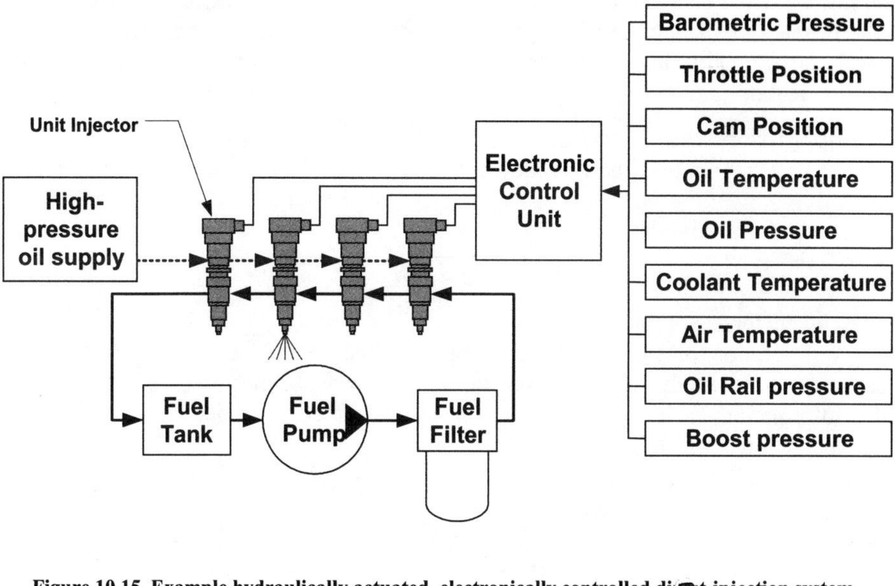

Figure 10.15. Example hydraulically actuated, electronically controlled direct-injection system.

Safety lighting provides illumination to allow the vehicle to be seen and to provide signals that can be seen and interpreted by operators of other road vehicles when the vehicle is operated on-road. This lighting includes beacon lights, tail lights, turn signal lighting, and marker lighting. ASAE S279.11 (ASAE, 2001) specifies the functions and configurations of safety lighting and markings for applications in North America. ASAE S279.11 specifies that agricultural vehicles will include two headlamps, one red tail lamp and two amber flashing lamps. The positioning, brightness, color, and special requirements for these lamps are specified and refer to applicable SAE standards. SAE 1029 and ISO 12509 provide similar recommendations for construction equipment (SAE, 1996b; ISO 1995). Similar standards are adopted into regulations in Europe and by countries worldwide. Exporting vehicles to foreign countries requires that safety lighting be configured to meet the regulations in each country. The process of confirmation that configured vehicles meet the particular country regulations for where it will be exported is known as homologation.

Work lighting illuminates surfaces that are critical to be seen during operation in the dark. It is not unusual to include one or more floodlighting circuits for night operation. Typical designs have included incandescent lamps. Halogen lamps are replacing incandescent lamps due to their longer life and higher efficiency. Some high-intensity discharge systems are now offered as optional equipment. Efficiencies of incandescent lamps are 10-18 lm/W, and halogen lamps are 22-26 lm/W, while gas discharge systems can operate at 85 lm/W (Bauer, 1999).

Power Distribution

The typical power distribution system follows the schematic presented in Figure 10.1. Power within the vehicle is normally distributed from the battery to a fuse or circuit breaker system and then individual wires carry current to the individual loads. The primary challenge after laying out a distribution system is sizing fuses or circuit breakers, sizing wiring, and selecting connectors to assure a safe, efficient, and economical design.

Wiring, Fusing, and Connections

Multiple factors must be considered when selecting cable sizes. The major factors are:

- Mechanical strength
- Temperature rise
- Voltage drop
- Connector requirements

Configuration of the circuit must be specified, including the required current capacity, cable lengths, connector positions, and physical exposure requirements. With this information, each of the potential constraints can be examined and the wire sized and insulation selected to meet all of the constraints.

Mechanical strength requirements constrain the minimum size that may be selected for a cable. Minimum cable sizes of 0.8 mm^2 in harnesses or 1 mm^2 in areas susceptible to physical damage are recommended to prevent failure due to inadequate mechanical strength (SAE, 1998a). Any other special physical requirements should be

Table 10.3. Cable sizes based on thermal circuit protection (SAE J1614).

70% Load Operating Current, A	Circuit Breaker Rating, A	Smallest Acceptable Cable Size, mm^2
5.2	7.5	0.8
7	10	1
10.5	15	2
14	20	3
21	30	5

considered in a design. SAE specifies wire characteristics, including types, conductor size, abrasion resistance, fluid resistance, bend strength, etc., in SAE J1128 for primary wiring and in SAE J1127 for battery cable (SAE, 1995a; SAE 1995b).

To control maximum temperature rise in a cable during a short-circuit condition, cable must be sized for circuit protection. The primary function of the circuit protection device is to protect the cable from failure during a fault. The circuit protection device may also provide some protection for the load device, but this must be a secondary function. The temperature of the circuit protection device must rise faster than the cable temperature. Table 10.3 is a portion of the requirements given in SAE 1614, which specifies methods for sizing cabling on off-highway vehicles. The wire sizes are specified based on calculation of the rate of temperature rise in cables. The operating current is set lower than the circuit breaker rating to prevent the breaker from tripping under the design load.

Requirements for temperature rise of the cable and voltage drop across the cable must also be within design constraints. SAE J1614 (SAE 1998b) specifies sizing methods for both steady and intermittent loads and shows a derivation of the equations for the calculations. For steady loads, and a 30°C rise for cables bundled in a harness or 10°C rise of a single cable in free air, the minimum cable sizes have been calculated and a partial extract is shown in Table 10.4. For intermittent loads of 90 s or less, Equation 10.3 may be used. It estimates the cable size based on a 40°C ambient temperature and allowing the wire to rise to 150°C through resistive heating.

$$\frac{i}{A} = 13.82 \tag{10.3}$$

where

i = current, A

A = conductor area, mm^2

Table 10.4. Steady-state thermal capacity (SAE J1614).

Cable Size, mm^2	Rating, A
0.8	7.5
1	10
2	14
3	20
5	29

Voltage drop in the cable should be evaluated and may be found using Ohm's Law to calculate voltage drop for the current requirement. Resistivity is defined in Equation 10.4 and can be used to calculate the resistance of a wire. Table 10.5 provides basic resistivity information for materials typically used in electrical and electronic circuits. Specific resistance, r_s, is commonly used to calculate voltage drops and is resistivity per unit length of wire as shown in equation 10.5.

$$R = \rho \frac{l}{A} \qquad 10.4$$

$$r_s = \frac{R}{l} = \frac{\rho}{A} \qquad 10.5$$

where

R = resistance, Ω
ρ = resistivity, Ω m
l = length, m
A = area, m^2
r_s = specific resistance, Ω m^{-1}

Resistivities may be compensated for temperature using

$$r_t = r_{base}\,[1 + \alpha\,(T - T_{base})] \qquad 10.6$$

where

r_t = electrical resistivity at temperature T
r_{base} = electrical resistivity at T_{base}
α = temperature coefficient of electrical resistance
T = temperature of the material
T_{base} = temperature of the material at which the base electrical resistivity, r_{base}, is given

Table 10.5. Steady-state thermal capacity.

Material	Electrical Resistivity[a] Ωm × 10^9 or μΩmm	Temperature Coefficient of Electrical Resistance, K^{-1}	Reference
Pure copper	16.7	0.00404	Fink and Christensen, 1996
Copper electrolytic tough pitch	17.1	0.00397	Fink and Christensen, 1996
Nickel	78.0	0.006	Lide, 2001
Gold	24.4	0.0034	Lide, 2001
Steel	104.0	0.005	Lide, 2001

[a] Values given at 20°C.

SAE J1614 details the specific resistance for nominal wire sizes. Minimum wire size to meet a voltage drop constraint may be selected given the acceptable voltage drop, the current requirement, and the cable length. Specific resistance for the cable may be calculated using Equation 10.7 as given in SAE J1614. A voltage drop constraint must be set based on the particular situation. In some cases specific constraints are recommended as in the case of battery cables supplying starters (see the section above on starters) (SAE, 1996a).

$$r_s = \frac{\Delta V}{Li} \times 10^6 \qquad 10.7$$

where

r_s = specific resistance, μΩ/mm
ΔV = acceptable voltage drop, V
L = cable length, mm
i = current, A

Table 10.6 shows specific resistances for a partial selection of wire sizes as given in SAE J1614. A wire size with equal or lower resistance to the allowable limits should be selected to meet the voltage drop constraint for the portion of the circuit being considered. Voltage drop should be calculated for the complete circuit including the conductors, whether cable or a frame return, connectors, and other components in the circuit.

Voltage drop and temperature rise in connectors are an issue and should also be considered in a distribution system design. SAE J1614 also makes recommendation regarding contact resistance and specifies that contact resistance in connectors should not exceed 0.020 Ω after endurance testing. This value can be used in estimating voltage drops in connectors.

Table 10.6. Specific resistance of copper cables (SAE J1614).

Cable Size, mm^2	Specific Resistance, μΩ/mm
0.8	21.496
1	14.370
2	9.088
3	5.676
5	3.576

Implement Connection

The primary objective of an off-road vehicle is generally to power some type of implement attached to the vehicle. Supply of lighting and electrical power to that implement is an important function. A connector is currently specified by ASAE in S279.11 for electrical connection of equipment at the hitch of agricultural tractors (ASAE, 2001). This connector provides definition for pins to drive flashing directional signals, work lights, and tail lamps. Two of the pins on the connector are defined as “auxiliary” and no power is provided for general implement use. The connector generally conforms to SAE J560 (SAE 1998) with the exception of the definition of the auxiliary pins. A new connector has been defined for hitches as a part of ISO 11783 (described further below) that includes power for implements, electronic control units, and a communications bus, which could be used for conveying lighting messages.

10.3 Electronic Systems

Manufacturers of agricultural equipment are increasingly turning to electronics to provide products with improved functionality, productivity, and performance to customers. Electronic content in agricultural equipment has increased and further increases are likely. Designers of off-highway vehicles need to have a basic understanding of electronic systems. Specialists are typically employed to design and develop the electronic components of vehicles, but a basic understanding of the systems is necessary to allow electronic components to be integrated and work effectively.

10.3.1 Electronic Control Units

Electronic control units combined with sensing and actuation elements make up the fundamental components of modern electronic systems. Some modules exist that are composed of only discrete electronic components, but microcontroller-based (a microprocessor with integrated peripherals) systems are being developed to replace even the simplest of these systems. Microcontroller-based systems offer greater functionality and can have costs that rival systems using discrete-based designs.

ECUs are composed primarily of microcontrollers, electronics to interface the microcontroller to inputs and outputs, input power conditioning, enclosures, and connectors. These computer-based modules run a program that has been stored in them by the manufacturer, and have the capability to input digital (ON or OFF) and analog (continuously variable) signals, which are converted to numerical values for use in the program. Calculations and decisions can be made based on the values of the input parameters. The ECU can then produce outputs, generally in a digital form to cause an action, for example, to set the position of a hydraulic valve or turn a motor ON or OFF. The stored program within the ECU is normally placed into either a UV erasable, programmable, read-only memory (EPROM) or an electrically erasable, programmable, read-only memory (EEPROM or Flash memory). The latter can allow update of the program within the ECU while on-board the vehicle.

ECUs have limited computational speed and this characteristic is revealed in the rate at which ECUs can input information, process the information, and produce an output. These three steps form the process elements of a control cycle in a controller. In general, typical ECUs used today do not process a control cycle faster than 1ms with 10 ms being more typical. For most control systems, these sample rates are effectively continuous in time. Care must be taken in applying ECUs in systems requiring high sample rates, for example in fuel control during an injection cycle, etc.

Signal conditioning is employed to convert analog inputs in the form of voltages and currents to voltages in the range appropriate for analog to digital (A/D) converters. A/D converters, either a part of the microcontroller or as a peripheral device, convert voltage inputs into binary numbers used within the microcontroller. The resolution of each number depends on the number of binary digits the A/D can provide. Typical A/D converters provide 8-, 10-, or 12-bit resolutions and these resolutions would provide numbers with maximum numerical resolutions of 1 part in 256, 1024, and 4096 respectively.

Analog signals may be in the form of a voltage level referenced to ground (the negative battery terminal) or may be a voltage difference without reference to ground. The former is known as a single-ended signal, and the latter as a double-ended signal. Both types of signals are usually conditioned by processing through an amplifier before being provided to the A/D converter. Double-ended signals require special amplifiers designed to condition differential signals. A/D converters are typically designed with multiple channels and the number of analog signals and their types are critical in the determination, if the A/D in a particular ECU is adequate.

ECUs provide for digital (ON or OFF) I/O capabilities, but some interface circuitry is usually required. The interface is normally provided within the ECU and is designed to suit each input. Many sensing devices produce digital signals that require little signal conditioning and can be directly input to the ECU. Figure 10.16 shows a typical circuit employing a "pull-up" resistor to convert a switch closure input to a logic signal. The logic signal is normally supplied as a voltage and either the supply voltage (12 V in the example) or the signal to the ECU is processed to convert the signal to the logic levels require by the microcontroller. This level is generally a nominal 0 and 5 V signal. Arcing during the switch closure or opening can result in "bouncing" of the signal between low and high logic levels. In systems where switch closures are counted, some type of de-bouncing circuit is used.

Digital outputs are typically used to drive valves, actuators, and indicators. Many different types of circuits are used. A common driver circuit for a digital output would drive a solenoid or motor coil. Figure 10.17 shows an example of a low-side driver circuit for this application. The power-MOSFET (metal oxide semi-conductor field effect transistor) acts as a switch, becoming conductive when the signal from the ECU is high (nominally 5 V) and open or non-conductive when the signal from the ECU is low (nominally 0 V). The MOSFET may be used rather than a conventional bi-polar

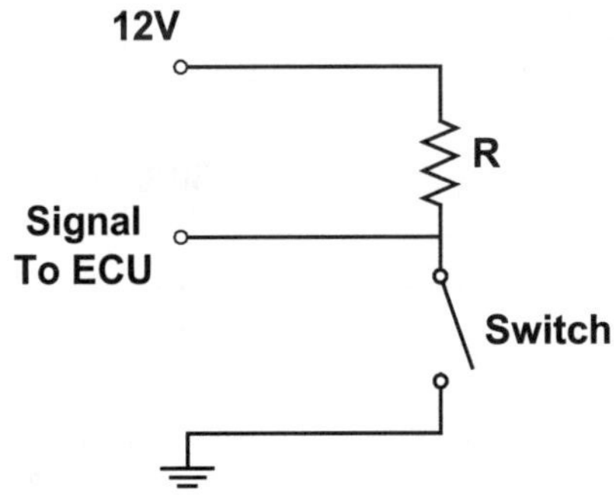

Figure 10.16. Example circuit for conversion of a switch closure to a digital signal.

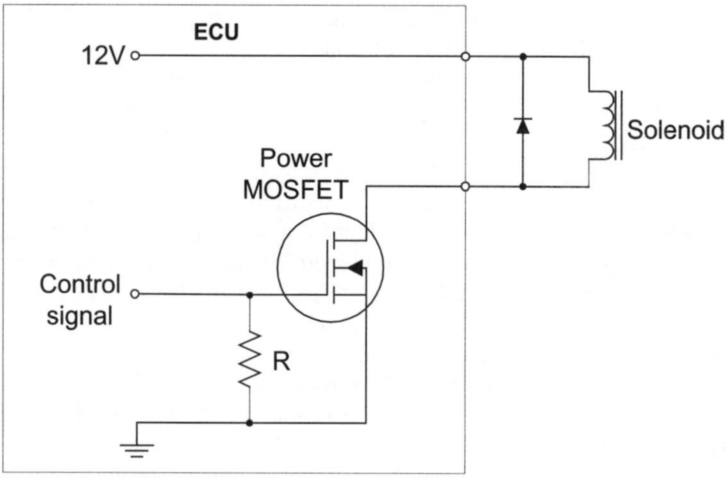

Figure 10.17. Low-side solenoid driver circuit.

junction transistor (BJT) because of its high "ON" conductance. The diode is used to control the voltage spike and dissipate current generated by the collapsing magnetic field in the solenoid coil when the solenoid is turned off. The low-side driver circuit has the feature that the high side of the solenoid is always powered. This can be a safety issue if faults occur and an alternative high-side design is also often used.

The circuit in Figure 10.17 is also commonly used to produce an analog output. This is done by driving the circuit at an appropriate frequency with a "PWM" or pulse-width modulated signal. The duty cycle, the percentage of time that the signal is in the "ON" state over a single cycle, is controlled. Figure 10.18 shows an example of PWM signals at three "ON" duty cycles. The frequency is normally set at a level high enough that pulses are averaged by the system being driven and not observed in the response.

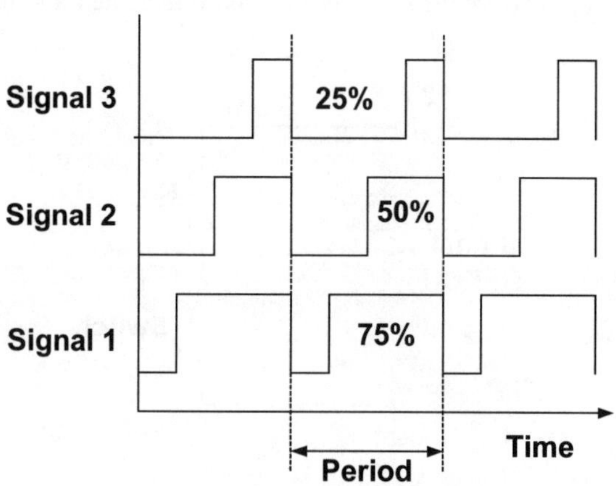

Figure 10.18. Pulse-width modulated signals.

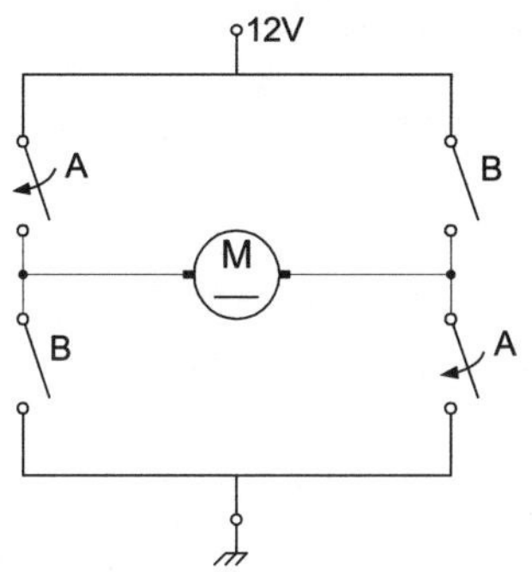

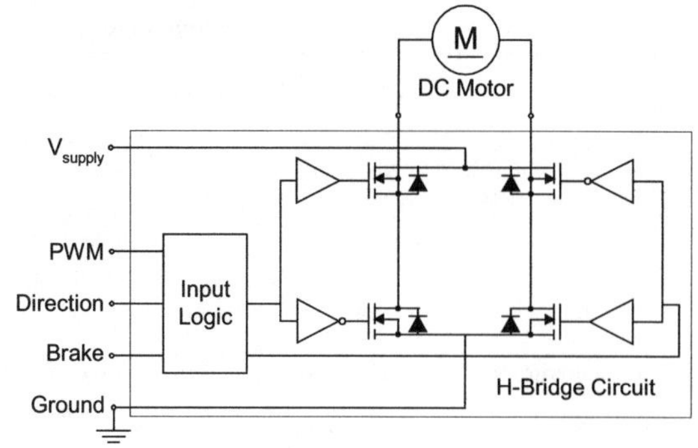

Figure 10.19. Schematic model of an H-bridge circuit (left) and an integrated circuit implementation (right).

When applied to a DC motor for example, the amount of power delivered to the motor is approximately proportional to the duty cycle and motor speed can be varied by varying the duty cycle. Use of PWM is the dominant method used to produce an analog output primarily because of the low cost of components and simple efficient implementation with microcontrollers. Other methods including the use of D/A converters exist but are not commonly used on off-road applications.

Figure 10.19 shows an H-bridge circuit typically used for bi-directional control of DC motors. The switches in this device are typically implemented similarly to the solenoid driver in Figure 10.17. When the switches marked "A" are closed, the motor runs in one direction, and the other direction when the "B" switches are closed. This type of circuit can also be used for PWM speed control in two directions.

10.3.2 Sensors

Sensors are a critical component in embedded electronics. Inputs to the embedded electronics including those provided directly by the operator are typically provided by sensors. The most common sensed parameters and sensor types on off-road vehicles include those outlined in Table 10.7.

The common methods used in sensors to transmit a signal from a sensor to an ECU are through analog, frequency, or a digital signal. Voltage or current analog signals are common where the voltage or current levels are proportional to the parameter being measured. Sensors that produce a voltage signal are usually powered and produce a voltage signal on a third conductor. Devices that produce current generally derive power from the current supplied externally. These devices usually regulate the current flow proportional to the parameter being measured. These devices are desirable where it is important to eliminate the effect of voltage drop through connectors or wiring, for example, where the sensor is a long distance from the ECU. The signal is current-based and is not a function of voltage drop in the circuit. Another method that allows elimination of the interference of voltage drops across sensor circuits is to use frequency to transmit a signal. Frequency signals are commonly a square wave with nominal voltage levels of 0 and 5 volts. The frequency of the square wave conveys the

Table 10.7. Common sensors for off-road vehicles.

Parameter	Sensor Type	Signal Type	Typical Application
Pressure	Piezoresistive	Single-ended voltage	Oil pressure
	Solid-state micro-machined monolithic silicon piezoresistive	Single-ended voltage	Engine manifold pressure, oil pressure
Temperature	Thermistor	Single-ended voltage	Water temperature, oil temperature
	Solid-state silicon junction devices	Single-ended voltage	Water temperature, air temperature
	Solid-state silicon junction devices	Serial digital	Water temperature, air temperature
Rotational position	Potentiometric	Single-ended voltage	Accelerator pedal position
Rotational speed	Incremental encoder	Frequency	Engine speed, wheel speed
Linear position	Potentiometric	Single-ended voltage	Engine fuel pump rack position
Linear speed	Doppler RADAR	Frequency	Ground speed
Level	Potentiometric	Single-ended voltage	Fuel level
Force	Strain-gauge load-cell	Double-ended voltage	Draft sensing

value of the signal. This type of signal is insensitive to voltage drops and electrical noise. Frequency signals are also convenient for use with microcontrollers as the signal can be input through a digital input port and simple counting over time can be used to reduce the signal to a numerical value.

Double-ended voltage signals are typically produced by sensors using elements in a bridge arrangement. Figure 10.20 illustrates a bridge with resistive elements. Changes in the resistance of any of the elements causes a voltage difference at the output. The advantage of the bridge is that equal changes in adjacent elements result in no change in output. This effect can be used to compensate for temperature or other interferences. The adjacent elements are both exposed to the interference, temperature for example, and only one of the pair is exposed to the effect of interest, strain for example. The temperature effect is nullified in this example. These devices are typically used in strain-based transducers including pressure and draft. An important characteristic of a double-ended signal is that neither of the output conductors is at ground potential and the information in the output signal is the voltage difference between the conductors.

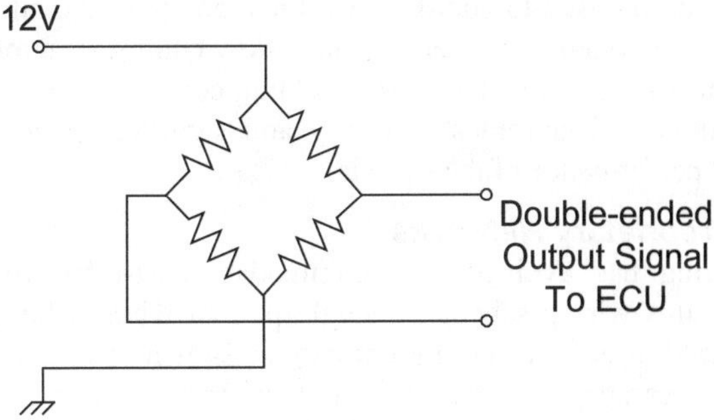

Figure 10.20. Double-ended signal from a bridge-type sensor.

Smart sensors and actuators are an alternative to using conventional methods for sending signals between ECUs and sensors. Smart sensors and actuators employ microcontrollers directly in the sensor. The microcontroller converts the sensor signal into a digital form within the sensor. The sensor data are then sent as a serial digital bit-stream to the ECU similar to that shown in figure 10.21. The reverse process occurs in a smart actuator. The devices sending and receiving the information typically use timing to allow interpretation of the signals as bits. A start bit indicates the beginning of data of known length and each bit is read at a fixed time after the start bit. Stop bits are used to enforce space between the messages so the next message may be distinguished from the last. Most smart sensors are designed to operate on a network as described in the next section.

10.3.3 Controller Area Networks

A natural consequence of adding electronic components to agricultural equipment has been the realization of the advantages of allowing the components to communicate with each other. A hitch controller on a tractor, for example, may communicate with a transmission and engine controller to allow optimized performance. Electronic

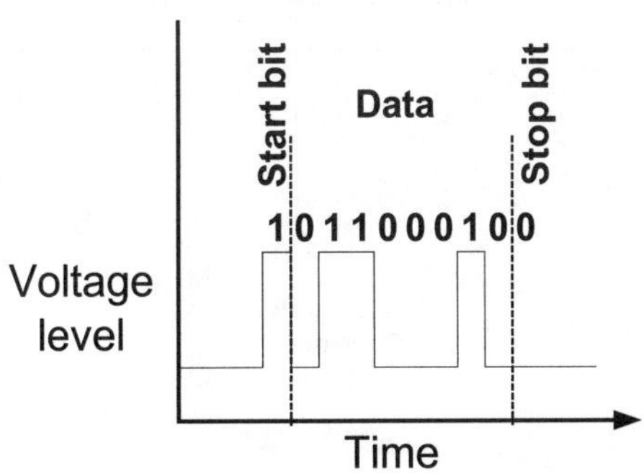

Figure 10.21. Data encoded in a serial bit-stream.

communications can be used to coordinate vehicle components, allow information to be shared among components of a vehicle, and allow control systems to be distributed across components of a vehicle. The cost of adding communications is a small part of the cost of stand-alone electronics, but may add significantly to the functionality, productivity, and performance of the vehicle.

Introduction/Proprietary Networks

Multiplex wiring has evolved to accommodate cost-effective communications among ECUs. In this wiring scheme, a single pair of wires–a bus–is shared among controllers and used to carry a serial bit-stream as shown in Figure 10.22. Groups of bits are sent as messages with the first bits of the message forming a message identifier and the latter composing data. The protocol embedded into the ECUs requires the ECU to check the bus to assure no other ECU is using it before transmitting. The strategy works because enough free time exists on the bus for all of the ECUs to pass their messages without significant delays.

Multiplexed wiring systems based on proprietary designs have been used in off-road equipment for many years. Early examples include the Chrysler Collision Detection (a CCD-based network) used on John Deere equipment. The Deere 7000 series tractor introduced in 1992 incorporated a network that had as many as five ECUs controlling various aspects of the tractor. Deere has used this network in various models of their equipment. New Holland has used a CAN-based network on their Genesis™ series tractors in 1994 (Young, 1994). The Controller Area Network (CAN) protocol was developed by Bosch (1991) for use in automobiles. Genesis™ uses four ECUs to handle the right-hand console, instrument cluster, transmission control, and draft control. Caterpillar's Challenger™ 75 and 85 series tractors used a SAE J1587 data link (Luebbering and Smith, 1993). Early applications of networks in implements have also been reported and highlight the need for integration of implement and tractor communications. Flexi-Coil used an SAE 1708-based network on their air seeder monitor and control system in 1993 (Weisberg et al., 1993). Flexi-Coil's system included cab and remote implement ECUs with provisions for as many as 18 ECUs. The application of network-based control systems for product application was

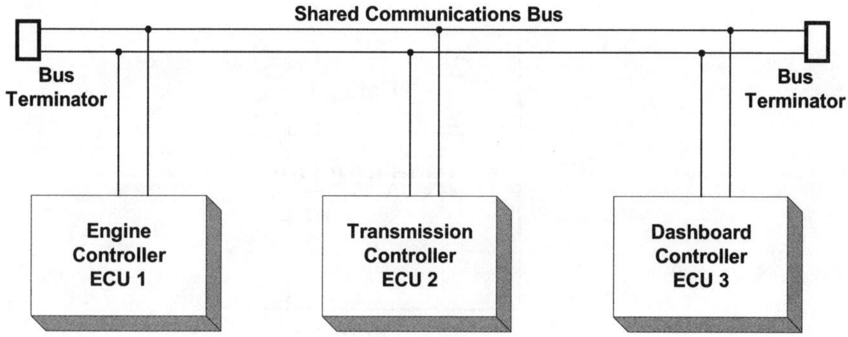

Figure 10.22. Multiplexed wiring.

patented by Ag-Chem Equipment Company, Inc. in 1995 (Monson et al. 1995) and was later introduced in their Falcon™ series application systems. In Germany during the early 1990s a standardized agricultural bus, LBS, was introduced. This bus was standardized as DIN 9684 and uses CAN version 1.2 (Auernhammer, 1983; Schueller, 1988). LBS is the basis of the network integrated into the Fendt Vario™ tractor introduced in the late 1990s.

ISO 11783/SAE J1939

The need for standardized communication networks for off-road vehicles became obvious once networks became popular. Standardized networks allow different manufacturers' ECUs to communicate and function on a common network. This need exists where engine, transmission, hitch controllers, etc. from different manufacturers are integrated into a single vehicle. In addition, the need in agricultural systems to connect different implements from various manufacturers to tractors of different manufacturers with a common network obviated a similar standardization requirement. SAE J1939 was developed to support component level communications in on-road and off-road vehicles (Stepper, 1993). The German DIN 9684 was developed to support agricultural implement to tractor-mounted display and control communications. An international standard, ISO 11783 evolved to support comprehensive standardized networks for agricultural tractor and implement systems. The various parts of ISO 11783 are derived from SAE J1939 and the German DIN 9684 standard. ISO 11783 relies on SAE J1939-derived components for the basic communications structure with application components (virtual terminal and task control) largely derived from DIN 9684. ISO 11783 is applicable in agricultural applications while construction and other applications are supported through SAE J1939 (Stone et al., 1999). ISO 11783 is described below and the description can serve as a basis for understanding the many proprietary networks. Further description of other types of networks can be found elsewhere (Stone, 1987).

ISO 11783 and SAE J1939 standardize a multiplex wiring system as described above, based on the CAN version 2 protocol. This protocol uses a prioritized arbitration process to allow messages access to the bus. When two messages are sent at the same time, their identifiers are imposed bit-serially onto the bus. The bus must be designed to allow dominant bits to overwhelm recessive bits when both are applied simultaneously by different ECUs. No conflict occurs as long as the ECUs are sending the same bits, but when one sends a recessive bit while the other sends a dominant bit, the bus state is dominant. The ECU sending the recessive bit must sense the bus is at a dominant state when the bit was sent and must cease transmitting the message at that time and retry the next time the bus becomes idle. This strategy allows more dominant identifiers, those with a lower value, to have a higher priority on the bus. To allow this feature to work properly, CAN protocol controllers synchronize messages at the beginning of each transmission to assure bits are aligned. The result is that ISO 11783 provides a communication system where ECUs share a communications link and messages at any point in time are allowed access to the bus based on their priority.

10.3.4 ISO 11783 Overview

ISO 11783 has been written to support applications of networks in agricultural equipment. The standard supports both self-propelled systems and tractor-implement systems. A tractor-implement model is assumed throughout the documents, with the recognition that the same or a simpler design can be used in self-propelled systems. Figure 10.23 shows in schematic form a simplified ISO 11783 network superimposed on an agricultural tractor and implement background. A network with no master controller has been defined. The network is composed of two communication busses, a tractor bus and an implement bus. The implement bus spans the tractor, crosses the hitch, and spans implements. The implement is shown in this schematic with an implement sub-network. The busses are interconnected with network interconnection ECUs, the tractor ECU and -one labeled "implement ECU and implement bridge." The characteristics of the tractor ECU are specifically described in ISO 11783 Part 9. A task controller and management computer gateway and virtual terminal (labeled "VT") are shown connected to the implement bus. The virtual terminal is described in Part 6 and the task controller and management computer gateway is described in Part 10 of ISO 11783. The task controller is an ECU, which normally resides on the tractor and is used to provide commands to implements to accomplish some task. An example might be to provide the commands of a prescription in a precision farming operation. The management computer gateway portion of the task controller and management computer gateway contains an interface that is compatible with the management computer and allows data to be exchanged between the task controller and the management computer. Standardized communications are defined between the task controller and implements and between the task controller interface and applications software on the management computer. The interface between management computer and task controller is not standardized.

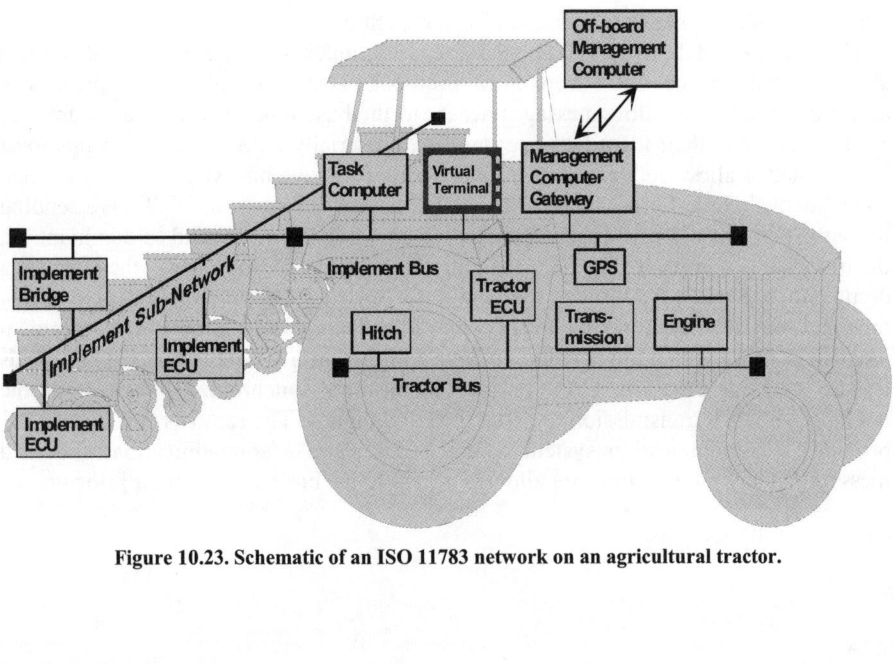

Figure 10.23. Schematic of an ISO 11783 network on an agricultural tractor.

The network has messages defined to allow communications between any of the components illustrated in Figure 10.23. An example might be communication between the task controller and the GPS ECU. Navigational messages are defined and allow positional information to be received by the task controller. In the same sense, messages are defined to allow the engine ECU to provide a current torque curve to the transmission. Information sharing and control messages are supported. Some messages are defined with repetition rates of 100 messages per second. This type of message utilizes approximately 5% of the bus capacity, with conservative maximum average bus use targeted at approximately 35%. Many messages are defined with various repetition rates, and careful planning is necessary to prevent overuse of the available bus capacity. The tractor ECU provides filtering of messages between the tractor and implement bus. This filtering is necessary to prevent heavy traffic on either bus from overloading the other. Support of precision farming applications across the implement bus has been included in ISO 11783 as well as support for implement and tractor coordination.

Flexible expansion of the communications in ISO 11783 has been implemented. The network supports the use of proprietary communications simultaneous with standardized messages. Manufacturers are free to implement enhanced control and information systems within the proprietary support in the standard. A process has also been included in the standard to accommodate requests to expand the message set beyond that currently defined.

10.3.5 Components

A basis for understanding ISO 11783 networks can be gained by examining the components that comprise a network. Logical and physical components of the network are described below.

Wiring and Connectors (The Physical Layer)

A *twisted quad* cabling system was developed especially for ISO 11783 networks. Selection of a bit rate to be carried in the cabling system had considerable influence on the design. 125 K bits/s was considered roughly the fastest rate that could be handled without a shielded cable while producing acceptable EMC (electro-magnetic compatibility) performance. A 250 K bits/s shielded twisted pair specification was available in the SAE J1939 documents (SAE J1939-11), but shielding was regarded as unacceptable by manufacturers. DIN 9684 included a 50 K bits/s unshielded design, but the bit rate was considered too low for the applications anticipated by manufacturers. An unshielded 250 K bit/s design with carefully selected voltage slope (dv/dt) and current control in the data lines was proposed by Deere. This design was proven and was adopted as Part 2 of the 11783.

The twisted quad cabling system uses four wires enclosed in a jacket as shown in cross-section in Figure 10.24. Two of the wires are used to carry data, CAN_H and CAN_L, and two (TBC_PWR and TBC_RTN) are used only to provide power to terminators at the end of the bus as shown in Figure 10.22. The terminator requirements and the method of powering the terminators are rigorously specified in the standard. The TBC (terminating bias circuit) lines are included in the cable at all points, but are not connected within an ECU.

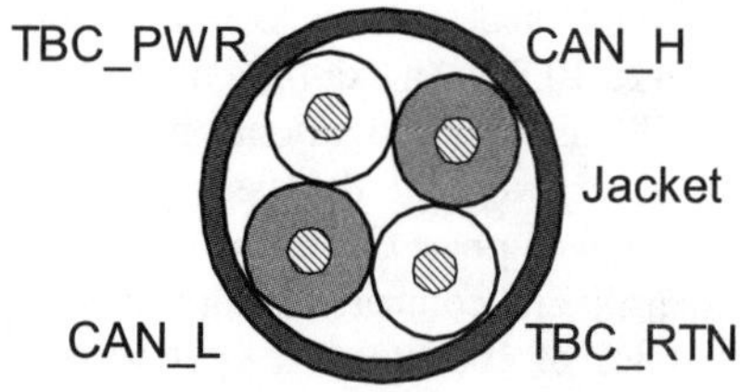

Figure 10.24. Cross-section of the twisted quad network cable specified in ISO 11783 (ISO, 2002).

The Part 2 specification has been written to allow use of ISO 11898 integrated circuit bus drivers to be used in ECUs. These integrated circuits are readily available at a reasonable cost. This feature provides the opportunity for ECUs designed to meet SAE J1939-11 to be connected and operate within an ISO 11783 bus system, though voltage slope control must be set within these ECUs to prevent EMC problems.

Termination of the bus at both ends is a requirement. This presents some problem at the hitches of a tractor. When an implement is unhitched and there are ECUs operating on the tractor portion of the implement bus, unhitching could result in the tractor portion of the bus being unterminated. A special automatic terminating bus breakaway connector has been developed to solve this problem and is specified in the Part 2 document. This connector is designed for the tractor at hitching points, and automatically applies termination when an implement is unplugged. This problem also exists where several implements are hitched together and requires implements that provide a hitch for other implements to also provide an automatic terminating hitch connector.

The Part 2 document specifies three standard connectors; bus breakaway, diagnostics, and bus in-cab connectors. Figure 10.25, taken directly from Part 2, shows a connector-use map. The physical specifications of the bus in-cab connector are given. It can be used for connecting components to the bus in the cab of a tractor, for example, virtual terminals and task controllers. A diagnostics connector is also specified in Part 2 and is provided for diagnosis of both the tractor bus (if it exists) and the implement bus.

The selection of a bus topology as shown in Figure 10.23 places a restriction on some applications. The maximum length of a single segment of the bus is 40 meters. ECUs may be connected to the bus at any point (not closer than 0.1 meter of each other), but the length from the bus to the ECU must not exceed 1 meter. This topology prevents configuring the network as a "T" or cross. An example might be on an implement where the bus traverses it from front to back, and there is a need to extend the bus more than 0.6 meters side to side. In this case either a serpentine arrangement of the bus or a network interconnection device must be used with an implement sub-network as shown in Figure 10.23.

Constraints exist for the number of ECUs that may be connected on a single segment of the bus. A maximum of 30 ECUs may be connected on a single segment. Multiple segments may be interconnected with bridges allowing up to 254 ECUs in a system.

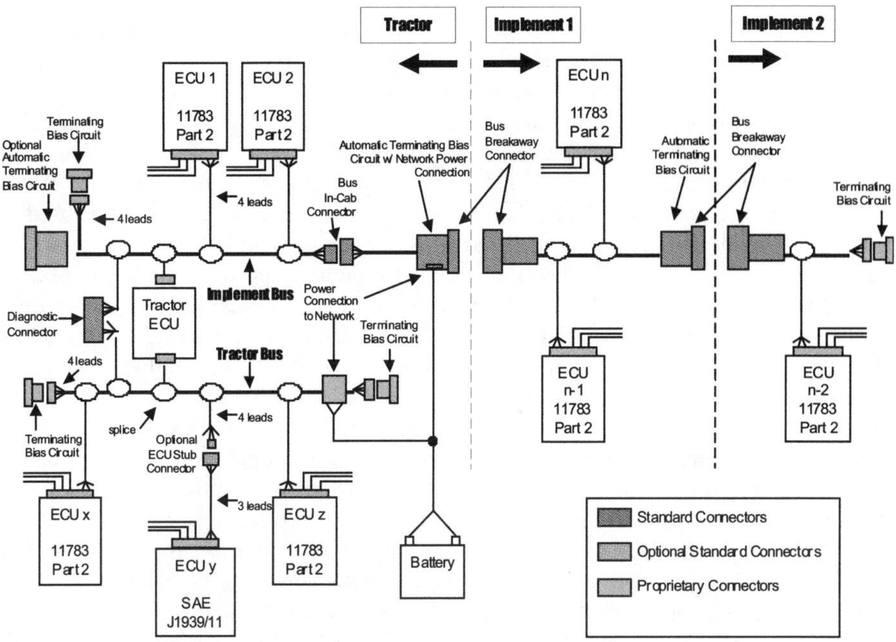

Figure 10.25. Connector use within ISO 11783 Part 2 (ISO, 2002).

Message Structure (The Data Link Layer—29-Bit CAN)

ISO 11783 is based on the use of the CAN 2.0b 29-bit protocol (Bosch, 1991). This protocol is designed to send bits serially as described earlier. A single frame or collection of bits sent by a CAN controller is shown in Figure 10.26 and consists of an identifier and a data field. Many additional bits (not shown in Figure 10.26) are defined in the frame for use in the CAN protocol controller, including cyclical redundancy check bits, which are used to allow receivers to determine if the data frame sent was received without bit errors. Undetected errors are confined in the CAN protocol to a probability of less than 4.7×10^{-11} multiplied by the message error rate.

ISO 11783 defines the interpretation of the 29 bits in the identifier of CAN frames as well as the interpretation of the data. Two types of identifier structures, or protocol data units (PDU), are defined. Figure 10.27 shows a schematic of the definition of the identifier bits for both types of identifier structures. In both types of PDUs, the least

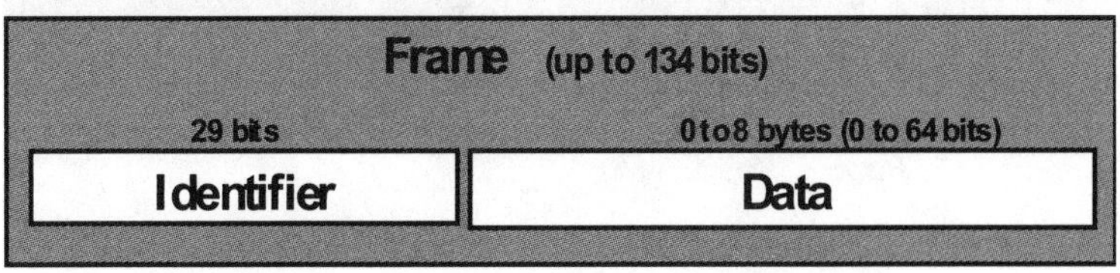

Figure 10.26. Components of a CAN frame.

significant 8 bits define a "source address." This value is a physical address of the ECU sending a message. The first three most significant bits of the identifier are defined as independent priority bits. Recommendations are provided for the values of these bits in the standard, but may be adjusted by a manufacturer in a particular application. The difference between the two PDU types is the inclusion of a destination address in PDU type 1. This message type allows the message to be sent to a particular ECU based on physical address. Addresses 0 through 253 may be used by an ECU, while 254 (the null address) must be unused and 255 as a destination indicates a message to all ECUs (global). The remaining portion of the identifier in each PDU is used to identify the parameter group that is contained in the data field, that is, the content of the data field that may be defined to contain multiple parameters. The PDU type 2 messages do not use a destination address and use this field (group extension) to create an expanded number of parameter group identifiers. The R (reserved), G (page), PDU format, and if PDU type 2, the group extension field, are combined to compute a "parameter group number," a unique numeric identifier for each group of parameters that may be identified and contained in the data field.
ISO 11783 messages are defined to allow any parameter group to be sent from any ECU. The inclusion of a source address in the identifier is used to guarantee uniqueness of the identifiers in the system, a requirement of CAN. This requires that addresses of ECUs in the system be set to unique values.

General purpose messages are defined in ISO 11783 to allow a request to be made for a particular parameter group. The use of the remote transmission request (RTR) feature of CAN is not defined for ISO 11783. A general purpose message is also defined to allow acknowledgment or negative acknowledgment of a message. The use of acknowledgment is defined for each message.

Messages in ISO 11783 are normally composed of a single CAN frame, but can be composed of multiple frames. Two types of multi-frame messages (transport protocol) are defined: 1) a broadcast announce mode message, where an initial frame is sent announcing the specifications on the frames to follow, followed by the rest of the frames, and 2) a connection mode message which is sent to a specific destination and allows the receiver to control the flow of messages being sent.

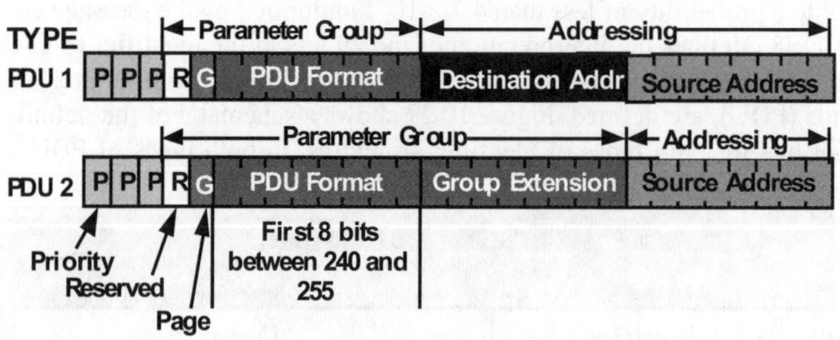

Figure 10.27. Interpretation of the identifier of CAN frames in ISO 11783.

Interconnection Structure (The Network Layer)

Though not a requirement, most ISO 11783 systems will have interconnected bus segments. The tractor-implement system shown in Figure 10.23 is an example. A bridge must be used if transparent communications are to occur on interconnected ISO 11783 bus segments. Bridges use a protocol controller to connect to each segment and pass messages between the segments. A repeater, which may not be used, simply echoes the electrical signal from one segment to the other. Limitations on bit timing in ISO 11783 prevent use of repeaters. The tractor ECU in an ISO 11783 network may perform as a bridge, but normally has special requirements defined in the Part 5 document.

ISO 11783 defines filtering capabilities for network interconnection ECUs. Provisions are made for these ECUs to prevent messages from being passed from one sub-network to another. Part 4 defines the structure and makes provisions that allow messages to be used to configure the filtering of messages. This capability is important in controlling network loading of bus segments. An example can be seen in Figure 10.23 on tractors where a tractor bus and an implement bus coexist. The tractor bus is likely to have heavy loading with engine, transmission, and hitch messages. Most of these are not of interest on the implement bus and can be filtered from the implement bus by the tractor ECU. In the same sense, messages to control set points or seeding rate on a seeder are generally not of interest on the tractor bus. Traffic partitioning performed by network interconnection devices can be used to control bus loading. Generally, manufacturers will need to configure network interconnection ECUs to optimize performance of their systems.

Some constraints exist on the way network segments may be interconnected and on the timing in network interconnection ECUs. Any segment may not be connected to another segment in more than one place. This precaution prevents loops in the network and the associated duplicate messages.

Addressing, NAMEing, and Initialization

ISO 11783 Part 5 includes requirements for a unique NAME to be included within each ECU. The NAME must be unique within a system and is a 64-bit value defined as shown in Figure 10.28. The NAME is divided into two distinct parts, an upper 32 bits that is used to provide a functional name and a lower 32 bits that provides a unique code based on the ECU manufacturer and an identity number. Manufacturers must obtain a manufacturers code by request to the ISO 11783 working group (ISO/TC23/SC19/WG1) in order to build ECUs compatible with the standard. Manufacturers may then assign unique identity numbers to each of the ECUs made.

In the initial design of the network, a requirement for an address was desirable to allow peer-to-peer operation but provide a coordinated method to assure unique message identifiers. An 8-bit address was selected, allowing 254 ECUs to be connected in a network. Simple assignment of addresses to all of the possible ECUs based on their type appeared impossible, particularly for implement systems. Many components in a network can have an address assigned, a primary engine, or a primary transmission for example, but ECUs that are temporarily connected would potentially have conflicting addresses assigned. This problem was managed by providing both

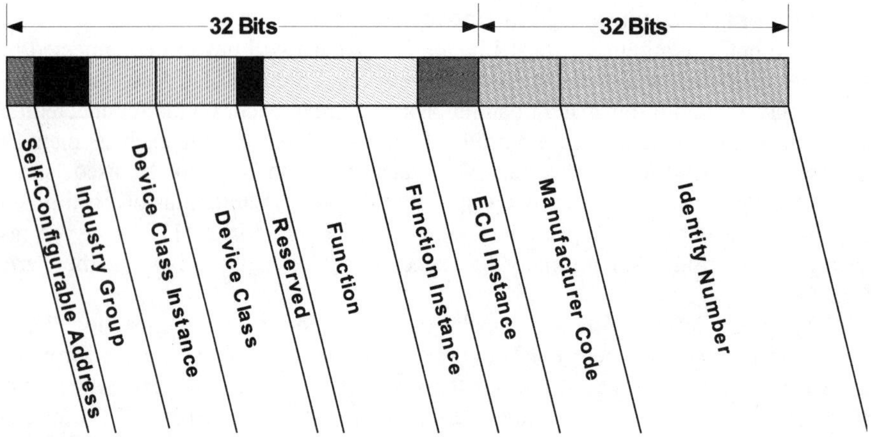

Figure 10.28. ISO 11783 NAME structure.

self and non-self configuring ECUs. ECUs in the network attempt to claim an address upon power-up. In the case of ECUs that are self-configurable, if they happen to claim a used address, an arbitration process occurs and the ECU with the lower valued NAME retains the address. Non-self-configuring ECUs always win arbitration with a self-configuring ECU because the NAME includes a self-configuring bit that assures self-configuring ECUs have higher valued NAMEs. Non-self-configuring ECUs are expected to be configured with a tool during configuration of the vehicle and conflicts are resolved at that time. Agricultural implements must be equipped with self-configuring ECUs

The upper 32 bits of the NAME have capacity for functional naming of an ECU. These fields include an industry group field that is set to agriculture for agricultural equipment. States are provided for truck and bus, forestry, construction, and marine industries. The device class field is used to identify implements and the tractor and other similar systems. An instance field is provided for device class, allowing multiple instances of implements. A Function field with capability for multiple instances is provided, which may be used to define particular functions associated with ECUs. An ECU instance field is provided for cases where a single function instance may be split among several ECUs. Functions are not currently defined in ISO 11783, but an example of one type of function might be pressure control on a sprayer. It is possible that several instances of this function could occur and that several ECUs might be used in a single-pressure control system.

Virtual Terminal

The virtual terminal (VT) is an operator interface device provided to allow display of information to operators and to allow operators to provide input information. VTs are designed to be slaves of ECUs on the network. An ECU may secure service from a VT and then be able to display its screens and retrieve operator information for its purposes. The ECU will not necessarily be aware of other ECUs using the terminal, that is, the VT appears to be exclusively dedicated from an ECU's perspective (ISO, 2003). From an operator's perspective, the VT may be switched to display one ECU's information or another or both if the VT supports that capability. An example of the application of a VT would be with a sprayer. A sprayer could secure use of the terminal and display spray rate and pressure. When the sprayer's panel is active on the VT screen, it has soft keys associated with the panel as shown on the bottom of the screen in Figure 10.29. The operator can switch to other panels that may be those of other implements.

The VT supports downloading of masks used to define panels displayed on the VT screen as well as alarm displays and soft key definitions for menus. Functions are also provided to allow masks to be loaded from or saved to some form of mass storage within the terminal. The ECU can simply instruct the terminal to load masks from mass storage and then select them to display them.

Masks in a VT can contain output fields that are used to display on the screen and input fields that are used to retrieve data from an operator. Numeric data can be displayed on the VT by selecting a field for update, sending the data to the particular output field, and selecting the field for end of update. The VT can format the data for display. In a similar sense, input from the operator can be obtained by selecting a field for input. When the operator has completed the input, data will be automatically sent to an ECU by the VT. Display from multiple ECUs can be coordinated from a single ECU.

The VT specification supports both text and graphic displays. Graphic functions are included for line drawing as well as for higher level functions including bar and dial gauges. Bitmap graphic elements may also be defined and displayed.

Task Controlling

ISO 11783 supports a task-control application. A task controller is contained within an ECU on the implement bus in the system. Commands may be loaded into a task controller from a management computer before a field operation and then the commands are delivered to a controlled device, an implement for example, during the field operation. Task controllers support three modes of command delivery: time-

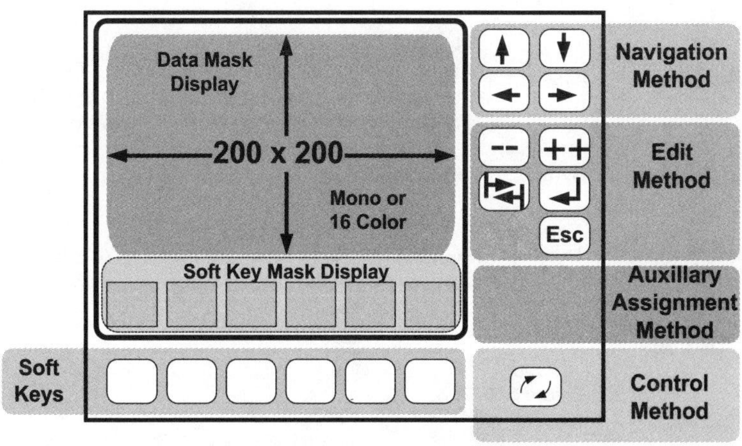

Figure 10.29. Schematic of a virtual terminal (ISO, 2003).

based, distance-based, and position-based. A common application of a task controller will be for use in precision farming systems. In that application, prescriptions created on a management computer can be transferred to the task controller. The task controller can then deliver the prescription to an implement as needed based on a position measured by an onboard GPS system. Task controllers also support the capability to log actual data during the field application and then transfer that data back to the management computer.

A message was created in ISO 11783 Part 7 to allow commands to pass from task controllers to implements and from implements to task controllers. Figure 10.30 summarizes the field definitions in the message. The identifier contains a value in the R, G, and PDU format fields that identifies the data field as the process data message. The message contains both source and destination address, allowing it to be sent to a particular ECU. If sent from a task controller to an implement, the message would be sent to the lowest instance of the function being controlled within that implement. A single process variable is sent in this message in the four-byte process variable data field. The selector indicates the data format of the variable, the type, and a modifier, a qualification regarding the variable. The data type indicates whether the process variable is an actual value or a set point and whether the message is a request or a response (commands are sent as a response). The count number field allows specification of a particular element within the implement or may be used to select all elements. This may be a particular row, or bin, depending on the process variable. The implement-type and position field allows selection of a particular implement function, for example, a seeding function versus a fertilizer application function within an implement. The position field allows selection of a particular instance or mounting

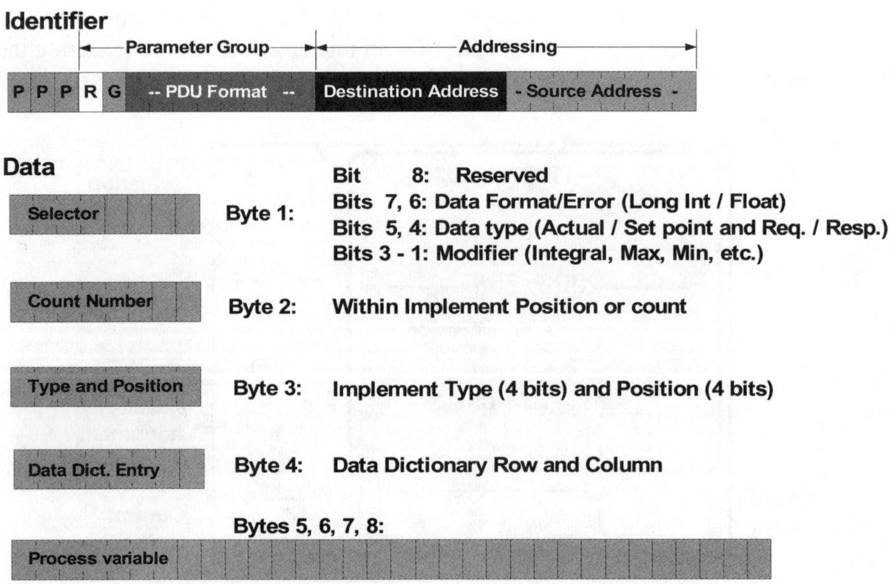

Figure 10.30. Field definitions for the process data message.

Table 10.8. Basic messages included in Part 7.

Message Title	Normal Source
Time and date	Tractor ECU
Wheel-based speed and distance	Tractor ECU
Ground-based speed and distance	Tractor ECU
GPS position and status data	GPS/Navigation ECU
Attitude (bearing, pitch, roll, altitude)	GPS/Navigation ECU
Hitch status (position and draft)	Tractor ECU
Power take-off status	Tractor ECU
Auxiliary valve status	Tractor ECU
Hitch and PTO commands	Implement ECU
Auxiliary valve commands	Implement ECU
Lighting	Tractor ECU
Process data	Task controller/Implement ECU
ECU power status and extension	Tractor/Implement ECU

position of that function. The data dictionary field identifies the particular process variable that is a function of the particular implement type.

The process data message allows task controllers to send commands and query implements regarding their current set points and actual operating points. The same message allows implements to send current set points and actual operating points.

Tractor and Tractor ECU Messages

Messages have been developed to allow basic information to be available on the implement bus. These messages have been included in ISO 11783 Part 7 as listed in Table 10.8. Most of these messages are sent repetitively at some rate fixed in the document and can be monitored by ECUs needing them on the implement bus. Others are sent on request, for example, time and date, which may be requested using the request message described earlier. An example of the use of these messages would be in an implement where seeding rate is being controlled. This implement could monitor speed and distance information and use that to regulate seeding rate.

ISO 11783 Part 9 specifies the requirements for tractors and tractor ECUs for the network. Tractors are classified with regard to their network capabilities. Table 10.9 summarizes the three different classifications and the messages that must be supported by tractors of different classes. Classification 1 provides basic information and lighting control. Classification 2 additionally provides more complete speed and draft information. Classification 3 adds the ability for an implement system to command the hitch, PTO and auxiliary hydraulic valves.

Messages on the Tractor Bus

A set of messages is available primarily for use on the tractor bus. These messages include an extensive set for power train control and information as well as messages supporting service logging. This message set is defined in ISO 11783 Part 8 and is equivalent to SAE J1939-71. An example of these messages is the engine configuration message. This message communicates the current torque curve of the engine and could be used by an implement to optimize power use. A forage harvester

Table 10.9. Message requirements for different tractor classifications.

Message Scope	Classification 1	Classification 2 (In addition to Classification 1)	Classification 3 (In addition to Classification 2)
Time and date		Time and date	
Power management	Key switch state		
	Maximum time of tractor power		
	Maintain power request		
Speed information	Wheel-based implement speed	Ground-based distance	
	Ground-based implement speed	Ground-based direction	
	Engine speed	Wheel-based distance	
		Wheel-based direction	
Hitch	Rear hitch position	Rear draft	Rear hitch position command
	Rear implement in-work indication		
PTO	Rear PTO output shaft speed		Rear PTO output shaft speed set point command
	Rear PTO output shaft engagement		Rear PTO output engagement command
Auxiliary valves		Estimated or measured auxiliary valve status	Auxiliary valves command
Lighting	Left turn signal lights	Full implement (non-tractor) lighting message set	
	Right turn signal lights		
	Marker light		
	Left stop light		
	Right stop light		
	Implement rear flood light		
Language	Language specific parameter storage in tractor ECU or VT		
Identification	Component ID		
	Software ID		

could monitor engine power use with the Electronic Engine Control #1 message, and use the TC1 message to request a transmission gear settings to optimize forward speed and engine efficiency. This type of control system impacts safe operation of the machine and would require agreement between the forage harvester manufacturer and the tractor manufacturer. Tractors will likely be equipped with security mechanisms to prevent unauthorized command of critical functions on the tractor. This type of control could just as well be used in a self-propelled combine harvester to optimize performance.

Diagnostics

Currently, diagnostics are not defined in ISO 11783, but the working group is discussing options. ISO 11783 does define a standard diagnostic connector (identical to SAE J1939) that provides connections to both the tractor and implement bus. This provides a standard physical connection point for data loggers and diagnostic tools. The VT also provides input/output capability that can be used to display and retrieve operator information for diagnostic purposes. SAE J1939 includes a diagnostic capability that is designed for use among network-based ECUs and can be used with diagnostic tools (Stepper et. al., 1995). There is no agreement at this point regarding whether to include this or a similar capability in ISO 11783. Initial diagnostics are likely to be supported through proprietary diagnostic tools or the VT.

10.3.6 Design Strategies for Implementation of ISO 11783

Guidelines for network development have been given by Young (1994) and example designs have been published (Stone, 1988). The guidelines by Young are:

1. Locate ECUs where concentrations of inputs and outputs exist.
2. Minimize the number of wires crossing critical boundaries.
3. Connect sensors or actuators to the closest module.
4. Locate ECUs so that critical closed-loop control is not performed over the network.
5. Condition, scale, and diagnose sensor or actuator information at the module to which they are connected.
6. Transmit information over the network in engineering units.
7. Make no assumptions about hardware or operator interface components connected to ECUs.
8. Broadcast data at a fixed rate.
9. Do not incorporate emerging standards until they are fully defined.

These guidelines are in general consistent with the current definitions in ISO 11783 and are suitable for examining addition of messages and parameters in the system. Some expansion and additions to Young's original guidelines are appropriate for ISO 11783 systems.

Location of ECUs to accommodate concentrations of input and output signals requires a survey of input and output signals in a proposed design. The survey should include identification of the signal, its physical location, its temporal frequency and resolution, and its magnitudinal resolution and range requirements. Identification of magnitudinal resolution and range of signals should include classification of signals into digital or analog. Naming or identification of the locations about the vehicle should be done carefully before the survey is initiated, with the understanding that once defined, locations may be split or combined to allow allocation of signals to ECUs. Signals should be classified as set points or actual values, and as a measured or a status value. In the later case, an ECU might contain inputs where the status of a switch might be measured, for example a switch to turn the PTO on or off. The same controller might also contain state data indicating whether the PTO is on or off. Both signals may need to be communicated on a network and may not have the same value, for example, the PTO switch is on, but the PTO state is off because it may be inhibited for some reason. This is less a problem for actual values than with set points since the measured signal of an actual value is typically the same as the state signal.

Once a signal survey is complete, the signals should be compared to parameters that have already been defined in ISO 11783. Where possible, parameters should be communicated through messages that have been standardized. Signals that haven't been defined in ISO 11783 but appear to be generally needed by other manufacturers should be considered for standardization. A request to the ISO TC23/SC19/WG1 should be made to include these parameters and group them into messages. Those signals that must be communicated and are not candidates for standardization should be assigned to proprietary parameters and grouped into proprietary messages.

Identification of the ECUs needed in a system can be done after a signal survey is complete. An initial set of ECUs should be proposed for the locations about the vehicle and signals assigned to the ECUs by location. The signal count and total frequency of throughput should be constrained at each ECU to the reasonable capabilities that can be provided by an ECU.

Once the messages have been identified and the ECUs are known, the ECUs should be examined for computational capacity. The algorithms that must be executed within each ECU, the frequency at which the algorithm must be computed, and the memory space it will occupy must be determined. In addition, the load on the ECU to manage the network traffic it must handle must be determined. Both loads must be totaled and compared to the ECUs computational capacity. A re-allocation of signals to ECUs may be necessary, and once done, the load computation process repeated until a reasonable design is found.

Network load must also be examined as a part of the design. Total network loads can be calculated or simulated based on the number of messages, their lengths, and frequency of transmission. The network load of each segment in the network should be calculated. A conservative target maximum load on each segment is approximately 35%. Larger loads should be considered carefully with regard to the impact on latency of messages. ECUs that are communications partners and contribute large loads may be partitioned to a separate sub-network or combined to eliminate the network traffic.

Initialization processes of ECUs must be considered in the network design. This is particularly true for ECUs on implements or those communicating with implements. ECUs that may be disconnected and reconnected to the network without the use of a tool to readjust the address of the ECU should be configured to perform self-configuring addressing. In addition, in cases where more than one instance of these ECUs can exist on the network, some method must be provided to set the instance fields in the network NAME of the device. This problem exists with agricultural implements. Consider a planter that may be used alone or may be hitched side by side with several other planters. The planter may be manufactured and programmed initially to have a device class of "planter," and an instance of "1." When two planters are connected together, some method must be provided by the manufacturer to set the instances of planter contained in the NAME in the ECUs. Several techniques are available, including a requirement that the components be connected and powered in sequence the first time they are used together. If this process is used, software must be included in a planter to detect other planters and to set its instance accordingly. In addition, some method must also be provided to allow the instance to be reset when the planter is configured differently. No specification is made in ISO 11783 regarding how instance setting is to be done, just that it must be done. Manufacturers must include resolution of this issue in their designs.

10.3.7 Fault Management

Analysis and control of potential faults in vehicle design is a normal part of the design process. The use of an ISO 11783 network introduces opportunities for failures. Some that should be considered in an analysis include breaks in or shorts of one or more of the conductors in the communications bus. The Part 2 document identifies many of the potential failures in the bus wiring and the potential effect on communications. It is possible to continue communications in some cases where single bus lines are shorted or broken. In addition, the detection of the failure of the communications bus is being considered by bus driver manufacturers. Failures in communications can also be detected in the ECU through the CAN protocol controller. These devices typically provide indications of errors that occur in transmission and reception of messages. CAN includes an error recovery protocol that automatically re-attempts to send a message when a transmission error occurs. A mechanism is provided to prevent this from occurring indefinitely.

The probability of an undetected error can be calculated in CAN-based networks based on the techniques used in protocol controllers for fault confinement. The probability of an undetected error in a CAN data link is $4.7x10^{-11}$ multiplied by the message error rate. If errors were detected at rates of 10 per second, the probability of a single undetected error occurring in a 10,000 hour life of a vehicle would be less than one chance in a thousand. Two conclusions can be drawn, first, that as long as errors do not occur at a high rate, it is unlikely that a single undetected error will occur within the life of a vehicle, and second, software should be designed to detect recurring errors in communications and to indicate a fault when the error rate is determined to be too high.

While CAN is very effective in detecting errors in messages received it does not provide a robust mechanism to determine if a message has been completely missed or not sent in the first place. ISO 11783 provides a definition of messages and that many of the messages are to be sent repetitively at a specified rate. It is up to designers to manage a failure to receive messages in an appropriate fashion.

Information that may be used in failure mode and effect analysis (FMEA) is provided in Part 2 and in the Bosch CAN 2.0 Specification (Bosch, 1991). The Part 5 document defines requirements regarding ECU operation during power drop-outs. ECUs must retain their information regarding network structure and continue to operate normally after a 10 ms power drop-out. No other significant requirements are made within the standard regarding actions that must be taken with a communications failure. Appropriate measures should be designed into software of ECUs to provide reasonable operation in the event of a communications failure.

10.4 System Environmental Concerns

Electronics are exposed to a harsh environment in off-road vehicle applications and significant care must be taken in design to assure reliability of these systems. Holloway (1987) reviewed the issues and requirements for electronics exposed to typical off-road environments. He categorized the environmental conditions into four areas:

- Atmospheric or climatic conditions including temperature, moisture, dust, and salt.
- Chemical conditions including fertilizer, fuels, and lubricants.
- Mechanical conditions including vibration and mechanical shock.
- Electrical conditions including steady-state electrical characteristics, transient electrical characteristics, electrostatic discharge, and electromagnetic compatibility.

ASAE EPP 455 provides recommendations regarding environmental robustness for electrical and electronic components (ASAE, 2000). An international standard, ISO 15003, is being developed to address the same issues. Table 10.10 lists a selection of electrical conditions specified by EPP 455.

Most of the specifications require little additional explanation, but give some measure of the rigor required to assure adequate robustness for off-road use. Operation without a battery is an issue because some vehicles are sometimes started with a battery and used without the battery in place. This situation removes the capacitance and load that the battery normally provides and exposes electronic components to the ripple produced by the alternator at the possible alternator frequencies. Load dump occurs when there is a sudden interruption of a load in the system. At that point, the alternator may be operating at full capacity and some time is required for the voltage regulator to react and for the current being generated to be dissipated. Load dump results in a large positive spike in voltage at the power supply to ECUs and other electronic components and their design must accommodate the transient. Alternator

Table 10.10. Selected supply voltage conditions from ASAE EPP 455 (ASAE, 2000).

Condition	Requirement
Operating voltage (DC)	Level 1: 9-16 V, Level 2: 10.5-16 V
Over-voltage	26 V for 5 min
Reverse polarity	-26 for 5 min
Short-circuit protection	Short with supply at 16 V for 5 min
Starting voltage	Test at 5.3 V after ramping from 0 V
Operation with no battery	Apply $6 + \lvert 12.6 \sin(2\pi ft)\rvert$V, f = 500 Hz to 1.5 kHz for 5 min
Load dump	Apply $14 + 106e^{-t/0.188}$ for 5 test cycles
Alternator field decay	Apply $-90e^{-t/0.038}$ from + 14 V for 60 cycles

field decay occurs when the engine is shutoff. A negative voltage spike occurs as the alternator field is disconnected. Reverse bias diodes, current blocking diodes, zener diodes, and transient absorption devices are commonly used to condition power used in ECUs and other electronics modules.

10.4.1 Electromagnetic Compatibility

The term *electromagnetic compatibility* encompasses the issues of radio frequency electronic emissions as well susceptibility to radio frequency electronic emissions. ASAE EPP 455 provides specifications regarding both emissions and susceptibility. These tests are typically run with the complete vehicle but may be run with individual modules. The ASAE EPP 455 identifies the critical FCC regulations and SAE recommendations. Many countries and some states have regulations regarding electromagnetic compatibility and significant efforts must be expended to assure vehicles meet the applicable regulations.

Management of EMC is a necessary part of design of electronic systems. ISO 11783-based systems may add somewhat to the effort that must be made to manage EMC. ISO 11783 systems include a communications bus that conventional systems may not. On the other hand, dedicated wiring that might be used for communications in past systems is eliminated. Electromagnetic emissions and susceptibility of the communications bus of ISO 11783 systems have been studied extensively. Systems built to conform to ISO 11783 Part 2 have been demonstrated to meet current EMC requirements in the US and in Europe. Application of conventional EMC control practices as well as following the recommendations in the Part 2 document is likely to produce a system that will meet current EMC requirements. Testing to confirm the EMC performance of ISO 11783 systems is necessary.

10.5 Chapter Summary

Electrical and electronic systems cover a broad spectrum of design issues in off-road equipment. These systems must be closely integrated into an off-road vehicle to allow the performance, functionality, and cost constraints to be met with a vehicle design. The material provided here is introductory, but provides those involved with off-road vehicles an overview and reference information to allow more in-depth work to be undertaken.

Homework Problems

10.1 Size the alternator for a vehicle with a 12 V electrical system and with the following specifications. Assume a 5 kW starter will be required by the engine and that the typical operating cycle for the vehicle is 30 min.

Long-term or Permanently Connected Loads	Current, A
Fuel pump	7
Electronic control units (ECUs)	
Engine controller	1
Transmission controller	2
Dashboard controller	1
Hitch controller	2
Work lights	25
Radio	2
Tail lamps	2
Flashing beacon lamp	3

Short-term or Intermittent Loads	Use Factor	Current, A
Ventilation blower	0.5	8
Windscreen wipers	0.25	6
Turn signal lamps	0.1	3
Implement actuators	0.2	40

10.2 Determine if an alternator with the following specifications will be adequate to provide electrical power for a vehicle with 600 W of long-term and permanently connected loads and 400 W of short term and intermittently connected loads. Assume the engine driving the alternator has an idle speed of 800 rpm, a normal operating speed of 2000 rpm and a maximum speed of 2500 rpm.

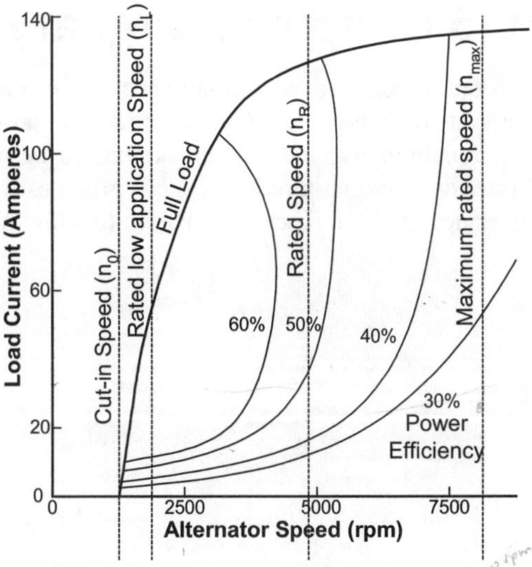

10.3 An electromagnetic noise is being received in the AM radio of an off-road vehicle, and is being blamed on the alternator. The frequency of the noise is suspected to be generated by the alternator. Estimate the frequency of the noise if the alternator speed is 5000 rpm.

10.4 Given a 12 V battery with a cold cranking current of 200 A, estimate the current that would be required to recharge the battery in a four-hour period following a starting load that was at 100% of the batteries cold cranking current.

10.5 Given the following engine and starter torque curves, determine if the starter described here will turn this engine fast enough to satisfactorily start the engine. Assume the starter drives the engine through a pinion geared at 12 starter revolutions to one flywheel revolution.

Engine data		Starter data	
Engine speed, rpm	Engine torque, Nm	Starter speed, rpm	Starter torque, Nm
50	135	0	22.5
75	210	900	15
100	180	1800	10
125	135	2700	7.5

10.6 Determine the voltage across a starter while the starter is drawing 100A from a 12 V system and is supplied by a 3/0 conductor from the battery to the starter with the same size conductor returning from the starter to the battery. Assume the distance from the battery to the starter is 3 m. Assume 0.02 ohm resistances in the battery connections and in the starter connections. Recalculate the voltage drop for connector resistances of 0.0005 ohms.

10.7 An electric screw-jack is supplied with current via a 10 m supply cable from the circuit breaker (thermal) and has a 10 m return cable. The cables are bundled with other conductors in a wiring harness. The harness includes connectors on the supply and return at the screw-jack and connectors for the supply and return cables at the circuit breaker. The screw-jack requires 12 amperes while operating and can be expected to operate infrequently for 30 s intervals. Size the conductor for a voltage drop of no more than 1 V and connector resistances of 0.001 ohms.

10.8 Estimate the number of undetected errors that could occur in the life of a vehicle with an ISO 11783 network. Assume the network has noisy connections and has 10 ECUs, each of which is generating one error in every other message that it sends. Assume each ECU is sending 100 messages per second. Assume the life of the vehicle is 15,000 hours.

10.9 Estimate the bus utilization that would occur for a message repeated at a rate of 100 per second (10 ms) for a message with 8 bytes of data on an ISO 11783 network.

References and Suggested Readings

Ackerson, E., J.H. English, and G.W. Harr. 1990. The application of electrical charging and cranking systems for agricultural equipment, ASAE Distinguished Lecture Series No. 15. St. Joseph MI: ASAE.

ASAE. 2001. Lighting and marking of agricultural field equipment on highways. ASAE Standard S279.11. *ASAE Standards,* 48th ed. St., Joseph, MI: ASAE.

ASAE. 2000. Environmental considerations in development of mobile agricultural electrical/electronic components. ASAE Standard EPP 455. *ASAE Standards,* 47th ed. St. Joseph, MI: ASAE.

Auernhammer, H. 1983. Die elektronische Schnittstelle Schlepper-Gerät. In Landwirtschaftliches BUS-System-LBS. Arbeitspapier 196, Kuratorium für Technik und Bauwesen in der Landwirtschaft e. V. (KTBL), Darmstat, Germany.

Bauer, H. (ed.) 1999. *Automotive Electrics and Electronics,* 3rd ed. Stuttgart, Germany: Robert Bosch GmbH.

Bosch, R. 1991. CAN Specification, Version 2.0. Stuttgart, Germany: Robert Bosch GmbH.

Denton, T. 1997. Automobile Electrical and Electronic systems. Edward Arnold: London, UK. Available through SAE, Warrendale, PA.

Fink, D.G., and D. Christensen (eds.) 1996. *Electronics Engineers' Handbook,* 4th ed. New York, NY: McGraw Hill, Inc.

Holloway, G.A. 1987. The environment of electrical/electronic components on agricultural equipment. ASAE Distinguished Lecture Series No. 12. St. Joseph, MI: ASAE.

ISO. 1995. Earth-moving machinery—Lighting, signalling and marking lights, and reflex-reflector devices. ISO Standard 12509. Geneva, Switzerland: ISO.

ISO. 2002. Tractors and machinery for agriculture and forestry—Serial control and communications data network—Part 2: Physical layer. ISO Standard 11783. Geneva, Switzerland: ISO.

ISO. 2003. Tractors and machinery for agriculture and forestry—Serial control and communications data network—Part 6: Virtual terminal. ISO Standard 11783. Geneva, Switzerland: ISO.

Lide, D.R. (ed.) 2001. *Handbook of Chemistry and Physics,* 81st ed. Cleveland, OH: CRC Press Inc.

Luebbering, B.L., and V.R. Smith. 1993. Engine electronics technology. SAE Technical Paper No. 932404. Warrendale, PA: SAE.

Monson, R.J., and E.M. Dahlen. 1995. Mobile control system responsive to land area maps. U.S. Patent 5,453,924.

SAE. 1984. Storage batteries for off-road work machines. SAE Standard J930. *SAE Handbook*. Warrendale, PA: SAE.

SAE. 1985. Unauthorized starting or movement of machines. SAE Standard J1085. *SAE Handbook*. Warrendale, PA: SAE.

SAE. 1987. Electrical charging systems for construction and industrial machinery. SAE Standard J180. *SAE Handbook*. Warrendale, PA: SAE.

SAE. 1994. Storage batteries. SAE Standard J537. *SAE Handbook*. Warrendale, PA: SAE.

SAE. 1995a. Low-tension primary cable. SAE Standard J1128. *SAE Handbook*. Warrendale, PA: SAE.

SAE. 1995b. Battery cable. SAE Standard J1127. *SAE Handbook*. Warrendale, PA: SAE.

SAE. 1996a. Voltage drop for starting motor circuits. SAE Standard J541. *SAE Handbook.* Warrendale, PA: SAE.

SAE. 1996b. Lighting and marking of construction, earth-moving machinery. SAE Standard J1029. *SAE Handbook.* Warrendale, PA: SAE.

SAE. 1997. Shielding of starter system energization. SAE Standard J1493. *SAE Handbook.* Warrendale, PA: SAE.

SAE. 1998a. Seven conductor electrical connector for truck-trailer jumper cable. *SAE Handbook.* Warrendale, PA: SAE.

SAE. 1998b. Wiring distribution systems for construction, agricultural, and off-road work machines. SAE Standard J1614. *SAE Handbook.* Warrendale, PA: SAE.

SAE. 1999. Road vehicles—Alternators with regulators—Test methods and general requirements. SAE Standard J56. *SAE Handbook.* Warrendale, PA: SAE.

Schueller, J.K. 1988. A proposed scheme for integrated tractor-implement communications and control. ASAE Paper No. 881533. St. Joseph, MI: ASAE.

Stepper, M.R. 1993. J1939 high speed serial communication, the next generation network for heavy duty machines. SAE Paper No. 931809. Warrendale, PA: SAE.

Stepper, M.R., S.R. Butler, and G.G. Zhu. 1995. On-board diagnostics, a heavy duty perspective. SAE Paper No. 951947. Warrendale, PA: SAE.

Stone, M.L. 1987. Applications of networks in agricultural controls and instrumentation. ASAE Paper No. 873005. St. Joseph, MI: ASAE.

Stone, M.L. 1988. Applications of distributed control systems on combines. ASAE Paper No. 881600. St. Joseph, MI: ASAE.

Stone, M.L., K.D. McKee, C.W. Formwalt, and R.K. Benneweis, 1999. ISO 11783: An electronic communications protocol for agricultural equipment. ASAE Distinguished Lecture Series No. 23. St. Joseph, MI: ASAE.

USDA. 1954. Development of the agricultural tractor in the United States, Part I: Up to 1919 inclusive. USDA Information Series No. 107. Beltsville, MD: USDA. .

Weisberg, P., R.K. Benneweis, and M. Bloom. 1993. Air seeder monitor/controller communications system. SAE Paper No. 932430. Warrendale, PA: SAE.

Young, S.C. 1994. Electronic control system for GENESIS™ 70 series tractors. SAE Paper No. 941789. Warrendale, PA: SAE.

CHAPTER 11

HYDRAULIC SYSTEMS

11.1 Introduction

Hydraulic systems came into widespread use in off-road vehicles following World War II. These systems removed the need for the vehicle operator to have great physical strength. Using the hydraulic system, the operator can easily maneuver heavy vehicle attachments such as front-end loaders, rear-mounted blades, back hoes, etc. The same hydraulic system can also provide power for hydraulic brakes and power steering. Through use of hydraulic motors, the hydraulic system can transmit power more conveniently than with mechanical drives.

In this chapter, the reader will learn the basic principles of hydraulic components and systems. The National Fluid Power Association (NFPA) standard symbols will be introduced as a way of describing the logic of hydraulic circuits. While most of the chapter will relate to steady-state behavior of hydraulic systems, the reader will also be introduced to techniques for modeling the transient behavior of such systems. Finally, the reader will be introduced to mechatronics as applied to hydraulic systems. The microprocessors incorporated into mechatronic devices provide flexibility and sophistication of control that is not feasible with mechanical control of hydraulic systems.

11.2 Basic Principles of Hydraulic Systems

Liquids have no shape of their own, but will flow to acquire the shape of their container. Hydraulic fluids are virtually incompressible at the pressures used in hydraulic systems. Liquids transmit pressure equally in all directions. The importance of these principles is illustrated in Figure 11.1, which is a schematic illustration of a hydraulic jack. By applying a downward force on the small piston, the jack can be made to lift a heavy mass on the large piston. The liquid fills the entire volume between the two pistons and, because it is incompressible, some liquid must move into the large piston chamber when the small piston is pushed downward. The pressure beneath the small piston is equal to F_1/A_1, i.e., the force on the small piston divided by the area of the small piston. The same pressure acts on the large piston because the liquid transmits pressure equally in all directions. Because $A_2 > A_1$, the force on the large piston is much larger than the force on the small piston. The small piston in Figure 11.1 can be considered to be a simple *hydraulic pump*, while the large piston can be considered to be a *hydraulic actuator*.

The jack described in Figure 11.1 is an example of a *positive displacement* system. It has positive displacement because each movement of the small piston displaces a definite quantity of liquid and forces a corresponding movement of the large piston. The movement of the large piston is independent of the load on it. In contrast, suppose the small piston was replaced with a powered turbine. The turbine might be capable of generating enough pressure to move the actuator (the large piston) but, as the load pressure increases, more fluid bypasses the turbine blades and less fluid is moved to the actuator. Thus, the actuator speed depends on the amount of load on the actuator. Such speed dependence on load is characteristic of a system with *non-positive displacement*. All hydraulic systems used on off-road vehicles are of the positive displacement type.

The final important principle of hydraulic systems is illustrated in Figure 11.2. Any flow of liquid through a pipe or orifice is accompanied by a reduction in liquid pressure. Figure 11.2a shows that the pressure is constant throughout the system because the liquid is not flowing. Figure 11.2b shows that the pressure is lower to the right of the orifice because the liquid is flowing from left to right.

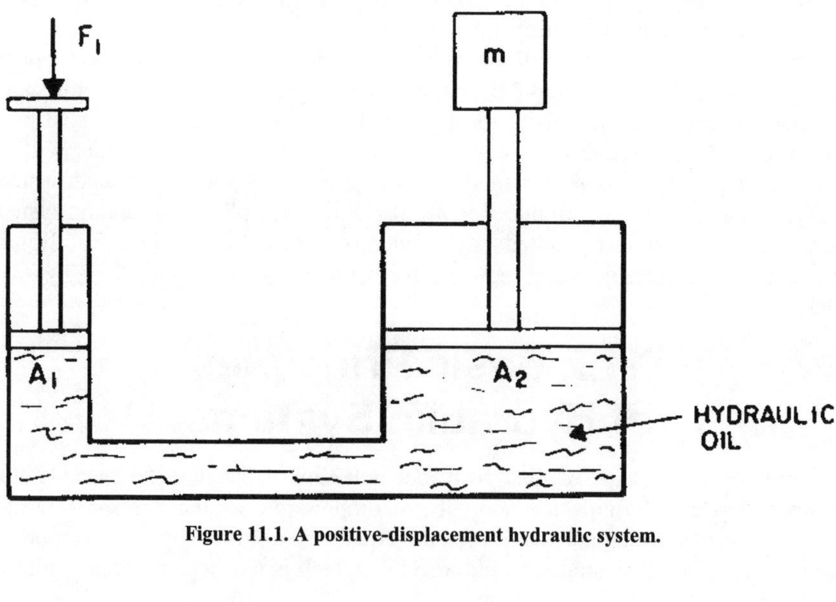

Figure 11.1. A positive-displacement hydraulic system.

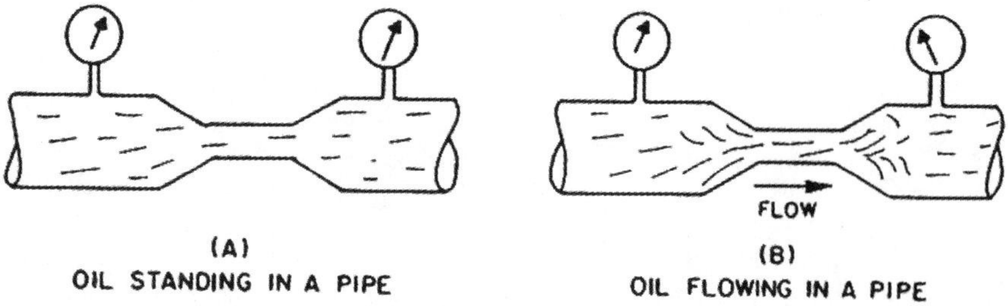

Figure 11.2. Pressure drops occurring as oil flows through a pipe.

11.3 Standard Symbols

The logic of a hydraulic component can be conveyed through a cutaway drawing, but preparation of such drawings would be laborious. A Joint Industry Conference of the fluid power industry developed a set of JIC symbols to convey the logic of hydraulic circuits just as the symbols for batteries, resistors, capacitors, etc., convey the logic of electric circuits. Hydraulic circuit symbols were later standardized by the NFPA and by the International Standards Organization (ISO). Figure 11.3 shows some of the most widely used symbols, which were designed to be as self-explanatory as possible. The small arrow on the pump symbol indicates pressurized oil is being forced out of the pump. The small arrow on the motor symbol indicates pressurized oil is being forced into the motor. Drawing an arrow through a component indicates the component is variable rather than fixed. For example, the pump of Figure 11.3d has fixed displacement but a variable-displacement pump would have an arrow drawn through the pump symbol. Similarly, the throttling valve in Figure 11.3l has an orifice of constant diameter; drawing an arrow through the symbol would indicate an orifice with adjustable diameter.

Although most hydraulic systems have only one reservoir, the symbol can be used at more than one place on a drawing of a hydraulic circuit to eliminate the need to draw numerous return lines; this is analogous to the drawing of electrical circuits, where the electrical ground symbol can be used at many places in a circuit drawing. The reservoir and ground symbols have another analogous relationship: just as the ground symbol indicates a point of zero voltage, the reservoir is usually at zero gage pressure, i.e., at atmospheric pressure.

The three valves at the bottom of Figure 11.3 control the direction of oil flow and thus are called *directional control valves* (DCVs). Each of the valves shown is a four-port device, i.e., each has four connections to the hydraulic circuit. Ports P and T are for connection to the pump and the tank, i.e., the reservoir. Ports A and B are the work ports for connection to the actuator or circuit to be controlled by the valve. Each of the three DCVs is also a three-position valve with the valve spool shown in the centered position. The valve spool can be slid to the right to align the left-most box with the external ports, or to the left to align the right-most box with the external ports. Each of the DCVs has a different type of center, as will be discussed further in the Section 11.7.3.

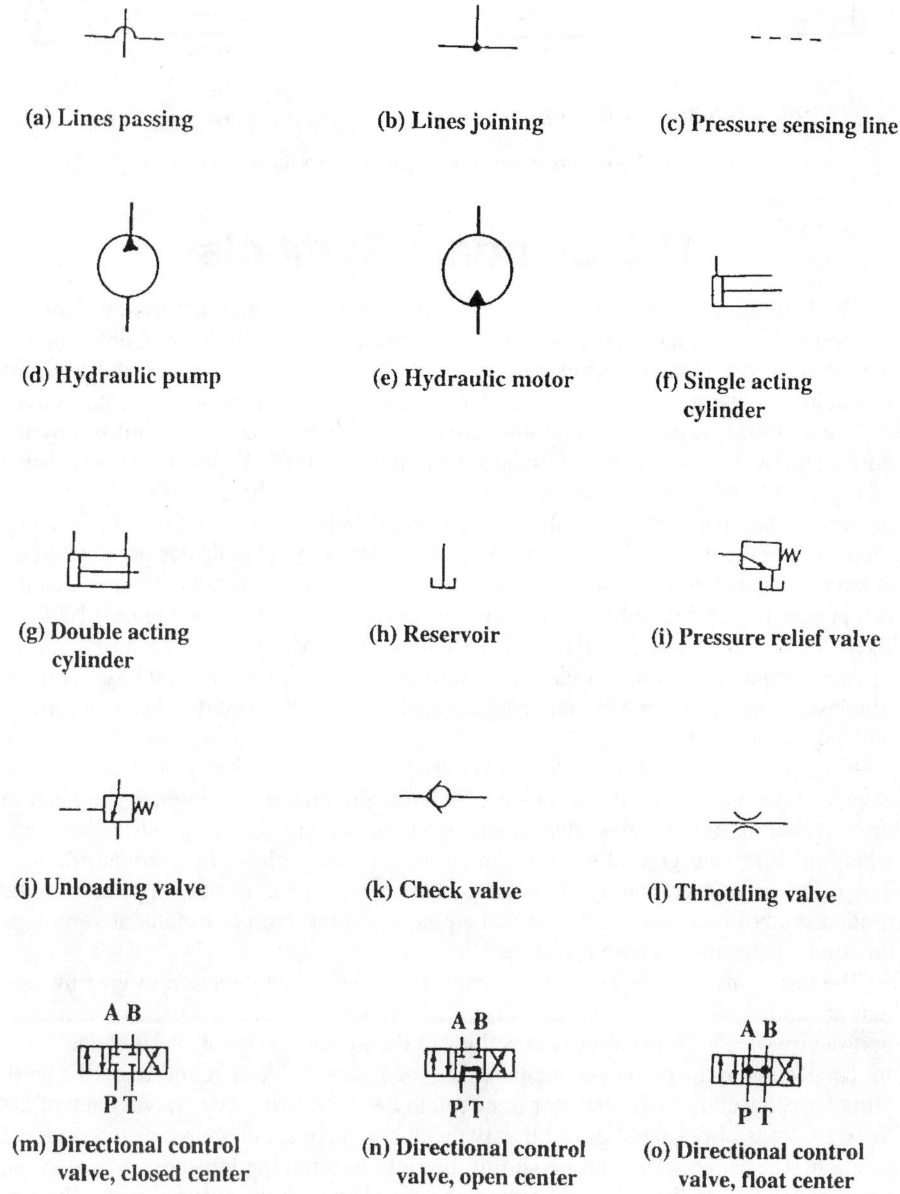

Figure 11.3. Frequently used hydraulic schematic symbols.

11.4 Hydraulic Pumps

The three basic types of hydraulic pumps are *gear pumps*, *vane pumps,* and *piston pumps*. In each case, the pump converts mechanical power at the pump shaft into hydraulic power at the pump outlet.

11.4.1 Gear and Vane Pumps

The gear pump illustrated in Figure 11.4 and the vane pump illustrated in Figure 11.5 are examples of *fixed-displacement pumps*. The schematic symbol for each of these pumps has a small arrow indicating that oil is being forced out of the pump. The pump displacement is the theoretical volume of oil the pump can deliver per revolution of the pump shaft. In the gear pump of Figure 11.4, the displacement is equal to the volume of space between two gear teeth and the housing, multiplied by the total number of gear teeth on both gears. The top gear in Figure 11.4 is driven by an external shaft (note the locking key in the top shaft) and, in turn, drives the lower gear. Oil flowing into the inlet port on the left is carried to the outlet port in the tooth spaces; the meshing of the gears blocks the tooth spaces and prevents the oil from flowing back to the inlet port through the middle. In Figure 11.5, the rotor turns counterclockwise (ccw); centrifugal force keeps the vanes in the rotor slots in contact with the housing and oil is carried from the inlet to the outlet in the spaces between the moving vanes. The displacement of both the gear and the vane pumps is fixed when the pumps are manufactured and cannot be changed. Thus, these pumps are said to have fixed displacement.

11.4.2 Axial Piston Pumps

There are two types of piston pumps: axial and radial piston pumps. A radial piston pump has pistons that move perpendicular to the pump axis. The pump shown in Figure 11.6 is called an axial piston pump because the pistons move parallel to the axis of rotation. The pistons are carried in a rotating cylinder barrel. As the piston shoes slide along the cam plate, the angularity between the drive shaft and the barrel forces the pistons to reciprocate in their bores. As the pistons move out of their bores on the left side of the barrel, oil is drawn in through the inlet port and the valve plate slot on the left. Along the right side, as the pistons move back into their bores, oil is forced out through the valve plate slot and outlet port on the right. This pump has fixed displacement because the angle between the drive shaft and barrel is fixed. The displacement is equal to the area of each piston times the piston stroke times the number of pistons. Similar pumps are available in which the angularity can be controlled to vary the pump displacement. The latter pumps are called *variable-displacement pumps*. If the pump of Figure 11.6 had variable displacement, the schematic symbol would be drawn with an arrow through it.

Figure 11.7 illustrates an axial piston pump with variable displacement. The arrow through the schematic symbol indicates variable displacement. The swash plate can be tilted to control the displacement. Because of the heavy forces on the swash plate, internal hydraulic cylinders are used to control the tilt. With the swash plate tilted as shown and with shaft rotation as indicated by the arrow, the pump would be at

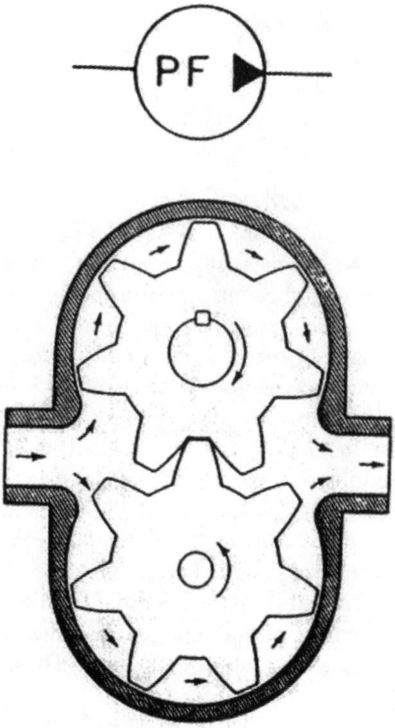

Figure 11.4. A gear pump.

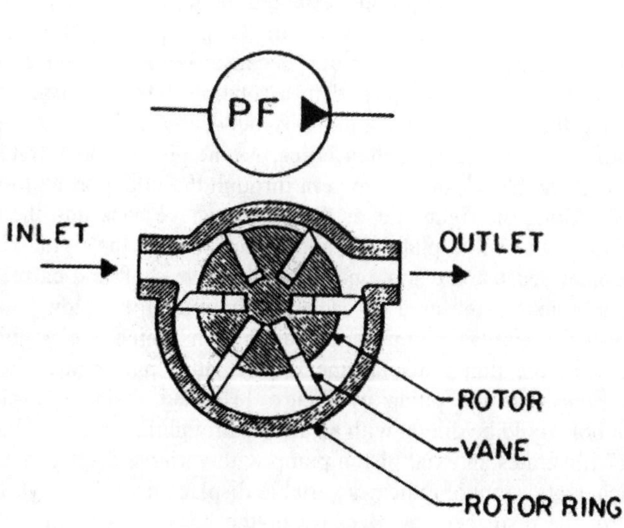

Figure 11.5. A vane pump.

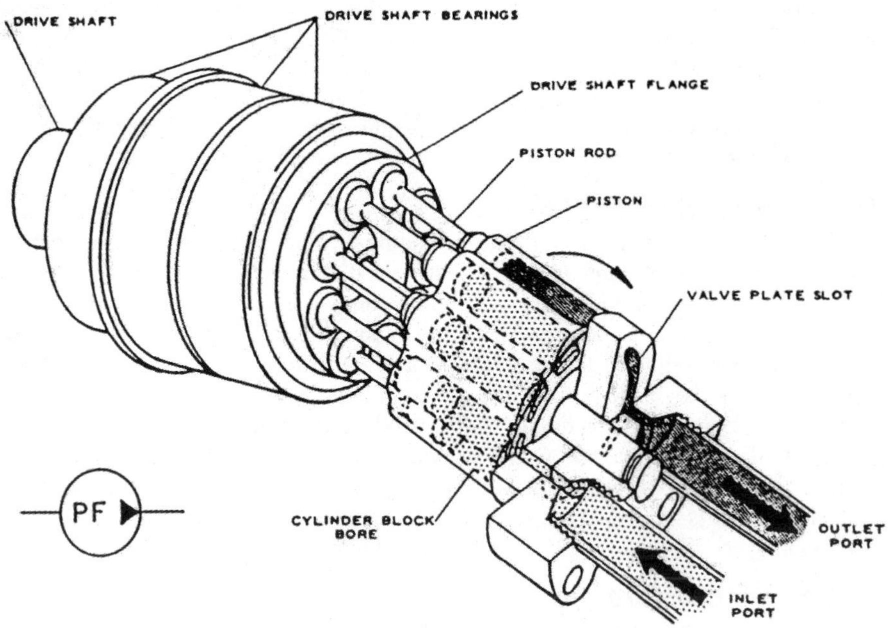

Figure 11.6. A fixed-displacement axial piston pump.

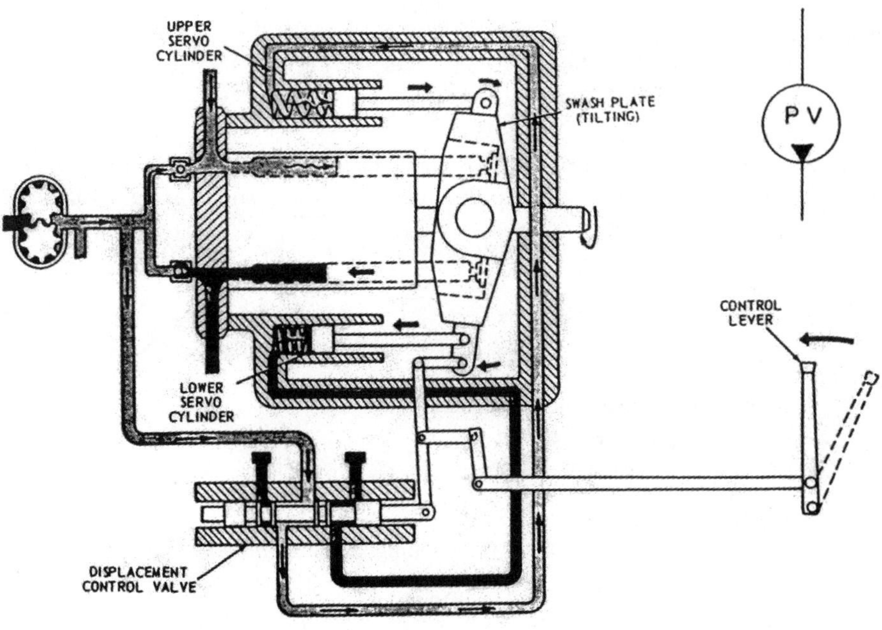

Figure 11.7. A variable-displacement axial piston pump.

maximum displacement to deliver oil through a valve plate slot and port on the side of the pump facing out of the page. With the swash plate tilted full ccw, the pump would be at maximum displacement to deliver oil through a valve plate slot and port on the side of the pump facing into the page, i.e., the oil flow direction would be reversed. If the swash plate were at zero tilt, i.e., perpendicular to the drive shaft, the pump displacement would be zero.

11.4.3 Pump Seal Protection

All of the above pumps have a shaft extending outside the pump housing for input of mechanical power. A *shaft seal* must be provided to prevent leakage of oil along the shaft. Pumps must be provided with seal protection, since any pressure buildup on the seal could force it from its housing, resulting in an oil leak. The protection consists of a vent to convey oil from the seal area to a low-pressure area. In a *unidirectional pump,* in which one of the ports is always the inlet port, the seal can be vented to the inlet port. Reversing the direction of shaft rotation and oil flow in such a pump would not be feasible because the seal would pop out of its housing. In pumps like the one in Figure 11.7, neither port is always at low pressure and cannot be used to vent the seal. Instead, the seal is vented to the reservoir via an external line, often referred to as a case drain.

11.4.4 Pump Delivery

The theoretical delivery from any hydraulic pump can be calculated by

$$Q_{pt} = \frac{D_p N_p}{1000} \tag{11.1}$$

where

Q_{pt} = theoretical pump delivery, L/min
D_p = pump displacement, cm^3/rev
N_p = pump speed, rpm

Pump delivery is actually less than the theoretical amount because of internal leakage from the outlet port back to the inlet port. Thus, pumps have a *volumetric efficiency*, defined as

$$e_{pv} = \frac{Q_{pa}}{Q_{pt}} = \frac{Q_{pt} - Q_{pl}}{Q_{pt}} \tag{11.2}$$

where

e_{pv} = pump volumetric efficiency, decimal
Q_{pa} = actual pump delivery, L/min
Q_{pl} = pump internal leakage, L/min

11.4.5 Pump Torque

The theoretical torque required to turn a pump shaft can be calculated using

$$T_{pt} = \frac{\Delta p\, D_p}{2\pi} \tag{11.3}$$

where

T_{pt} = theoretical pump torque, N·m

Δp = pressure rise across the pump, MPa

Internal friction causes the actual torque demand to be greater than theoretical. Thus, the pump has a *torque efficiency*, defined as

$$e_{pt} = \frac{T_{pt}}{T_{pa}} = \frac{T_{pt}}{T_{pt} + T_{pf}} \tag{11.4}$$

where

e_{pt} = pump torque efficiency, decimal

T_{pa} = actual pump torque, N·m

T_{pf} = pump friction torque, N·m

11.4.6 Pump Power

Hydraulic power is the product of pressure and flow rate. Across a circuit with the same flow in as out, the power change with units correctly converted is

$$P_h = \frac{Q\,\Delta p}{60} \tag{11.5}$$

where

P_h = hydraulic power, kW

Q = flow through device, L/min

Δp = pressure rise or drop across device or circuit, MPa

If there is a pressure rise across the device, as in a pump, then the hydraulic power is positive, i.e., the device produces hydraulic power. If there is a pressure drop across the device, as in a hydraulic motor or an orifice, then the hydraulic power is negative, i.e., the device converts hydraulic power into useful work and/or heat.

The actual pump power is the power delivered to the pump shaft by the prime mover (usually an engine or an electric motor) and can be calculated using Equation 2.4. The actual power into the pump is greater than the theoretical (hydraulic) power output because of pump friction and internal leakage. The *pump power efficiency* is

$$e_{pp} = \frac{P_{ph}}{P_{ps}} = e_{pv}e_{pt} \tag{11.6}$$

where

e_{pp} = pump power efficiency, decimal

P_{ph} = hydraulic power from pump, kW

P_{ps} = shaft power into pump, kW

It can be shown that the power efficiency of a pump is equal to the product of its volumetric and torque efficiencies.

11.4.7 Pump Efficiencies

Figure 11.8 illustrates pump efficiencies for a particular pump but the curve shapes are typical. The efficiencies are plotted versus a dimensionless parameter, oil viscosity times pump speed divided by pressure rise, as shown on the horizontal axis. Efficiency Equations 11.2, 11.4, and 11.6 help to explain the shapes of the efficiency curves on Figure 11.8.

As shown in Equation 11.2, the volumetric efficiency of a pump declines with increasing internal leakage. The leakage is driven by the pressure difference between the outlet and inlet ports of the pump. If there were no pressure difference, the internal leakage would be zero and Equation 11.2 shows that the volumetric efficiency would be 1.0, i.e., 100%. Internal leakage increases and volumetric efficiency declines as the pressure differential increases. For any given pressure difference, the volumetric efficiency is also affected by pump speed. At a very low pump speed, the leakage flow may be as large as the theoretical pump delivery and, as shown by Equation 11.2, the volumetric efficiency would then be zero. As pump speed increases for a constant pressure differential, the leakage flow becomes smaller relative to the theoretical flow and the volumetric efficiency increases. Thus, the volumetric efficiency approaches zero at very low pump speeds when the outlet pressure is high, but can approach 1.0 at high pump speeds when the outlet pressure is low.

There is always some internal friction in a pump. When the pump outlet pressure is zero, the theoretical torque is also zero and Equation 11.4 shows that the pump torque efficiency would be zero. As the pump outlet pressure rises, the friction torque becomes smaller relative to the theoretical torque and the torque efficiency rises. Usually, for a given outlet pressure, the friction torque increases and the torque efficiency declines with increasing pump speed.

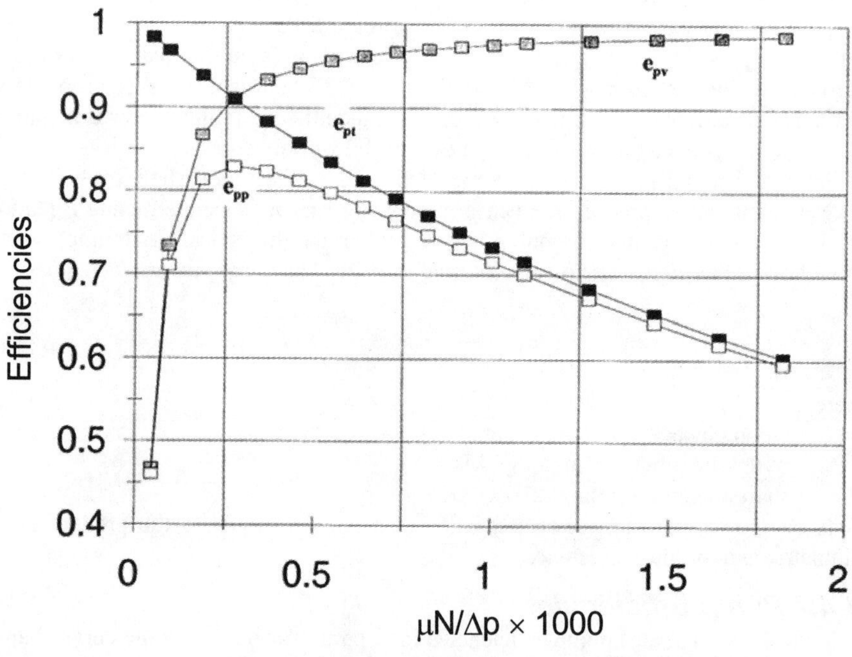

Figure 11.8. Typical pump efficiency relationships.

The power efficiency is the product of the volumetric and torque efficiencies (Equation 11.6). Thus, the power efficiency must be zero at zero pump speed and high outlet pressure because the volumetric efficiency is zero. At high pump speed and low outlet pressure, the power efficiency also approaches zero because the torque efficiency approaches zero. There is always some combination of pump speed and outlet pressure that yields maximum power efficiency. The power efficiency may also remain high over some range of speed and pressure combinations above and below the optimum combination. It is the responsibility of the hydraulic circuit designer to select a pump that can operate at high power efficiency over the pump speed and pressure combinations likely to be encountered by the hydraulic circuit.

11.5 Hydraulic Actuators

A *hydraulic actuator* is a device for converting hydraulic power into mechanical power. There are two types of actuators, rotary and linear. Rotary actuators are called *hydraulic motors*, while linear actuators are called *hydraulic cylinders*.

11.5.1 Hydraulic Motors

Hydraulic motors are similar in appearance to hydraulic pumps. If care is taken to avoid damaging the seal, pumps and motors can often be used interchangeably. For example, if a gear pump is unidirectional because the pump seal is vented to the input port, the same device can be used as a hydraulic motor if the direction of rotation is reversed so that the port to which the seal is vented becomes the outlet port.

The theoretical speed of a hydraulic motor is calculated using a variation of Equation 11.1, i.e.,

$$N_{mt} = \frac{1000\,Q_{ma}}{D_m} \tag{11.7}$$

where

N_{mt} = theoretical motor speed, rpm
Q_{ma} = liquid flow rate into the motor, L/min
D_m = motor displacement, cm^3/rev

Internal leakage from the inlet port to the outlet port causes the actual motor speed to be less than the theoretical speed. The volumetric efficiency of a motor is defined as

$$e_{mv} = \frac{N_{ma}}{N_{mt}} = \frac{Q_{ma}}{Q_{ma} + Q_{ml}} \tag{11.8}$$

where

e_{mv} = motor volumetric efficiency, decimal
N_{ma} = actual speed of motor, rev/min
Q_{ml} = internal leakage in motor, L/min

The equation for calculating the theoretical torque produced by a motor is similar to Equation 11.3, except that the p-subscripts are replaced by m, i.e.,

$$T_{mt} = \frac{\Delta p\, D_m}{2\pi} \tag{11.9}$$

where

T_{mt} = theoretical torque from motor, N·m
Δp = pressure drop across motor, MPa

Internal friction causes the actual torque production to be less than the theoretical torque. Motor torque efficiency is

$$e_{mt} = \frac{T_{ma}}{T_{mt}} = \frac{T_{mt} - T_{mf}}{T_{mt}} \tag{11.10}$$

where

e_{mt} = motor torque efficiency, decimal
T_{ma} = actual motor torque, N·m
T_{ma} = motor friction torque, N·m

Equation 11.5 is valid for calculating the hydraulic power into a motor, given the actual flow into the motor and the pressure drop across the motor. The actual shaft power out of the motor can be calculated using Equation 2.4. An equation similar to Equation 11.6 can be used calculate the power efficiency of a hydraulic motor, i.e.:

$$e_{mp} = \frac{P_{ms}}{P_{mh}} = e_{mv} e_{mt} \tag{11.11}$$

where

e_{mp} = power efficiency of motor, decimal
P_{ms} = shaft power into motor, kW
P_{mh} = hydraulic power into motor, kW

The volumetric, torque, and power efficiencies of a hydraulic motor vary in a manner similar to that illustrated in Figure 11.8 for hydraulic pumps. Most of the discussion of Section 11.4.7 for pumps also applies to motors because the efficiency equations for motors are similar to those for pumps.

11.5.2 Hydraulic Cylinders

A cutaway of a double-acting hydraulic cylinder is shown in Figure 11.9. When oil is forced into the port on the left, the cylinder is forced to extend while expelling oil from the port on the right. Conversely, the cylinder retracts and expels oil from the port on the left when oil is forced into the port on the right. The force generated by a cylinder is

$$F_c = \frac{p_1 A_1 - p_2 A_2}{10} \tag{11.12}$$

where

F_c = force exerted by the cylinder rod, kN
A_1 = area of piston face, cm^2
A_2 = area of piston face minus area of rod, cm^2
p_1 = pressure acting on A_1, MPa
p_2 = pressure acting on A_2, Mpa

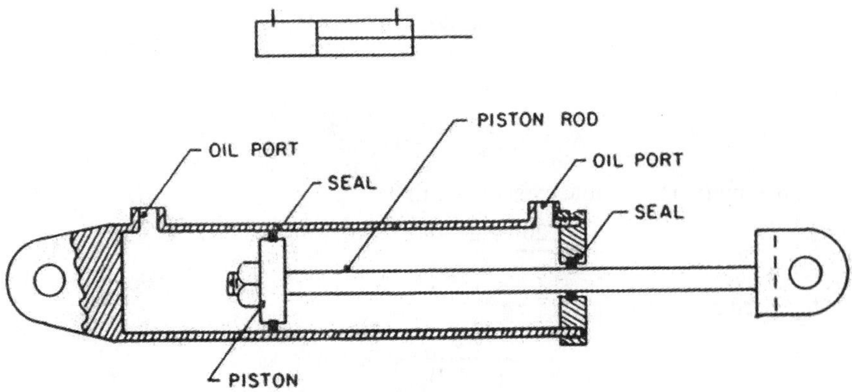

Figure 11.9. A double-acting hydraulic cylinder.

When $p_1A_1 > p_2A_2$, the cylinder extends and F_c is positive. When $p_1A_1 < p_2A_2$, F_c is negative and the cylinder retracts.

The cylinder speed can be calculated using the following equation:

$$v_c = \frac{Q}{6A} \qquad (11.13)$$

where

v_c = cylinder speed, m/s

A = area on which inflowing oil acts, cm^2

The factor 6 is a units factor to allow the use of more convenient units in the equation. Example Problem 11.1 illustrates the performance of a double-acting cylinder.

Example Problem 11.1.

The double-acting cylinder as in Figure 11.9 has a bore of 6.5 cm and a rod diameter of 2.5 cm. A pump supplies 80 L/min of oil to the cylinder at a maximum pressure of 20 MPa. If the left cylinder port is connected to the pump while the port on the right is connected to the reservoir,

(a) What is the maximum load the cylinder can move while extending?

(b) How fast will it extend?

(c) What will be the flow rate of oil returning to the reservoir?

If the port connections are reversed to retract the cylinder,

(d) What is the maximum load the cylinder can move while retracting?

(e) How fast will it retract?

(f) What will be the flow rate of oil returning to the reservoir?

To simplify the problem, assume line pressure drops caused by the flowing fluids are negligible.

Solution

The areas on which the oil acts are

$$A_1 = \frac{\pi(6.5)^2}{4} = 33.2\ \text{cm}^2$$

and

$$A_2 = 33.2 - \frac{\pi(2.5)^2}{4} = 28.3\ \text{cm}^2$$

(a) The maximum force while extending will be

$$F_c = \frac{20(33.2) - 0(28.3)}{10} = 66.4\ \text{kN}$$

(b) The speed of extension will be

$$v_c = \frac{80}{6(33.2)} = 0.402\ \text{m/s}$$

(c) The return rate of oil to the reservoir will be

$$Q = 6\, v_c A_2 = 6(0.402)(28.3) = 68.3\ \text{L/min}$$

Note that the pump will withdraw 80 L/min from the reservoir to extend the cylinder, but the cylinder will return only 68.3 L/min to the reservoir. The reservoir must have sufficient capacity to make up the difference.

The reader can verify that the answers to parts (d), (e), and (f) are:

(d) F_c = -56.6 kN
(e) v_c = 0.471 m/s
(f) Q = 93.8 L/min

Because $A_1 > A_2$, the cylinder can lift less force but moves faster while retracting and returns more oil to the reservoir than it receives from the pump. The difference in extension and retraction speeds can cause difficulties in applications such as power steering. If a single cylinder were used, the vehicle steering response would not be the same to both the left and the right. In such applications, a double-rod cylinder may be used, i.e., the rod extends from both ends of the cylinder. The effective area of both sides is thus the piston area minus the rod area.

The reader may notice that no mention was made of internal leakage or friction in cylinders. Internal leakage is generally so small compared to the flow rate into the cylinder that it has negligible effect on the cylinder speed. Likewise, internal friction has negligible effect on the cylinder force. In other words, the volumetric and force efficiencies of a hydraulic cylinder are assumed to be equal to 1.0.

Single-acting cylinders are used in some hydraulic applications. The cylinder of Figure 11.9 could be converted to a single-acting cylinder by disconnecting the hose and installing a breather in the port on the right to keep out dirt. An external load must supply the force to cause a single-acting cylinder to retract. The analysis of the extension of a single-acting cylinder is identical to the procedure followed in Example Problem 11.1, except that the cylinder is not returning any oil to the reservoir when the cylinder is extending. While retracting, the cylinder does return oil to the reservoir.

11.6 Hydraulic Orifices

There may be occasions in hydraulic circuit design when it is necessary to restrict flow to some segment of the circuit and/or to create a pressure differential. Both of these goals can be achieved by use of a hydraulic orifice. Equation 11.14, the standard orifice equation, can be used to calculate the flow rate corresponding to a given pressure drop across an orifice. If means are available to measure the pressure drop, the orifice can be used as a flow meter. Alternatively, by solving Equation 11.14 for pressure drop, one can calculate the pressure drop resulting from any given flow through the orifice.

$$Q = 2.68\, C_d A_o \sqrt{\frac{\Delta p}{\rho}} \tag{11.14}$$

where

Q = flow through orifice, L/min
C_d = orifice coefficient, dimensionless
A_o = cross sectional area of orifice, mm^2
Δp = pressure drop across orifice, MPa
ρ = fluid density, kg/L

The constant, 2.68, is a units constant to allow use of more convenient units in the equation. The orifice coefficient varies with the Reynolds number that, for flow through an orifice, is defined as

$$Re = \frac{10^3\, \rho\, v_o d_o}{\mu} \tag{11.15}$$

where

Re =Reynolds number, dimensionless
ρ = fluid density, kg/L
v_o = fluid velocity through orifice, m/s
d_o = orifice diameter, mm
μ = dynamic viscosity of fluid, mPa·s

The units constant, 10^3, was inserted to allow use of more convenient units in the equation. The reader can verify that Equation 11.15 does produce a dimensionless quantity. An equation similar to Equation 11.13 can be used to calculate the flow velocity through the orifice from the flow rate and orifice diameter. Figure 11.10 shows an approximate relationship between orifice coefficient and Reynolds number for a sharp-edged orifice. For high Reynolds numbers, i.e., $Re > 2500$, the orifice coefficient is often assumed to be equal to 0.6.

Many practical orifices are not circular. For such orifices, the effective diameter can be calculated using the following equation:

$$d_{eff} = \frac{A_o}{L_{co}} \tag{11.16}$$

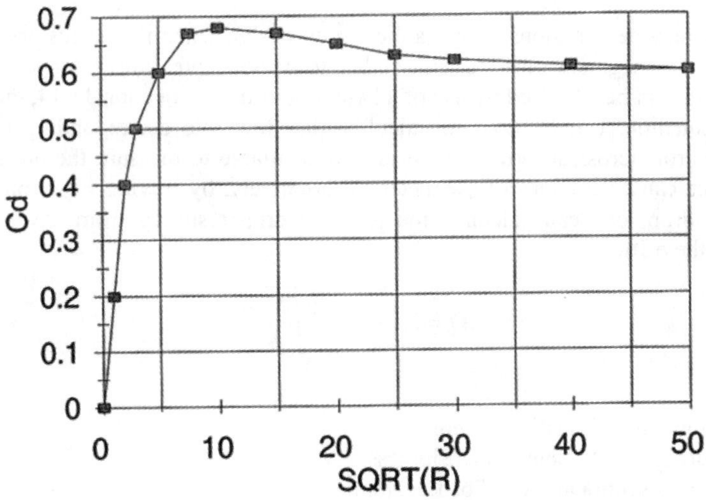

Figure 11.10. Sharp-edged orifice coefficient as affected by Reynolds number.

where

d_{eff} = effective orifice diameter, mm

A_o = cross sectional area of orifice, mm^2

L_{co} = length of orifice circumference, mm

The reader can verify that, for a circular orifice, the effective diameter is equal to the actual diameter of the orifice.

The reader will note that Figure 11.10 gives approximate C_d values for an ideal, sharp-edged orifice. Many valves form orifices that are used to control flow and the orifices are not ideal, i.e., Figure 11.10 does not describe their orifice coefficients. In such cases, Equation 11.14 can be used with experimental data to calculate C_d values. The experimenter takes simultaneous measurements of flow and pressure drop for various orifice areas, then uses Equation 11.14 to calculate the corresponding values for C_d.

11.7 Hydraulic Valves

Valves are used in hydraulic circuits to control pressure, volume flow rate, and direction of flow. Accordingly, valves are classified as pressure, volume, or directional control valves. Flow control valve is a commonly used alternate name for volume control valve.

11.7.1 Pressure Control Valves

The most common type of pressure control valve is the *pressure relief valve*, which is used to limit the pressure in a hydraulic circuit to a safe level. In a hydraulic circuit in which flow is supplied by a fixed-displacement pump, for example, the pump may continue to produce flow even when an actuator is stalled and incapable of accepting flow. In the absence of a pressure relief valve, the pressure would climb rapidly until the circuit ruptured at some point and provided an escape path for the flow. A pressure relief valve prevents such ruptures by providing a flow path back to the reservoir when the pressure reaches the pressure setting of the relief valve. A *direct-acting pressure relief valve* is illustrated in Figure 11.11. The inlet port is normally teed into the line from the pump to the directional control valve. When the *cracking pressure* is reached, i.e., when the pressure is high enough to lift the ball from the seat and compress the spring, oil can flow from the inlet port to the outlet port. As Figure 11.12 illustrates, the pressure drop across the relief valve increases with the flow rate through it. The increase in pressure drop from the cracking pressure to the full-flow pressure is called *pressure override*. The flow through the relief valve causes a substantial power loss, which can be calculated using Equation 11.5. Because the lost power is converted to heat, it is important to minimize such power losses. One way of doing so is by use of a *pilot-operated pressure relief valve*, as illustrated in Figure 11.13. The small relief valve (3) opens when the cracking pressure is reached and allows oil to flow from the small drain at the top to the reservoir. The resulting flow through passage (1) causes a pressure differential across the piston (6), and the pressure imbalance across the piston causes it to move upward to compress the large spring (5), thus opening a passage to the large outlet. As a result, the pilot-operated pressure relief valve has a much smaller pressure override than a direct-acting pressure relief valve, as shown in Figure 11.12.

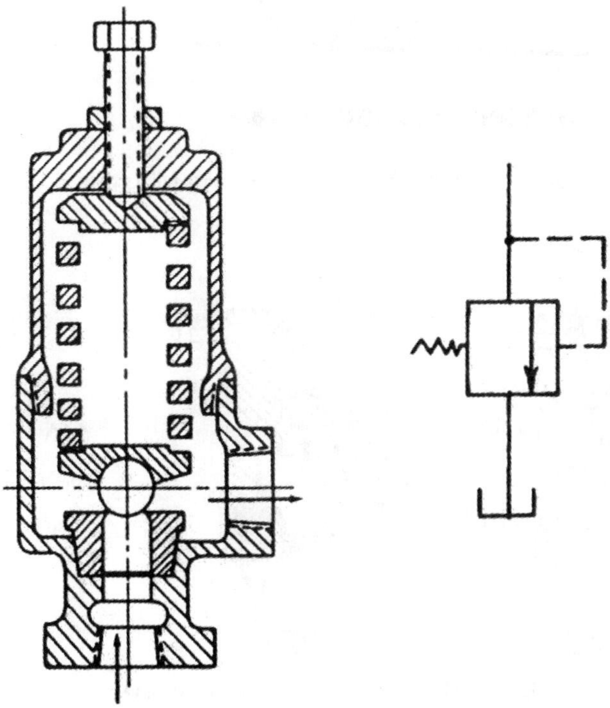

Figure 11.11. A direct-acting pressure relief valve.

An unloading valve, illustrated in Figure 11.14, is used to unload the pump when the pressure at some point in a hydraulic circuit reaches a desired level. When the pressure at the sensing port reaches that level, the plunger is pushed back against the spring until the groove in the plunger aligns with the inlet and outlet passages, thus allowing the pump to discharge freely to the reservoir. The unloading valve could be used as a pressure-relief valve by connecting the sensing port to the pump port.

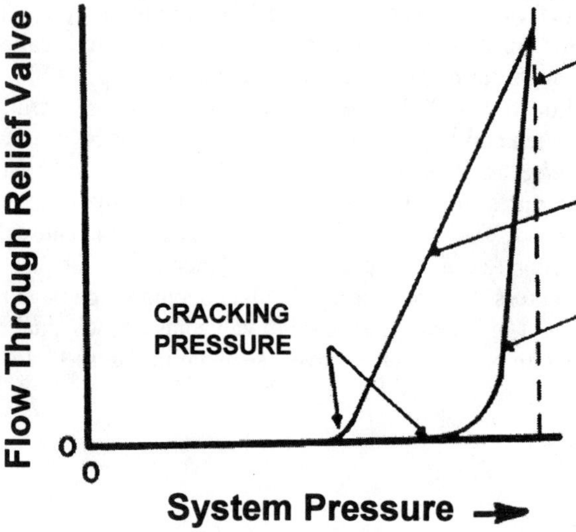

Figure 11.12. Typical characteristics of pressure relief valves.

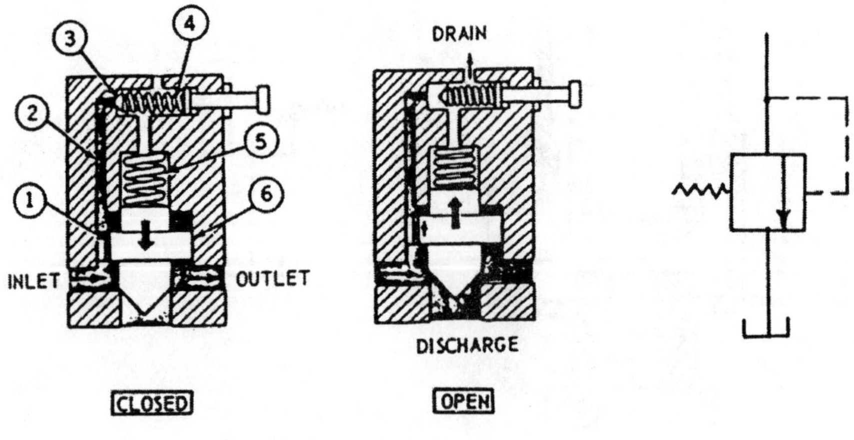

Figure 11.13. A pilot-operated pressure relief valve.

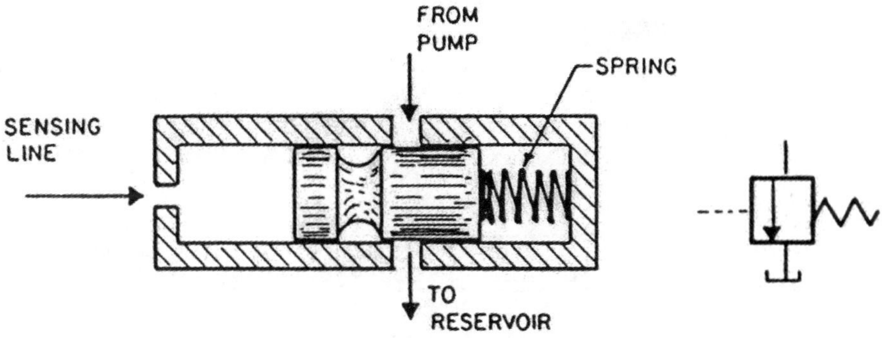

Figure 11.14. An unloading valve.

11.7.2 Volume Control Valves

Two types of volume control valves are shown in Figures 11.15 and 11.16. The *pressure-compensated throttling valve* regulates flow to the outlet port regardless of pressure variations in the downstream circuit. If pressure at the outlet port decreases, resulting in a momentary increase in pressure drop across the orifice and increased flow, the pressure change causes the spool to move to the right to further restrict and limit the flow. Conversely, if the outlet pressure rises, the spool moves to the left. The hand knob permits the user to adjust the orifice size to set the metered flow rate. Note the schematic symbol, which includes PC to indicate pressure compensation and an arrow to indicate that the metered flow rate can be adjusted. A simple hand valve could also meter flow but it would not be pressure compensated, i.e., the flow rate would be affected by pressure variations in the circuit.

Note that the throttling valve of Figure 11.15 would not be suitable if the inlet flow was supplied by a fixed-displacement pump. If the pump was supplying too much flow, the spool would move to the right to block the flow, causing even higher pressure at the inlet and causing the spool to move even further to the right. The throttling valve would close completely, forcing the pump to discharge through a relief valve with the consequent loss of hydraulic power. The *flow divider valve* of Figure 11.16 can be used with a fixed-displacement pump. Note that when the spool in the flow divider valve moves to the right to reduce flow to the outlet port, a bypass port opens to pass the excess flow. The flow divider valve is also pressure compensated. The flow divider valve is also called a *priority valve*. Suppose, for example, that 20 L/min of flow was needed for the vehicle power steering system, which was connected to the outlet port. If the pump was supplying 70 L/min, then 50 L/min would be routed to the bypass port to be used for other hydraulic functions. If the pump speed decreased until the pump was supplying only 25 L/min, the power steering system would still receive 20 L/min, but the bypass port would pass only 5 L/min. As this example illustrates, the outlet port receives first priority for the flow entering the valve.

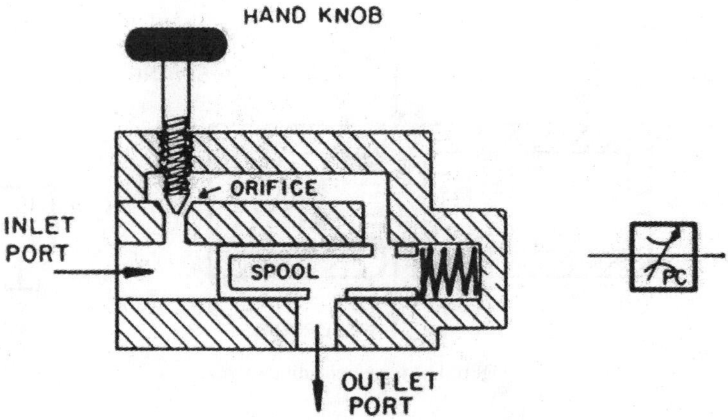

Figure 11.15. An adjustable pressure-compensated throttling valve.

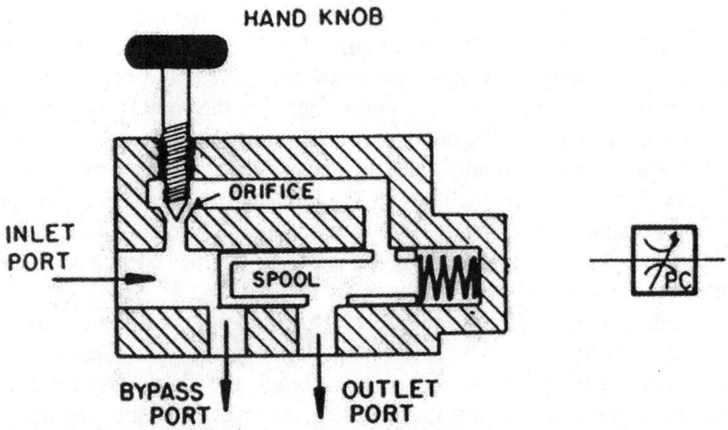

Figure 11.16. An adjustable pressure-compensated flow divider valve.

11.7.3 Directional Control Valves

Figure 11.17 illustrates a *directional control valve* (DCV). This DCV is a four-port device, i.e., it has four connections to the hydraulic circuit. One port connects to the pump, another to the reservoir (tank) and there are two work ports to connect to the circuit to be controlled. This is also a three-position valve, i.e., the spool has three possible positions, left, centered or right. Figure 11.17 also illustrates a closed-center valve, i.e., all ports are blocked when the spool is centered. If a hydraulic cylinder was connected to the work ports, the cylinder might extend when the DCV spool was moved to the left, hold in position when the spool was centered, and retract when the

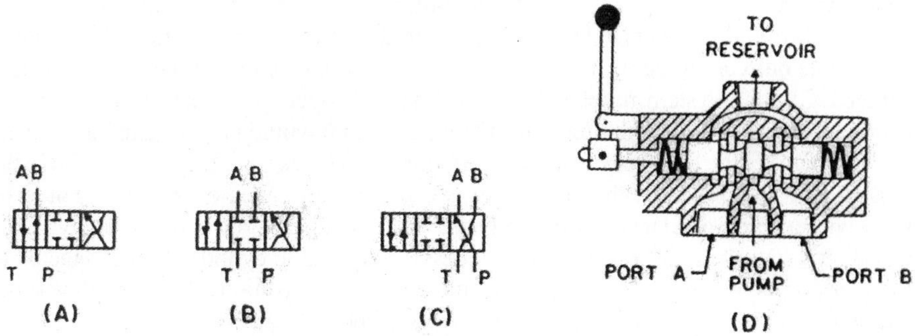

Figure 11.17. A closed-center directional control valve.

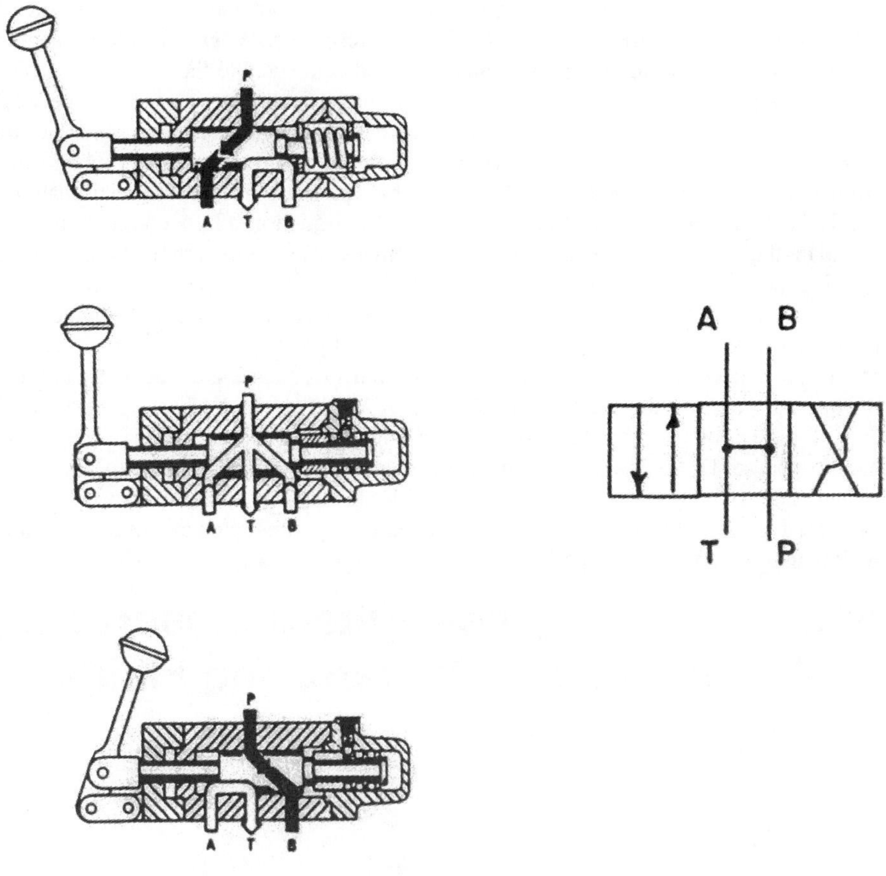

Figure 11.18. An open-center directional control valve.

spool was moved to the right. A variable-displacement pump is ordinarily used to supply oil to a closed-center DCV; then, when the DCV spool is centered, the pump reduces its delivery to zero. If a fixed-displacement pump supplied the oil to a closed-center DCV, the system relief valve would have to open when the DCV spool was centered, and all of the hydraulic power from the pump would be converted into heat.

An open-center DCV is illustrated in Figure 11.18. Its operation differs from the closed-center DCV only when the spool is centered. The open-center DCV could be used with a fixed-displacement pump because the pump could discharge freely to the reservoir when the DCV spool was centered. If a hydraulic cylinder was connected to the work ports, however, the cylinder would not hold in position when the DCV spool was centered, i.e., the cylinder position would be free to float.

A tandem-center DCV is illustrated in Figure 11.3n. When the spool is centered, the work ports are blocked but the pump can discharge freely to the reservoir. The valve of Figure 11.19 is also a tandem-center DCV, but it has two spools for control of two separate hydraulic circuits. It would be classified as a six-port, three-position, tandem-center DCV. Studying the dual DCV will show that the pump can discharge freely to the reservoir when both spools are centered, but moving either spool off center will block the free passage of oil to the reservoir. When the free passage is blocked, oil pressure can build up to move the load connected to the actuators.

The schematic symbols for the DCVs may suggest that the passages to the work ports are either fully open or fully closed. However, through careful movement of the spool, the operator can control the volume of flow through the work ports and thus control the speed of the actuator. The flow versus pressure differential relationship in the DCV is governed by orifice Equation 11.14. Equation 11.16 could be used to calculate the effective diameter of the non-circular orifice formed by the spool and its housing.

There are a wide variety of DCVs available to the hydraulic circuit designer. They differ as to the number of ports, the number of spool positions, and as to the means of moving the spool. The DCVs of Figures 11.17 and 11.18 each show a handle to allow the operator to move the spool manually. The DCV of Figure 11.17 has springs at each end of the spool to keep the spool centered until the operator moves it using the handle. Some DCVs are equipped with solenoids to allow control of the spool movement from a location remote to the DCV. It is not feasible to show all of the possible DCVs in this textbook, but the three that were illustrated are among the most widely used in off-road equipment.

11.8 Hydraulic Lines, Filters, Reservoirs, Accumulators, Coolers, and Fluids

While not closely tied to the logic of a hydraulic circuit, hydraulic lines, filters, reservoirs, and fluids are essential to the proper functioning of the circuit. Hydraulic accumulators provide a means of potential energy storage. In many cases, oil coolers are necessary.

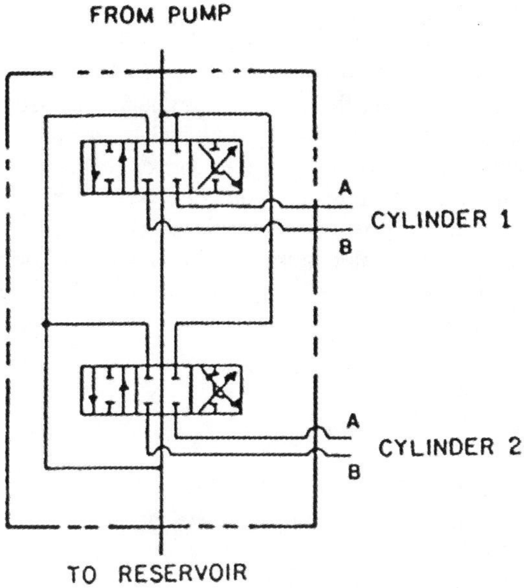

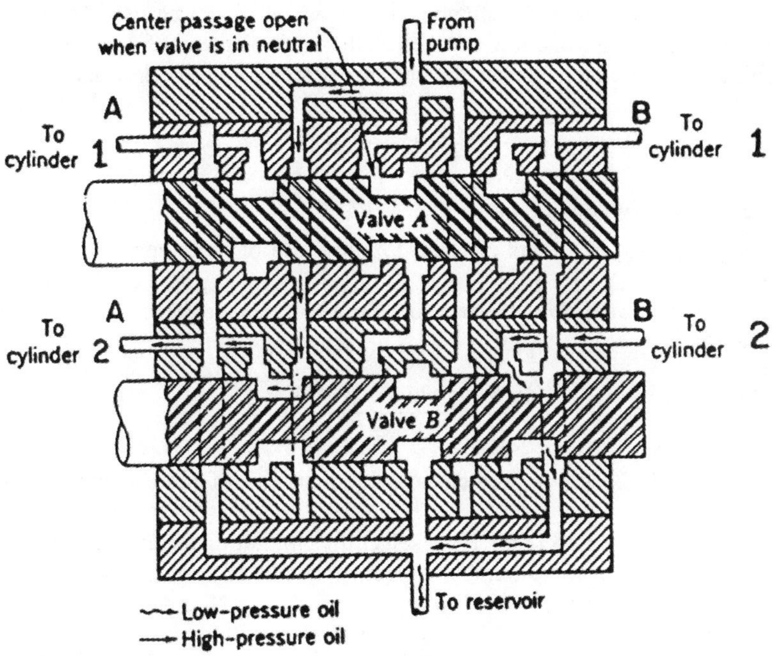

Figure 11.19. A tandem-center directional control valve stack.

11.8.1 Hydraulic Lines

Hydraulic lines, or conduits, are used to transfer hydraulic fluid between components. Typically, rigid lines are made from steel, while flexible lines are made from wire-reinforced rubber. In either case, there are two primary considerations in designing each line. The line must be strong enough to withstand the maximum pressure to which it will be subjected, and large enough to convey the hydraulic fluid without excessive pressure drop. Manufacturers of hydraulic hoses normally specify the limiting pressure rating of their hoses. For a line made of steel or other homogeneous material, the maximum allowable pressure is limited by the hoop stress, i.e., the stress that would cause rupture along a line parallel to the centerline of the conduit. The allowable pressure is

$$p_{max} = \frac{2\,t\,S_{des}}{d} \tag{11.17}$$

where

p_{max} = maximum allowable pressure, MPa
t = wall thickness of conduit, mm
d = conduit diameter, mm
S_{des} = design stress for conduit material, MPa

The steel hydraulic lines and rubber hoses used in off-road vehicles can be classified as smooth conduits. The flow in the lines can be either laminar or turbulent, depending upon the Reynolds number. Equation 11.15 can be used to calculate the Reynolds number for a line if d_o is taken as the inner diameter of the line. The Hagen-Poiseuille law is used to calculate the pressure drop for *laminar flow* in conduits, i.e.,

$$\frac{\Delta p}{L} = \frac{2.13\,\mu\,Q}{\pi\,d^4} \tag{11.18}$$

where L = length of conduit, m.

The constant, 2.13, is a units constant to allow use of more convenient units in the equation.

For fully *turbulent flow*, the pressure drop can be calculated using the following equation:

$$\frac{\Delta p}{L} = \frac{5.92\,\mu^{0.25}\rho^{0.75}Q^{1.75}}{d^{4.25}} \tag{11.19}$$

The constant, 5.92, is a units constant to allow use of more convenient units in Equation 11.19.

Use of Equation 11.18 or 11.19 to select the conduit diameter leads to an iterative solution, since the Reynolds number cannot be calculated until the conduit diameter is known. Both equations were used to plot Figure 11.20. Note that the flow is laminar for Reynolds numbers below 2500, fully turbulent for Reynolds numbers above 4000 and transitional for intermediate Reynolds numbers. Given the flow rate, the designer can use Figure 11.20 to select a pipe diameter and determine the approximate pressure drop per meter of conduit length, then use Equation 11.15 to calculate the Reynolds number. Knowing the Reynolds number, the designer can then choose the appropriate equation, either Equation 11.18 or Equation 11.19, to calculate the pressure drop more exactly. In hydraulic circuits on off-road vehicles, flow velocities in the conduits are usually high enough to produce turbulent flow.

Equations 11.18 and 11.19 are valid for straight conduits. Additional pressure losses result in conduits containing bends. The following equation can be used to calculate pressure drops in bends:

$$\Delta p = 0.139 K \rho \frac{Q^2}{A^2} \tag{11.20}$$

where

Δp = pressure drop, MPa
A = cross sectional area of conduit, mm^2
K = dimensionless factor from Fig. 11.21

The abscissa values in Figure 11.21 are ratios of bend radius over conduit inside diameter.

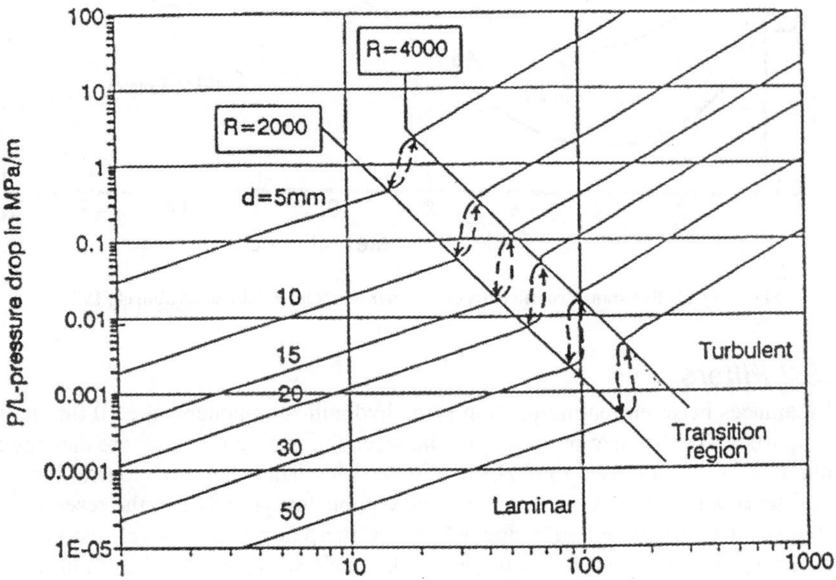

Figure 11.20. Pressure drops in hydraulic conduits for oil with specific gravity of 0.85 and dynamic viscosity of 27.6 mPa·s.

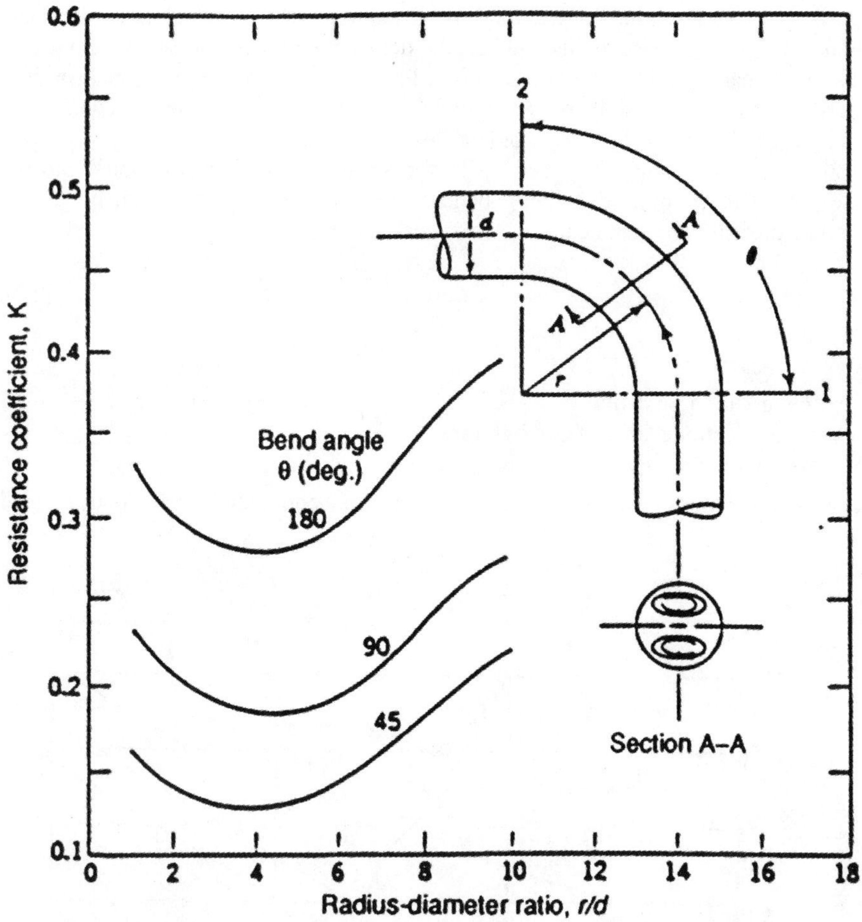

Figure 11.21. Resistance coefficients of pipe bends. (Reprinted from Taborek, 1959.)

11.8.2 Filters

Clearances between mating parts in some hydraulic components are 10 μm or less, and if particles of that size or larger pass between the mating parts, severe damage can result. Thus, filters are used to remove solid particles. There are three logical locations for a filter in a hydraulic circuit for off-road equipment: (a) between the reservoir and the pump inlet, (b) immediately downstream of the pump outlet, or (c) just upstream of the reservoir return port. Location (a) is seldom chosen as the sole filtration solution because the pressure drop across the reservoir could cause the fluid pressure at the pump inlet to be sub-atmospheric and cause *cavitation* in the pump. With cavitation, vapor bubbles form in the inlet port then implode while going through the pump to the high-pressure outlet port. Implosions near a metal surface can pull metal particles from the surface and damage the pump. If location (b) is used, the filter housing must withstand the maximum system pressure and thus a more expensive filter is needed. Location (c) is often chosen for the filter. To prevent large particles (150 μm or larger) from entering the pump, a strainer or porous filter is usually placed on the reservoir withdrawal tube. The filter in location (c) filters out the smaller particles, i.e., down to 2 μm in size.

The International Standards Organization (ISO) has developed an oil cleanliness code to aid in filter selection. ISO 4406 is based on two range numbers representing counts of 5 and 15 μm particles per 100 ml of sample fluid. The smaller size was thought to be representative of fine silt present in the fluid and the larger size was indicative of wear contaminants present. The ISO class number converts the number of particles per 100 ml of sample into class codes. The first class is 0 to 2 particles per 100 ml of sample. In the remaining classes, the upper class limit is double the lower class limit. For example, Class 2 is for 2 to 4 particles per 100 ml, Class 12 is for 2000 to 4000 particles per 100 ml, while Class 13 is for 4000 to 8000 particles per 100 ml.

The filtration ratio (β_{fr}) is

$$\beta_{fr}(x) = \frac{\text{No. of particles upstream}}{\text{No. of particles downstream}}$$

where x = particle size. The filter manufacturer assigns the β_{fr} values for each particle size by adding particles of that size upstream of the filter and then sampling the fluid downstream for particles of that size.

The filter efficiency is given by

$$e_{fil} = \frac{\beta_{fr} - 1}{\beta_{fr}}$$

A $\beta_{fr}(x) = 2$ is called the nominal rating and corresponds to removal of 50% of the particles of size x. A $\beta_{fr}(x) = 75$ is often called the absolute rating. It corresponds to 98.7% removal efficiency and higher values are difficult to verify statistically.

11.8.3 Reservoirs

Every hydraulic system includes a *reservoir* to supply hydraulic fluid to the pump and to provide storage for fluid returning from the hydraulic circuit. The reservoir must have sufficient volume to allow the returning fluid sufficient resident time to cool and to allow air to escape before the fluid re-enters the pump. If the reservoir cannot provide sufficient cooling, an oil cooler may be needed. The line supplying the pump must be below the fluid level in the reservoir to prevent air entry into the line. Although the return line could be above the fluid level, it is normally also below fluid level in the reservoir to prevent *air entrainment* and foaming of the fluid. The reservoir designer can use careful placement of the two reservoir ports and baffles, if necessary, to prevent the returning fluid from immediate entry into the pump port; otherwise, the fluid would not have time to cool. Finally, reservoirs normally operate at atmospheric pressure and thus are vented to the atmosphere. The vent must have a filter to prevent dust entry into the reservoir.

11.8.4 Hydraulic Accumulators

A cross-sectional view of a hydraulic accumulator is shown in Figure 11.22, along with the NFPA symbol for an accumulator. An inert gas above the diaphragm is compressed when hydraulic fluid is forced into the space below the diaphragm. The ideal gas law describes the compression and re-expansion of the inert gas. The compressed gas represents potential energy that can be re-converted into hydraulic energy when needed. For example, the stored energy could be used for emergency powering of power brakes or power steering during engine failure. Because the compressed gas provides cushioning, an accumulator can also be used as a shock absorber to reduce maximum stresses when the system is subjected to unusual loads.

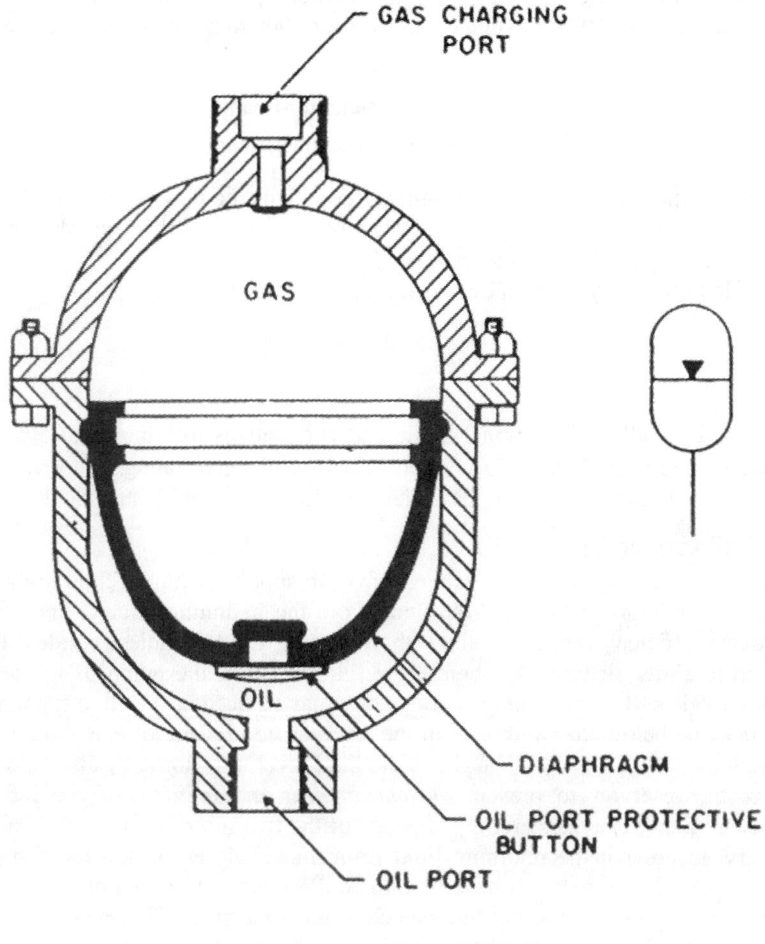

Figure 11.22. A hydraulic accumulator.

11.8.5 Oil Coolers

If the reservoir volume is too small to allow sufficient cooling of the hydraulic fluid, an *oil cooler* may be used. Typically, the oil cooler is a liquid-to-liquid heat exchanger that transfers heat from the hydraulic fluid to the engine coolant. Alternatively, a liquid to air intercooler can be used in which the heat exchanger transfers heat from the hydraulic fluid to the ambient air.

11.8.6 Hydraulic Fluids

The most important property of a *hydraulic fluid* is its viscosity. Manufacturers generally recommend fluid viscosities between 12 and 48 mPa·s at operating temperature. Oil viscosity is highly dependent on temperature, but the reduction in viscosity at high temperatures is less for oils with a high viscosity index. The relationship between viscosity and temperature for a petroleum-based hydraulic fluid was given by Equation 5.12. Viscosity control is important because pump and motor efficiencies depend on viscosity, as discussed in Section 11.4.7 and 11.5.1.

The density of a hydraulic fluid varies with both temperature and pressure, according to the following linearized equation:

$$\frac{\rho}{\rho_s} = 1 + \frac{1}{\beta_e}(p - p_s) - \alpha(T - T_s) \tag{11.21}$$

where

ρ = fluid density, kg/L
p = pressure, MPa
T = temperature, °C
ρ_s, p_s, and T_s = reference values at some standard condition

The *bulk modulus*, β_f, of the hydraulic fluid is defined as

$$\beta_f = \rho_s\left(\frac{\partial p}{\partial \rho}\right)_T$$

The thermal expansion coefficient, α, is defined as

$$\alpha = \frac{1}{\rho_s}\left(\frac{\partial \rho}{\partial T}\right)_p$$

The effective bulk modulus of petroleum is the change in pressure associated with a change in volume of a given mass of fluid; its value is above 1500 MPa for petroleum-based hydraulic fluids, i.e., the fluid is virtually incompressible. In practice, small amounts of entrained air reduce can appreciably decrease the effective value of β_f. The thermal expansion coefficient for petroleum-based fluids is about 0.9×10^{-3}/°C.

Petroleum-based hydraulic fluids are subject to oxidation. The oxidation rate doubles for every 10°C increase in temperature, but the rate is very low for temperatures below 60°C. Additives are also used in hydraulic fluids. On off-road vehicles, the transmission gear case is often used as the reservoir for the hydraulic system. Then the hydraulic fluid must also lubricate the transmission. Vehicle manufacturers formulate special fluids to serve this dual purpose. An anti-foaming additive is usually used to reduce the surface tension of the fluid and thus reduce foaming. A rust inhibitor may be used to protect metal surfaces from rusting. An anti-wear additive may be used to reduce wear at points in the hydraulic pump or motors that may be subject to boundary lubrication. Finally, an extreme pressure additive may be used to reduce wear at the high pressure points at which the transmission gear teeth mesh.

11.9 Types of Hydraulic Systems

Most off-road vehicles have *live hydraulic power*, i.e., the hydraulic pump is driven directly by the engine so that hydraulic power will be available whenever the engine is running. When no hydraulic power is needed, the hydraulic system is said to be in *standby*. When the system is in standby, any power input to the pump is converted to heat. Thus, means must be found to reduce the standby power to near zero. As Equation 11.5 shows, there are three different methods to reduce standby power to near zero and each of these has led to a different type of hydraulic system. In an *open-center* (OC) system, the pump pressure is reduced to near zero during standby. In a *pressure-compensated* (PC) system, the flow is reduced to near zero during standby. In a *pressure-flow-compensated* (PFC) system, both the pressure and flow are close to zero during standby. PFC systems are also called *load-sensing* (LS) systems. In the discussions that follow, each type of system is illustrated with two DCVs for use in controlling up to two actuators simultaneously. In practice, more than two DCVs could be used; only two were shown for simplicity.

11.9.1 Open-Center Systems

An OC hydraulic system is illustrated in Figure 11.23. The pressure-flow relationship for the OC circuit is shown in Figure 11.24. The system has a twin-spool DCV for control of two hydraulic cylinders. When both spools are centered, the open-center DCVs permit the pump to discharge to the reservoir at near-zero pressure, i.e., the system is operating at Point A in Figure 11.24. When either DCV spool is moved off center, the open passage to the reservoir is blocked, thus allowing pressure to build to move the actuator. With increasing actuator load, the system moves toward Point B in Figure 11.24. The circuit has a relief valve to protect against excess pressure if an actuator stalls. At Point B on Figure 11.24, system pressure has reached the cracking pressure of the relief valve and flow begins to divert to the reservoir. At Point C, all of the oil is being diverted to the reservoir before reaching the DCV.

The OC system can encounter difficulty when trying to power two actuators simultaneously. Suppose, for example, that the relief valve cracking pressure was 30 MPa, cylinder A required 25 MPa of pressure to move its load, and cylinder B needed 20 MPa of pressure to move its load. In this example, when both DCV spools were moved off center, the pressure would rise to 20 MPa; cylinder B would move but cylinder A would be stalled until cylinder B reached the end of its stroke. Then the pressure could rise to 25 MPa to move cylinder A. This action is called *sequencing*, since the actuators move sequentially rather than simultaneously. Flow divider valves

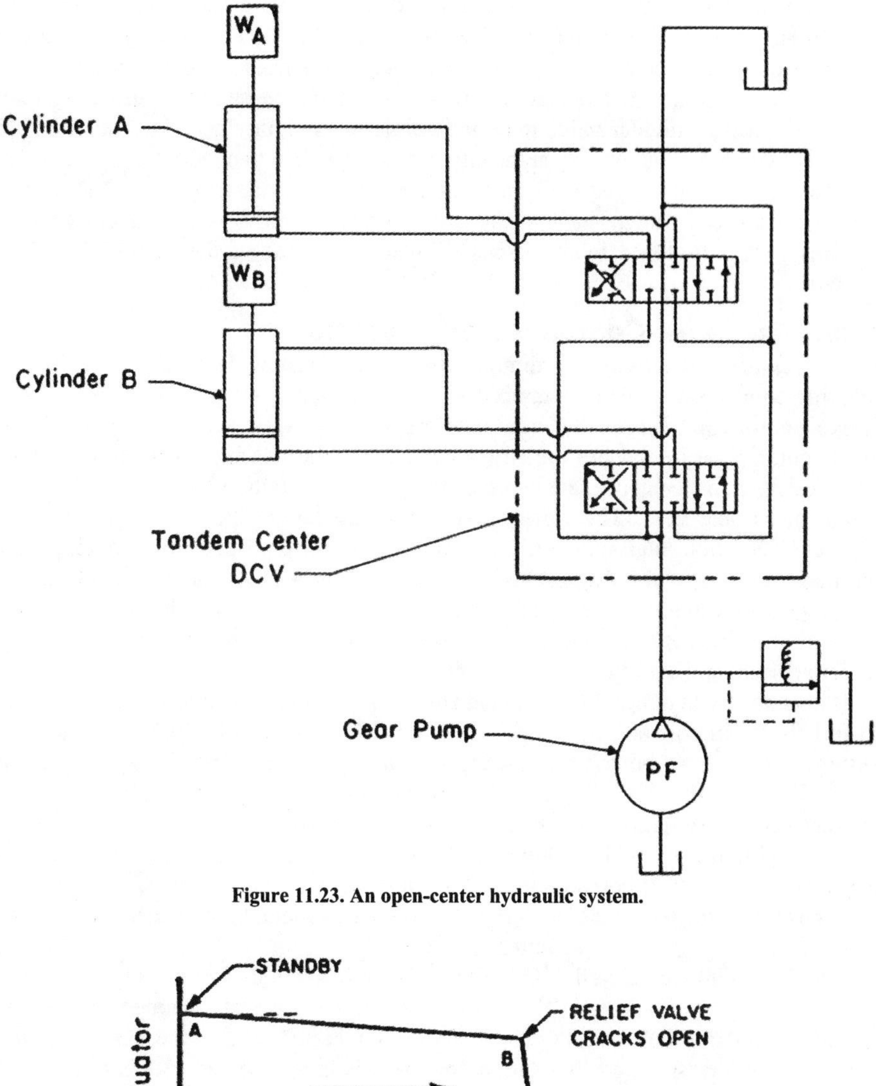

Figure 11.23. An open-center hydraulic system.

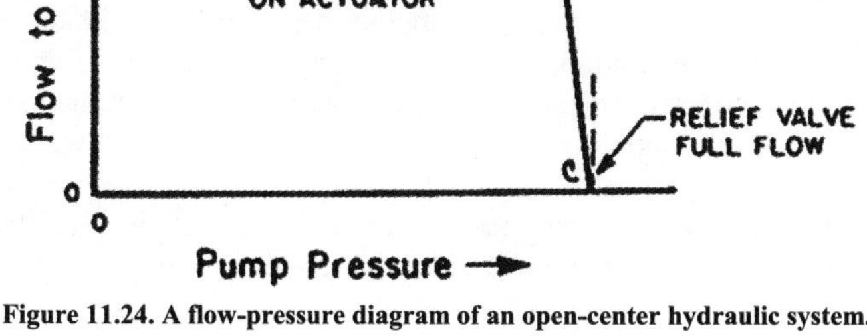

Figure 11.24. A flow-pressure diagram of an open-center hydraulic system.

are available to prevent sequencing, but at the cost of additional expense and power loss. In the above example, the pressure could rise to 25 MPa to move both cylinders simultaneously, but the flow divider valve would experience a 5 MPa pressure drop in feeding oil to cylinder B. The resulting power loss could be calculated using Equation 11.5. Also, a flow divider valve may be feasible for dividing the flow when a circuit has only two DCVs, but not when the circuit has more than two DCVs.

The OC systems worked well on vehicles with one actuator and are still used on some vehicles. OC systems have the advantage of using an inexpensive gear pump. PC systems were introduced in the 1960s to overcome some of the limitations of OC systems.

11.9.2 Pressure-Compensated Systems

A pressure-compensated hydraulic system is illustrated in Figure 11.25. A pressure-compensated piston pump is designed to maintain constant output pressure. A stroke control valve senses the outlet pressure and causes the pump stroke to decrease as the outlet pressure reaches the target level. When the system is in standby with the DCV valve spools centered, the pump cannot discharge to the reservoir and the pump discharge is near zero. The flow-pressure diagram for the PC system is shown in Figure 11.26. Note that standby is at Point A, i.e., at high pressure but zero flow. Normal operation of the PC system is between Points A and B on Figure 11.26. The stroke control valve needs a small decrease in pressure to cause the pump to move toward full stroke. If the circuit demand is not excessive, the pump can maintain high outlet pressure and avoid actuator sequencing.

If actuators with a high flow demand are connected to the circuit, the pump may go to full stroke and operate between Points B and C on Figure 11.26. In this case, the system has the same behavior as an OC system and actuator sequencing may occur. When PC systems first replaced OC systems, system pressure levels were raised appreciably. As Equation 11.5 shows, the necessary hydraulic power could then be attained with much smaller flow rates. Actuator speed is maintained by use of cylinders with smaller bores compared to those used on OC systems. Use of lower flow rates and smaller-bore cylinders helps to restrict operation between Points A and B on Figure 11.26 and thus prevent sequencing.

The PC circuit works well with hydraulic cylinders, but difficulty can arise when a hydraulic motor is connected. If the motor displacement is small enough to cause the system to operate between Points A and B on Figure 11.26, the flow is highly dependent on pressure and thus the motor speed is highly dependent upon the motor torque load. Even small variations in torque load can cause substantial variations in motor speed. A solution to this problem is to select a motor with large enough displacement to cause the PC system to operate between Points B and C on Figure 11.26. In this region of operation, the flow is nearly independent of pressure and the motor can maintain constant speed. If the operator attempts to simultaneously operate a second actuator with a higher pressure demand than the motor, however, the second actuator will stall.

Maintaining high standby pressure is a disadvantage of the PC system. Load-sensing systems were developed to overcome this disadvantage.

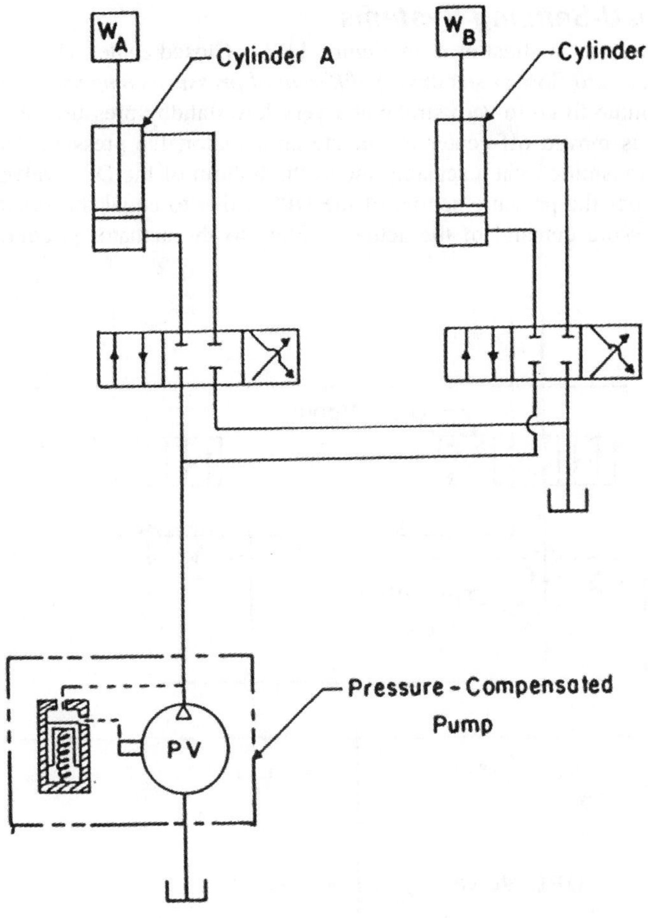

Figure 11.25. A pressure-compensated hydraulic system.

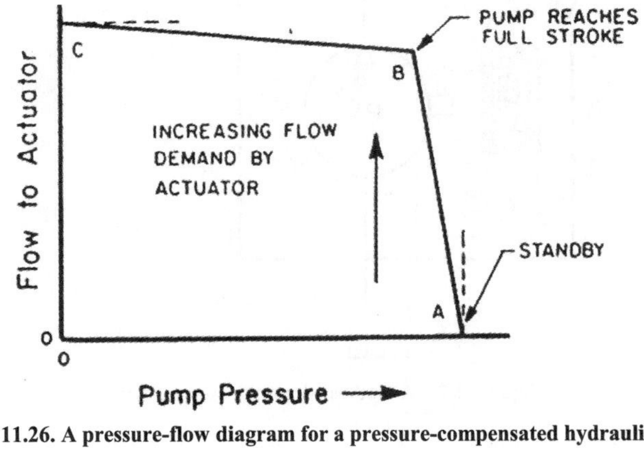

Figure 11.26. A pressure-flow diagram for a pressure-compensated hydraulic system.

11.9.3 Load-Sensing Systems

An LS system is illustrated in Figure 11.27. Closed-center DCVs are used to maintain near zero flow at standby. A *differential pressure compensator* (DPC) *valve* causes the pump to go to zero stroke at a very low standby pressure. When only one DCV spool is moved off center to operate an actuator, the pressure demand of the actuator is transmitted via a sensing line to the bottom of the DPC valve; the sensed pressure boosts the pressure setting of the DPC valve to equal the *standby pressure* plus the pressure demand of the actuator. Flow to the actuator is controlled by an

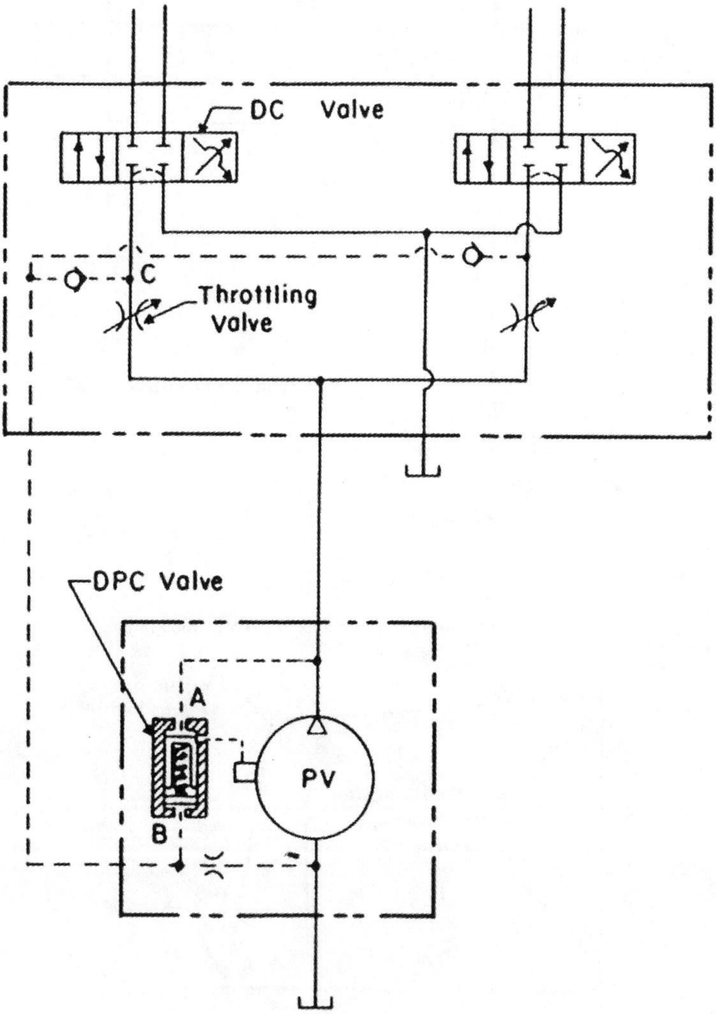

Figure 11.27. A load-sensing hydraulic system.

adjustable throttling valve similar to the one illustrated in Figure 11.15. The pressure drop across the throttling valve is always equal to the pump standby pressure. The flow-pressure diagram for the LS system would be similar to Figure 11.26, except that standby would be close to the origin of the graph. System pressure is controlled by actuator demand, up to a maximum safe pressure level at which point a stroke-control valve causes the pump to go to zero stroke. Flow is controlled by the setting of the throttling valve but cannot exceed the pump capacity at full stroke.

Next, consider the case when two DCV valve spools are moved off center to deliver oil to two actuators simultaneously. Assume a different pressure demand for each actuator. Because of the check valves shown near the throttling valves, only the higher of the pressure demands is transmitted to the DPC valve. Thus, both actuators can move simultaneously and sequencing is avoided if the total flow demand is not excessive. However, the throttling valve serving the actuator with the lowest pressure demand will experience a pressure drop larger than the standby pressure and thus will cause some hydraulic power loss.

The designer of an LS system has a dilemma in choosing the capacity of the throttling valves. For situations in which only one actuator is being used, it seems desirable for the throttling valve capacity to equal the pump discharge at full stroke; if not, the full pump capacity could not be utilized. If two actuators are used and both throttling valves are set to the wide-open position, the pump will quickly go to full stroke and the LS system will operate between Points B and C on Figure 11.26, that is, just like an open-center system. In that situation, actuator sequencing can occur. Thus, to prevent sequencing in an LS system, it is the vehicle operator's responsibility to set the throttling valves such that their total flow capacity is a little less than the pump capacity at full stroke.

Hydraulic motors operate well on an LS system. The system pressure is controlled by the motor torque, while the motor flow and thus the motor speed are controlled by the setting of the throttling valve. An LS system uses more expensive components than the OC or PC systems, but has sufficient advantages to make it the system of choice on many off-road vehicles.

11.10 Introduction to Transient Analysis of Hydraulic Systems

Analyzing transient behavior of a hydraulic circuit usually involves complex computer simulations. However, a simple example will illustrate some of the techniques involved. Imagine a pressure-compensated axial piston pump is connected to the pump port of the DCV shown in Figure 11.3(m). Imagine one port of a hydraulic motor is connected to work port A and the other to work port B. Port T of the DCV is connected to the reservoir. Imagine a solenoid is used to shift the DCV spool to the right to connect the pump to work port A. For simplicity, assume the pump or a pump-accumulator combination can maintain constant pressure, p_p, at the pump port.

We imagine the simulation to begin when the solenoid is energized to shift the DCV spool. The spool movement is controlled by the spool dynamics, i.e.,

$$\ddot{x}_s = \frac{1}{m_s}(F_s - F_f) \tag{11.22}$$

where

x_s = spool displacement; the double dot indicates second derivative of displacement with respect to time, i.e., spool acceleration

m_s = mass of spool

F_s = solenoid force, which could vary with time

F_f = spool friction force, which varies with spool velocity

Spool displacement is calculated by two integrations of Equation 11.22. Spool movement creates two orifices in the DCV, one connecting port P to port A and the other connecting port B to port T. For simplicity, we will assume these orifice areas are equal and are given by

$$A_{vo} = f(x_s) \tag{11.23}$$

The function, f (x_s), must be determined from the geometry of the DCV.

The DCV spool movement allows flow to enter the line connected to work port A. For simplicity, we assume pressure is equal throughout this line. The bulk modulus equation, Equation 7.2, can be differentiated to obtain the following equation for line A, the line connected to work port A:

$$\dot{p}_A = \frac{\beta_{fA}}{V_A}(Q_{Ain} - Q_{Aout}) \tag{11.24}$$

where

p_A = pressure in line A and the dot represents time derivative.

β_{fA} = effective bulk modulus of the fluid in line A

V_A = internal volume of line A

Q_{Ain} = instantaneous flow into line A

Q_{Aout} = instantaneous flow out of line A

The same equation would apply to line B, except that subscripts B would be substituted for subscripts A. The pressure in each line is calculated by integrating each respective equation.

The motor has an external load, T_L, which is assumed to be some known function of time. The motor will accelerate if the motor torque from Equation 11.9 is greater than T_L or decelerate if the converse is true. The motor acceleration or deceleration is given by the following equation:

$$I_m \dot{N}_m = \frac{D_m e_{mt}}{2\pi}(p_A - p_B) - T_L \tag{11.25}$$

Here, I_m is the mass moment of inertia of the moving parts of the motor plus any moving parts driven by the motor. The dot over the motor speed indicates motor acceleration and the motor speed are calculated by integrating Equation 11.24.

The following three equations can be used to calculate fluid flows in the circuit:

$$Q_{Aout} = Q_{Bin} = \frac{D_m N_m}{1000\, e_{mv}} \tag{11.26}$$

and

$$Q_{Ain} = 2.68\, A_{vo} C_d \sqrt{\frac{(p_p - p_A)}{\rho}} \tag{11.27}$$

and

$$Q_{Bout} = 2.68\, A_{vo} C_d \sqrt{\frac{(p_B - 0)}{\rho}} \tag{11.28}$$

In numerical simulation software, the integrator outputs (x_s, p_A, p_B, and N_m) are assumed to be known. Then sufficient information is available to calculate the valve orifice area, motor acceleration, line pressure derivatives, and all flow rates to complete the simulation. Each integrator requires specification of an initial condition. For example, the valve spool could initially be centered ($x_{so} = 0$), the motor starting speed could be assumed to be zero, and the initial line pressures could be assumed to be zero gage pressure. When the simulation began, each of these quantities would begin varying with time as calculated by the simulation.

Even in this simple simulation, some complications arise. One of these has to do with the valve spool dynamics. Another term is actually needed in Equation 11.21 to model the effect of the spool reaching the end of its travel. Because of the elasticity of the spool and the valve housing, the spool typically bounces at the end of its travel before settling to a stop. While the spool is bouncing, the valve orifice area is oscillating in size. Another complication is that the pressure-compensated pump typically does not maintain unvarying outlet pressure. A more accurate simulation would include equations to model the exact pump performance. Despite such complications, modeling of the transient behavior of hydraulic circuits can be very valuable to the circuit designer and can help the designer to foresee and avoid problems in the performance of the circuit.

11.11 Mechatronics and System Control

Many functions accomplished by hydraulics were previously accomplished by mechanical devices in earlier off-road vehicles. However, the great flexibility of hydraulic systems has allowed them to replace mechanical systems in many cases. Hydraulic systems and especially mechatronic systems allow much more precise and flexible control than mechanical systems.

11.11.1 An Introduction to Mechatronics

Electrohydraulic valves have been used in hydraulic circuits for many years to allow control of the valves at locations remote from the valves. As discussed in Chapter 10, microprocessors are becoming ever more widely used in the control of off-road vehicles. Microprocessor control of electrohydraulic valves is one example of the emerging field of *mechatronics*. The microprocessors in mechatronic devices allow such devices unique capabilities and great flexibility. Figure 11.28 is an example of a prototype mechatronic valve.

Through microprocessor control, the valve in Figure 11.28 can perform many functions in a hydraulic circuit. The valve has four ports, as in an ordinary, single-spool DCV. Five proportional cartridge valves can block or open any of the possible passages between the four ports. A pressure transducer in each port allows the microprocessor to determine the pressure in each port. Cartridge valve 5 is normally open but can be closed on signal from the microprocessor. The other four cartridge valves are normally closed but can be opened on signal from the microprocessor. Although the mechatronics valve of Figure 11.28 can perform many functions, only a few will be discussed.

The mechatronics valve removes the distinction between open-center and closed-center DCVs. The computer keeps cartridge valve 5 closed to simulate a closed-center valve. For the mechatronics valve to function in an open-center circuit, the computer keeps cartridge valve 5 open until actuator action is needed. For example, to extend a hydraulic cylinder, the computer might open cartridge valves 1 and 4, then smoothly close cartridge valve 5 to cause a smooth start to cylinder extension. When the pressure in work port A indicated the cylinder had reached the end of its stroke, the computer would close cartridge valves 1 and 4 while opening cartridge valve 5. The reader can deduce which cartridge valves the computer would need to control to cause the cylinder to retract.

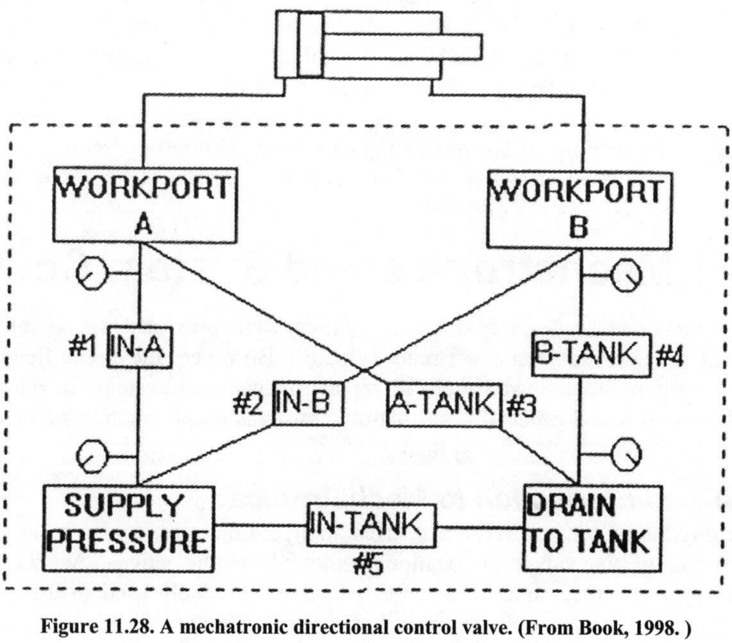

Figure 11.28. A mechatronic directional control valve. (From Book, 1998.)

The mechatronic valve of Figure 11.28 has built-in pressure relief. If the pressure in the supply pressure port became too high, for example, the computer could cause cartridge valve 5 to open to provide a flow path to the tank. Compared to the ordinary relief valves discussed in Section 11.7.1, the mechatronics relief valve could have near zero pressure override.

In addition to providing relief against excessively high pressures, the mechatronics valve can provide protection against sub-atmospheric pressures and cavitation. Consider the circuit that was discussed in Section 11.10. Further consider the situation if the hydraulic motor was powering a high-inertia load at high speed when the DCV was closed to stop the motor. Momentarily, the external load would drive the motor, causing it to function as a pump to remove fluid from line A and pump fluid into line B. With the DCV closed, there would be no inflow to line A or outflow from line B. The pressure in line A would fall below atmospheric pressure and could possibly cause collapse of the line. If the mechantronic valve was used in place of the DCV, the computer could sense the approach of sub-atmospheric pressure and open cartridge valve 3 to allow oil from the reservoir to relieve the vacuum. Note also that the mechatronics valve could also close cartridge valve 4 in a controlled way to stop the hydraulic motor without causing excessively high pressures in line B.

11.11.2 System Control

Many functions are powered and controlled by hydraulics on off-road vehicles and space does not permit a discussion of all of them. Instead, we will discuss only one control function, the control of the hitch used to link implements to tractors. Many types of hitches have been used in the past, but the *three-point hitch* illustrated in Figure 11.29 has become the standard hitch. Figure 11.30 shows three-point hitch connections on an implement. Interchangeability is achieved through standardization of three-point hitches, i.e., it is possible for implements from one manufacturer to be connected to another manufacturer's tractors. As shown in Table 11.1, there are four different three-point hitches. A Category I hitch can be used for tractors with 15 to 35 kW of drawbar power; Category II is for 35 to 75 kW of drawbar power, Category III for 60 to 168 kW, and Category IV is for 135 to 300 kW. There is currently no three-point hitch available for tractors with more than 300 kW of drawbar power.

Standard terminology is used to describe the three-point hitch. The points at which the hitch links attach to the tractor are called *link points*. The points at which the hitch links attach to the implement are called *hitch points*. The key elements of hitch standardization relate to the geometry of the lower hitch studs and the upper hitch pin (Figure 11.30). The mast height is also standardized.

Figure 11.31 is an example of a mechanical system for controlling a three-point hitch. A single-acting hydraulic cylinder rotates lift arms for raising the two lower links. The weight of the attached implement provides the force for lowering the hitch. In Figure 11.31, the operator moves the position control handle to the left to lower the hitch or to the right to raise the hitch. Moving the position control handle displaces the spool in the DCV that controls the hydraulic lift cylinder. The mechanical linkage senses the resulting movement of the hitch and moves the spool to stop oil flow to the cylinder when the hitch has reached the desired position. Thus, because of the

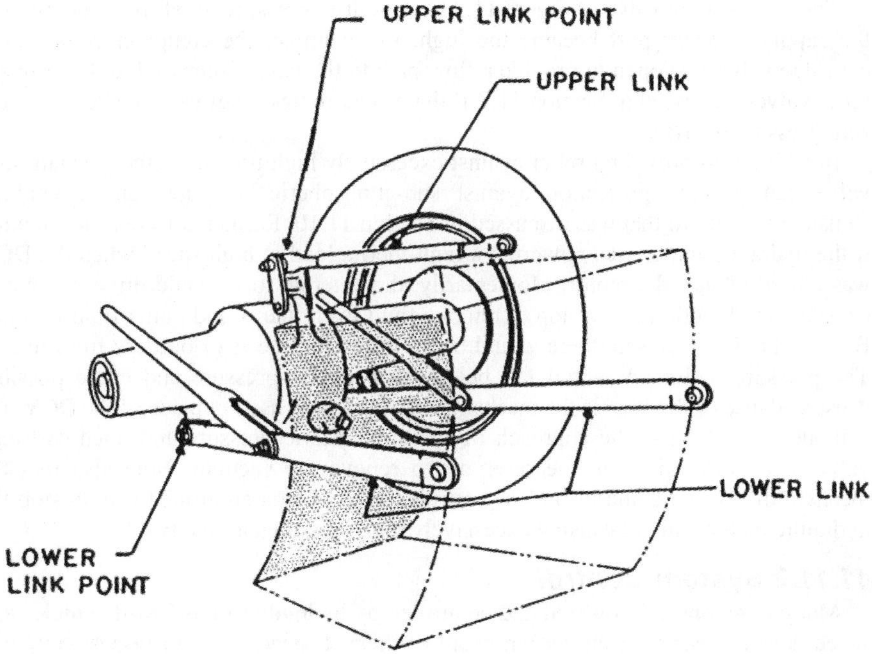

Figure 11.29. A three-point hitch for a tractor.

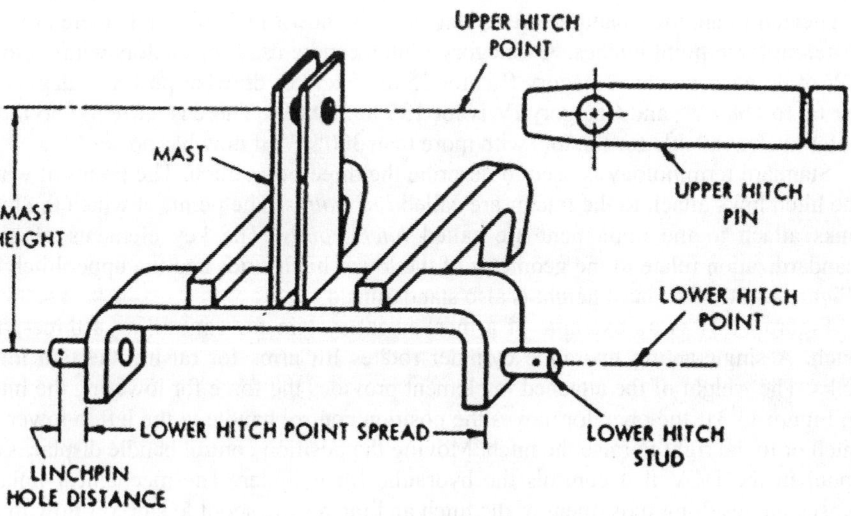

Figure 11.30. Three-point hitch connections on an implement.

	Category I				Category II				Category III[#]				Category IV[#]			
	Millimeter		Inches		Millimeter		Inches		Millimeter		Inches		Millimeter		Inches	
	Min	Max	Min	Max	Min	Max	Min	Max	Min	Max	Min	Max	Min	Max	Min	Max
UPPER HITCH POINT																
Width inside	44.5	-	1.75		52.3	-	2.06	-	52.3	-	2.06	-	65	-	2.56	-
Width outside	-	85.9	-	3.38	-	95.3	-	3.75	-	95.3	-	3.75	-	132	-	5.20
Clearance radius for upper link*†	57.2	-	2.25	-	57.2		2.25	-	57.2	-	2.25	-	76.2	-	3.00	-
Hitch pin hole diameter	19.3	19.56	0.76	0.77	26.65	25.91	1.01	1.02	32.0	32.26	1.26	1.27	45.2	45.5	1.78	1.79
LOWER HITCH POINT																
Stud diameter	21.84	22.10	0.86	0.87	28.19	28.45	1.11	1.12	36.32	36.58	1.43	1.44	49.7	50.8	1.96	2.00
Linchpin hole distance†	38.86		1.53		46.52		1.91		48.52	-	1.91	-	68		2.68	-
Linchpin hole diameter	11.68	12.19	0.46	0.48	11.68	12.19	0.46	0.48	11.68	12.19	0.46	0.48	17.5	18	0.69	0.71
Lower hitch point spread	681.0	684.3	26.81	26.94	822.5	825.5	32.38	32.50	963.7	966.7	37.94	38.06	1165	1168	45.87	45.99
Clearance radius for lower link*†	63.5		2.50		73.2		2.88		82.6	-	3.25		82.6	-	3.25	-
Implement encroachment in front of lower hitch point if implement extends laterally behind tire		12.7		0.5		12.7		0.5		12.7	-	0.5	-	12.7		0.5
IMPLEMENT MAST HEIGHT†§	457		18		483		19		559		22		686		27	

*Some tractors with quick-attachable connectors require 140 mm (5.50 in.) space for clearance above the upper hitch point and below the lower hitch points.

†Refer to standard for attachment of implements to agricultural wheel tractors equipped with quick-attaching coupler for three-point free link hitch. Tractors equipped with a standard quick-attaching coupler for the three-point free-link hitch require an auxiliary attaching pin on the implement mast located 76 mm (3.0 in.) for categories I and III, and 102 mm (4.0 in.) for Category II, below the standard upper hitch point. To facilitate attachment and detachment of the implement a clearance zone must be maintained 76 mm (3.0 in.) rearward from and extending 104 mm (4.10 in.) above and 216 mm (8.50 in.) below this pin. In addition, a clearance zone must be maintained 94 mm (3.70 in.) rearward from and extending 25 mm (1.0 in.) above and 211 mm (8.30 in.) below each lower hitch point.
To facilitate car attachment and detachment of the implement with tractors equipped with a standard Category IV-N or IV quick-attaching coupler for the three-point free-link hitch, a clearance zone must be maintained 85 mm (3.35 in.) rearward from and extending 120 mm (4.75 in.) above and 252 mm (9.92 in.) below the standard upper hitch point on the implement mast. In addition, a clearance zone must be maintained 94 mm (3.7 in.) rearward from and extending 32 mm (1.26 in.) above and 272 mm (10.71 in.) below each lower hitch point.

§The mast height is not necessarily a mechanical dimension on the implement itself. It is a figure used in design and if properly used for design of both implement and tractor, a well performing interchangeable implement and tractor combination will be achieved. This standard makes it possible to produce tractors and implements that will give good performance in any combination; therefore, consideration to hitch geometry is essential. This makes it desirable to establish a standard mast height and a standard mast adjustment within a working range, because these items influence the position of hitch points that are common to both the implement and the tractor. Mast height is one of the essential factors in establishing the virtual hitch point of the free-link system, draft signal for the draft-responsive system, loads on the linkage and hitch points, changes in implement pitch corresponding to changes in working depth, implement pitch when the implement is in transport position, clearance of the implement with the tractor, especially in transport position and clearance of the hitch links with the implement or with the tractor, especially in the transport position. When an implement mast height is made different than standard to accomplish some specific performance feature, care should be exercised to insure that the desired performance is secured with tractors likely to operate the implement.

‡Some Category II tractors are designed to accommodate a 559 mm (22 in.) mast height for optimum performance. In the design of implements for use on these tractors care should be taken to investigate the need for providing a 559 mm (22 in.) mast height. Tractors that are designed only for use with 559 mm (22 in.) mast height must be properly identified.

#See Section 3 for Category IV-N or III-N, Narrow Hitch, dimensions.

Table 11.1. Dimensions for four categories of three-point hitches.

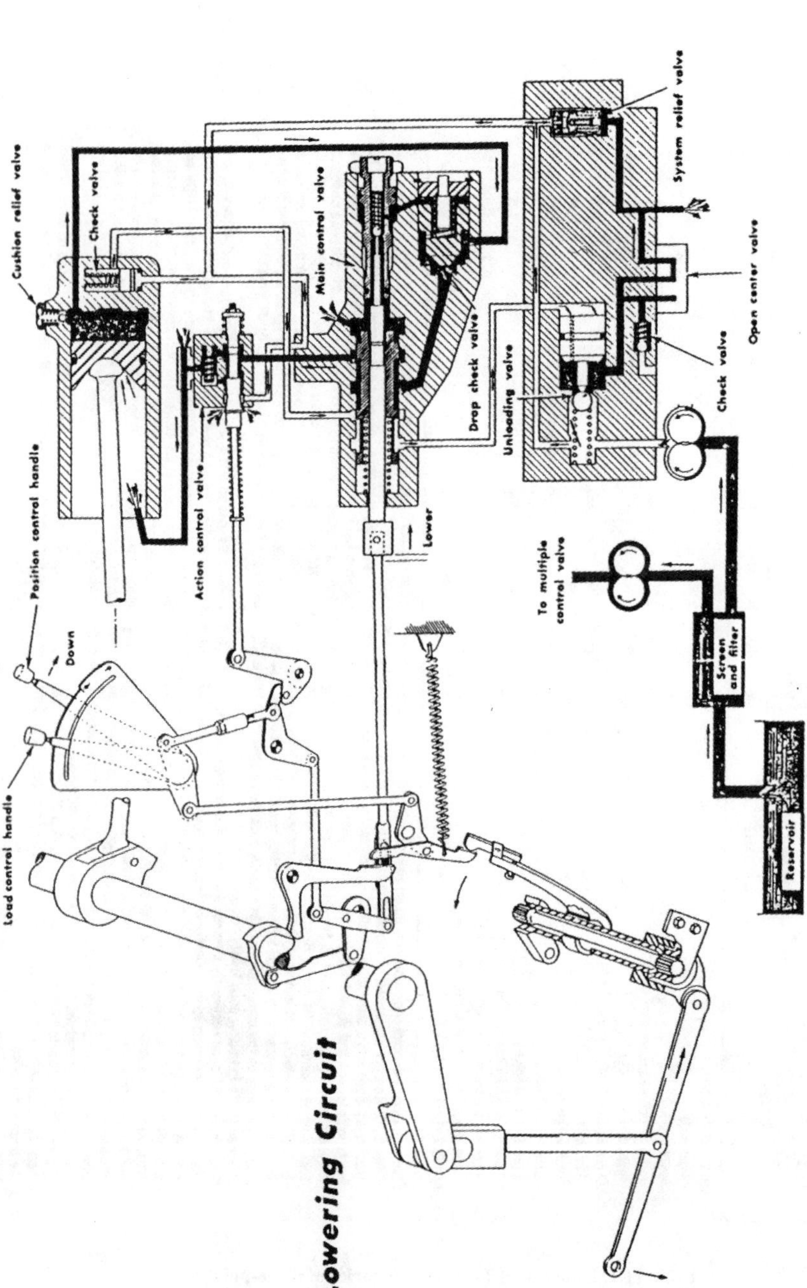

Figure 11.31. Mechanical control of a three-point hitch.

feedback linkage, the hitch automatically moves to a height proportional to the setting of the position control handle.

The earliest use of three-point hitches was for mounting a plow to a tractor. Undulating soil surfaces caused difficulties during plowing with a fully mounted plow. When the tractor front wheels encountered an incline or the tractor rear wheels encountered a swale, the plow depth could increase until the tractor engine approached stall. Operators found they could avoid shifting to a lower gear by raising the hitch to reduce the plowing depth somewhat, thus reducing the load on the tractor. The operator would lower the hitch when the soil surface became level again. Tractor designers automated this practice by adding a load control lever and linkage. In Figure 11.31, the lower link points are attached to arms on a torsionally rigid torque tube. One end of a torsion bar is attached to the torque tube, while the other end of the torsion bar is anchored to the vehicle frame. With increasing pull on the lower links, the torsion bar allows increasing rotation of the torque tube. An arm near the center of the torque tube moves to control the DCV actuating linkage as the torque tube rotates. An increasing load signals the DCV to raise the hitch while a decreasing load signals the DCV to lower the hitch. The operator uses the load control handle to set the desired load on the hitch. Note that the system of Figure 11.31 has two control levers, one for load control and one for position control. The two controls together control the hitch. Pulling the load control lever to the rear increases the hitch sensitivity to implement draft, which can result in erratic plowing depth in variable soils. Conversely, moving the draft control lever forward decreases hitch sensitivity to draft and provides for more constant plowing depth on level ground. The three-point hitch enhances weight transfer from the front to the rear wheels of the tractor, thus improving traction and reducing the need for tractor ballast.

While mechanical linkages were able to control three-point hitches successfully, the complex linkages were expensive to manufacture and were subject to wear. Thus, such linkages are being replaced by mechatronic devices. In the mechatronic equivalent of Figure 11.31, each control handle is replaced by a very small handle that sets the position of a potentiometer. Another potentiometer senses the rotation of the shaft that rotates the lower link lift arms. The lower hitch links are linked to the tractor by special pins that produced an electric signal proportional to the shear in the pins. The mechanically controlled DCV is replaced by an electrically controlled DCV. For position control, the microprocessor senses the setting of the position-control handle, compares it with the current hitch position, and signals the DCV to raise or lower the hitch as needed. For load control, the microprocessor senses the setting of the load-control lever, compares it with the load as measured by the lower link pins, and raises or lowers the hitch as needed. In addition to having fewer moving parts subject to wear, the mechatronic system offers enhanced control of the hitch. For example, while controlling load, the microprocessor might restrict the raising and lowering of the hitch to programmed limits to avoid excessive variation in tillage depth.

11.12 Chapter Summary

In this chapter, basic principles were discussed to show that the hydraulic systems used on off-road vehicles have positive displacement. Except for variations in volumetric efficiency, the flow produced by a hydraulic pump is independent of the pressure at the pump outlet port. While the pump is capable of moving hydraulic fluid under high pressure, it is the circuit connected to the pump that actually causes the pressure.

Standard schematic symbols were introduced as a convenient way of describing the logic of a hydraulic circuit. The symbols are analogous to the symbols for electric resistors, capacitors, etc., that are used to convey the logic of an electrical circuit.

Hydraulic pumps provide the means for converting mechanical power into hydraulic power. Equations were presented for calculating the pump flow delivery, the pump torque requirement, and the hydraulic power output of a pump. The volumetric, torque, and power efficiencies of a pump were also discussed.

Hydraulic actuators were presented as a means for converting hydraulic power back into mechanical power. Hydraulic motors convert hydraulic power into rotary power, while hydraulic cylinders convert hydraulic power into linear power. Double-acting hydraulic cylinders were shown to behave differently while extending compared to retracting. Single-acting cylinders are also available; the hydraulic system can force them to extend, but an external load must be supplied to cause them to retract when oil is released from the cylinder.

Hydraulic orifice theory was presented as a means to analyze the orifices found in hydraulic valves. Valves allow control of system pressure, flow volume, and direction of flow. Three types of directional control valves (DCVs) were discussed: closed-center, float-center and open-center DCVs. The DCVs are also classified as to their number of ports for connection to the circuit and by the number of possible spool positions.

Equations were presented to calculate allowable pressure in hydraulic lines and to select line sizes of sufficient diameter to avoid excessive pressure loss. Filters and reservoirs were also discussed as necessary elements of a hydraulic system. If the reservoir has insufficient volume to prevent overheating of the oil, an oil cooler may be needed to prevent overheating. Hydraulic accumulators were discussed as a means for storing potential hydraulic energy. Also, important characteristics of hydraulic fluids were discussed.

The rationale for three different types of hydraulic systems was presented. OC systems can use an inexpensive gear pump, but when more than one actuator is used simultaneously, OC systems are subject to sequencing. PC systems use a more expensive piston pump and can avoid actuator sequencing except if the flow demand takes the pump to full stroke. LS systems have become the system of choice on most new off-road vehicles. With LS systems, the pressure rises to meet the demand of the actuator with the highest pressure demand. Adjustable flow control valves control the flow rate to each actuator. Sequencing is avoided unless the total demand of all flow control valves exceeds the pump capacity at full stroke.

While most of the chapter was devoted to steady-state analysis of hydraulic components and systems, the reader was given a brief introduction to transient analysis. Transient analysis involves the solution of differential equations to calculate DCV spool displacement, line pressures, and actuator speed. Software is available to the designer to simulate transient hydraulic behavior using a digital computer.

When applied to hydraulics, mechatronics uses a combination of hydraulic components, electrical components, and computer control to achieve flexible and sophisticated control of hydraulic systems. A mechatronic DCV was discussed to illustrate some advantages of the mechatronic approach. Finally, the three-point hitch was presented as an example of the use of hydraulics to achieve control of a vehicle function. The terminology of three-point hitches was presented, as were the four categories of hitches. The mechanical approach to hitch position and load control was presented and then the more modern mechatronic approach to such control was presented.

Homework Problems

11.1 In Figure 11.1, let $A_1 = 3$ cm^2 and $A_2 = 30$ cm^2. If the load on the larger piston is 7.5 kN,

(a) Calculate the pressure in the hydraulic oil.

(b) Calculate the force needed on the small piston to hold it in place.

(c) If the small piston moves down 1 cm, how far does the large piston raise? Note that energy is conserved, i.e., if there is no energy loss in the system, the product of input force times distance moved must be the same for both pistons.

11.2 Rework Problem 11.1, but let $A_2 = 40$ cm^2.

11.3 A hydraulic pump has a displacement of 80 cm^3/rev. When the pump is running at 1000 rpm and the outlet pressure is 20 MPa, the pump delivers 77 L/min of flow.

(a) Calculate the theoretical delivery.

(b) Calculate the internal leakage flow.

(c) Assuming the pump outlet pressure and leakage flow remain constant, calculate and plot the pump volumetric efficiency versus pump speeds for pump speeds from 40 to 1000 rpm.

11.4 Rework Problem 11.3, except the pump delivers 75 L/min at 1000 rpm.

11.5 The pump of Problem 11.3 requires 268 N·m of shaft torque at 20 MPa outlet pressure.

(a) Calculate the theoretical torque.

(b) Calculate the friction torque.

(c) Assuming the pump speed and the internal friction remain constant, calculate and plot the pump torque efficiency versus outlet pressure for outlet pressures from 1 to 20 MPa.

11.6 Rework Problem 11.5, except the pump requires 270 N·m of shaft torque at 20 MPa outlet pressure.

11.7 Prove that the power efficiency of a pump is equal to the product of its volumetric and torque efficiencies.

11.8 Assume that the internal leakage flow in a hydraulic motor obeys Equation 11.18, which can be simplified to $Q_{ml} = C_l\ \Delta p/\mu$. A motor with displacement of 60 cm^3/rev runs at 880 rpm while receiving 55 L/min of flow when the inlet pressure is 15 MPa and the oil viscosity is 25 mPa·s.

(a) Calculate the theoretical motor speed.

(b) Calculate the motor volumetric efficiency.

(c) Calculate the leakage flow.

(d) Calculate the constant, C_l, in the leakage flow equation.

(e) Finally, calculate and plot the volumetric efficiency versus pressure as the pressure varies from 0 to 15 MPa.

11.9 Rework Problem 11.8, but assume the motor receives 57 L/min of flow.

11.10 Rework Problem 11.8, except in part (e) calculate and plot the volumetric efficiency versus oil viscosity as the viscosity varies from 12 to 48 mPa·s.

11.11 Assume that, in a gear-type hydraulic motor, the friction torque obeys the equation $T_{mf} = C_f \mu N_m$. When a motor with displacement of 60 cm^3/rev has input pressure of 15 MPa, it produces 130 N·m of torque while running at 800 rpm. The oil viscosity is 25 mPa·s.

(a) Calculate the theoretical torque.

(b) Calculate the torque efficiency.

(c) Calculate the constant, C_f, in the friction equation.

(d) Assuming the speed and oil viscosity remain constant, calculate and plot the motor torque efficiency versus pressure for pressures up to 15 MPa.

11.12 Rework Problem 11.11, except in part (d) calculate and plot motor torque efficiency versus oil viscosity for viscosity ranging from 12 to 48 mPa·s.

11.13 A double-acting hydraulic cylinder has a 7.5 cm bore, a 20 cm stroke, and the rod diameter is 2.5 cm.

(a) Calculate the inflow rate of oil that is required to extend the cylinder at a rate of 0.35 m/s.

(b) Calculate the input pressure needed to move a load of 90 kN.

(c) Calculate the flow rate of oil returning to the reservoir from the rod end of the cylinder.

11.14 Rework Problem 11.13, except let the cylinder be retracting and returning oil from the end of the cylinder opposite the rod.

11.15 The flow rate is 40 L/min through an orifice 5 mm in diameter. The fluid density is 0.85 kg/L and the viscosity is 27 mPa·s.

(a) Calculate the flow velocity through the orifice.

(b) Calculate the Reynolds number.

(c) Estimate the orifice coefficient from Figure 11.10; if the Reynolds number is off the chart, assume the orifice coefficient remains constant for Reynolds numbers greater than 2500.

(d) Finally, calculate the pressure drop across the orifice.

11.16 Rework Problem 11.15, except let the flow rate be 70 L/min.

11.17 Rework Problem 11.15, except let the flow rate vary from zero to 70 L/min and plot the pressure drop versus flow rate.

11.18 Assume that a steel conduit has a maximum design stress of 200 MPa. Calculate and plot the maximum allowable pressure versus wall thickness for wall thickness ranging from 0 to 5 mm; plot curves for conduit diameters of 10, 20, 30, 40, and 50 mm.

11.19 A hydraulic fluid has a density of 0.85 kg/L and a viscosity of 27.6 mPa·s.

(a) Select a conduit diameter if the flow rate is 90 L/min and the pressure drop is to be no higher than 0.1 MPa/m.

(b) Calculate the Reynolds number and (c) the exact pressure drop for this flow.

11.20 Rework Problem 11.19, except use a flow rate of 9 L/min.

11.21 In Figure 11.23, 12 MPa of pressure are required to move the load on cylinder A, 16 MPa of pressure are required to move the load on cylinder B, the cracking pressure of the relief valve is 20 MPa, and the pump delivers 60 L/min of flow. Both cylinders have a 7.5 cm bore and a 20.0 cm stroke. If the operator simultaneously moves both DCV spools to try to extend both cylinders and continues to hold the spools in the off-center position, plot the pump pressure as a function of time. Measure time from the instant both spools are moved off center when the pressure increases from zero gage pressure. For each pressure change, briefly explain the cause of the change.

11.22 Rework Problem 11.21, except let cylinder A be compressing a linear spring whose spring rate is 400 kN/m.

11.23 Rework Problem 11.21, except the cylinders are connected to the DCV in the LS circuit of Figure 11.27, the pump has a maximum delivery of 60 L/min, and each throttling valve is set to pass 28 L/min.

11.24 Rework Problem 11.21, except the cylinders are connected to the DCV in the LS circuit of Figure 11.27, the pump has a maximum delivery of 60 L/min, and each throttling valve is set to pass 35 L/min.

References and Suggested Readings

ASAE. 1997a. Application of hydraulic remote control cylinders to agricultural tractors and trailing-type implements. ASAE/ANSI Standard S201.4. *ASAE Standards,* 44th ed. St. Joseph, MI: ASAE.

ASAE. 1997b. Three-point free-link attachment for hitching implements to agricultural wheel tractors. ASAE Standard S217.11 (SAE Standard J715). *ASAE Standards,* 44th ed. St. Joseph, MI: ASAE.

ASAE. 1997c. Attachment of implements to agricultural wheel tractors equipped with quick-attaching coupler. ASAE Standard S278.6 (SAE Standard J909). *ASAE Standards,* 44th ed. St. Joseph, MI: ASAE.

ASAE. 1997d. Category "0" three-point free-link attachment for hitching implements to lawn and garden ride-on tractors. ASAE Standard S320.1. *ASAE Standards,* 44th ed. St. Joseph, MI: ASAE.

ASAE. 1997e. Test procedure for measuring hydraulic lift capacity on agricultural tractors equipped with three-point hitch. ASAE Standard S349.2 (SAE Standard J283). *ASAE Standards,* 44th ed. St. Joseph, MI: ASAE.

ASAE. 1997f. Dimensions for cylindrical hydraulic couplers for agricultural tractors. ASAE Standard S366.1 (SAE Standard J1036, ISO Standard 5675). *ASAE Standards,* 44th ed. St. Joseph, MI: ASAE.

ASAE. 1997g. Hydraulic pressure available on agricultural tractors for remote use with implements. ASAE Standard S489. *ASAE Standards,* 44th ed. St. Joseph, MI: ASAE.

Book, R. 1998. Programmable hydraulic valve. Ph.D. dissertation, University of Illinois at Urbana-Champaign.

Book, R., and C.E. Goering. 1997. Load sensing hydraulic system simulation. *Applied Engineering in Agriculture* 13(1): 17-25.

Book, R., and C.E. Goering. 1999. Programmable electrohydraulic valve. SAE Paper No. 1999-01-2852. Warrendale, PA: SAE.

Frankenfield, T., and P. Stavrou. 1993. Developing trends in hydraulics tied to electronic controls. *Control Engineering* 40(5): 62-66.

ISO. 1999. Hydraulic fluid power solid contaminants code. ISO Standard 4406. Milwaukee, WI: National Fluid Power Assn.

Keller, G.R. 1978. Hydraulic System Analysis. Penton/IPC. Cleveland, OH.

Kim, S.D, H.S. Cho and C.O. Lee. 1988. Stability analysis of a load-sensing hydraulic system. In *Proceedings of the Institution of Mechanical Engineers, Part A: Power and Process Engineering* 202(A2): 79-88.

Luomaranta, M.K., M.J. Vilenius and I. Ahonoja. 1994. Microcomputer makes a flexible electrohydraulic valve. Mechatronics Spells Profitability, International Conference on Machine Automation, 805-815. Tampere, Finland, 15-18 February.

Mai, V.K.L., and P. Dransfield. 1989. Load-sensitive pump control performance. *Mechanical Engineering Transactions* 143(3): 149-157. Australia: Institution of Engineers.

Merritt, H.E. 1967. *Hydraulic Control Systems*. New York, NY: John Wiley and Sons.

Taborek, J.J. 1959. Fundamentals of line flow. *Machine Design Magazine,* 16 April 1959.

Vaughan, N.D., and J.B. Gamble. 1996. The modeling and control of a proportional solenoid valve. *Transactions of the ASME, Journal of Dynamic Systems, Measurement and Control* 118(1): 120-125.

Zeiger, G., and A. Akers. 1986. Dynamic analysis of an axial piston pump swashplate control. In *Proceedings of the Institution of Mechanical Engineers,* Part C (200 CI): 49-58.

CHAPTER 12

DRIVE TRAINS

12.1 Introduction

The vehicle drive train includes all of the power transmission elements between the engine and the drive wheels. It may also include a power take-off (PTO) drive for delivering mechanical power to a vehicle attachment. The drive train serves the following functions:

- Transmit power from the engine to the drive wheels and the PTO shaft.
- Provide means for smoothly engaging the engine power during startup.
- Transform the engine torque and speed to meet demands of the load.
- Provide means for reversing the direction of travel of the vehicle.
- Provide means for smoothly stopping the vehicle.

An overview of a power train for a rear-wheel drive vehicle is shown in Figure 12.1. The drive train includes a friction clutch, a transmission, a differential, final drives and brakes. A PTO may also be included.

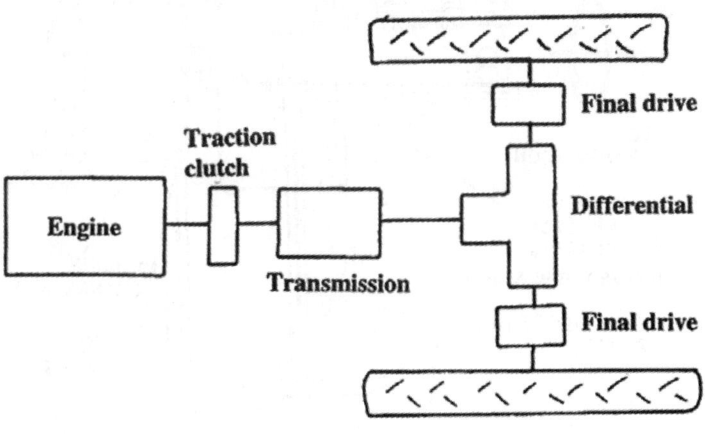

Figure 12.1. A complete vehicle drive train.

In this chapter, the reader will be introduced to procedures for designing transmission elements as well as the rationale for including some of the elements in the power train.

12.2 Friction Clutches and Brakes

Friction clutches and brakes must have the ability to absorb energy during slippage. A clutch slips during engagement to provide a smooth startup of the vehicle. A brake slips during engagement to provide a smooth stopping of the vehicle.

12.2.1 Clutches

The purpose of the *clutch* is to provide smooth engagement of the engine to the load and to permit interruption of power flow. Virtually all vehicle clutches are *disk clutches*. Figure 12.2 illustrates a *spring-loaded clutch*, i.e., mechanical springs squeeze a *friction disk* between pressure plates that rotate with the flywheel. The friction disk is spline-connected to the output shaft.

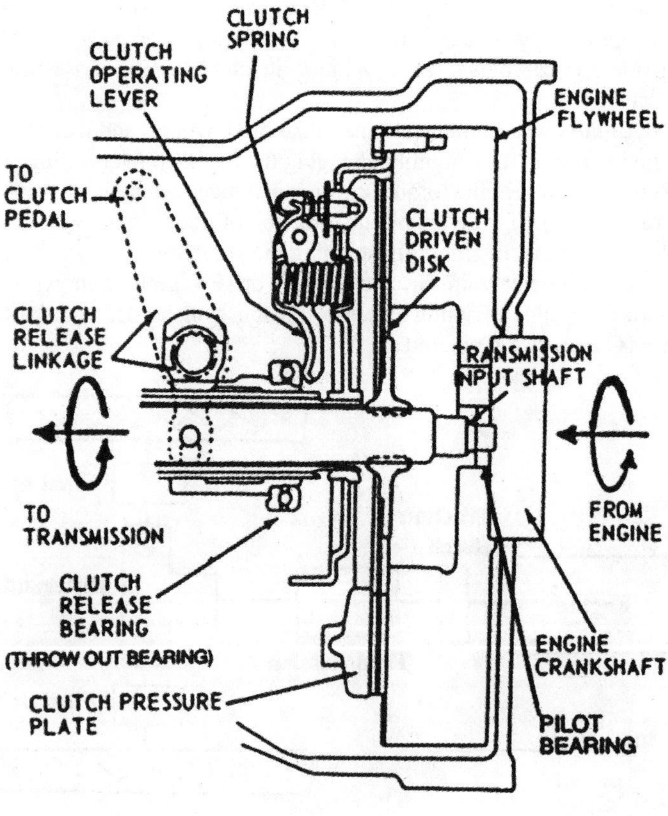

Figure 12.2. A spring-loaded dry clutch. (Courtesy of Deere and Co.)

Clutch actuating mechanisms could be mechanical linkages. Typically, actuation of a traction clutch would be through a pedal operated by the operator's left foot. On some vehicles, the foot pedal operates a master hydraulic cylinder that, in turn, drives a slave cylinder to move the disengaging collar, thus eliminating the mechanical linkage.

The clutch illustrated in Figure 12.2 is a *dry clutch* suitable for vehicles of low to moderate power output. In heavier-duty vehicles, a dry clutch may not be able to dissipate the heat that is generated during clutch engagement and a *wet clutch* is used. The wet clutch shown in Figure 12.3 is a dual clutch. The clutch on the right, with two clutch disks (No. 10 in the figure) interspersed between three pressure plates, is the traction clutch. The clutch on the left, also with two clutch disks (No. 7) interspersed between three pressure plates, controls the PTO. The inner shaft (No. 34) goes to the transmission, while the outer shaft drives the PTO shaft (No. 33) through an idler gear (No. 31). Multiple disks are required because the oil cooling reduces the coefficient of friction between the disks and pressure plates. A hydraulic actuator (No. 4) pushes clutch levers (No. 5) to engage the traction clutch. Another hydraulic actuator (No. 28) pushes clutch levers (No. 25) to engage the PTO clutch. In this wet clutch, a foot pedal operates a valve that directs oil into or allows release of oil from the actuating cylinder of the traction clutch. Usually, the clutch valve that regulates oil flow into or out of the actuating cylinder of the PTO clutch is controlled by a hand lever.

The torque-transmitting capacity of a clutch can be calculated by

$$T_c = f\, F_c\, r_m\, n_s \tag{12.1}$$

where

T_c = torque capacity, N·m
f = coefficient of friction
F_c = clamping force provided by clutch springs, kN
r_m = mean radius of clutch, mm
n_s = number of torque-transmitting surfaces = 2 times the number of disks.

The torque capacity of the clutch should be two to three times the peak torque of the engine to allow some reserve capacity. The torque reserve allows quick lockup and minimum slippage after lockup when the clutch is engaged. The mean radius of the clutch is

$$r_m = \frac{D_o^3 - D_i^3}{3(D_o^2 - D_i^2)} \tag{12.2}$$

where

D_o = outer diameter of clutch disks, mm
D_i = inner diameter of clutch disks, mm

Coefficients of friction of some common clutch disk materials are given in Table 12.1. There are two classes of materials, depending on whether a *dry clutch* or a *wet clutch* is used.

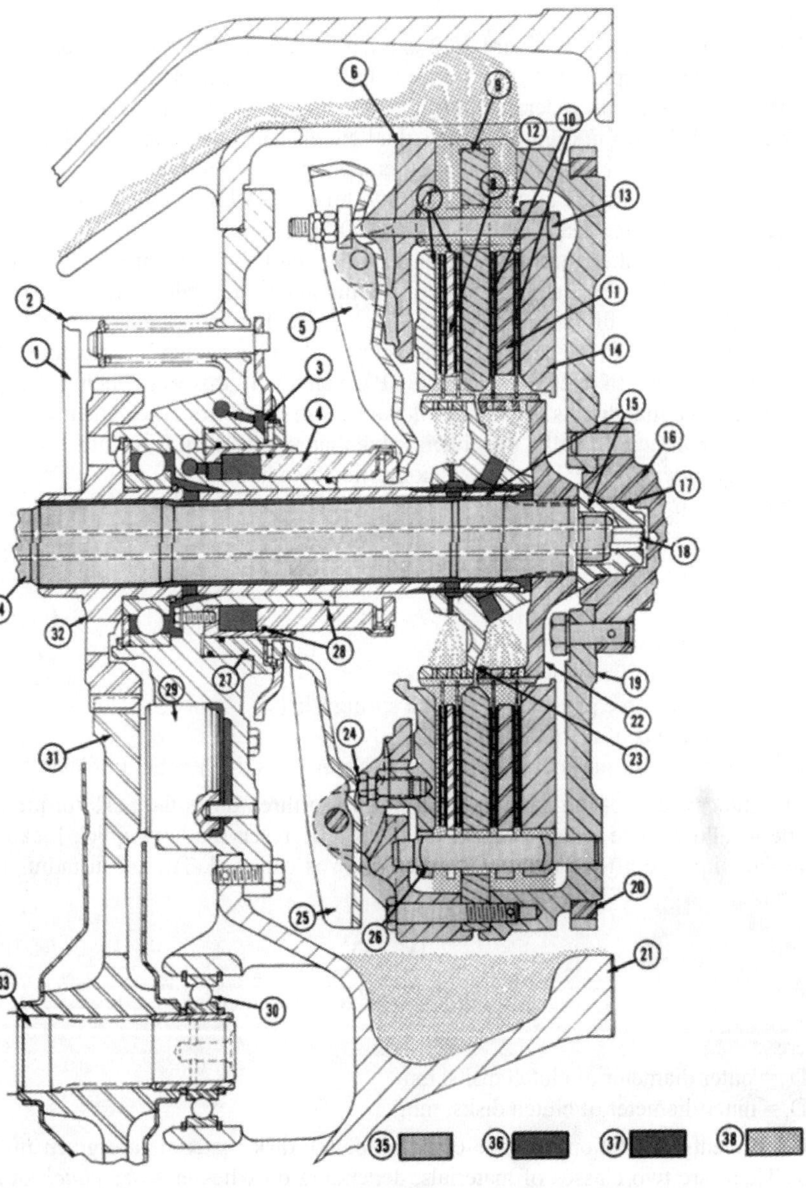

Figure 12.3. A hydraulically operated wet clutch. (Courtesy of Deere and Co.)

Table 12.1. Design data for clutches and brakes.

Material	Friction Coefficient, f	Max. Facing Pressure, N/mm^2	Cooling Oil Flow, L/s/m^2	Max. Energy Rate, W/mm^2	Max. Bulk Temperature, °C
Dry operation					
Organic	0.2 – 0.3	0.1 – 0.3	Dry	0.5 – 0.8	150
Cerametallic	0.3 – 0.4	0.7 – 0.9	Dry	0.8 – 2.3	200 – 250
Wet operation					
Paper	0.09 – 0.13	0.8 – 2.5	6 – 30	0.8 – 1.2	230 – 280
Molded	0.08 – 0.10	1.5 – 2.5	2.5 – 30	1.0 – 1.2	230 – 280
Filled fluorocarbon	0.08 – 0.10	1.5 – 2.5	2.5 – 12	1.0 – 3.0	250 – 300
Sintered metallic	0.04 – 0.09	1.2 – 3.5	2.5 – 12	1.0 – 2.0	300

The heat generated in slippage can be estimated as

$$Q = \frac{\pi}{30}(T_s\, N_s\, t_s) \tag{12.3}$$

where

Q = heat generated, J

T_s = average torque being transmitted while clutch is slipping, N·m

N_s = maximum speed of pressure plate relative to clutch disks, rpm

t_s = duration of slip, s

The temperature rise can be estimated by

$$\theta - \theta_o = \frac{Q}{m_c C_p} \tag{12.4}$$

where

θ = clutch temperature after engagement, °C

θ_o = clutch temperature before engagement, °C

m_c = mass of heat absorbing parts of clutch, kg

C_p = specific heat of heat absorbing elements, J/(kg °C)

Nonmetallic friction disks will not absorb much heat; therefore the moving metal parts of the clutch must provide the heat absorption. The equation for calculating the specific rate of heat generation is a modification of Equation 12.3:

$$E_r = \frac{2\pi\, T_s\, N_s}{60\, A_c} \tag{12.5}$$

where

E_r = specific rate of heat generation, W/mm^2

A_c = combined area of all friction surfaces, mm^2

It is important to limit the rate of heat generation in a clutch. The specific rate of heat generation must be less than the maximum energy rate and the clutch temperature must be less than the maximum bulk temperature given in Table 12.1 for the material used in the clutch.

12.2.2 Brakes

Originally, *drum brakes* were used to stop vehicles. *Disk brakes* have become more popular in recent years. Disk brakes include a disk that is attached to the rotating axle of each wheel and a *brake caliper* that is attached to the non-rotating axle housing. Each caliper contains a pair of brake pads that straddle the *brake disk*. When the brakes are actuated, the caliper squeezes the brake pads together and forces them to grip the rotating disk. Equation 12.1 can be used to calculate the stopping torque. As with clutches, the brake design should provide for a torque reserve of two to three. The effective radius, r_m, is the radial distance from the center of the axle to the center of the brake pads. Equation 12.3 can be used to calculate the heat generated as the brake shoes absorb energy in bringing the vehicle to a halt. Equation 12.4 can be used to calculate the temperature rise and Equation 12.5 can be used to calculate the specific rate of heat generation in the braking process. As with a clutch, the specific rate of heat generation must be within the limit given in Table 12.1 for the material used in the brake pads.

12.3 Transmissions and Load Matching

A transmission serves a number of functions. It provides for forward travel, reverse travel, or neutral. Also, it provides a means for transforming the engine torque and speed to levels that are suitable for the external load on the vehicle. Figure 12.4 illustrates the use of the transmission to transform torque and speed. The torque and speed of the engine are transformed to those at the vehicle axles through use of the following equations:

$$N_A = \frac{N_e}{G_{pt}} \tag{12.6}$$

and

$$T_A = G_{pt} e_{pt} T_e \tag{12.7}$$

where

N_e = engine speed, rpm
N_A = axle speed, rpm
T_e = engine torque, N·m
T_A = axle torque, N·m
$e_{pt} = e_T\, e_D\, e_{FD}$ = power train efficiency, product of transmission (e_T), differential (e_D), and final drive (e_{FD})efficiencies
$G_{pt} = G_T\, G_D\, G_{FD}$ = power train gear (speed) ratio, product of speed ratios of transmission (G_T), differential (G_D), and final drive (G_{FD})

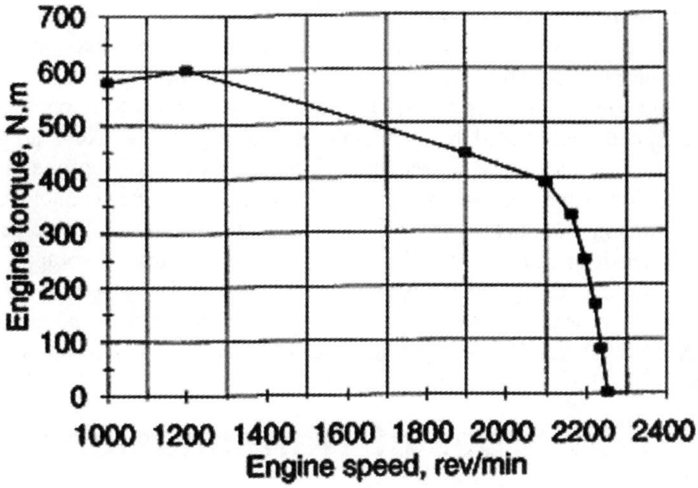

(A)

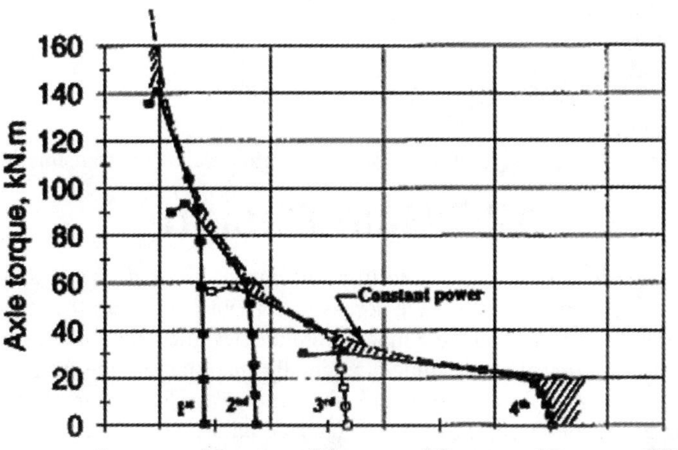

(B)

Figure 12.4. Torque-speed curves of (a) an engine and (b) vehicle rear axle.

The final drive ratio in some vehicles is 1.0, i.e., the drive axles are connected directly to the differential side gears (G_D = 1.0). Usually, G_D and G_{FD} are nonadjustable in a vehicle, but the operator can change G_T by shifting the transmission. The power train transforms the engine torque-speed curve to a series of torque-speed curves at the axles, with one axle curve for each speed ratio, G_T, in the transmission. Figure 12.4b shows the axle torque-speed curves for a four-speed transmission. When the transmission is in a specific gear, the axles can operate only on the torque-speed curve for that gear (if the engine is at the maximum governor setting) or at points below the curve (if the engine is at reduced governor setting). The dashed line on Figure 12.4b is a constant power line; it shows the combinations of torque and speed that would be attainable if the engine was kept at governor's maximum at all times. The cross-hatched areas show axle torque and speed combinations for which engine power is sufficient but which cannot be reached because the transmission lacks a suitable gear ratio for those areas. The vehicle is most productive when it operates on the dashed constant-power line. Two approaches are used in transmission design to improve the efficiency and productivity of vehicles. The first is to increase the number of available speed ratios, thereby reducing the areas that are cross-hatched. The second is to design for rapid shifting so that less momentum is lost during a gear shift.

12.4 Types of Transmissions

The following list of types of transmissions is in order of increasing vehicle productivity and convenience in changing transmission speed ratios:

1. Sliding gear
2. Constant-mesh
3. Synchromesh
4. Power-shift
5. Continuously Variable Transmission (CVT)
6. Hydrokinetic

A CVT or a hydrokinetic drive may be integrated into a gear-type transmission.

12.5 Gear Design

Most transmissions rely on gears to transmit power. *Gear design* is a specialized activity. Manufacturers generally have access to computer software that encompasses gear design theory and, with suitable inputs from the designer, can produce output that drives numerically controlled machines to actually manufacture the gears. However, many manufacturers still use hobs and shavers to produce gears based on the computer-aided designs. Because of the specialized nature of gear design, only a brief overview will be given to acquaint the reader with the terminology and basic considerations in gear design.

Spur gears or *helical gears* are meshed from parallel shafts. The teeth on spur gears are parallel to the axis of rotation, while helical gear teeth are at an angle to the axis of rotation. Helical gears gradually transfer the load from one tooth to the next and are much quieter than spur gears.

Gear teeth have an *involute profile*. The profile is one that is generated by a point on a string as the string unwraps from a cylinder (see Figure 12.5). A perfect involute profile would transmit constant rotational velocity. However, imperfections in machining and in heat treating cause profile errors, which result in impact stresses and noise. Some of the details and terminology of involute gears are illustrated in Figures 12.6 and 12.7.

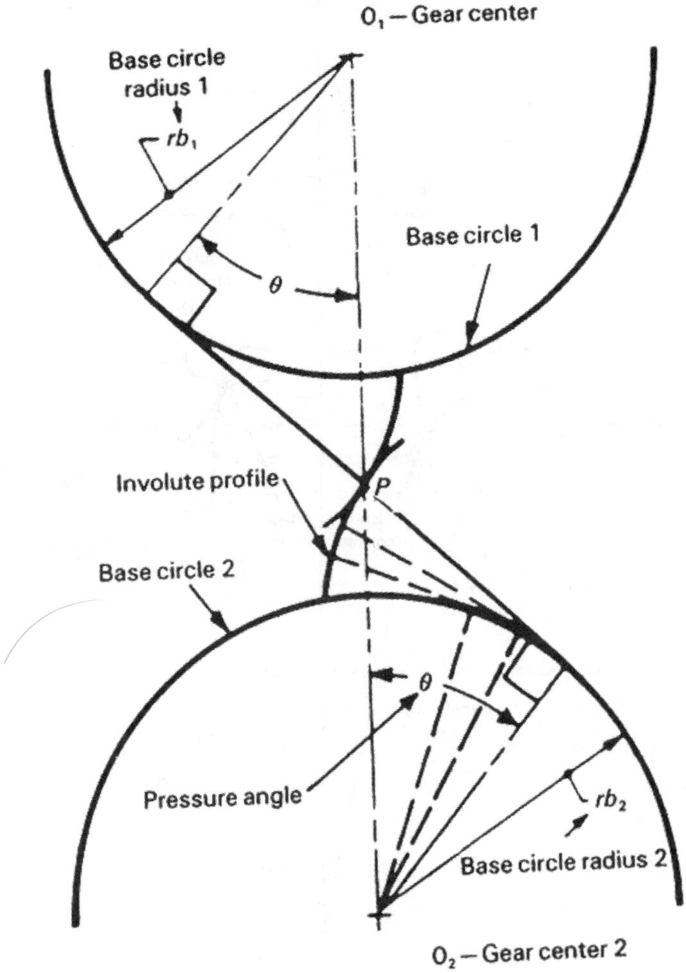

Figure 12.5. Involute gear tooth profile. (From Browning, 1978.)

Bevel gears connect shafts with intersecting axes. While the most common application is for perpendicular shafts, there are applications for bevel gears in which the shafts form angles less than or greater than 90°. Straight bevel gears are the bevel gear equivalent of spur gears. Thus, they transfer load abruptly from tooth to tooth and are noisy. The curved teeth on *spiral bevel gears* transfer load gradually from tooth to tooth and thus are much quieter than straight bevel gears. *Hypoid bevel gears* are similar to spiral bevel gears, except that the shafts do not intersect. Hypoid bevel gears are stronger and can be used for greater speed ratios than the spiral bevel gear, but the greater tooth sliding of hypoid bevel gears make them less efficient than spiral bevel gears.

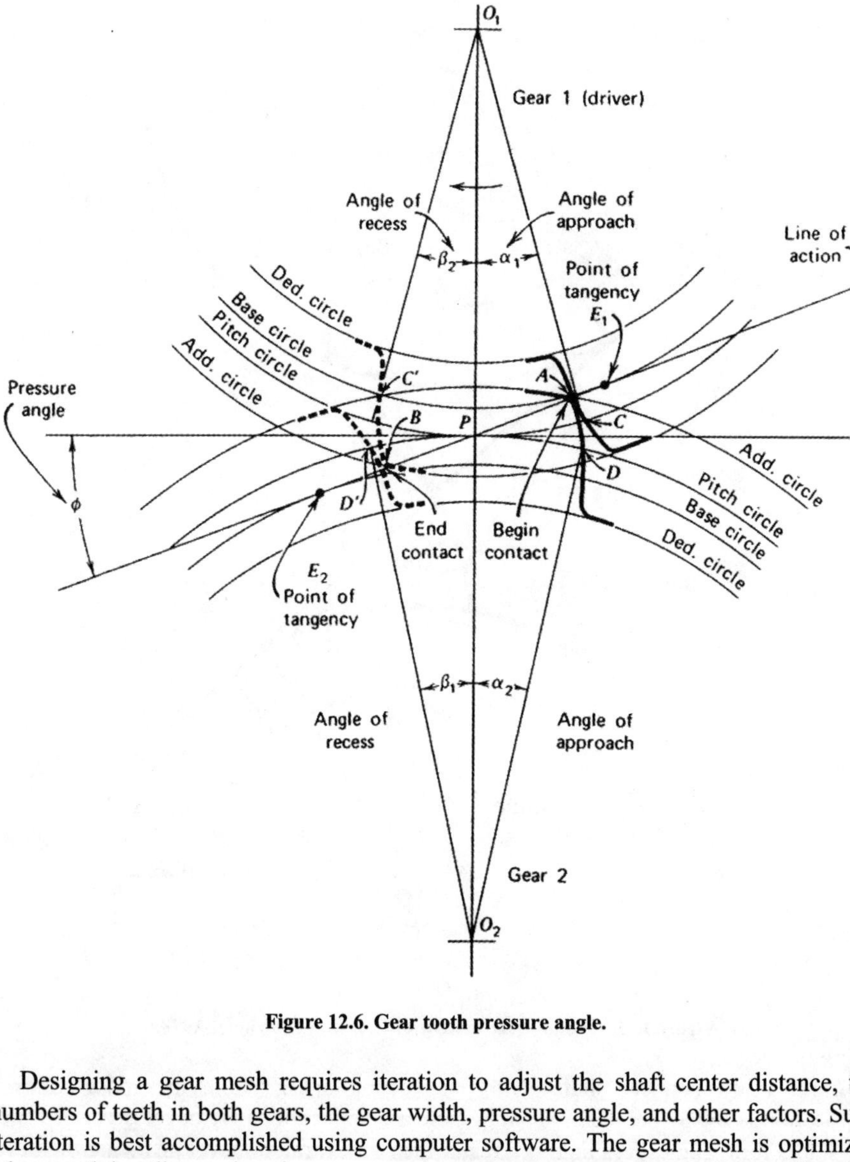

Figure 12.6. Gear tooth pressure angle.

Designing a gear mesh requires iteration to adjust the shaft center distance, the numbers of teeth in both gears, the gear width, pressure angle, and other factors. Such iteration is best accomplished using computer software. The gear mesh is optimized when tooth bending and surface compressive stresses are near the maximum allowable unless reduced to reduce noise. Usually, the desired gear ratio is known. When the nominal tooth size (pitch) and center distance have been selected, the pitch diameter and number of teeth can be calculated. For spur gears and helical gears, the pitch diameter of the pinion (the smaller gear) is given by

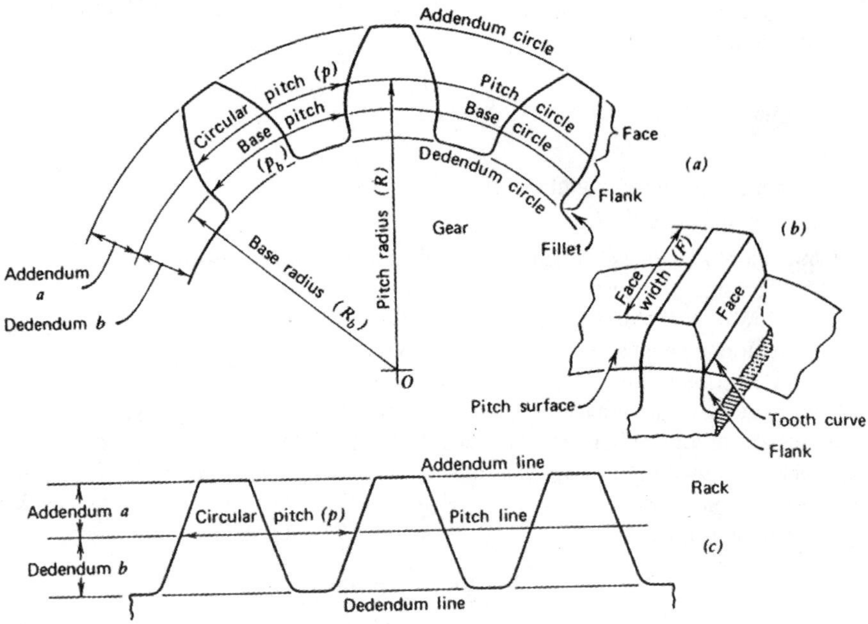

Figure 12.7. Gear tooth terminology.

$$\frac{D_p}{D_c} = \frac{2}{1+G} \tag{12.8}$$

where

D_p = pitch diameter of pinion, mm
D_c = shaft center distance, mm
G = speed ratio, greater than one

The same equation can be used to calculate the pitch diameter of the larger gear if G is replaced by 1/G. The following equation gives the relationship between the gear module, the pitch diameter and the number of teeth on the gear:

$$m = \frac{D_G}{n} \tag{12.9}$$

where

m = tooth size module, mm
D_G = pitch diameter of the gear
n = number of teeth on the gear

For helical gears, m is the module in the transverse plane and is related to the normal or cutter module by

$$m = \frac{m_n}{\cos \psi} \tag{12.10}$$

where

m_n = normal or cutter module, mm

ψ = the helix angle

Typical tooth modules for gears in a vehicle drive train are as follows:

Transmission gears	4 to 5 mm
Power-shift planetary gears	1.5 to 3.5 mm
Spiral bevel gears	8 to 12 mm
Final drive gears	5 to 7 mm

Gear cutters of nominal size and pressure angle are usually used, but center distances and tooth numbers are often chosen such that the *nominal pitch diameters* do not touch, as shown in Figure 12.5. Thus, the *working pitch diameter* is different from the nominal pitch diameter used in Equation 12.9. The corresponding working pressure angle is given by

$$\phi_w = \cos^{-1} \frac{m(n_1 + n_2) \cos \phi_t}{2\, D_c} \tag{12.11}$$

where

ϕ_w = working pressure angle

m = gear module in the transverse plane, from Equation 12.10

n_1, n_2 = numbers of teeth on the gears in the mesh

ϕ_t = the transverse pressure angle = $\tan^{-1}(\tan \phi_n / \cos \psi)$

ϕ_n = nominal or pressure cutter angle

Gears in many off-road vehicles require high-quality materials and manufacturing to withstand the very high stresses encountered in service. The tooth surfaces must be very hard to withstand the compressive interfacial stresses, but must also be fatigue-resistant to endure the cyclic loading as the teeth move into and out of mesh. The required properties are obtained by using a medium carbon or alloy steel and then quenching or induction hardening to obtain a very hard surface with a fatigue-resistant core. A surface hardness of Rockwell C 56 to 60 is generally specified, which gives a yield strength of at least 2000 MPa. The gear teeth must be designed for stresses below the endurance limit for fatigue, because a gear running at only 1000 rpm can go through 1,000,000 stress cycles in only 20 hours.

Gears must be lubricated to reduce friction and provide cooling. An *extreme pressure additive* must be used to prevent surface welding. The extreme interfacial pressures prevent hydrodynamic lubrication. Instead, elastohydrodynamic lubrication occurs in which the oil viscosity increases so much that the lubricant resembles cold asphalt. Such lubrication allows transmission gears to mesh with very little indication of mating surface contact. However, spiral bevel gears generally show a high polish, while final drive gears show some contact and wear.

The sliding contact between gear teeth is an important cause of power loss in transmissions. Power losses are typically 1% to 2% for external gears and 0.5% to 1.5% for internal gears. Thus, an average efficiency of 98.5% for external gear sets and 99% for internal gear sets is often used.

12.6 Sliding Gear, Constant-Mesh, and Synchromesh Transmissions

The first four types of transmissions listed in Section 12.4 are all manually shifted transmissions. The first three types include two or more parallel shafts, as shown in Figure 12.8. The gears can be permanently fixed on one of the shafts. In the sliding gear transmission, the moveable gears are spline-connected to their shaft. The moveable gears are slid axially to move out of or into mesh with fixed gears on the other shaft. Meshing of spur gears, such as those shown in Figure 12.8, is noisy. Noise can be reduced by designing a stiff housing with curved surfaces to minimize noise amplification, and by providing uniform and low-stiffness gear teeth to minimize cyclic force generation. Helical-cut gears, such as those used in constant-mesh transmissions, are quieter but cannot be slid axially. Thus, the gears are kept in constant-mesh with mating gears on the other shaft, but one set of gears (those on the lower shaft in Figure 12.8) are free to rotate on their shaft. Collars (or sliding couplings) are spline-connected to their shaft and can be slid forward or backward to latch to an adjacent gear. In Figure 12.8, for example, sliding coupling number 1 forward would latch it to the 55-tooth gear, and power would flow from the main shaft through the meshing of the 28-tooth gear with the 55-tooth gear. When gears with fixed centers are meshed, the speed ratio can be calculated by remembering that the product of the number of gear teeth times the gear rotational speed is constant for both gears in the mesh. Thus, the speed ratio of the mesh is given by

$$G_m = \frac{N_{in}}{N_{out}} = \frac{n_{out}}{n_{in}} \tag{12.12}$$

where

G_m = speed ratio of the mesh
N_{in} = speed of the input gear
N_{out} = speed of the output gear
n_{in} = number of teeth on the input gear
n_{out} = number of teeth on the output gear

The torque ratio for the mesh is

$$\frac{T_{out}}{T_{in}} = e_m\, G_m \tag{12.13}$$

where

T_{in}, T_{out} = input and output torques, respectively
e_m = efficiency of the mesh

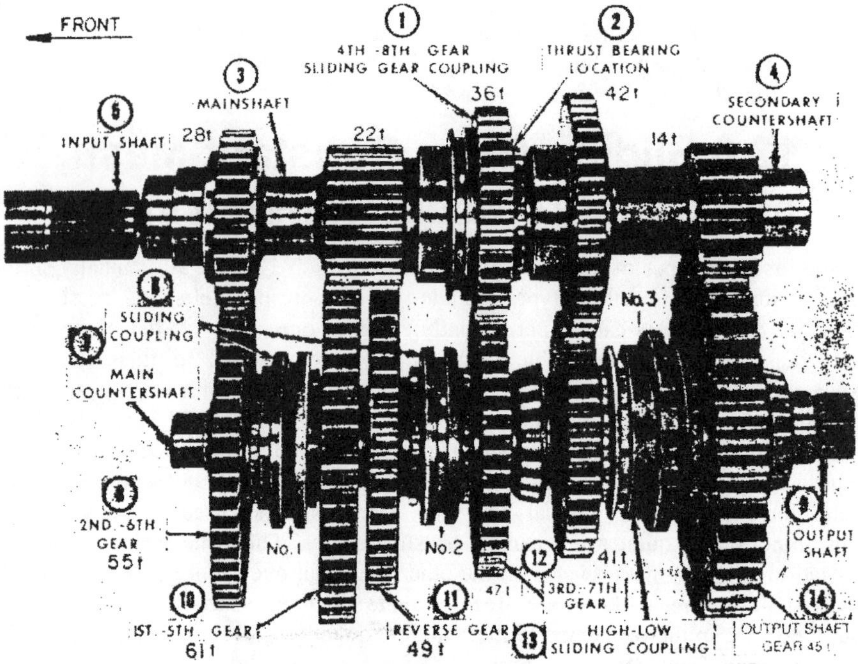

			Shift Code				
	Shift	Range	Slider Positions			Sliding	
Gear	Lever	Lever	No.1	No.2	No.3	Gear	Key
R_1	1(F)	F	C	F	B	F	
1	2	F	B	C	B	F	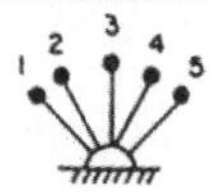
2	3	F	F	C	B	F	
3	4	F	C	B	B	F	
4	5(B)	F	C	C	B	B	SHIFT LEVER
R_2	1	B	C	F	F	F	
5	2	B	B	C	F	F	
6	3	B	F	C	F	F	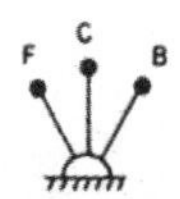
7	4	B	C	B	F	F	
8	5	B	C	C	F	B	
N	X	C	X	X	C	X	RANGE LEVER

Key: F = forward position; C = centered position; B = back position; X = irrelevant position.

Note: Mainshaft does not continue through to secondary countershaft. 36t gear is spline connected to mainshaft; when back, it couples to secondary countershaft. Reverse gear is driven from 22t gear through an idler gear (not shown). Main countershaft does not continue through to output shaft. Output shaft gear (45t) rotates freely on output shaft. 41t gear is keyed to main countershaft. Four other gears rotate freely on main countershaft. High-low sliding coupling is spline connected to output shaft. Two sliding couplings are spline connected to main countershaft.

Figure 12.8. A constant-mesh transmission.

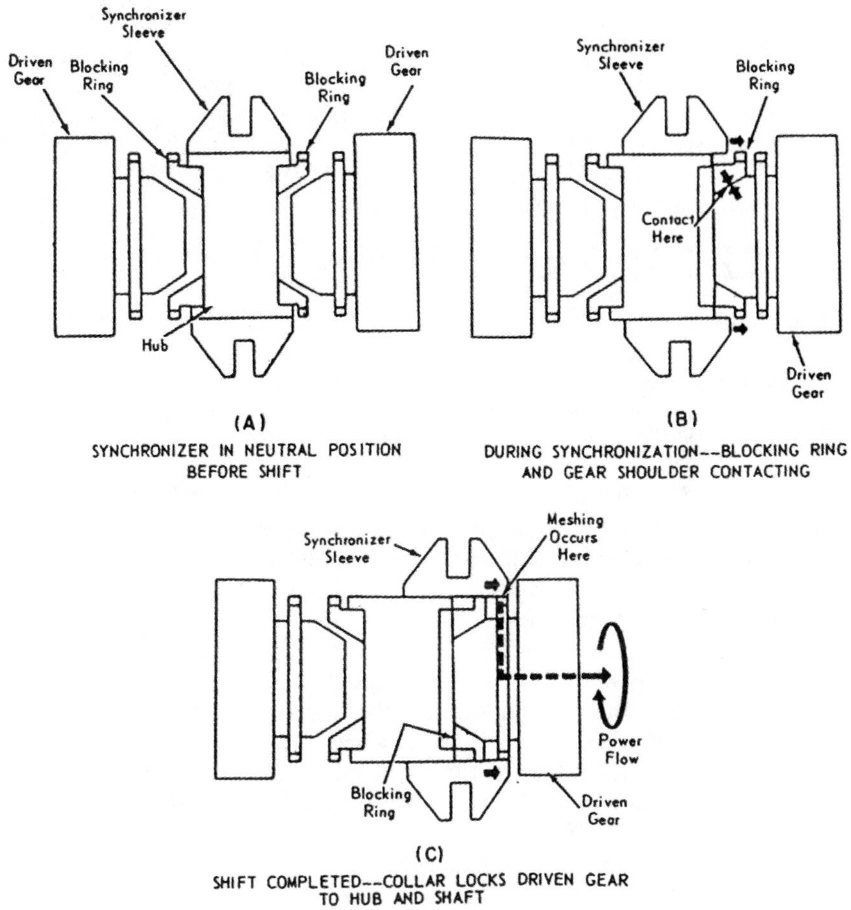

Figure 12.9. A transmission synchronizer.

The sliding gear and constant-mesh transmissions are unsynchronized and shifting is slow. For example, consider shifting the transmission in Figure 12.8 from first to second gear by moving the number 1 sliding coupling forward to unlatch from the 61-tooth gear and to latch with the 55-tooth gear. The operator could easily slide the coupling forward to unlatch from the 61-tooth gear after releasing the clutch. However, the main shaft would immediately begin decelerating because it has insufficient momentum torque to overcome bearing friction. The main countershaft is still connected to the drive wheels through the drive train and vehicle momentum will keep that shaft turning with little loss of speed. Thus, the coupling will be rotating at a different speed than the 55-tooth gear and it will be difficult to latch them together. It almost always is necessary to bring the vehicle to a complete stop to complete the shift. The synchromesh transmission overcomes this loss of productivity by replacing the sliding couplings with synchronizers.

The principle of a synchronizer is shown in Figure 12.9. The synchronizer is spline-connected to the shaft, while the adjacent gears rotate freely on the shaft. Note the slot in the synchronizer sleeve for insertion of a shifting fork. When the operator begins sliding the sleeve toward the adjacent gear, conical friction surfaces come into contact to equalize the speeds of the sleeve and the gear. The sleeve is then moved further to latch to the gear before the clutch is re-engaged. The friction surfaces need not transmit high torque, because latching occurs before the traction clutch is re-engaged.

12.7 Power-Shift Transmissions

Power-shift transmissions can be shifted with virtually no interruption in power flow to the drive wheels. Hydraulically operated clutches and brakes within the transmission are used to accomplish shifting, so there is no need for a traction clutch. The two principal types of power-shift transmissions are the countershaft type and the planetary type.

12.7.1 Countershaft-Type, Power-Shift Transmissions

Figure 12.10a shows a countershaft-type transmission used as a high-low unit, while Figure 12.10b shows a shuttle (reversing) transmission. In both cases, the input shaft drives the drum on the right end of the input shaft. The left drum in Figure 12.10a is driven through gear meshes G-1 to G-2 and G-3 to G-4; thus, depending on the numbers of teeth on the gears, the left drum can turn faster or slower than the right drum. The unit in Figure 12.10a is for an underdrive, so the left drum turns slower than the right drum. The disks in both clutches are spline-connected to the output shaft. When the left clutch is engaged, as shown in Figure 12.10a, power flows through the two gear meshes and the left clutch, and the output shaft turns slower than the input shaft. The speed ratio can be calculated when the numbers of teeth on the gears is known. When the right clutch is engaged, the output shaft turns at the same speed as the input shaft. When both clutches are disengaged, the transmission is in neutral. When both clutches are engaged, the transmission is locked, as for a parking brake. Figure 12.10b works on the same principle, except that an extra gear (G-5) has been added to cause the left drum to rotate opposite in direction to the right drum. Thus, the vehicle reverses in direction when the left clutch is engaged.

Power-shift transmissions are subject to the same power losses per gear mesh as was discussed in Section 12.5. In addition, when hydraulic power is used to engage or disengage the clutches and brakes in a power-shift transmission, that hydraulic power can be charged as a power loss to the transmission. Consequently, power-shift transmissions offer more convenient speed changing than manual-shift transmissions, but also absorb more of the input power than manual-shift transmissions.

12.7.2 Planetary Gear Sets

A *planetary gear set* is illustrated in Figure 12.11. In an analogy to our solar system, the *planet gears* rotate about a *sun gear* while being carried on a *planet carrier*. The planets are also in mesh with a *ring gear*. The planetary gear set cannot transmit power unless two elements are locked together or one of the elements is held

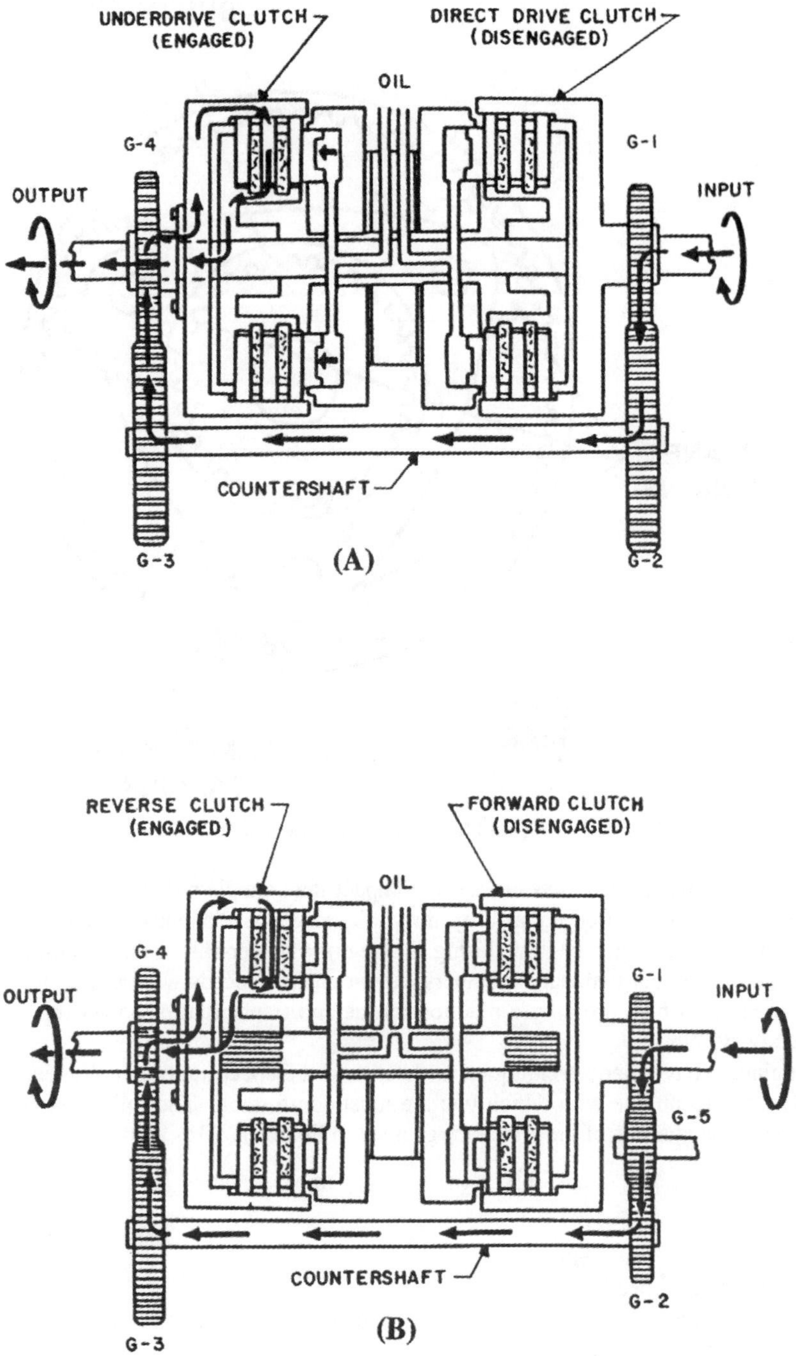

Figure 12.10. Countershaft-type, power-shift transmissions for (a) high-low and (b) reversing.

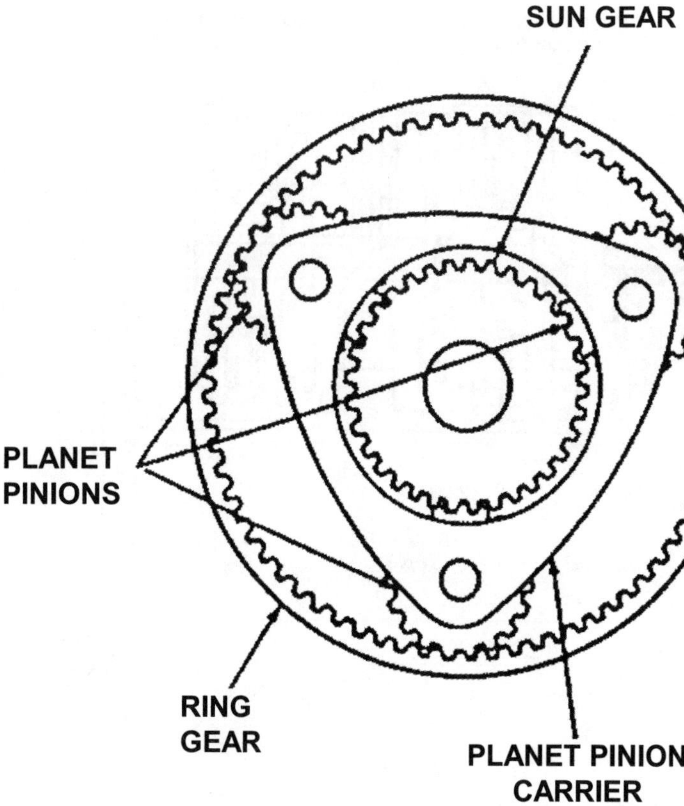

Figure 12.11. A planetary gear set.

stationary. Planetary gear sets are more compact than simple reduction gears because the gear loads are shared equally by three planet gears. Planetary gear sets help to maintain continuous power flow during shifting of a power-shift transmission because the input and output planetary members cannot change speeds without accelerating a third member, which provides a reaction torque and transmittal of power through the planetary system.

Engineers have many techniques to determine the speed ratio of a planetary gear set. Use the technique with which you are most comfortable. The following equation also gives speed ratios of the planetary gear set in Figure 12.11:

$$n_s N_s = (n_s + n_r) N_{pc} - n_r N_r \tag{12.14}$$

where

n = number of gear teeth

N = rotational speed, rpm

r, s, pc = subscripts referring to the ring gear, sun gear and planet carrier, respectively.

The speed ratio of a transmission is the input speed divided by the output speed (Equation 12.6). A frequent application of the planetary gear set of Figure 12.11 is to hold the ring gear stationary while putting power into the sun gear and taking power out on the planet carrier. The speed ratio of such a drive can be calculated from Equation 12.14 by setting $N_r = 0$ and solving for N_s/N_{pc}. The result is

$$G_t = \frac{N_s}{N_{pc}} = \frac{n_s + n_r}{n_s} \tag{12.15}$$

Another alternative is to drive the sun gear and take power out from the ring gear while holding the planet carrier stationary. Note that when the resulting equation is solved for N_s/N_r the result is negative, indicating that the ring gear turns opposite in direction to the sun gear. Conversely, a positive ratio indicates the input and output elements rotate in the same direction.

The efficiency of a planetary gear set can be analyzed using engineering principles. To illustrate, we will determine the efficiency of the planetary gear set of Figure 12.11 when the input is to the sun gear, the output is from the planet carrier, and the ring gear is being held. We start with the basic definition of efficiency as applied to the gear set, i.e.,

$$e_t = \frac{\text{power out}}{\text{power in}} = \frac{T_{pc}\, N_{pc}}{T_s\, N_s} \tag{12.16}$$

where

e_t = efficiency of the transmission

T_{pc}, T_s = torque of planet carrier and sun gear, N·m

The speed ratio, N_s/N_{pc}, is given by Equation 12.15. The torque ratio can be calculated by following the power through the gear set as if the planet carrier was fixed and noting torque reductions due to the inefficiencies in the meshes encountered. Thus, the planet torque is

$$T_p = \frac{n_p}{n_s} e_{sp}\, T_s \tag{12.17}$$

where e_{sp} = efficiency of the sun to planet mesh.

Likewise, the ring gear torque is

$$T_r = \frac{n_r}{n_p} e_{pr}\, T_p \tag{12.18}$$

where e_{pr} = efficiency of the planet to ring gear mesh.

Next, from a static equilibrium of the external torques on the planetary gear set, we note that the torques on the sun and ring gears are in the same direction and opposite to the torque on the planet gear, resulting in

$$T_{pc} = T_s + T_r \tag{12.19}$$

Finally, combining Equations 12.15 through 12.19 and simplifying gives the following equation for the efficiency, e_t, of the planetary gear set when input is to the sun, output is from the planet carrier, and the ring gear is being held:

$$e_t = \frac{1 + e_{sp}\, e_{pr} \dfrac{n_r}{n_s}}{1 + \dfrac{n_r}{n_s}} \tag{12.20}$$

The same principles can be applied to calculate the efficiencies of other planetary gear drives.

Certain principles must be observed if a planetary gear set is to assemble and function properly. The number of teeth on the planet gears is not important as long as the tooth profile will mesh properly with the teeth of the sun and ring gears. Note the following relationship:

$$\frac{n_s + n_r}{\gamma_p} = \text{Integer} \tag{12.21}$$

where γ_p = number of equally spaced planet gears.

Compound planetary gear transmissions, to be discussed in the next section, are much more complex except when the number of teeth in all the sun and ring gears is divisible by the number of planet gears. Simple inspection will show that proper assembly will be obtained when this is the case. There may be other combinations that will assemble properly, but the correct indexing of the planet gears must be found by trial and error methods as discussed by Meyers (1965, in Suggested Readings and References).

12.7.3 Compound Planetary Transmissions

The planet carrier in a *compound planetary transmission* carries two sets of planets of differing size: one set meshes with a sun gear and the other meshes with a ring gear that is offset from the sun gear. Compound planetary gear sets may also include two sun gears and/or two ring gears. The transmission illustrated in Figure 12.12 includes a compound planetary in the input section of Figure 12.12a, another in the center section of Figure 12.12b, and a third in the output section of Figure 12.12c. Taken together, the three sections form a full power-shift transmission. Numerous power-shift transmissions have been and are available on the market. The transmission in Figure 12.12 was included only to illustrate the calculation of speed ratios and the possibility of gear duplication.

The input section of Figure 12.12a has two concentric output shafts. The inner shaft is driven when clutch C_1 is engaged, while the outer shaft is driven when clutch C_2 is engaged. Connected to outer shaft C_2 are the disk portion of clutch C_{Lo} and the planet carrier of a compound planetary gear set. The smaller planet gears roll in a ring gear that drives the output shaft of the compound planetary gear set. The larger planets roll on a sun gear that is connected to the high-low drum. When clutch C_{Lo} is engaged, the

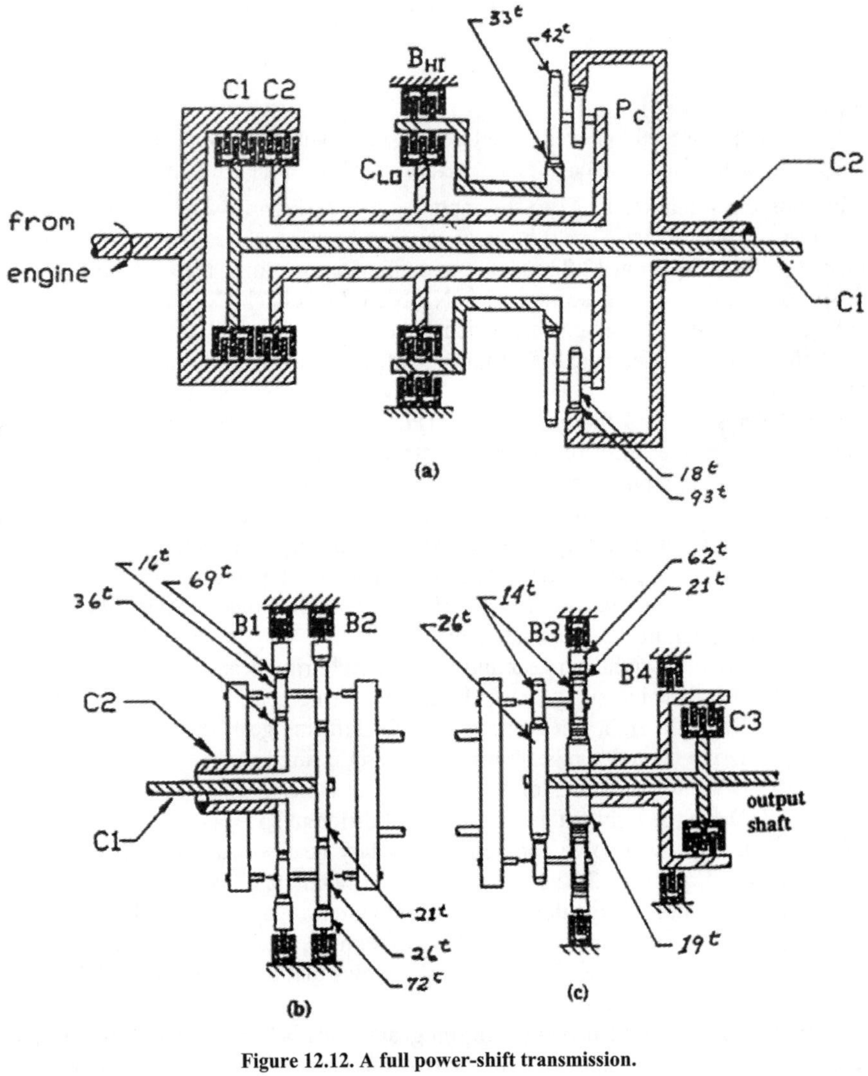

Figure 12.12. A full power-shift transmission.

sun gear is forced to rotate at the same speed as the planet carrier, thus the planet gears cannot rotate on their axes, the ring gear must rotate at the same speed as the planet carrier, and the output ring gear turns at the same speed as the engine. When brake B_{Hi} is engaged, the sun gear is held stationary, the larger planet gears roll on the sun gear, and the ring gear is driven by the smaller planet gears. Application of the principles discussed in Section 12.7.2 leads to the following equation for calculating the speed ratio of a compound planetary drive:

$$N_s n_s = N_{pc}(n_s + n_r \frac{n_{ps}}{n_{pr}}) - N_r n_r \frac{n_{ps}}{n_{pr}} \tag{12.22}$$

where

n_{ps} = number of teeth on each planet gear in mesh with the sun gear
n_{pr} = number of teeth on each planet gear in mesh with the ring gear

In applying Equation 12.22 to the input section of Figure 12.12 when C_2 and B_{Hi} are engaged, notice that $N_s = 0$ because the sun gear is held stationary by engaging brake B_{Hi}. Then Equation 12.22 reduces to a simpler equation which can be solved for the speed ratio of the input section, $G_i = N_{pc}/N_r$. When clutch C_2 is engaged, the outer shaft (C_2) is driven at engine speed when clutch C_{Lo} is engaged, but with speed ratio G_i when brake B_{Hi} is engaged. The inner shaft (C_1) is driven at engine speed when clutch C_1 is engaged.

The center section of the transmission (Figure 12.12b) also contains a compound planetary. Notice that this compound planetary contains two sun gears and two ring gears in addition to a planet carrier with two sets of planet gears. The input is through the sun gear on the right if clutch C_1 is engaged, or through the sun gear on the left if clutch C_2 is engaged. The ring gear on the right can be held stationary by engaging brake B_2, or the ring gear on the left can be held by engaging brake B_1. Output is always through the planet carrier. Six different combinations are possible from the center section of Figure 12.12b. They are:

I. Drive the right sun gear and hold the right ring gear.
II. Drive the left sun gear and hold the left ring gear.
III. Drive the right sun gear and hold the left ring gear.
IV. Drive the left sun gear and hold the right ring gear.
V. Drive the right and left sun gears at the same speed.
VI. Drive the right and left sun gears at different speeds.

Cases I through IV involve compound planetary gear sets in which a ring gear is held stationary. Equation 12.22 can be adapted to calculate the speed ratios by setting $N_r = 0$. In Cases I and II, Equation 12.22 is used with $n_{ps} = n_{pr}$. Conversely, in Cases III and IV, n_{ps} is not equal to n_{pr}. In Case III, for example, $N_r = 0$, $n_s = 21$, $n_r = 69$, $n_{ps} = 26$, and $n_{pr} = 16$. In that case, Equation 12.22 shows that the speed ratio of the center section is $G_c = N_s/N_{pc} = 5.34$.

Cases V and VI involve driving both sun gears while both ring gears free wheel and the planet carrier is the output. When the two sun gears are driven at the same speed, the planets cannot roll on the sun gears and the speed ratio of the center section is $G_c = 1.0$. The input and center sections must work together when both sun gears are driven at different speeds. A kinematic analysis shows the combined ratio for the two sections can be calculated using the following equation:

$$G_i G_c = \frac{N_{ss}}{N_{pc}} = \frac{1 - \frac{n_{fs}\, n_{ps}}{n_{ss}\, n_{pf}}}{1 - G_{fs} \frac{n_{fs}\, n_{ps}}{n_{ss}\, n_{pf}}} \tag{12.23}$$

where

G_iG_c = combined speed ratio
N_{ss} = speed of slower sun gear in rpm
$G_{fs} = N_{fs}/N_{ss}$ = ratio, faster sun to slower sun
n_{fs} = number of teeth on faster rotating sun gear
n_{ss} = number of teeth on slower rotating sun gear
n_{pf} = number of teeth on planet in mesh with faster sun gear
n_{ps} = number of teeth on planet in mesh with slower sun gear

If brake B_{Hi} is engaged (Case VI), the left sun is driven faster than the sun on the right and G_{fs} is greater than one.

Combining the transmission sections of Figures 12.12a and 12.12b provides a total of eight possible speed ratios. There are two ratios each for Cases II and IV (depending on whether clutch C_{Lo} or brake B_{Hi} is engaged) and one ratio for each of the other four cases. Up to this point, no reverse has been provided.

The output section in Figure 12.12c is always driven by the planet carrier. Unlike the previous compound planetary gear sets, two sets of planet gears on the carrier are carried on separate shafts arranged such that one set of planet gears meshes with the other. The larger set of planet gears on the right roll in a ring gear and on the right sun gear, as in a standard planetary gear set. The smaller set of planet gears mesh with the larger planets and also with the sun gear on the left. The output shaft is connected to the sun gear on the left. The right sun gear is connected to a drum to which clutch C_3 and brake B_4 are attached. In direct drive, when clutch C_3 is engaged, the two sun gears are forced to rotate at the same speed. Consequently, the output shaft is forced to rotate at the same speed as the planet carrier. Overdrive is achieved by engaging brake B_4, thus holding the right sun gear stationary. The planet carrier forces the larger planet gears to roll in the right sun gear, driving the smaller planet gears, which in turn drive the left sun gear in the same direction as the planet carrier.

Equation 12.22 does not hold for the compound planetary in the output section because the paired planet gears run on separate shafts. However, the following equation can be used to calculate the speed ratio in overdrive:

$$G_o = \frac{N_{pc}}{N_{sl}} = \frac{n_{sl}}{n_{sl} + n_{sr}} \tag{12.24}$$

where

G_o = speed ratio of output section
N_{sl} = speed of sun gear on the left in rpm
n_{sl} = number of teeth on sun gear on the left
n_{sr} = number of teeth on sun gear on the right

When brake B_3 is engaged, the ring gear is held stationary and the planet carrier forces the large planet gears to roll in the stationary ring gear. The larger planets drive the smaller planets that, in turn, drive the sun gear on the left. Again, Equation 12.22 does not hold but the following equation can be used to calculate the speed ratio:

$$G_o = \frac{N_{pc}}{N_{sl}} = \frac{n_{sl}}{n_{sl} - n_r} \tag{12.25}$$

In the usual case, $n_r > n_{sl}$, thus G_o is negative, and the output direction is opposite the input direction, i.e., the vehicle is in reverse.

In the output section, engaging brake B_3 provides eight possible reverse speeds, one for each of the eight possible speeds of the planet carrier of the output section. Depending on whether clutch C_3 or brake B_4 are engaged, two possible ratios are provided for each of the eight possible speeds of the planet carrier for a total of sixteen possible forward speeds for the entire transmission. Table 12.2 summarizes the use of the transmission in Figure 12.12 to provide a power-shift transmission with 15 forward speeds and four reverse speeds. The manufacturer decided that only four of the possible eight reverse speeds were needed in the transmission. Also, one of the 16 possible forward speeds was not used because it nearly duplicated 5th gear. Transmission designers strive to avoid such gear duplication, but it sometimes occurs. Note that locking C3 and B4 simultaneously locks the output shaft to the transmission case, thus providing a vehicle parking brake. Two neutrals, NR and NF are provided to allow the vehicle to be brought to a halt before the parking brake is engaged.

Calculating speed ratios for a full power-shift transmission may seem to be a daunting task, but it is simply a matter of breaking the problem down into smaller steps and then combining the results, as illustrated in Example Problem 12.1.

Table 12.2 Element engagements and resulting speed ratios in the power-shift transmission illustrated in Figure 12.12.

Gear	C1	C2	C_{Lo}	B_{Hi}	B1	B2	B3	C3	B4	R_t
4R		X		X		X	X			-1.40
3R		X	X		X		X			-2.11
2R	X			X*		X	X			-3.20
1R	X		X*		X		X			-4.58
NR										
Park								X	X	
NF										
1	X			X*	X			X		6.34
2	X			X*		X		X		4.43
3	X		X*		X				X	3.66
4		X	X		X			X		2.92
5		X		X	X			X		2.53
6		X	X			X		X		2.23
7		X		X		X		X		1.94
8		X	X		X				X	1.69
9		X		X	X				X	1.46
10		X	X			X			X	1.29
11		X		X		X			X	1.12
12	X	X	X					X		1.00
13	X	X		X				X		0.81
14	X	X	X						X	0.578
15	X	X		X					X	0.467
Unused	X					X			X	2.56

X indicates the element is engaged for the gear indicated.

X* indicates the element not needed but engaged for convenience.

Example Problem 12.1

Calculate the speed ratio of the power-shift transmission in Figure 12.12 in 11th and 15th gears.

Solution

Clutch C_2 and brake B_{Hi} are engaged in either 11th or 15th gears (see Table 12.2). When those two elements are engaged and the input section drives only one of the sun gears in the center section, the ratio is

$$G_i = N_{pc} / N_r = n_r / [n_r + n_s\, n_{pr} / n_{ps}] = 93 / [93 + 33\ (18 / 42)] = 0.868$$

The input section is an overdrive in this case, that is, the speed is increased in the input section.

In 11th gear, the ring gear of the input section (Figure 12.12a) drives the right sun of the center section (Figure 12.12b), while the left ring gear of the center section is held stationary. The speed ratio of the center section is

$$G_c = N_s / N_{pc} = 1 + n_r\, n_{ps} / [n_s\, n_{pr}] = 1 + (72)(16) / [(36)\ (26)] = 2.23$$

In 15th gear, the left sun gear in the center section is driven at a faster speed than the sun gear on the right. Equation 12.23 is used to calculate the combined speed ratio of the input and center sections. Notice that, when B_{Hi} is engaged, $G_{fs} = 1 / G_i = 1 / 0.868 = 1.152$. Also, $n_{fs} = 36$, $n_{ss} = 21$, $n_{pf} = 16$, and $n_{ps} = 26$. Grouping these tooth numbers in the combination needed in Equation 12.23 gives

$$[(36)\ (26)] / [(21)\ (16)] = 2.788$$

Then the combined speed ratio of the input and center sections is

$$G_iG_c = [1 - 2.788] / [1 - 2.788\ (1.152)] = 0.808$$

In the output section, the planet carrier is driven, the right sun gear is held stationary, and the left sun gear drives the output shaft. Equation 12.24 can be used to calculate the speed ratio, as follows:

$$G_o = N_{pc}/N_{sl} = n_{sl}/[n_{sl} + n_{sr}] = 26/[26 + 19] = 0.578$$

The output section functions as an overdrive in this case, that is, the speed is increased in the output section.

The speed ratio for the entire transmission, G_t, is the product of the speed ratios of the three sections. In 11th gear, the ratio of the entire transmission is

$$G_t = G_i\, G_c\, G_o = (0.868)\ (2.23)\ (0.578) = 1.12$$

In 15th gear, the transmission ratio is

$$G_t = G_i\, G_c\, G_o = (0.808)\ (0.578) = 0.467$$

The transmission is an overdrive in 15th gear, that is, the output shaft rotates faster than the input shaft.

12.7.4 Electro-Hydraulic Shifting

Imbedded hydraulic actuators are used to engage the various clutches and brakes in power-shift transmissions. For some power-shift transmissions, mechanically operated valves are used to direct pressurized hydraulic fluid to the appropriate actuators as indicated, for example, in Table 12.2. For some power-shift transmissions, solenoid valves are used to direct the fluid. When two-state, i.e., off-on solenoid valves are used, accumulation and restriction valves are needed to smooth the shifting action. However, fluid viscosity and density changes resulting from temperature changes tend to degrade the consistency of shifting, resulting in operator discomfort and early driveline failures.

A recent innovation, the electro-proportional pressure-reducing (EPPR) valve, was designed to provide smooth, consistent shifting without the need for accumulation and restriction valves. A cross sectional view of an EPPR valve is shown in Figure 12.13. One such valve is provided for each clutch or brake in a power-shift transmission. Note the three external o-ring seals on the EPPR valve. The portion of the valve to the right of the right-most o-ring is immersed in port 1, the pump port. The portion of the valve between the center and right-most o-rings is immersed in port 2, the clutch (or work) port. The portion of the valve between the center and the left-most o-rings is immersed in port 3, the tank port. A pilot operated sliding spool determines the connectivity between the ports. Pilot operation is initiated by supplying pulse-width-modulated (PWM) electrical current to an electrical coil in the left end of the EPPR valve.

In the de-energized mode, a bias spring at the right end of the sliding spool holds the spool in the left-most position illustrated in Figure 12.13. Port 2 is connected to port 3 to reduce the work port pressure to near zero to keep the corresponding clutch or brake disengaged. Oil from Port 1 flows via a cross passage and central passage in the sliding spool, through a filter screen and orifice to the open ball valve, and through threads on the outer edge of the spool housing to Port 3. The oil experiences some pressure drop in passing through the threads but, if the pressure in Port 3 is kept small enough, the pressure in the area of the ball housing will be less than the 100 kPa needed to compress the bias spring to move the sliding spool to the right.

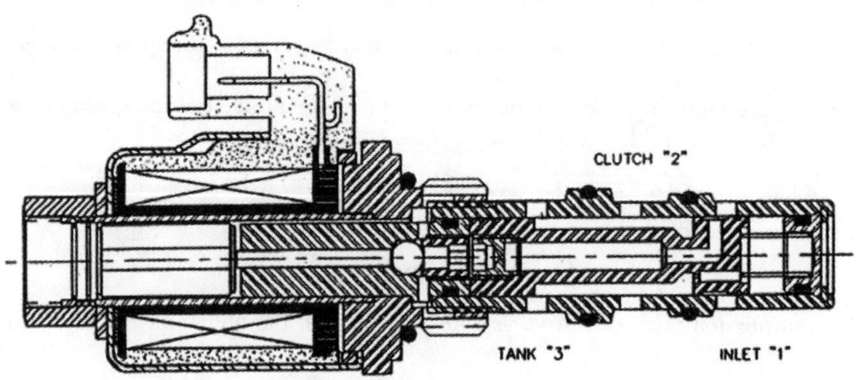

Figure 12.13. An electro-proportional pressure-reducing valve. (From Harms and Guse, 2000.)

Actuation of the EPPR valve begins when the duty cycle of the PWM signal to the coil is increased, causing the steel pin to the left of the ball valve to push the valve onto its seat. The resulting pressure buildup in the pilot chamber to the right of the ball valve overcomes the bias spring and causes the sliding spool to move to the right. With sufficient movement, the sliding spool disconnects the work port from the tank port and connects the work port to the pump port. The resulting pressure buildup in the work port engages the clutch or brake whose cylinder is connected to the work port. Note that the amount of sliding spool movement, and thus the pressure level in the work port, varies with the duty cycle of the PWM signal delivered to the coil. This is because the degree of openness of the ball valve varies with the force balance between the hydraulic force supplied by the pressurized oil flowing through the center of the sliding spool and the electrical force supplied by the coil in moving the steel pin against the ball. When the duty cycle of the PWM signal reaches 100%, the ball valve is fully closed and the pressure in the pilot chamber rises to essentially the same level as the pressure in the pump port.

Each hydraulic actuator operating the clutches and brakes functions like a spring-return, single-acting cylinder. The speed of engagement is controlled by the rate of pressure increase to the cylinder, while the speed of disengagement is controlled by the rate of pressure decrease. Typically, about 167 kPa of pressure is required to initiate movement of the cylinder and 230 to 330 kPa is required to reach the *kiss point* (the point at which the clutch or brake begins to engage). The slippage of the clutch or brake that begins with engagement is essentially stopped when the pressure reaches 1 MPa. Typically, the pressure then builds to about 2 MPa to provide for reserve capacity. If the actuating cylinder is empty when pressurization begins, significant time can elapse before the clutch or brake engages. Thus, the conduits should be arranged to allow gravity to keep the cylinders full between actuations. Also, with careful control, the PWM signal can be brought to 100% duty cycle for a short time before each actuation to allow rapid filling, but then reduced before the cylinder engages the clutch or brake. Finally, between actuations, the EPPR can keep the cylinder pressurized to a pressure just below the kiss point. When actuation is desired, the EPPR can ramp up the pressure to provide a controlled engagement of the clutch or brake. Likewise, pressure can be ramped down to provide a controlled disengagement. Such careful control of the EPPR is accomplished by a microprocessor. Using a small lever attached to an indicating device (a potentiometer or multi-position switch), the vehicle operator can signal the microprocessor as to a desired gear change. Upon reading the indicating device, the microprocessor consults a table such as Table 12.2 to determine which clutches and brakes are to be de-energized to move out of the gear that is being left, and which clutches and brakes are to be energized to move into the new desired gear. The microprocessor also handles the rates of de-pressurization and pressurizations to cause a smooth shift.

The discussion above focused on the use of EPPR valves to control a power-shift transmission. However, the EPPR valve could also be used to control the engagement and disengagement of a hydraulically operated wet clutch.

12.7.5 Practical Power Shifting

A primary advantage of a full-range power-shift transmission is that speed ratios can be changed with essentially no interruption of power. A traction clutch is not needed for a full-range power-shift transmission. However, a clutch pedal is usually provided for safety reasons. In an emergency, an operator will instinctively reach for the clutch pedal to stop the vehicle. The clutch pedal need not operate a conventional clutch; it could rather shift the transmission into neutral, for example, by releasing both clutches C_1 and C_2 in Figure 12.12.

Full-range power-shift transmissions also have disadvantages. Considerable energy is used in the hydraulic shifting of the various clutches and brakes and because of friction in the numerous gear meshes in the transmission. The efficiency, e_t, of a full-range power-shift transmission is typically below 85%. Such transmissions also are expensive to manufacture.

Tractor manufacturers have recognized that full-range power shifting may not be essential. As an option, they may provide part-range power shifting by placing a synchromesh transmission in series with a power-shift element. The traction clutch must be disengaged to shift the synchromesh transmission. Power shifting within each manually selected gear can be accomplished without disengaging the tractor clutch. The part-range power-shift transmission is less expensive and its efficiency approaches 90%.

12.8 Continuously Variable Transmissions

The transmissions discussed up to this time all provided a set of discrete gear ratios. The *continuously variable transmission* (CVT) does not have discrete gear ratios; rather, the speed ratio is continuously variable. Considerable research and development work is underway on mechanical CVTs, and at least one is on the market. However, the most widely available CVT for heavy-duty off-road vehicles is the *hydrostatic transmission*. Figure 12.14 illustrates the concept of a hydrostatic transmission. Power input is to a pump, while power output is from a hydraulic motor.

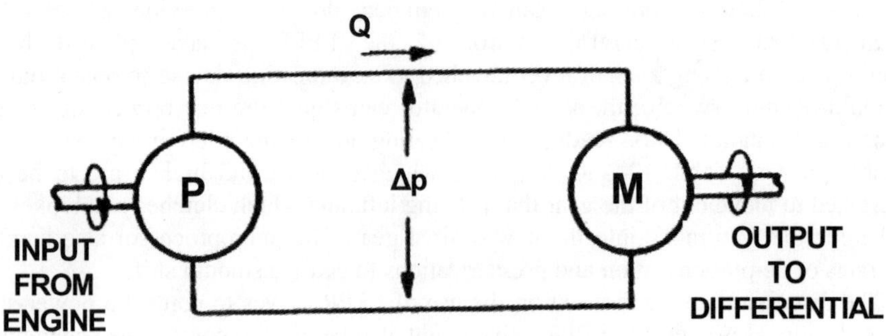

Figure 12.14. A hydrostatic transmission.

All hydrostatic transmissions can be classified into one of the four following types:

1. Fixed-displacement pump and fixed-displacement motor
2. Variable-displacement pump and fixed-displacement motor,
3. Fixed-displacement pump and variable-displacement motor
4. Variable-displacement pump and variable-displacement motor

The following equations govern the input-output relationships of a hydrostatic transmission. The speed relationship is

$$N_m = (e_{mv} e_{pv}) \frac{D_p}{D_m} N_p \tag{12.26}$$

The following two equations govern the torque relationship:

$$T_m = (e_{pt} e_{mt}) \frac{D_m}{D_p} T_p \tag{12.27}$$

and

$$T_m = \frac{e_{mt} D_m \, \Delta p}{2\pi} \tag{12.28}$$

where

N_p = pump (input) speed, rpm
N_m = motor (output) speed, rpm
e_{pv} = volumetric efficiency of pump, decimal
e_{mv} = volumetric efficiency of motor, decimal
e_{pt} = torque efficiency of pump, decimal
e_{mt} = torque efficiency of motor, decimal
D_p = displacement of pump, cm^3/rev
D_m = displacement of motor, cm^3/rev
Δp = pressure difference across pump or motor, MPa
T_p = pump (input) torque, N·m
T_m = motor (output) torque, N·m

12.8.1 Type 1 Hydrostatic Transmissions

The Type 1 hydrostatic transmission provides no possibility for changing the output speed unless part of the pump flow is diverted away from the motor; thus, the Type 1 transmission is seldom used.

12.8.2 Type 2 Hydrostatic Transmissions

The Type 2 hydrostatic transmission provides excellent control of speed and direction, as illustrated in Figure 12.15. The pump displacement can be varied from full delivery in one direction to zero delivery to full delivery in the reverse direction. The Type 2 transmission has the disadvantage of being a *constant torque transmission*. For maximum power-transmitting capability at full speed, the transmission must be operated at the maximum allowable pressure difference, Δp.

When D_p is reduced to reduce the output speed, the output torque is not increased, as Equation 12.28 shows. Thus, the power-transmitting capability declines at lower speeds with the Type 2 transmission. If such a transmission were to be used on a heavy-duty vehicle, a very large pump and motor would be needed to provide adequate power at low speeds. Type 2 transmissions are commonly used in lawn tractors, where power demands are limited. Heavy-duty off-road vehicles have need for *constant power transmissions*. All of the gear-type transmissions discussed earlier are constant power transmissions.

12.8.3 Type 3 Hydrostatic Transmissions

The Type 3 hydrostatic transmission is a constant power transmission. D_m must be increased to reduce the output speed (Equation 12.25), but increasing D_m increases the torque-transmitting capability (Equation 12.26 or 12.27) without increasing Δp. However, as Figure 12.16 indicates, the Type 3 transmission has poor speed-changing characteristics. Allowing the swash plate in the motor to cross over zero displacement would cause a sudden reversal from maximum speed in one direction to maximum speed in the opposite direction. Such abrupt changes cannot be permitted, so variable-displacement motors are designed to prevent zero displacement. The swash plate is blocked from reaching neutral, and thus the Type 3 transmission cannot provide reversing of direction of travel. Also, a very large motor displacement would be required to obtain very slow speeds.

12.8.4 Practical Hydrostatic Transmissions

The Type 4 hydrostatic transmission is used in heavy-duty applications, such as in the transmission of a tractor. The variable-displacement pump provides excellent speed control and reversing. When the pump displacement is reduced to reduce speed,

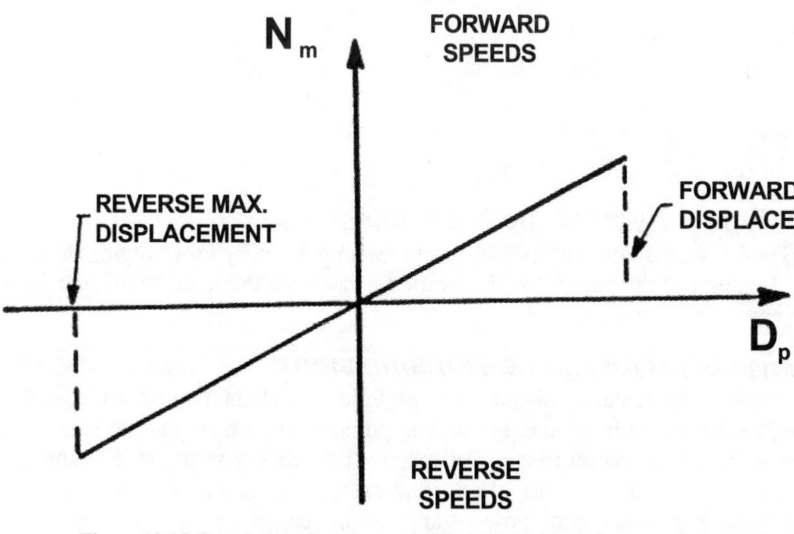

Figure 12.15. Speed-changing characteristics of a hydrostatic transmission with variable-displacement pump and fixed-displacement motor.

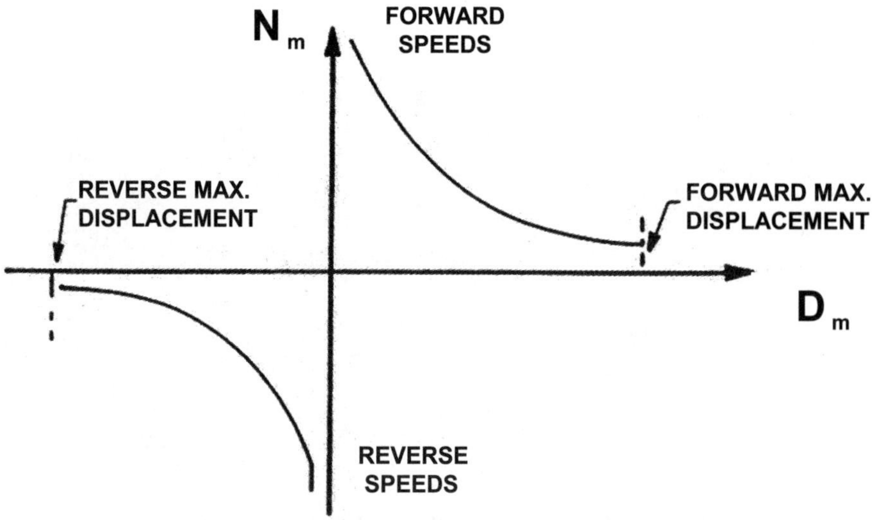

Figure 12.16. Speed-changing characteristics of a hydrostatic transmission with fixed-displacement pump and variable-displacement motor.

the motor displacement is simultaneously increased to provide some increase in torque capacity. Although the torque increase at low speeds is less than for a Type 3 hydrostatic transmission, it must only be sufficient to provide enough torque to reach the traction limit of the drive wheels.

Theoretically, the hydrostatic transmission of Figure 12.14 could cover the entire speed range. However, volumetric efficiencies of the pump and motor approach zero at low speeds, while the torque efficiencies approach zero at high speeds. The power efficiency of the entire transmission can remain high only over a limited range of speeds. Thus, as illustrated in Figure 12.17, a hydrostatic transmission is used in series with a gear-type transmission in tractors. A speed range is selected by shifting the gear transmission, and the hydrostatic transmission is continually variable within that range. A single lever is provided to set the displacement of the pump and motor simultaneously. Once that lever is set, the transmission behaves exactly like a gear transmission, i.e., it follows a torque-speed relationship similar to that for one of the gears in Figure 12.4. Since the hydrostatic transmission is infinitely variable, however, the operator can choose from an infinite variety of ratios and thus eliminate the cross hatched areas that are inaccessible in Figure 12.4. Poor power efficiency has limited the widespread adoption of hydrostatic transmissions in tractors. It can be shown that the overall power efficiency of a hydrostatic transmission is $e_t = e_{pt}\, e_{pm}\, e_{mv}\, e_{mt}$. If each of these efficiencies could be maintained at 0.95, the power efficiency of the entire transmission would be only 0.81. However, hydrostatic transmissions have been adopted in applications where the speed and convenience they offer in speed changing offsets the lower power efficiency.

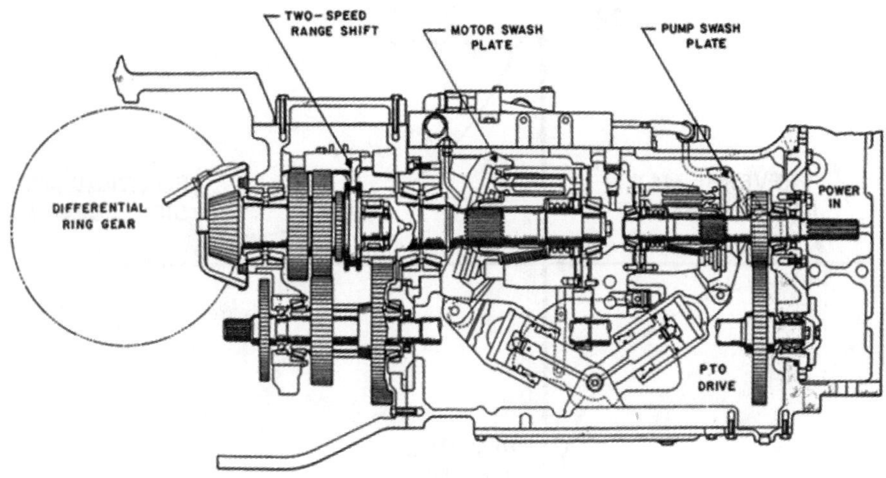

Figure 12.17. A heavy-duty hydrostatic transmission in series with a gear transmission.

Some manufacturers have developed *split-power* transmissions in which part of the power is transmitted hydrostatically while the majority of the power is transmitted through the planetary gear sets of a power-shift transmission. The hydrostatic element allows continuous variability of speed within each of the speed ratios of the power-shift transmission. Transmitting most of the power through the planetary gear sets helps to reduce power losses in the hydrostatic unit. The result is a CVT that is more efficient than one that uses only a hydrostatic unit.

12.9 Hydrokinetic Transmissions

A schematic diagram of a hydrokinetic transmission is shown in Figure 12.18. The input shaft drives an *impeller* that generates oil flow within the unit. Typically, the impeller blades are attached to the inner edge of the transmission case, and the entire case spins with the engine flywheel. The oil flows across a *turbine* that drives the output shaft. After leaving the turbine, the oil flows through a *stator*, i.e., a stationary set of vanes that change the flow direction of the oil. The torque reaction on the stator is effectively transferred to the turbine. Without the stator, the turbine torque could never be greater than the impeller torque but, with the stator, the transmission is able to provide a torque increase. Although the turbine has some effect on the impeller, the hydrokinetic transmission is more easily understood if we assume that effect is insignificant. Then, the impeller has a definite torque-speed curve of its own, as illustrated in Figure 12.19. The intersection of the impeller torque-speed curve with that of the engine defines an operating point that remains fixed unless the operator changes the governor setting. Both the input speed and input torque of the transmission remain essentially constant. Then, because the product of torque and speed is constant across the hydrokinetic transmission, the output speed is controlled by the output torque, as indicated in the following equation:

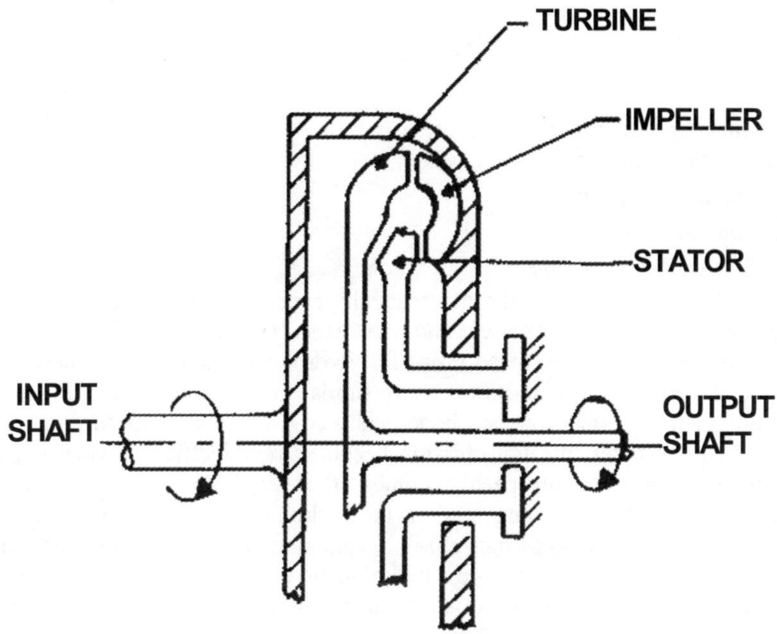

Figure 12.18. Schematic diagram of a hydrokinetic transmission.

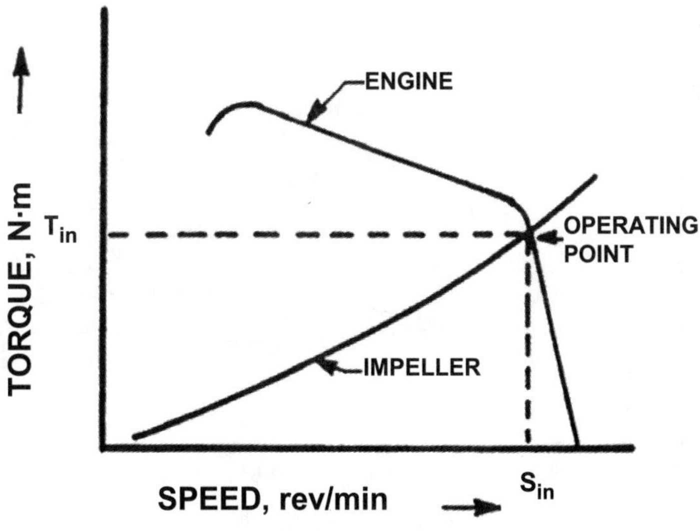

Figure 12.19. Torque-speed curves for a governed engine and a hydrokinetic transmission impeller.

$$N_{out} = \frac{T_{in} N_{in}}{e_t T_{out}} \tag{12.29}$$

where

N_{in} = input speed, rpm

N_{out} = output speed, rpm

T_{in} = input torque, N·m

T_{out} = output torque, N·m

e_t = transmission efficiency

The hydrokinetic transmission automatically reduces output speed if the torque load on the output shaft increases, and vice versa. Consequently, the hydrokinetic transmission has no definite speed ratio. The hydrokinetic transmission performance would be plotted as a constant-power curve similar to the dashed line in Figure 12.4. The hydrokinetic transmission can theoretically cover the entire speed range but, as Figure 12.20 illustrates, the transmission becomes very inefficient when the speed ratio (output speed over input speed) becomes either very low or very high. Thus, as in the case of the hydrostatic transmission, the hydrokinetic transmission is generally used in series with a gear-type transmission. Automatic transmissions in automobiles, trucks, busses, etc., typically include a hydrokinetic transmission in series with a two- or three-speed, planetary type power-shift transmission. No traction clutch is provided; the gear transmission is shifted automatically to keep the hydrokinetic transmission working at an efficient speed ratio. On heavy-duty tractors, the hydrokinetic

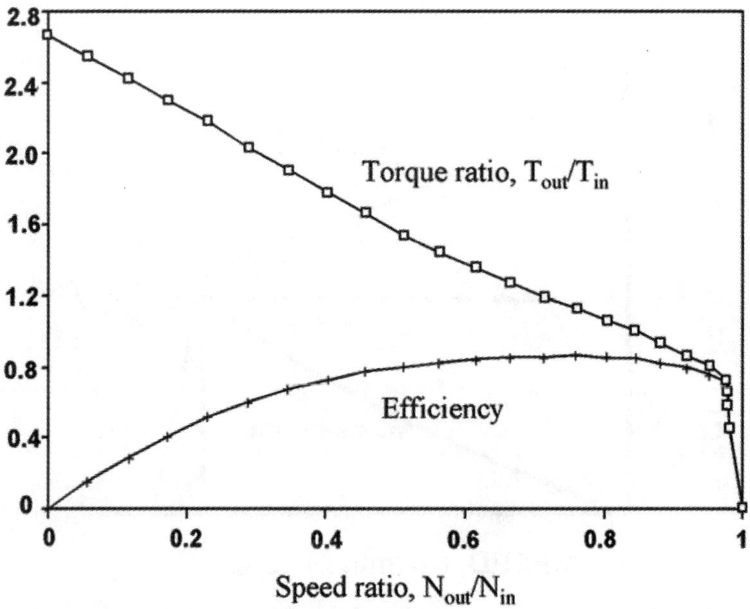

Figure 12.20. Typical performance curves for a hydrokinetic transmission.

transmission is most useful when exact speed control is unnecessary, as in doing heavy tillage work. When control of travel speed is important, provision must be made for locking out the hydrokinetic transmission to prevent the speed from changing in response to changes is torque load. A traction clutch is provided to allow manual shifting of the gear-type transmission when the hydrokinetic unit is locked out.

12.10 Comparison of Transmissions

The sliding gear and constant-mesh transmissions are simple and efficient in transmitting power, but are difficult to shift while the vehicle is in motion. The synchronizers in a synchromesh transmission improve productivity by allowing easy shifting on the go, but it is still necessary to release the traction clutch to shift gears. A full-range, power-shift transmission can be shifted into any of its speed ratios without releasing a traction clutch. Although the vehicle could be designed without a clutch pedal, one is usually included for safety reasons. In an emergency, most operators can release a traction clutch instinctively, but may not think fast enough to stop the tractor by using the shift lever. By allowing quicker shifting without releasing the traction clutch, the full-range power-shift transmission provides a further productivity boost over the synchromesh transmission. As a cost saving measure, a two-speed power-shift element is sometimes used in series with a synchromesh transmission. Power shifting without a clutch is then possible within any of the speed ranges of the synchromesh transmission.

Vehicle productivity is increased by providing more speed ratios, so that the operator has a better chance of selecting a ratio that will perfectly match the engine torque and speed to the needs of the load. The CVT theoretically allows the operator to accomplish perfect load matching by providing an infinite number of speed ratios. It remains the operator's responsibility to select the best ratio for load matching. Currently, the only CVT available for heavy-duty off-road vehicles is the hydrostatic transmission.

The hydrokinetic transmission has no definite speed ratio. It automatically adjusts its output speed in response to changes in output torque. Thus, it provides automatic load matching along a constant power curve. In agricultural tractors, it is best suited for heavy tillage work and other operations where travel speed is not critical. If travel speed is critical, the hydrokinetic transmission must be locked out.

12.11 Resonances in Transmissions and Computer Simulation

The shafts in transmissions undergo torsional deflection in response to applied torque, and therefore act as torsional springs. Some of the rotating masses in a transmission are large enough to have significant inertia. The torsional springs allow the rotating masses to oscillate torsionally at natural frequencies that are proportional to the square root of the torsional spring rate/inertia ratio. Torsional disturbances occur in a transmission due to engine harmonics, gear tooth meshing and other factors. The amplitude of oscillation of a simple spring-mass torsional pendulum is given by the following equation:

$$\theta = \frac{T(t)}{K - J\omega^2} \tag{12.30}$$

where

θ = amplitude of shaft oscillation, radians

$T(t)$ = amplitude of the torque oscillations, N·m

K = torsional spring rate, N·m /radian

J = moment of inertia of rotating masses, kg·m^2

ω = angular frequency of the torque fluctuations, radians/s

The angular natural frequency of the system, T_n, is

$$\omega_n = \sqrt{\frac{K}{J}} \tag{12.31}$$

Note that if the driving frequency, T, approaches the natural frequency, T_n, the torsional oscillation becomes very large. In practice, there are multiple natural frequencies in a transmission and if any of them are approached, oscillations can build to amplitudes that are annoying to the operator and even damaging to the transmission. The reader may consult books on vibration analysis for more extensive information on vibration.

Because of the complexity of the actual vibratory system, computer simulation is usually used to identify potential vibratory problems in transmissions and to test potential solutions. Transmission shafts are subject to torsional and bending stresses that must be resisted without failure and computer simulation is also used for stress analysis.

12.12 Differentials

The drive wheels of a vehicle negotiating a turn travel at different speeds, since the drive wheel most distant from the center of turn must travel farther and faster than the drive wheel nearest the center of turn. An exploded view of a *differential* is shown in Figure 12.21. The transmission drives the *bevel pinion*, which in turn drives the *ring gear* to provide a speed reduction and a 90 change in shaft direction. A *carrier assembly* attached to the ring gear carries four *differential pinion gears*. Because the differential pinion gears are free to rotate on their shafts, the two *side gears* can rotate at different speeds. The speed relationships for the differential are

$$\frac{N_L + N_R}{2} = N_{ave} = \frac{N_{in}}{G_d} \tag{12.32}$$

and

$$G_d = \frac{n_r}{n_{bp}} \tag{12.33}$$

where

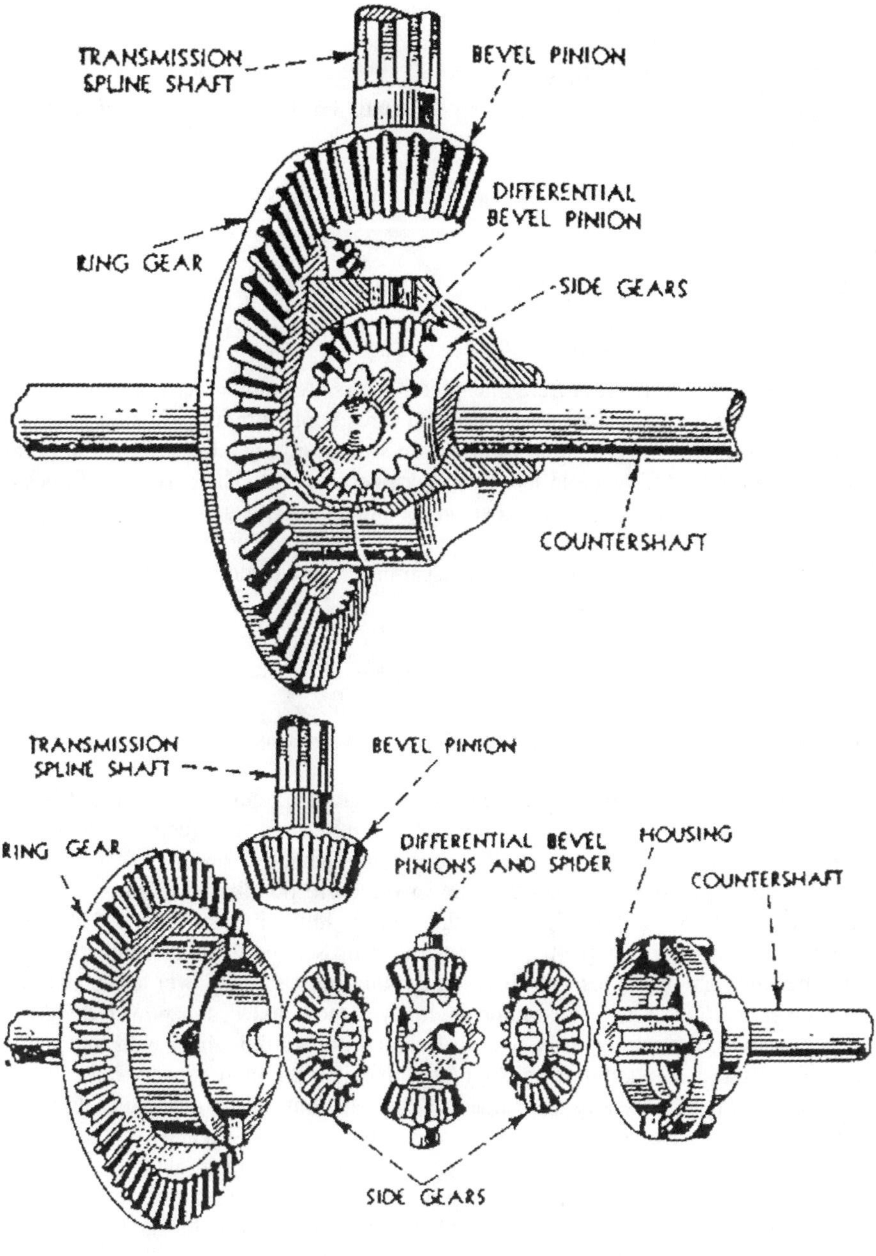

Figure 12.21. An exploded view of a differential.

N_L = speed of left side gear, rpm
N_R = speed of right side gear, rpm
G_d = speed ratio of differential
n_{bp} = number of teeth on bevel pinion
n_r = number of teeth on ring gear

Typically, G_d is in the range between 2.5 and 3.5 for differentials. The power relationship for the differential is

$$T_L N_L + T_R N_R = e_d\, T_{in}\, N_{in} \tag{12.34}$$

where

T_L = torque in left side gear, N·m
T_R = torque in right side gear, N·m
e_d = efficiency of differential

Consider any one of the differential pinion gears. Assuming zero bearing friction on the differential pinion shaft, the peripheral force applied by the left side gear must equal the peripheral force applied by the right side gear, since the differential pinion cannot sustain any difference in these two peripheral forces. The peripheral forces on the differential pinion are equal to those on the side gears and, since the two side gears have the same radius, they must also transmit the same torque. Thus, the law of differentials is that the two side gears must transmit equal torques. Then, since $T_L = T_R$, Equation 12.34 reduces to

$$T_L = T_R = \frac{e_d\, G_d\, T_{in}}{2} \tag{12.35}$$

The factor of 2 appears in the denominator because the ring gear torque is split, with half going to the left axle and half to the right.

A differential is needed for turning, but it degrades vehicle performance in two ways. First, the torque in either side gear is limited by the traction that can be developed by the attached wheel. If one driving wheel is on a sheet of ice, for example, virtually no torque can develop in the corresponding side gear. Thus, the maximum torque transmitted through either axle is limited to the torque that can be supported by the wheel with the poorest traction. Secondly, in dividing the input power between the two axles, the differential routes most of the power to the axle with the poorest traction. Since the torque is equal in both axles, the power divides in accordance with the axle speeds. The differential allows the wheel with the poorest traction to rotate faster and thus transmit more than half the input power. Since the two drive wheels are attached to the same vehicle and thus must translate at the same speed, any difference in power transmission between the wheel with the poorest traction and the wheel with the best traction is lost in wheel slippage.

A *differential lock* improves vehicle performance by locking one of the side gears to the carrier or ring gear. Then both axles are forced to rotate at the same speed. The differential pinions are no longer free to rotate on their shafts, and thus the torques need no longer be equal in the two side gears. Equation 12.34 still holds, but now the speeds of the two side gears are equal. Thus, the power divides between the two axles in accordance with the axle torques; the wheel with the best traction transmits the majority of the power. Since the differential lock must be released to turn the vehicle in a tight turn, a mechanism should be provided to release the differential lock whenever either of the turning brakes is engaged.

Some vehicles are equipped with a *limited-slip differential*. Such a differential allows speed differences between the two axles sufficient for vehicle turning. However, when the speed difference exceeds a certain limit, a centrifugally-operated device causes the differential to lock up. When the speed difference falls below the limit, the differential automatically unlocks. Limited-slip differentials thus prevent a vehicle from being stalled when one of the two wheels connected to the differential loses traction.

12.13 Final Drives

A typical *final drive* is shown in Figure 12.22. The purpose of the final drive is to provide a large reduction in speed and corresponding increase in torque. Final drives are a necessity in vehicles that have drive wheels of large radius; the combination of high tangential tire load caused by high drawbar pull and the large tire radius generates very high axle torques. By allowing all drive train parts between the engine

Figure 12.22. A planetary gear set used as a final drive.

and the final drive to operate at higher speed and lower torque, the final drive permits the power train to be of lighter construction, more compact and less expensive. Equations 12.15 and 12.20 can be used to calculate the speed ratio and efficiency of final drives of the type shown in Figure 12.22. Typically, the speed ratio for final drives is in the range from 4 to 5.

12.14 Steering Drives for Skid-Steer Vehicles

Most wheeled vehicles accomplish steering by causing one or more of the wheels to be angled relative to the main longitudinal axis of the vehicle. In contrast, *skid-steer vehicles* accomplish steering through differential speeds of the right and left traction devices. In negotiating a turn, at least some portion of the tractive device skids sidewise, thus giving rise to the name.

The wheels of the skid-steer loader of Figure 1.5 are driven hydrostatically and, typically, one variable-displacement pump drives the hydraulic motor that powers the left wheels while another such pump powers the motor that drives the right wheels. By appropriately varying the pump displacements, the wheels on the two sides can be made to rotate at the same speed, at different speeds, or even in opposite directions. In the latter case, the loader can turn about a point without translating forward or backward.

The tracks of a crawler tractor are driven mechanically. Figure 12.23 is a schematic illustration of the steering drive for a rubber-tracked agricultural vehicle. Shaft 1 receives power from the vehicle transmission. Shaft 7 powers the right track and shaft 14 powers the left track. The rotational speed of ring gear 19 is controllable by a hydraulic motor via the spur gear shown below ring gear 19. When the motor speed is zero, it can be shown that shafts 7 and 14 turn at the same speed and the vehicle moves in a straight line. When the motor turns clockwise (as shown by the arrow in Figure 12.23), the vehicle turns to the right, while turning the motor counterclockwise turns the vehicle to the left. The faster the motor turns in either direction, the steeper the angle of turn of the vehicle. If the motor turns sufficiently fast in either direction, the tractor can be made to turn about a point without translating, as was described for the skid-steer loader.

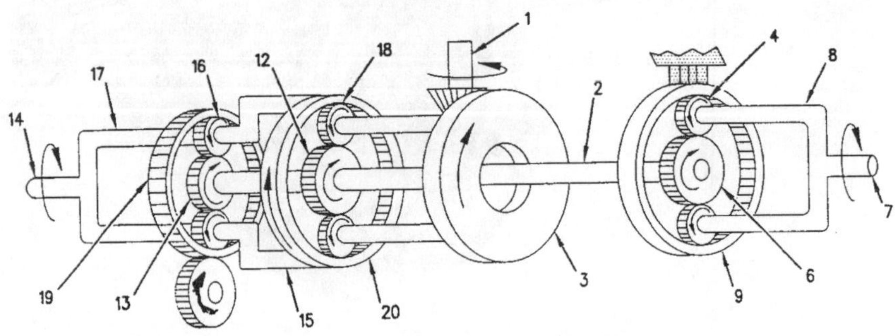

Figure 12.23. A differential steering drive. (Courtesy of Caterpillar, Inc.)

12.15 Power Take-Off Drives on Agricultural Tractors

A *power take-off* (PTO) drive provides a means for mechanically transmitting rotary power from a tractor to an implement. The original standard location for the PTO shaft was at the rear of the tractor, as shown in Figure 12.24. The PTO drive was first standardized by ASAE in 1926, and the standard has been updated periodically since then. The direction of rotation (clockwise as viewed from the rear), exact shaft dimensions, rotational speed, and approximate location are specified in the standard. The first standard specified a standard rotational speed of 540 rpm. As the power demand of implements increased, the 540 rpm PTO was no longer capable of transmitting enough power and a 1000 rpm PTO was developed. Later, for even greater power transmission, another 1000 rpm standard was developed, but with a larger shaft. The following standard PTO drives are now available (Figure 12.25):

Type 1: 540 rpm, 35 mm shaft for tractors up to 65 kW of PTO power

Type 2: 1000 rpm, 35 mm shaft for tractors of 45 to 120 kW of PTO power

Type 3: 1000 rpm, 45 mm shaft for tractors of 110 to 190 kW of PTO power

Some large, four-wheel drive tractors exceed all of the above power ranges and do not have a PTO drive. The splines on the two 35 mm PTO shafts differ to prevent the use of a 540 rpm implement with a 1000 rpm PTO drive. In addition to the rear PTO shaft, some tractors are now equipped with a front PTO shaft to power implements that are attached to the front of the tractor. If present, the front PTO is a Type 2 shaft and rotates clockwise as viewed when facing the projecting end.

Figure 12.24. A tractor power take-off drive.

Figure 12.25. The three types of PTO shafts.

Three different arrangements have been standardized for delivering engine power to the PTO drive. They are:

- Transmission-driven PTO
- Continuous-running PTO
- Independent PTO

As the name implies, the first type of PTO is driven from the transmission. Thus, disengaging the traction clutch interrupts power flow to the drive wheels and to the PTO shaft. Because it is sometimes inconvenient to stop the PTO every time the traction clutch is disengaged, the continuous-running PTO was developed. It has two separate clutches, but both are controlled by the same foot pedal. Initial movement of the pedal releases the traction clutch, and further movement releases the PTO clutch. Most new tractors now have an independent PTO, in which a hand lever or switch disengages a separate (from the traction clutch) PTO clutch.

12.16 Summary

The power train of a vehicle includes the traction clutch, transmission, and differential. If the vehicle is a tractor, the power train may also include final drives and a PTO drive. The power train serves the functions of transmitting power to the drive wheels and PTO shaft, disconnecting power, reversing directions, and transforming the engine torque and speed to meet demands of the load. In skid-steer vehicles, the power train also accomplishes vehicle steering.

Clutches permit interruption of power flow from the engine to the drive wheels. Hand-operated clutches are used on independent PTO drives on tractors and are designed to stay in either the engaged or disengaged position without operator attention. The traction clutch on small to medium-sized tractors is usually a spring-loaded dry clutch. Large tractors have a wet, hydraulically operated clutch. Clutches must be designed to transmit the required torque with at least 100% torque reserve and have sufficient capacity to avoid overheating.

Brakes provide a means for stopping a vehicle. The procedure for designing a brake is similar to that for designing a clutch, i.e., a clutch must absorb energy in starting a vehicle while a brake must absorb energy in stopping the vehicle. Drum brakes were used on nearly all early vehicles, but disk brakes are more common on recent vehicles.

Manual-shift transmissions prevent the tractor from operating at some pull-speed combinations that are achievable by the engine. Increasing the number of gear selections decreases the number of unusable pull-speed combinations. Use of synchronizers permits faster shifting for increased tractor productivity. Power shifting permits shifting without use of a traction clutch and with no apparent interruption of power. Continuously variable transmissions (CVTs) theoretically allow the operator to select any pull-speed combination permitted by the engine. The hydrostatic transmission is the most widely available CVT. A practical hydrostatic transmission for a heavy-duty vehicle includes a variable-displacement pump to allow full-range speed changing and reversing, and a variable-displacement motor to allow increased torque transmission at lower speeds. The hydrostatic transmission can operate over only a limited speed ratio at peak efficiency, so a selective gear transmission is used in series to permit operation in more efficient speed ranges. Recently, a hydrostatic transmission element has been integrated into a planetary type power-shift transmission to obtain a more-efficiency CVT. Hydrokinetic transmissions automatically adjust their speed ratios to match the load; the engine tends to operate at a fixed point on the torque-speed map, and thus the output load versus speed curve is a constant-power curve.

Differentials are necessary to allow drive wheels further from a center of turn to travel faster than those nearer the center of turn. A differential forces both axles to transmit equal torque. Thus, neither axle can transmit more torque than is permitted by the wheel with the poorest traction. Also, since the wheel with the poorest traction turns faster, over half of the power is transmitted to the wheel with the poorest traction. A differential lock can be used when the tractor is traveling straight ahead or on a slight curve. The differential lock forces both axles to rotate at the same speed and torque can be higher in the axle whose attached wheel has the best traction. As a result, over half of the power is delivered to the wheel with the best traction. The differential lock must be disengaged during a sharp turn.

Tractors are equipped with final drives to produce a large increase in torque and decrease in speed at the drive wheels. The large diameter of tractor drive wheels and the heavy loads pulled by tractors require that the final axles must turn at low speeds and be capable of transmitting high torque. Use of final drives permits all drive components nearer to the engine to turn at much higher speeds and carry much less torque. Thus, the final drives permit the remainder of the power train to be of lighter construction, more compact, and less expensive.

A power take-off (PTO) drive provides a means for transmitting rotary power from the tractor to an attached implement. The PTO drive was first standardized by ASAE in 1926. The standard specifies the approximate location of the PTO shaft, its direction and speed of rotation, and its exact dimensions. Three types of PTO shafts are available. Type 1 is a 540 rpm PTO, Type 2 is a 1000 rpm PTO, and Type 3 is a 1000 rpm PTO with a larger shaft. Three different types of PTO drives have been available, the transmission-driven PTO, the continuous-running PTO and the independent PTO. The trend on late model tractors is to equip them with an independent PTO.

Homework Problems

12.1 The data below are for a governed, CI engine.

Speed, RPM	Torque, N·m	Remarks
2363	0	High idle
2338	110	
2310	220	
2276	330	
2243	441	
2200	518	Governor's maximum
1978	585	
1850	639	Peak power
1755	673	
1542	689	
1321	689	Peak torque
1101	663	

Design a dry, cerametallic clutch for the engine whose torque speed data are given above. Use mid-range values for the clutch material properties, but design for 100% clutch reserve capacity. In Equation 12.2, assume that the outer diameter is 75 mm larger than the inner diameter.

(a) Specify the maximum clamping force that will not exceed the maximum facing pressure.

(b) Specify the clutch diameter.

(c) Specify the number of clutch disks required to provide the design capacity and keep the energy rate within allowable limits. Note that the number of clutch disks must be an integer.

12.2 Rework Problem 12.1, but design a wet clutch using filled fluorocarbon clutch material.

12.3 Rework Problem 12.1, but design a wet clutch using sintered metallic clutch material.

12.4 Design disk brakes to decelerate a 2 Mg vehicle from a speed of 20 km/hr at a rate of 3 m/s^2 if the radius from the axle center to the pad centers is 180 mm and the loaded radius of the wheels is 750 mm. Assume only the rear wheels are used for braking and each brake includes two pads; the brakes are located inboard of the final drive and thus the disks rotate at 4.5 times the wheel speed. Assume the brake material is the dry organic material in Table 12.1.

(a) Allowing 100% reserve capacity, calculate the required clamping force.

(b) Also calculate the required pad area to keep the specific rate of heat generation within allowable limits; use midrange values from Table 12.1.

(c) Using the pad area of part (b), calculate whether the actual facing pressure is within allowable limits.

12.5 Rework Problem 12.4, but use cerametallic brake linings.

12.6 The data below are speed ratios for the entire power train for the 15 forward speeds of the tractor whose engine torque-speed data were given in Problem 12.1.

Gear	$R_T R_D R_{FD}$	Gear	$R_T R_D R_{FD}$
1	355	9	82
2	248	10	72
3	205	11	63
4	163	12	56
5	142	13	45
6	125	14	32
7	108	15	26
8	94		

Plot a diagram similar to Figure 12.4b for this tractor using 1, 4, 7, 10 and 13th gears.

(a) Assuming $e_{pt} = 0.95$, plot and label an axle torque-speed curve for each gear.

(b) On the same graph, plot the constant power curve at the axles when the engine is at maximum power.

(c) If the transmission had only these gears, cross hatch the areas the engine is capable of serving but the transmission is incapable of reaching at full throttle setting.

12.7 Rework Problem 12.6, but use gears 2, 5, 8, 11, and 14.

12.8 Rework Problem 12.6, but use gears 3, 6, 9, 12 and 15.

12.9 When the transmission of Figure 12.8 is in 2nd gear

(a) Determine which gear meshes are transmitting the power through the transmission.

(b) Calculate the speed ratio of the transmission.

(c) Calculate the overall efficiency of the transmission assuming the efficiency of each mesh is 0.985. Hint: Use the explanations at the bottom of Figure 12.8 with the table to identify the active gear meshes; then calculate the ratio for each mesh using Equation 12.12 and multiply the ratios together to get the transmission ratio. Also multiply the individual mesh efficiencies together to get the transmission efficiency.

12.10 Rework Problem 12.9, except use 4th gear.

12.11 Rework Problem 12.9, except use 8th gear.

12.12 Assume the planetary gear set of Figure 12.11 has 15 teeth on the sun gear. Assume power input is to the sun gear, the ring gear is held fixed and power output is from the planet carrier.

(a) Plot the transmission speed ratio as the number of teeth on the ring gear varies from 30 to 50 teeth, i.e., plot speed ratio on the y-axis versus n_r/n_s on the x-axis.

(b) On another graph, plot the efficiency of the drive versus n_r/n_s if $e_{sp} = 0.985$ and $e_{pr} = 0.99$.

12.13 In Problem 12.12, identify the combinations of sun and ring gear tooth numbers that would satisfy Equation 12.20.

12.14 In Table 12.2, verify the transmission speed ratios for gears 1R, 1st and 3rd.

12.15 In Table 12.2, verify the transmission speed ratios for gears 2R, 5th and 7th.

12.16 In Table 12.2, verify the transmission speed ratios for gears 3R, 9th and 11th.

12.17 In Table 12.2, verify the transmission speed ratios for gears 4R, 13th and 15th.

12.18 Given that speed ratio is defined as input speed over output speed

(a) Derive an equation for the speed ratio of a hydrostatic transmission, starting with Equation 12.26.

(b) Plot the transmission speed ratio versus D_m/D_p for $0.01 < D_m/D_p < 100$ when $e_{mv} = e_{pv} = 0.98$.

(c) Add a second curve to the graph of part b, but use $e_{mv} = e_{pv} = 0.95$.

12.19 Assuming the efficiency of each gear mesh is 0.985, calculate the efficiency of the differential

(a) When both axles rotate at the same speed.

(b) When the two axles rotate at different speeds.

12.20 For the differential of Figure 12.21, assume $n_{bp} = 20$, $n_r = 60$, and the bevel gear turns at 550 rpm while transmitting 2800 N·m of torque. Use the differential efficiencies calculated in Problem 12.19 as appropriate.

(a) Assuming the vehicle is traveling straight ahead on a uniform, level surface, calculate the speed, torque and power at each side gear.

(b) Now assume the wheel driven by the left side gear encounters poorer traction, causing its torque-transmitting ability to drop 10% and also causing it to rotate 30% faster than the wheel driven by the right side gear. Assume the speed of the bevel pinion remains constant. Again, calculate the speed, torque and power at each side gear.

(c) By how many kW does the power in the left axle exceed that in the right axle? What is the fate of this power difference?

12.21 Rework Problem 12.20 but, in part b, assume the torque-transmitting ability drops 20% and the left wheel begins turning 50% faster than the right wheel.

12.22 Rework Problem 12.20, but assume the differential is locked by a differential lock.

12.23 Rework Problem 12.21, but assume the differential is locked by a differential lock.

References and Suggested Readings

ASAE. 1998. ASAE standard S205.2 Power take-off definitions and terminology for agricultural tractors. ASAE standard S205.2. *ASAE Standards,* 45th ed. St. Joseph, MI: ASAE.

ASAE. 2001. Front and rear power take-off for agricultural tractors. ASAE standard S203.11. *ASAE Standards,* 48th ed. St. Joseph, MI: ASAE.

Asmus, R.W., and W.R. Borghoff. 1968. Hydrostatic transmissions in farm and light industrial tractors. SAE Paper No. 680570. Warrendale, PA: SAE.

Browning, E.P. 1978. Design of agricultural tractor transmission elements. ASAE Distinguished Lecture Series No. 4. St. Joseph, MI: ASAE.

Drislane, E.W. 1965. Cermetalix clutch material. SAE Paper No. 650683. Warrendale, PA: SAE.

Dudley, D.W. 1965. Elastrohydrodynamic behavior observed in gear tooth action. Elastrohyrodynamic Lubrication. In *Proceedings of the Institution of Mechanical Engineers*, Part 3B (180): 206-214.

Graham, G.R. 1970. Friction materials for heavy duty oil cooled clutches. SAE Paper No. 700741. Warrendale, PA: SAE.

Harms, L., and W. Guse. 2000. Electric/hydraulic proportional control valves and their use in agricultural powershift transmissions. ASAE Distinguished Lecture Series No. 24. St. Joseph, MI: ASAE.

Kemper, I. And L. Elfes. 1981. A continuously variable traction drive for heavy duty agricultural and industrial applications. SAE Paper No. 810948. Warrendale, PA: SAE.

Koch, L.G. 1972. Power train-vehicle modeling to simulate shifting transients of off-highway vehicles. SAE Paper No. 720044. Warrendale, PA: SAE.

Meyers, W.I. 1965. Compound planetaries. *Machine Design* 37(20): 134.

Otten, R.L., W. von Wrisbert, D.E. Rieser, and R.L. Miller. 1993. Enhanced series of mid-size agricultural tractors. ASAE Paper No. 931571. St. Joseph, MI: ASAE.

Radzimovsky, E.I. 1956. A simplified approach for determining power losses and efficiency of planetary gear drives. *Machine Design* 28(3): 101.

Renius, K.T. 1984. New developments in tractor transmissions. *Grundlagen der Landtecknik* 34(3):132-142.

Wienkes, J.W., J.A. Miller and L. Becker. 1993. New John Deere tractors, 92 and 100 PTO hp. ASAE Paper No. 931569. St. Joseph, MI: ASAE.

CHAPTER 13

TRACTION AND TRANSPORT DEVICES

13.1 Introduction

Probably the most discernable visual differences between an on-road and off-road vehicle are the devices used to provide vertical support, propulsion, and steering for the vehicle. While on-road vehicles use pneumatic tires of a relatively narrow size range, off-road vehicles use a wide variety of such devices. These include much larger and multiple pneumatic tires, as well as steel tracks and, recently, rubber tracks. The off-road vehicle may be presented with significant mobility challenges in the form of soft soils and steep and/or rough terrain seldom encountered by an on-road vehicle. In addition, many off-road vehicles are required to pull or push significant external loads requiring tractive capability greatly in excess of that needed for self-propulsion. In most cases, the off-road vehicle is expected to fulfill its mission with minimal rutting and/or compaction of the terrain it traverses.

Although they may otherwise be very similar, a *traction device* is powered to provide tractive capability to the vehicle while a *transport device* is unpowered. It provides vertical support and, sometimes, steering capability. Although pneumatic tires are the most commonly used traction/transport devices for off-road vehicles, such devices can also include steel tracks, rubber tracks, and more exotic possibilities such as those used in hovercraft and walking machines. In this chapter, our attention will be limited primarily to pneumatic tires with some coverage of steel and rubber tracks.

The term traction/transport device doesn't fully describe the functions such devices must perform. They must also provide for vertical support and usually the development of lateral forces for steering the vehicle. All of these forces are generated

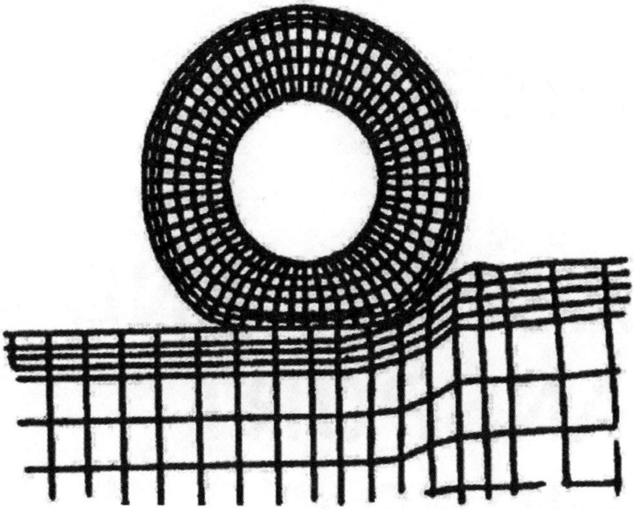

Figure 13.1. In a general off-road situation, both the tire and terrain deform. (From Aubel, 1993.)

by interaction of the traction/transport device with the terrain upon which the vehicle is operating. Due to the complex behavior of soil as an engineering material, an exact understanding of these interactions has yet to be developed although finite element models and increasingly sophisticated finite element software are showing increased promise for developing more accurate predictions. However, both the theoretical and experimental research conducted to date has provided a substantial basis for understanding the basic force generation behavior of vehicle-soil systems. This research has much of its roots in the need for improving the mobility of military vehicles and increasing the tractive performance of agricultural tractors. A specialized technical society, the International Society for Terrain-Vehicle Systems (ISTVS), has published much of the research in this area in the proceedings of its international and regional conferences and in its publication, the *Journal of Terramechanics*.

The forces transmitted to the vehicle through the contact patches of its traction and transport devices are of primary importance in determining the vehicle's traction performance as well as its ride and handling characteristics. This chapter provides an introduction to the basic principles of these forces to provide a better understanding of an off-road vehicle's static and dynamic behavior.

13.2 Vertical Support

13.2.1 Deformation

Figure 13.1 shows a deformable tire moving over a deformable terrain under an applied vertical load. For most off-road applications, the terrain must deform significantly to produce the stresses required to support the vertical load imposed on the tire. The weaker the terrain is, the greater this deformation. The tire also deforms depending primarily on its inflation pressure and to a lesser extent on its carcass stiffness. A general rule of thumb is that the average surface contact pressure is slightly higher than the tire's inflation pressure with the difference attributable to carcass stiffness. The mean contact pressure multiplied by the contact patch area must equal the applied vertical load. If the inflation pressure is decreased while the vertical load remains constant, the tire's deformation must increase to increase the contact patch area. Or, if the inflation pressure is held constant, an increase in vertical load must also be accompanied by an increase in tire deflection and contact patch area. This situation is in contrast to a vehicle (such as one with tracks) with a rigid undercarriage where an increase in vertical load on one of the support devices is accompanied by an increase in ground pressure.

13.2.2 Pressure Distribution

Figure 13.2 shows measured surface pressure distribution contours under a smooth tire that supports the same vertical load at two inflation pressures. Note the growth of the contact patch area as the inflation pressure is reduced and that the centers of the

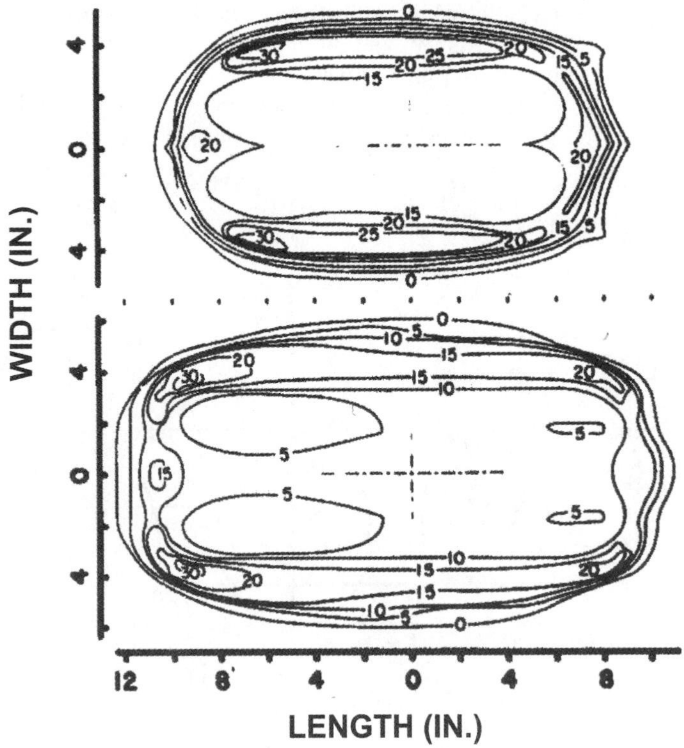

Figure 13.2. Measured surface pressure distribution (in psi) under a smooth tire for the same vertical load and at two different inflation pressures.
Top: 10 psi (70 kPa). Bottom: 6 psi (40 kPa). (From Vandenberg and Gill, 1962.)

contact patch have pressures approximately equal to the inflation pressure. The higher pressures at the edges indicate the influence of the stiffness of the tire sidewalls. From equilibrium considerations, the surface pressure distribution in the soil must be similar. Thus, the soil pressures immediately underneath the tire are also approximately equal to the tire inflation pressure

Figure 13.3 shows how soil pressure varies with depth for tires of different sizes and carrying different vertical loads, but all operating at the same inflation pressure. These soil pressures are difficult to measure experimentally, but the few experimental pressure measurements that have been conducted confirm the general trends shown by the calculated values. Note that the pressures just underneath the tires are somewhat higher than the inflation pressure. The pressure contours extend deeper into the soil profile with increased vertical load even though the surface pressures are similar. This effect is the basis for the rule of thumb that surface soil compaction is related to the inflation pressure while subsoil compaction is related to the total vertical load on the axle. Subsoil compaction is considered more critical because normal tillage and/or surface freeze-thaw cycles are generally not effective in alleviating it.

13.2.3 Load Capacity

The vertical load carrying capability of off-road tires is mainly determined on the basis of the volume and pressure of the air that they contain. Thus vertical load capacity increases with tire size and inflation pressure. The maximum inflation pressure is limited by the tire construction as expressed by either ply rating for *bias-ply tires*, the "star" rating for conventional-sized *radial-ply tires*, or the load index

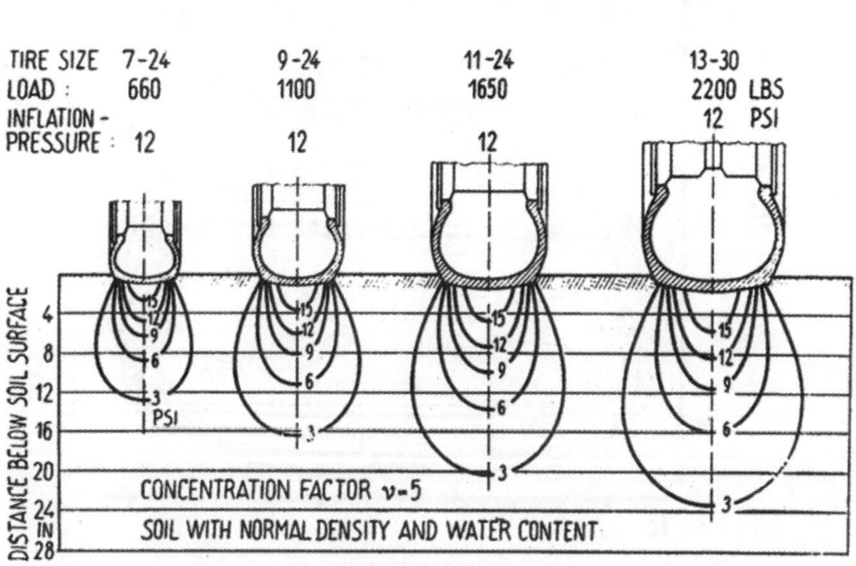

Figure 13.3. Calculated vertical pressures under tires of different sizes supporting different vertical loads but all at the same inflation pressure. (From Söhne, 1958.)

rating for metric-sized radial-ply tires. Maximum load ratings and tire dimensions are published by the tire manufacturers in the form of handbooks available in print and at the manufacturer's web site, as well as in the annual yearbooks of the Tire and Rim Association. Maximum vertical load ratings are given in these handbooks for a corresponding maximum inflation pressure at a given maximum travel speed. Supplemental load-inflation tables (also in the handbooks) show vertical load limits at lower inflation pressures. These ratings are also dependent on whether the tire will be used as a single tire or as a dual or triple. The dual or triple load ratings are lower than for a single because of the possibility that in some situations one of the dual or triple tires might bear part or the entire load normally split between the multiple tires on the axle. The load-inflation pressure tables basically provide the loads needed to maintain the same tire deflection as the inflation pressure is varied. These tables for radial-ply tires were extended in the early 1990s to allow for operation at lower pressures contingent on also lowering the vertical loads. Normally a range of tire sizes is available for use on a given vehicle.

Use of larger tires allows operation at lower pressures, improving tractive capability and reducing surface soil pressures and, hence, surface compaction. These lower but still correct pressures allowed rubber-tired tractors to operate at the same nominal ground pressures as the rubber tracked-tractors introduced in the mid 1980s.

13.2.4 Stiffness and Damping Characteristics

Because many off-road vehicles have no formal suspension system, the vertical spring and damping properties of the tires become important. A measure of the *vertical spring rate* is the slope of the tire's vertical static load-deflection relation at a given deflection and for a given inflation pressure. Alternatively, the spring and damping properties can be determined from free or forced vibration testing which can allow for rotation of the tire. Typical spring rates measured in this way are shown in Figure 13.4 for a range of inflation pressures and travel speeds. Note the spring rate increases with increasing inflation pressure but is relatively independent of travel speed. The use of liquid ballast in pneumatic tires substantially increases the spring rate over the air-filled condition.

Although not intentional, off-road tires also possess some *damping* or energy dissipation capability. Such damping levels are low (compared to the damping provided by a shock absorber in an on-road vehicle suspension) and have been found to decrease with travel speed, as shown in Figure 13.5. Damping levels are naturally kept low on off-road tires since the energy absorbed is dissipated in the form of heat, which is detrimental to the life of the tire. Energy dissipation concerns also limit the maximum allowable travel speeds of off-road tires.

From a vibration standpoint, both the off-road tire and the terrain have flexibility so that the stiffness and damping properties of the combination should be considered. A limited amount of such information currently available indicates that compared to the stiffness and damping values measured for the tire on a hard surface, operation in soil lowers the stiffness and increases the damping of the tire-terrain system.

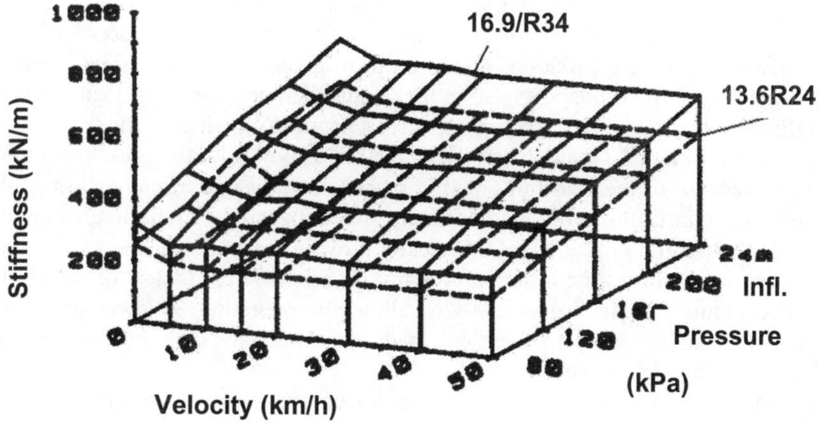

Figure 13.4. Measured vertical stiffness as a function of forward velocity and inflation pressure for two off-road tires. (From Kising and Göhlich, 1989.)

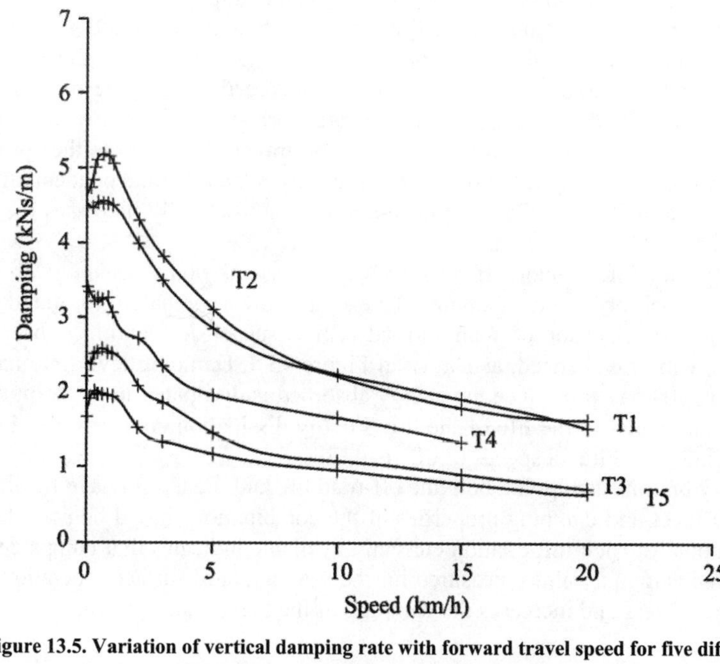

Figure 13.5. Variation of vertical damping rate with forward travel speed for five different off-road tires. (From Lines and Murphy, 1991.)

13.3 Traction

Off-road vehicles, such as the agricultural tractor, are often expected to provide a significant amount of tractive capability in excess of that required for self-propulsion on a hard flat surface. In fact, the word "tractor," derived from the term "traction engine," indicates the importance placed on developing traction for this type of off-road vehicle. Historically, considerable improvements in traction have been made starting with steel wheels and progressing to steel tracks, bias and then radial-ply pneumatic tires and, more recently, rubber tracks. Of the three principal ways (power take-off, hydraulic outlet, drawbar) of converting a tractor's engine power into useful work, the least efficient but most used method is pulling external loads through the drawbar.

13.3.1 Fundamental Mechanics Approach

Considerable insight into how a track or tire develops a tractive force on a deformable surface such as soil can be obtained from the experimental setup of Figure 13.6 and the plot of Figure 13.7. For several levels of the vertical load W, the maximum value of the shearing force F is determined. If we plot the maximum shearing force F_{max} versus the corresponding normal force W, we find in general that F_{max} does not approach zero as W approaches zero and that the change in maximum shearing force is linearly related to the change in the normal load or

$$F_{max} = Ac + W \tan \phi \tag{13.1}$$

Dividing by the area A,

$$\tau_{max} = \frac{F_{max}}{A} = c + \frac{W}{A} \tan \phi = c + p \tan \phi \tag{13.2}$$

or the maximum shear stress τ_{max} is equal to a constant c (termed the *soil's cohesion*) plus the normal stress p multiplied by the tangent of another constant ϕ (termed the *angle of internal friction*). Such a relation is referred to as a Mohr-Coulomb failure criterion. The shear test shown is one way of measuring these two soil properties. Soils high in clay content are said to be mainly cohesive while dry sands are said to be mainly frictional in nature. Cohesion reflects the tendency of the soil particles to stick together regardless of vertical load while the angle of internal friction measures the resistance of the soil particles as they interlock and slide over one another. Thus, tan ϕ can be thought of as the maximum friction coefficient obtainable on the soil. Most field soils exhibit both cohesive and frictional properties.

The above discussion relates to the maximum shear forces that can be developed. While important, the relation of shear force to shear displacement shown in Figure 13.6 provides further insight into how a track or tire develops a traction force along its entire length of ground contact within the contact patch. Although Figure 13.6 depicts a track shoe, the same concept holds true for a tire. At the leading edge of the contact patch, the shear displacement is zero and grows linearly as the track or tread element moves rearwardly through the contact patch reaching a maximum value at the rear of the contact patch.

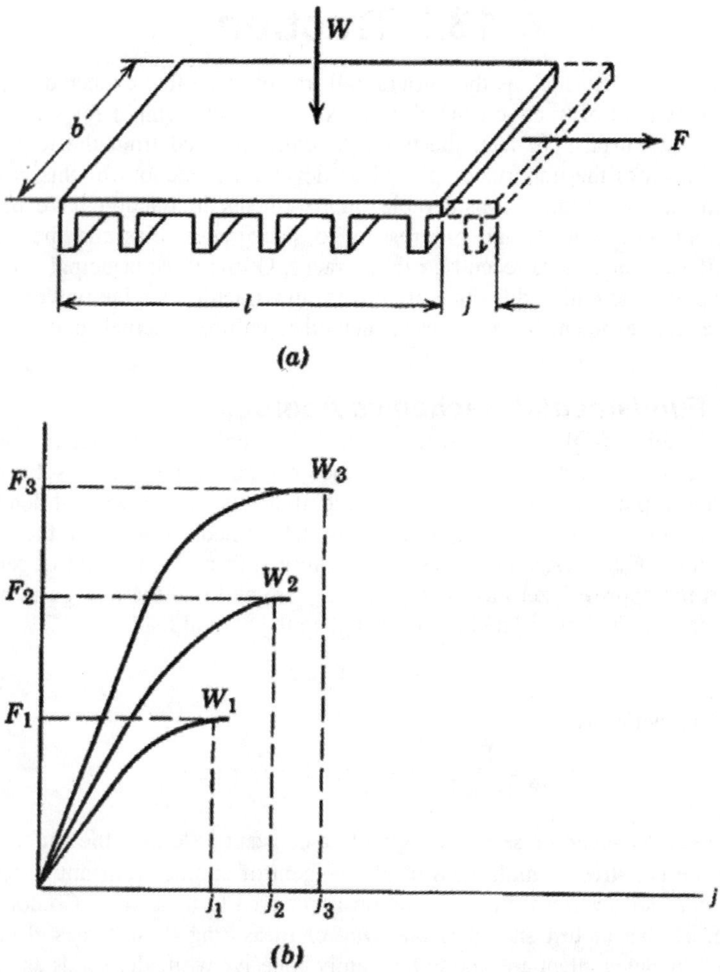

Figure 13.6. (a) Shear plate apparatus for measuring soil shear strength. (b) Nature of the shear force versus shear displacement relation for various vertical loads.

13.3.2 Shear Displacement and Slip

To determine the shear displacement, let us first define the slip of the tractive device as

$$\text{slip} = s = \frac{V_t - V_a}{V_t} = \frac{V_s}{V_t} \tag{13.3}$$

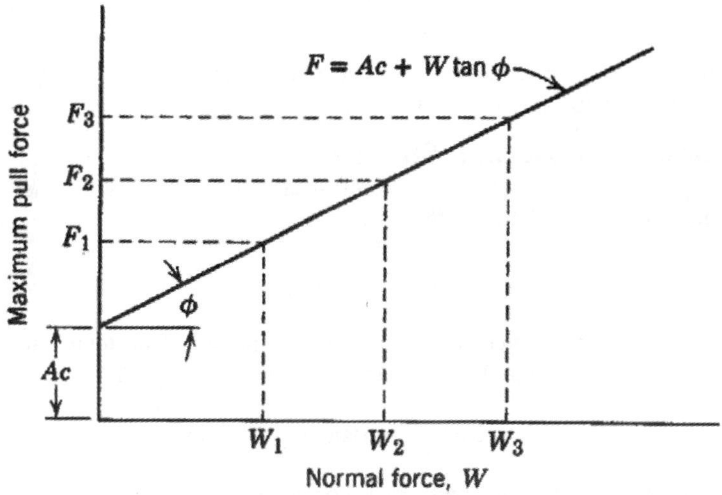

Figure 13.7. The soil parameters c and ϕ can be determined from a plot of the maximum shear force versus the normal force.

V_t is the theoretical forward travel speed of the vehicle if the rotational motion of the driving axle was perfectly transformed (zero slip) into forward motion. V_a is the actual forward travel speed obtained. Thus slip is the difference (V_s) between the theoretical and actual forward travel speeds normalized by the theoretical speed. A slip of zero indicates the theoretical and actual speeds are equal while a slip of 1 or 100% indicates the actual speed V_a has gone to zero and the traction device is spinning with no forward motion.

Referring to Figure 13.8, V_s can be thought of as the velocity at which the shear displacement of the soil takes place, which will be the same for every point on the track along its contact length with the ground. If t is time measured from when the shear displacement begins at the leading edge of the contact patch, the shear displacement j at time t is simply V_s t. If the position of a point along the length of the contact patch (starting from its leading edge) is denoted as x, then x = V_t t. Thus,

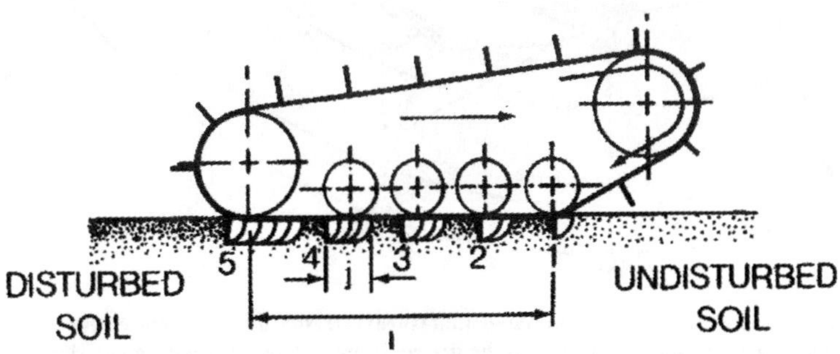

Figure 13.8. The shear displacement, j, varies linearly along the contact patch length. (From Bekker, 1956.)

$$j = V_s t = V_s \frac{x}{V_t} = sx \tag{13.4}$$

which simply states that the shear displacement varies linearly along the contact patch length depending on the slip of the traction device.

13.3.3 Shear and Normal Stresses

Referring to Figure 13.6, the shape of the shear stress-shear displacement relation suggests an exponential relation of the form

$$\tau = (c + p \tan\phi)(1 - e^{-j/K}) = \tau_{max}(1 - e^{-sx/K}) \tag{13.5}$$

where K is a constant related to the strength of the soil. The total tractive force developed along the length of the track can be found by integration

$$F = b\int_0^l \tau dx = b\int_0^l (c + p\tan\phi)(1 - e^{-sx/K})dx \tag{13.6}$$

where b is the width of the track and l is the contact length.

Figure 13.9 shows measured values for the normal and shear/tangential stress distributions on a pneumatic tire operating on soil.

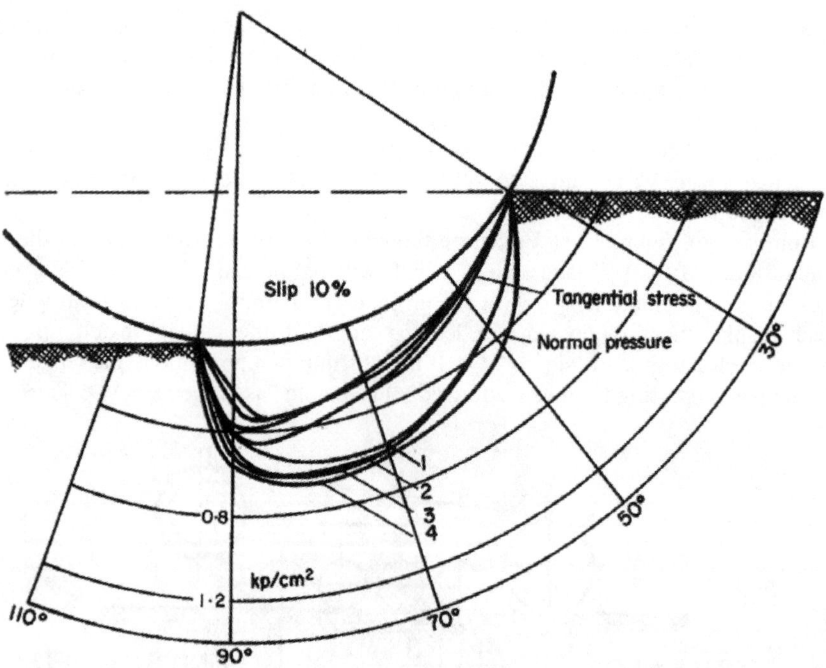

Figure 13.9. Measured normal and tangential (shear) stresses at the interface between a driven pneumatic tire and soil. Curve 1 is on the tire centerline while curve 4 is on one side. 1 kp/cm^2 is approximately equal to 100 kPa. (From Krick, 1969.)

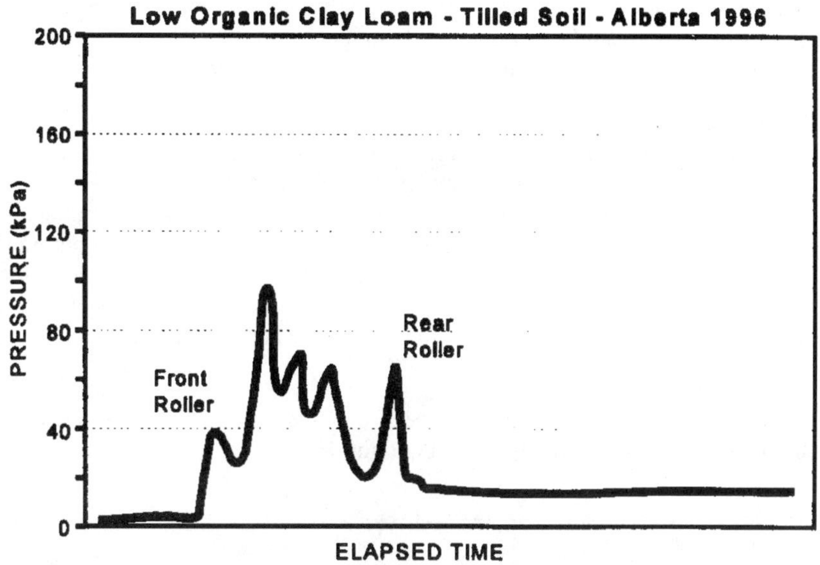

Figure 13.10. Measured normal pressure time history from a transducer buried in the soil as a rubber-tracked tractor pulling a load passes over it. (From Turner, Shell, and Zoz, 1997. Reprinted with permission from SAE Paper No. 972730, © Society of Automotive Engineers, Inc.)

The nominal normal pressure for a tracked vehicle is often calculated by dividing the total tractor weight by the total ground contact area. In practice, such a uniform ground pressure is not achieved, with pressure pulses seen underneath the drive sprocket, the idler, and each of the rollers (Figure 13.10).

The above discussion indicates the importance of contact length on the development of tractive force. Traction device designs that provide longer contact lengths are expected to provide higher tractive forces for the same level of slip. These predictions are borne out in reality when tracks are compared to tires and larger diameter tires are compared to those with smaller diameters.

However, the above discussion is a simplification of how a traction device develops a *propulsive force* through the creation of shear and normal stresses in the soil. Although a considerable amount of research has been done in analyzing traction from this basic stress concept, difficulties in accurately characterizing the stress-strain properties of soils have limited the usefulness of this approach. Recent developments in finite element analysis software programs and improved soil constitutive models are showing promise for realistically modeling both a deformable traction device (such as a rubber tire) and a deformable soil as they interact with each other.

13.3.4 Semi-Empirical Approach

Recognizing the limitations of the fundamental approach discussed above, other researchers have used a semi-empirical approach for predicting tractive and motion resistance forces. This approach uses empirically measured soil properties such as cohesion, angle of internal friction, and penetration resistance measured by shear and plate penetration tests. While useful for explaining some aspects of tractive device-soil interaction, the semi-empirical approach has had limited practical application.

13.3.5 Empirical Approach

Empirical methods using field and/or soil bin laboratory tests of traction devices either by themselves or as part of a complete vehicle are the most used technique for assessing tractive performance by both vehicle and traction device manufacturers. Figure 13.11 includes a plot of the typical quantities (input torque T and net tractive force H) that are measured in such tests and free body diagrams illustrating the forces acting on the wheel during different portions of the test. Although a wheel is shown and discussed here, the same concepts apply to a tracked traction device.

13.3.6 Slip

The axle torque and net tractive force are plotted as functions of *wheel slip*. Expanding on the definition of slip from Equation 13.3,

$$\text{slip} = \frac{V_t - V_a}{V_t} = 1 - \frac{V_a}{V_t} = 1 - \frac{V_a}{r\omega} \tag{13.7}$$

where

r = rolling radius of the wheel

ω = angular velocity of the wheel

All of the quantities in Equation 13.7 are well defined except for the rolling radius. The term *rolling radius* is defined in ASAE S296.4 as the distance traveled per revolution of the driving axle of a traction device under the *specified zero condition*, divided by 2π. The most commonly used zero condition is operation in a self-propelled condition on a hard surface. This definition results in the numerical value of the rolling radius lying between the tire's undeflected radius and its static loaded radius. Tire manufacturers routinely supply (in the tire data handbooks mentioned earlier) the rolling circumference of their tires from which the rolling radius can be calculated.

13.3.7 Traction Mechanics

When the input torque is zero (the towed wheel condition), an external force TF must be supplied to tow the wheel forward (Figure 13.11). As the input torque becomes positive, the wheel develops some traction force of its own and the necessary externally supplied force declines (the net tractive force H becomes less negative).

When the net tractive force becomes zero, the wheel is in the *self-propelled state*. Note this state occurs at a positive slip since zero slip was defined for self-propelled operation on a hard surface. As additional torque is supplied to the wheel, a positive net tractive force is produced and the driving wheel state is reached which continues until a slip of 1 is reached (the actual forward travel speed V_a is reduced to zero).

If a negative axle torque is supplied, the wheel becomes braked and creates a *retarding force* to forward travel. Typically traction tests are conducted only in the driving wheel region so that little information is available on the braking force

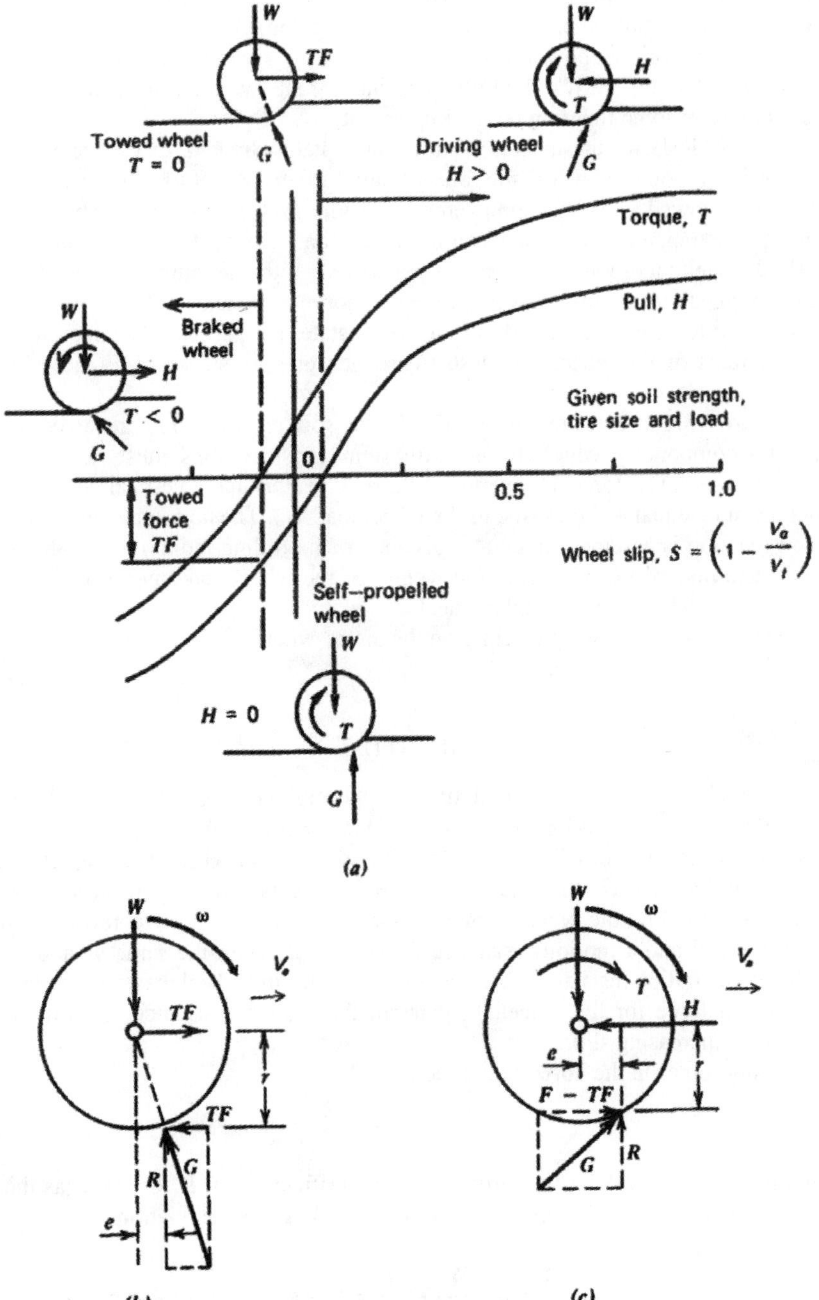

Figure 13.11. (a) Net traction (pull)-torque-slip relation for wheels on soil. (From Wismer and Luth, 1974 .) (b) Free body diagram of a towed wheel. (c) Free body diagram of a driving wheel.

properties of off-road tires. Another way to interpret the braked wheel region is to think of it as the region where the wheel might drive something on the vehicle. Before the introduction of the power take-off, many early agricultural implements were ground-driven in this way. Here the braking torque would represent the torque required to drive some function on the implement.

Figure 13.11 shows these three distinct force states: braked, driven, and driving. The transition point between the braked and driven wheel is the towed wheel condition. A towed wheel is unpowered; axle torque T is zero (neglecting bearing friction). The transition point between the driven and driving force states is the self-propelled wheel condition. For a self-propelled wheel, the net traction force H is zero with the applied torque simply overcoming the motion resistance of the wheel.

The towed force condition and driving wheel states are particularly important. Free body diagrams of the wheel for these two conditions are shown in Figures 13.11b and c.

In Figure 13.11b for the towed wheel, the soil reaction G is resolved into a horizontal component (which from equilibrium considerations must be equal and opposite to the towed force TF exerted at the axle center) and a vertical component R (which must be equal and opposite to the wheel load W). The horizontal component of the soil reaction is assumed to act at a distance r (the rolling radius) below the wheel center. Note that W includes both the weight of the wheel and any vertical reaction force from the vehicle on which the wheel is mounted.

Since there is no axle torque acting on the towed wheel,

$$\mathrm{TF}\,r - R\,e = 0 \tag{13.8}$$

or

$$e = (\mathrm{TF}/R)r = (\mathrm{TF}/W)r$$

The *motion resistance ratio* (ρ) is defined as the motion resistance force divided by the normal load on the traction device or $\rho = \mathrm{TF}/W$. Thus $e = \rho\, r$.

For the driving wheel (Figure 13.11c), the soil reaction G is again resolved into horizontal and vertical components. The horizontal component is again assumed to act at a distance r below the wheel center and is now divided into two forces: a gross traction force F and a motion resistance force TF. Although the same symbol TF is used here, the motion resistance force acting on the driving wheel may not be the same as the towed force for the wheel. In general, the motion resistance force increases slightly with increasing slip.

Summing forces in the horizontal direction,

$$H = F - \mathrm{TF} \tag{13.9}$$

Defining $\mu_g = F/R = F/W$ as the gross traction coefficient, $\mu = H/R = H/W$ as the net traction coefficient, and dividing Equation 13.9 by W, gives the relation

$$\frac{H}{W} = \mu = \frac{F}{W} - \frac{\mathrm{TF}}{W} = \mu_g - \rho \tag{13.10}$$

Summing the moments acting on the wheel,

$$T - (F - TF)r - R\,e = 0 \tag{13.11}$$

By using Equation 13.8,

$$T = F\,r$$

Thus the gross tractive force F is assumed equal to the wheel torque T divided by a moment arm equal to the rolling radius r. Note that in Figure 13.11, the two curves shown can be considered to represent plots of μ_g (= T/rW) and μ (= H/W) versus slip. The difference between these plots represents the motion resistance ratio ρ. In fact, in such a traction test, μ_g and μ are determined from measurements of axle torque and net tractive force respectively and ρ is determined from their difference.

13.3.8 Tractive Efficiency

So far, our analysis has been restricted to just the forces acting on the tractive device. To consider the efficiency of the tractive device in converting the axle input power into output power, the term *tractive efficiency* (TE) has been defined as

$$TE = \frac{\text{output power}}{\text{input power}} = \frac{HV_a}{T\omega} = \frac{HV_a}{Fr\frac{V_t}{r}} = \frac{(\frac{H}{W})V_a}{(\frac{F}{W})V_t} = \frac{\mu}{\mu_g}(1-s) = (1-\frac{\rho}{\mu_g})(1-s) \tag{13.12}$$

The term $(1-\frac{\rho}{\mu_g})$ can be considered the efficiency of converting the axle torque T into net traction H while the term $(1-s)$ represents the efficiency of converting the rotational input motion ω into a rectilinear output motion V_a.

The curves presented in Figure 13.11 represent a given soil strength, tire size and vertical load. As soil strength increases, the curves move upward and to the left; as soil strength decreases, they move downward and to the right as illustrated in Figure 13.12 for a variety of surface conditions.

13.3.9 Traction Testing

Figure 13.13 shows results from a typical traction test. Tractive efficiency (TE), net traction ratio (μ = NTR), and the motion resistance ratio (ρ = MRR) are plotted as a function of slip. As the scatter of the individual data points indicates, a certain amount of variability is always present in such tests necessitating the use of statistical methods to determine the "best fit" curves and analytical relations shown. Unlike the net traction ratio that increases with slip, the tractive efficiency shows a distinct peak. Such a peak is to be expected from Equation 13.12 because at low slips, most of the gross tractive force is going into overcoming motion resistance (the net tractive force H is near zero) while at high slips, the traction device has lost most of its forward motion (V_a is near zero.). Usually the net tractive force at the slip at which the tractive efficiency is a maximum is considerably less than the maximum net tractive force achievable at higher slips (and lower tractive efficiency). Ideally, the vehicle is operated nearer the point of maximum tractive efficiency with the necessary net traction force obtained by proper ballasting, rather than by operating at higher rates of slip.

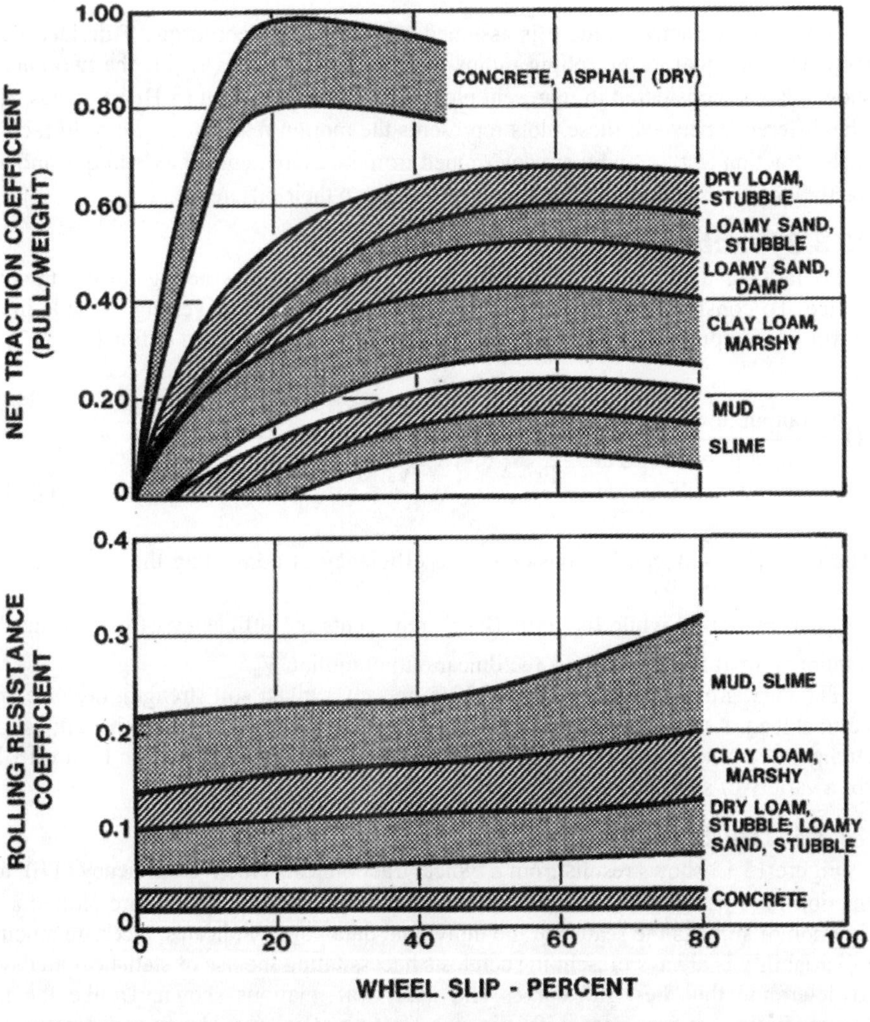

Figure 13.12. Net traction and motion resistance ratios as a function of slip for a variety of surface conditions. (From Wismer, 1982. Reprinted with permission from SAE Paper 820656, © 1982 Society of Automotive Engineers, Inc.)

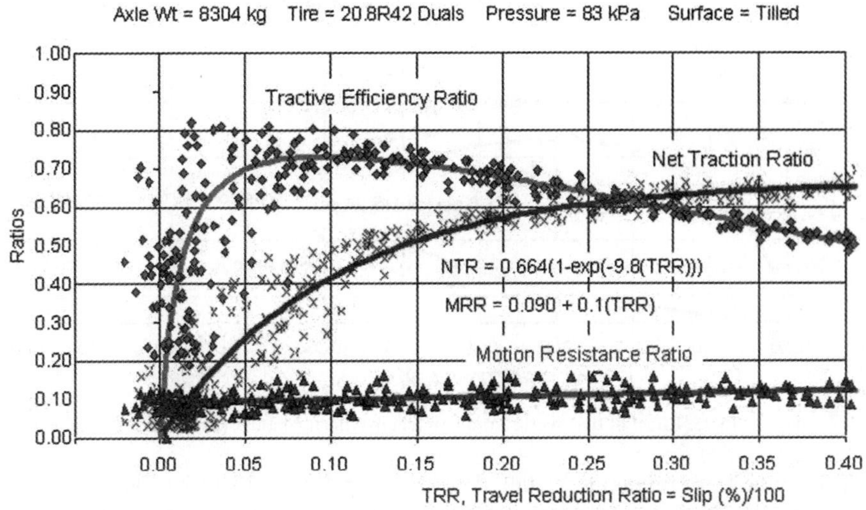

Figure 13.13. Tractive efficiency, net traction ratio, and motion resistance ratio versus slip results for a typical traction test. (From Zoz, Turner. and Shell, 2002.)

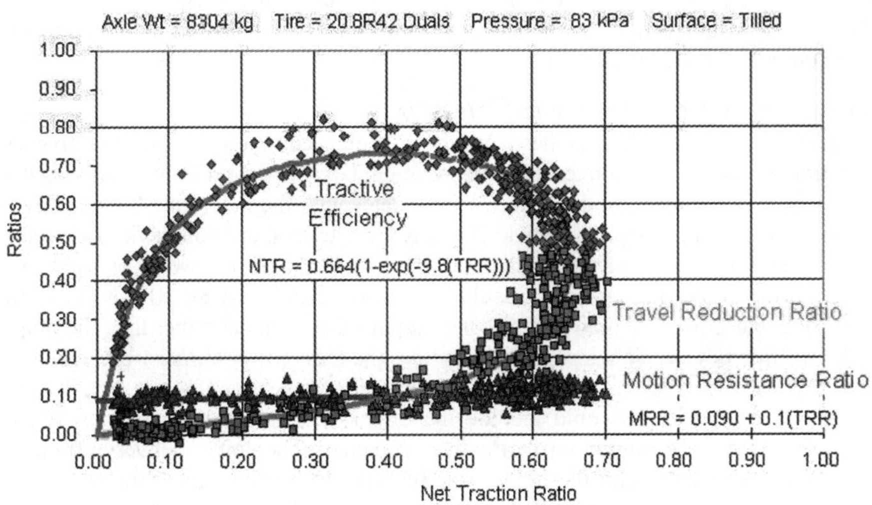

Figure 13.14. The traction test results from Figure 13.13 plotted as a function of net traction ratio. (From Zoz, Turner, and Shell, 2002.)

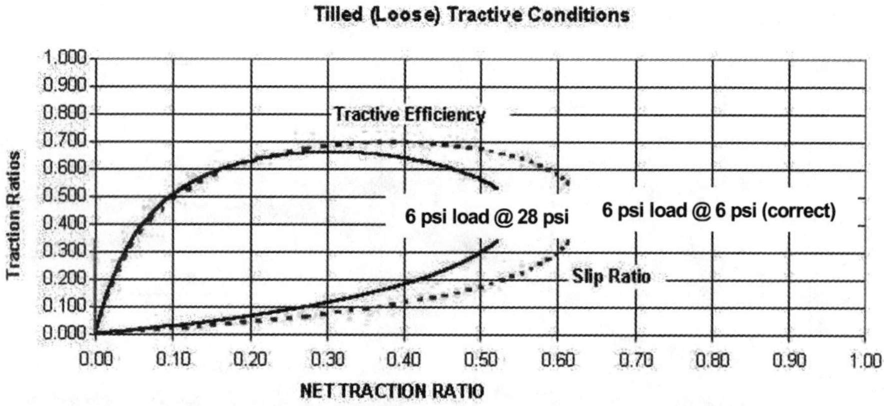

Figure 13.15. Effect of tire overinflation on tractive performance. (From Zoz, 1997.)

The method of presenting traction data shown in Figure 13.13 has been in use for many years. Recently, an alternative method for showing the same data has come into use. Figure 13.14 uses the same data as Figure 13.13, but the net traction ratio is the independent variable and slip is the dependent variable. This new method is intuitively appealing since during a complete vehicle traction test, the traction device is usually loaded down by requiring the development of an increased net tractive force. The result is an increased level of slip to develop this increased force. Thus, slip is more naturally thought of as a result, rather than as an input. This is especially true when analyzing a complete vehicle where a given draft load is assumed as an input and one of the outputs is the resulting slip.

13.3.10 Power Delivery Efficiency

Although tractive efficiency has historically been used to describe the effectiveness of a traction device in converting axle rotational power to rectilinear power, it is important to remember that this measure only deals with the individual traction device itself rather than the complete vehicle that may have several traction devices as well as unpowered transport devices. Sometimes the term "drawbar power" is used as a synonym for the output power in calculating tractive efficiency. Strictly speaking, the term "drawbar power" should relate to the output of the vehicle rather than the output of the traction device. To address this situation, the term *power delivery efficiency* has been proposed to measure how well the vehicle converts engine power into drawbar power. Such a measure would include the effect of additional power-train related losses that are not accounted for by tractive efficiency. The power delivery efficiency concept helps explain why the higher tractive efficiency of a rubber-tracked tractor may not translate into higher field work rates per unit of fuel consumed.

13.3.11 Tractive Performance

Comparisons of tractive performance are usually made on both a slip and tractive efficiency basis. For example, Figure 13.15 shows the effect on traction of over-

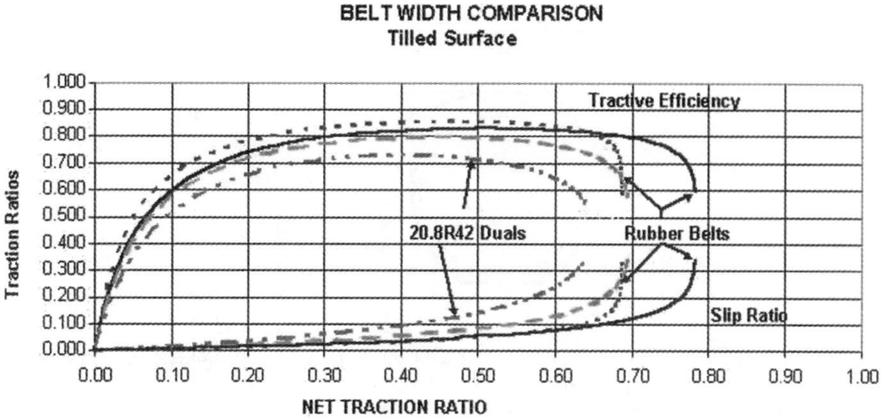

Figure 13.16. Comparison of tractive performance of three different width rubber tracks to a pneumatic tire on a tilled soil condition. (From Zoz, 1997.)

inflating a tire. The loads referred to in that figure are those from the load-inflation tables discussed earlier. Thus the curve labeled "6 psi load @ 6 psi" reflects the correct load for the tire. The "6 psi load @ 28 psi" curve reflects the effect of over-inflating the tire from 6 to 28 psi (40 to 190 kPa) while keeping the load recommended for the 6 psi pressure. The peak tractive efficiency is lowered by overinflation as is the maximum net traction ratio. For any given level of the net traction ratio, a higher slip is required by the over-inflated tire. A higher slip translates into a lower forward travel speed reducing the productivity of the vehicle using the traction device.

Figure 13.16 shows a comparison of three different belts widths for a rubber-tracked tractor compared to a rubber tire on a tilled soil. The rubber tracks exhibit higher tractive efficiencies and lower slips than the tire at a given net traction ratio. This performance difference was shown to narrow as the soil became firmer.

13.3.12 Dimensional Analysis

Traction data such as shown in Figure 13.13 are for a specific tire size, inflation pressure, and soil strength condition. Although useful, such data does not provide information on the changes in tractive performance that could be expected with a different tire size and/or different soil strength. In an attempt to consolidate such data, the technique of using dimensional analysis of traction data has been used in several studies to reduce the number of variables to a smaller number of dimensionless ratios.

The most widely used dimensional analysis approach to predicting off-road traction (incorporated into ASAE D497.4) makes use of the following ratios in addition to the gross tractive coefficient (μ_g), motion resistance ratio (ρ), net tractive coefficient (μ), and slip (s):

Wheel Numeric (C_n) = CI b d / W

Width-to-Diameter Ratio = b / d

Deflection Ratio = δ / h

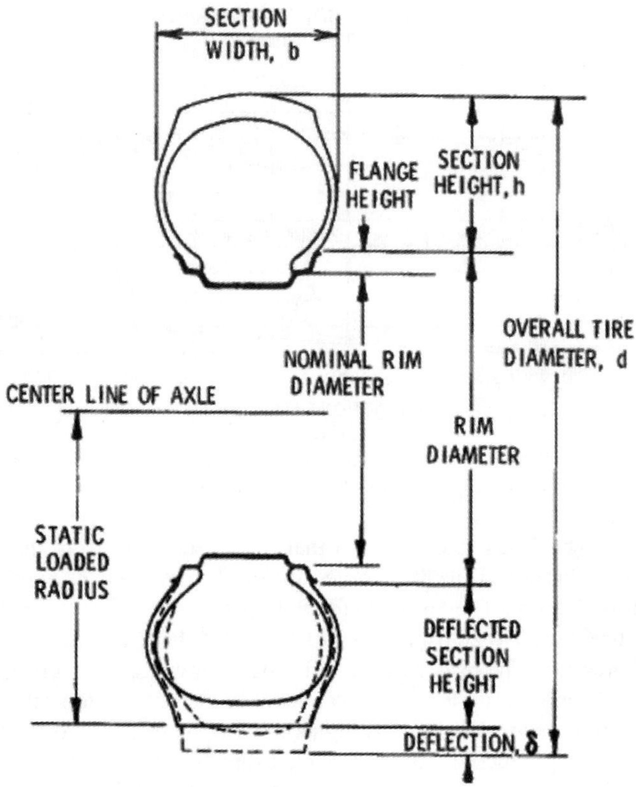

Figure 13.17. Tire Parameters. (From Brixius, 1987.)

13.3.13 Tire and Soil Parameters

Figure 13.17 illustrates the tire parameters (b, d, δ, and h) used. The section width b is the first number in a tire size designation (i.e., nominally 18.4 inches for an 18.4-38 tire). The overall unloaded diameter d can be obtained from the tire data handbooks available from off-road tire manufacturers. The tire deflection on a hard surface, δ, is equal to d/2 minus the static loaded radius. The static loaded radius for the tire's rated load and inflation pressure is also standard tire data from the tire data handbooks. The section height h is equal to half the difference between the overall unloaded diameter and the rim diameter. The rim diameter can in turn be estimated by adding 50 mm to the nominal rim diameter, which is the second number in a tire size designation (i.e. 38 inches for an 18.4-38 tire).

Soil strength is often represented by the cone index, CI, which is the average force per unit area required to force a cone-shaped probe vertically into the soil at a steady rate. The average before-traffic cone index for the top 150 mm layer of soil is used in the prediction equations that follow. ASAE Standards S313.3 and EP542 describe the soil cone penetrometer and procedures for its use. Some representative cone index values for a range of soil conditions are given in Table 13.1.

To further simplify the prediction equations, the above three dimensionless ratios were combined with the cone index into a single product termed the *mobility number* B_n:

$$B_n = \left(\frac{CIbd}{W}\right)\left(\frac{1+5\frac{\delta}{h}}{1+3\frac{b}{d}}\right) \tag{13.13}$$

13.3.14 Traction Prediction Equations

A number of experimental test results (each with varying values for CI, b, d, δ, h and W) were used with statistical curve fitting techniques to come up with the following two equations for bias-ply tires:

$$\mu_g = 0.88\,(1-e^{-0.1B_n})\,(1-e^{-7.5s}) + 0.04 \tag{13.14}$$

$$\rho = \frac{1.0}{B_n} + 0.04 + \frac{0.5s}{\sqrt{B_n}} \tag{13.15}$$

The following modified forms of the above two equations were suggested for radial-ply tires:

$$\mu_g = 0.88\,(1-e^{-0.1B_n})\,(1-e^{-9.5s}) + 0.0325 \tag{13.16}$$

$$\rho = \frac{0.9}{B_n} + 0.0325 + \frac{0.5s}{\sqrt{B_n}} \tag{13.17}$$

Note the similarity of Equations 13.5 and 13.14, which are both exponential functions of slip.

Table 13.1. Cone index values for a range of soil conditions. (From Brixius, 1987.)

Soil Class	Cone Index kN/m^2 (kPa)	Typical Operating Conditions
Soft or sandy soil	350	Rice harvest
	480	Disking on plowed ground or low-land logging
	700	Spring plowing or earthmoving on moist soil
Medium or tilled soil	850	Planting, field cultivating
	1000	Corn harvesting, fall plowing
	1200	Wheat harvesting
Firm	1750	Summer plowing, logging in dry season, earthmoving on dry, clay soil

Figure 13.18 is a plot of Equation 13.14 for several values of the mobility number B_n. The tractive performance of the wheel increases (higher μ_g at given slip) as the mobility number increases, which can reflect higher strength soil (greater CI), an increased tire deflection δ, a greater overall tire diameter d, etc.

Figure 13.19 is a plot of Equation 13.15 at zero slip (which can be considered to represent the towed wheel condition). The motion resistance ratio ρ is then a function of the mobility number B_n reaching a minimum value of 0.04 for the higher B_n values representing, for example, higher strength soils. B_n also increases as the deflection ratio δ/h increases (lower inflation pressures correspond to greater tire deflections, increased contact patch area, and less wheel sinkage) and as b/d decreases (a long [due to greater d], narrow [lower b] contact patch compacts a smaller soil volume). While lower inflation pressures reduce motion resistance on soft soils, lower pressures increase motion resistance on a hard surface. For vehicles that operate both on and off road (such as for transporting logs), the provision of a central tire inflation system has been shown to have definite benefits in optimizing overall performance. Such systems may also be of benefit whenever there is a substantial load change between the axles, as might happen when a mounted implement is raised for transport.

13.3.15 Improving Tractive Performance

Due to their importance in determining vehicle performance, the tractive characteristics of off-road tires have been investigated in much more detail than their vertical or lateral characteristics. Many of these investigations are reported in the references at the end of this chapter. Much of this work has been done at the United States Department of Agriculture's National Soil Dynamics Laboratory in Auburn, Alabama. Factors such as lug pattern, height and spacing have been found to be of secondary importance unless slippery conditions require the use of lugs to penetrate to a stronger soil layer below the surface. Radial construction (as opposed to bias) has been shown to provide improved tractive performance and lower motion resistance. Using larger volume tires to allow the use of lower but still correct inflation pressures has been shown to improve traction performance as well as improve flotation and minimize surface soil compaction. Similarly, increased drive wheel diameter provides the longer tire contact patches that have been shown to improve performance. Rubber tracks have been shown to outperform tires from a net tractive coefficient and tractive efficiency standpoint especially on tilled soil surfaces. All of these improvements in tractive efficiency are important to minimize the fuel consumption of off-road vehicles per unit of work output.

13.4 Lateral Forces

The handling or steering performance of off-road vehicles is also important to increase maneuverability and minimize the area and time required for making turns. Turning requires the development of yaw moments that may be generated through the development of *lateral forces* on a conventionally steered wheel or by creating differences in the tractive forces acting on two sides of a vehicle (skid steering).

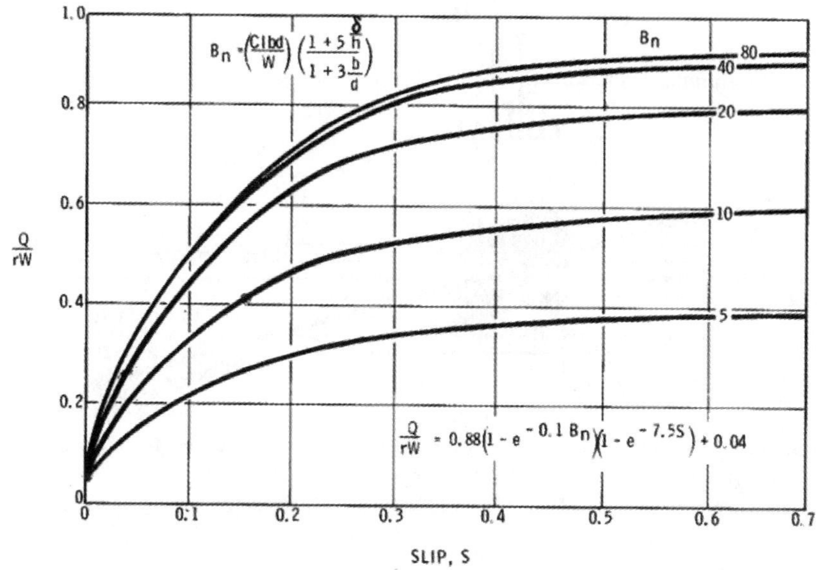

Figure 13.18. Gross tractive ratio as a function of slip and mobility number, B_n. Note the symbol Q refers to the axle torque T used in this book. (From Brixius, 1987.)

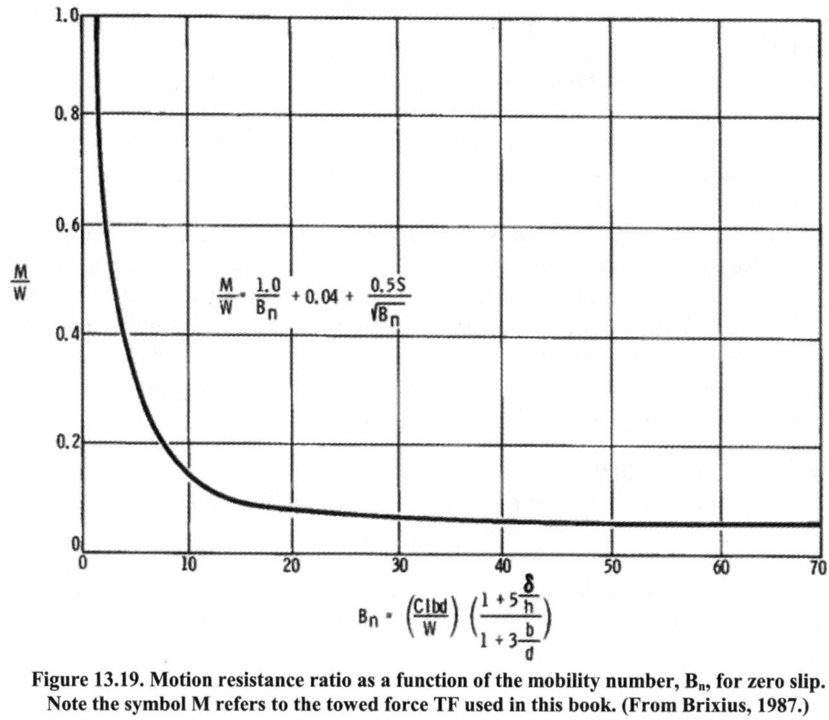

Figure 13.19. Motion resistance ratio as a function of the mobility number, B_n, for zero slip. Note the symbol M refers to the towed force TF used in this book. (From Brixius, 1987.)

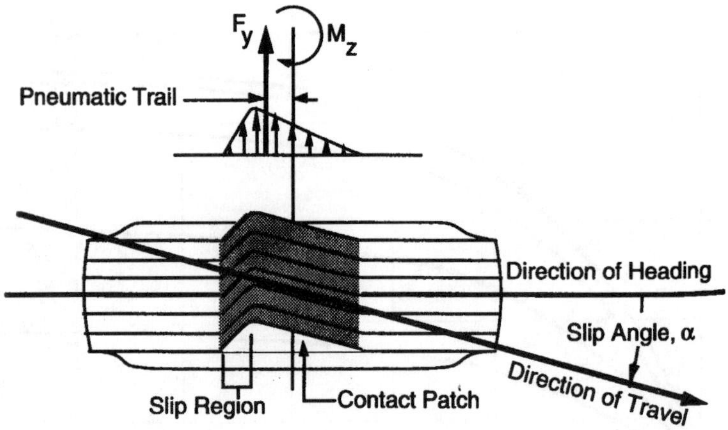

Figure 13.20. A pneumatic tire develops a lateral force when its heading and travel directions differ. (From Gillespie, 1992. Reprinted with permission from SAE Publication R-114, Fundamentals of Vehicle Dynamics, © 1992 Society of Automotive Engineers, Inc.)

13.4.1 Slip Angle

A lateral force is developed by a pneumatic tire whenever the direction the tire is traveling differs from the heading direction of the plane of the wheel itself. The lateral elasticity of the contact patch of the tire allows the lateral deflection shown exaggerated in Figure 13.20.

The difference in directions is called the *slip angle*, α. The resultant lateral force generated lies behind the wheel center creating a moment that seeks to align the heading direction with the travel direction (i.e., decrease the slip angle) and reduce the lateral distortion of the contact patch. This moment is thus called the self-aligning moment and helps to explain why a steered tire returns to the unsteered position when the steering input is removed. The self-aligning moment can be considered to be the product of the lateral force generated and an offset distance termed the pneumatic trail.

The lateral force developed by a tire at a given vertical load takes the form shown in Figure 13.21. The lateral force L varies nearly linearly with slip angle at small slip angles and reaches a maximum value asymptotically. The slope of the curve at the origin is termed the *cornering stiffness*, C_α. Thus, for small slip angles, $L = -C_\alpha \alpha$; i.e., a positive slip angle produces a negative lateral force.

13.4.2 Carpet Plots

Much less measured information exists on the lateral forces generated by off-road tires on both soil and hard surfaces as compared to automotive tires on hard surfaces. Figure 13.22 shows the lateral forces generated by an off-road radial tire on an asphalt surface. This plot is in the form of a *carpet plot*, a common method of plotting lateral force data for automotive tires. Three separate lateral force-slip angle curves are

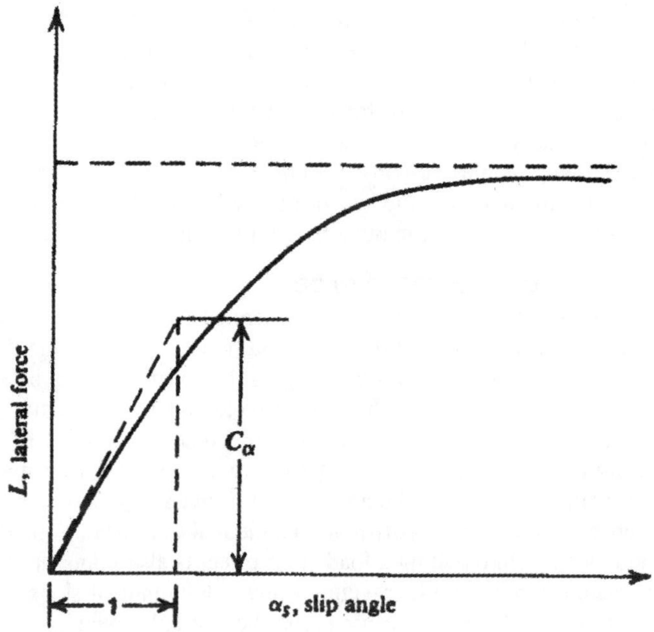

Figure 13.21. Lateral force-slip angle functional relationship.

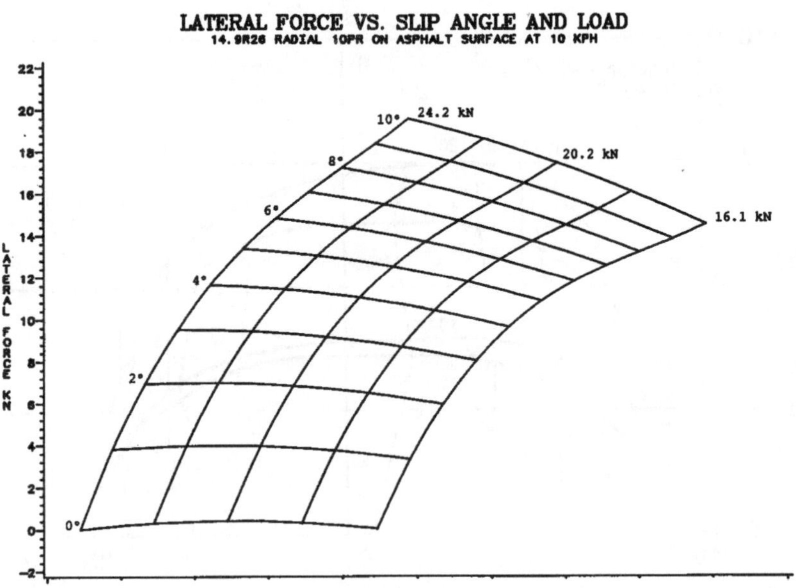

Figure 13.22. Carpet plot depiction of lateral forces on an off-road tire for various slip angles and three vertical loads. (Reprinted with permission of Firestone Agricultural Tire Division of Bridgestone/Firestone. For more details, see Janssen and Schuring, 1985.)

plotted on the same graph, each representing a different vertical load (16.1, 20.2, and 24.2 kN). The origins of the three curves are offset laterally (along the horizontal or slip angle axis) by a constant amount from each other. Then lateral force data points representing the same slip angle at different vertical loads are connected together by curve segments. Similarly the lateral force data points representing the same vertical load for different slip angles are also connected. The resulting plot somewhat resembles a carpet being tossed in the air with one edge fixed, hence the name carpet plot. Similar plots are made for automotive tires for the aligning moment.

13.4.3 Influence of Tractive Force

Since both tractive and lateral forces are generated in the tire's contact patch, it appears natural that the development of one would be influenced by the development of the other. This, termed the *friction ellipse effect*, has been found to be the case for both automotive and off-road tires where the development of a tractive or braking force reduces the lateral force generated at a given slip angle. Figure 13.23 illustrates the concept using measured data for an 18.4R38 tire. As the tractive or braking force generated by the tire increases, the lateral force at a given slip angle decreases. This effect can be encountered when an off-road vehicle makes a turn or operates along a contour of a side slope while pulling a load. This effect is also important in situations where a rear-wheel-drive vehicle is traveling down a longitudinal slope. Locking the brakes can result in the vehicle jack-knifing with the rear axle changing positions with the front axle. Locking the rear wheels drastically lowers their capability to generate the lateral forces required to keep the vehicle from sliding sideways.

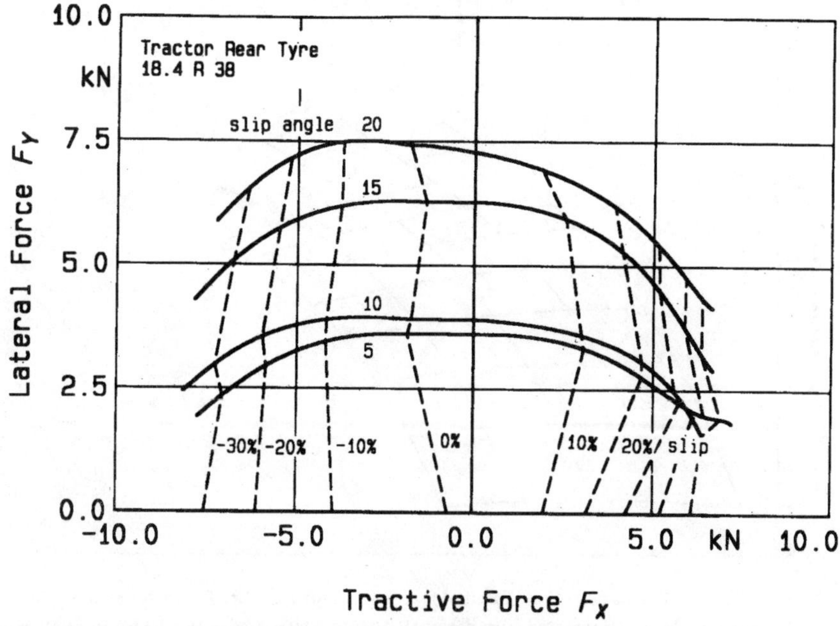

Figure 13.23. The lateral force on an off-road tire is reduced as a tractive or braking force is developed. (From Armbruster and Kutzbach, 1991.)

13.5 Summary

The forces for supporting, propelling, and steering an off-road vehicle are all generated in the contact patches of its tractive and transport devices. These usually are pneumatic rubber tires, steel tracks, or rubber tracks. The tractive and transport devices should provide sufficient flotation to the vehicle to enhance its mobility and minimize rutting and/or soil compaction. Unless the vehicle is suspended, the traction and transport devices may provide the major portion of isolating the vehicle chassis from shock and vibration through their spring and energy dissipation/damping properties. Because of the need to negotiate steep terrain and/or pull draft loads, the tractive performance of an off-road vehicle is of special importance for both mobility and efficiency. The generation of lateral forces by the tractive and transport devices allows the vehicle to make relatively sharp turns to enhance maneuverability.

Homework Problems

Since results from one problem may be needed as input data to the other problems, answers are given in parentheses at the end of each problem.

13.1 Using Equation 13.1, estimate the maximum tractive force of a tractor with two steel tracks each 360 mm wide by 1680 mm long. The weight of the tractor is 31.75 kN. Soil parameters are c = 3 kPa and $\phi = 27°$. (F_{max} = 19,806 N)

13.2 A series of soil shearing tests are made with a plate as shown in Figure 13.6a where l = b = 100 mm. As a result of these tests, a family of curves was obtained similar to those of Figure 13.6b, where F_1 = 105 N at W_1 = 200 N, F_2 = 265 N at W_2 = 600 N, and F_3 = 425 N at W_3 = 1000 N. Determine the cohesion c and angle of internal friction ϕ of the soil. (c = 2.5 kPa, $\phi = 21.8°$)

13.3 Janssen and Schuring, 1985, give the following relation for estimating the vertical load-deflection relation for agricultural off-road tires:

$$F = S_0Z + BZ^{1.6}$$

where

$S_0 = -3.25 + 0.12\,(SW) + 0.39\,(CONST) + 0.47\,(RIM) + 0.006\,(INF)$

Z = vertical deflection (cm)

F = vertical load (kN)

$B = (FL - S_0ZL)/ZL^{1.6}$

FL = tire rated load (kN)

ZL = tire rated deflection (cm)

= (overall tire diameter)/2 – (static loaded radius)

SW = nominal tire section width (inches)

CONST = 2 for bias tires, 3 for radial tires

RIM = 1 for nominal rim diameters < 30 inches

= 2 for nominal rim diameters ≥ 30 inches

INF = inflation pressure (kPa)

Estimate the spring rate at 80% of the rated deflection of an 18.4R38 tire for an inflation pressure of 138 kPa if the rated load is 26.61 kN at that pressure. The overall tire diameter is 1744 mm while the static loaded radius is 787 mm.

(3.62 kN/cm = 362 kN/m)

13.4 A linear spring (k) – mass (m) – damper (c) system analogy can be used to estimate the damping rate c of an off-road tire if measured data are not available. For such a system, the critical damping rate, c_c, is defined as $c_c = 2\sqrt{km}$ where k is the stiffness and m the mass. The damping ratio, ζ, is defined as $\zeta = c/c_c$. If the stiffness k of the tire is 300 kN/m, estimate the damping rate c of the tire if the total load on the tire is 10 kN and the damping ratio is 0.1. (Off-road tires typically exhibit damping ratios in the range of 0.03 to 0.1.) (c = 3.5 kN/m/s).

13.5 A mechanical front-wheel drive tractor weighing 91,597 N has its static weight divided so that 41,672 N is on the two front wheels and 49,925 N is on the four rear wheels. The front tires are 14.9-30 bias-ply tires (overall unloaded diameter d = 1410 mm with a static loaded radius of 650 mm). The rear tires are 18.4-42 bias-ply tires (overall unloaded diameter d = 1864 mm with a static loaded radius of 846 mm).

(a) Estimate the force and power required to tow the tractor at a speed of 8 km/h on a soil with a cone index of 800 kN/m^2. (6305 N, 14 kW)

(b) What would be the force required if the cone index were 400 kN/m^2?

(8946 N)

13.6 A single wheel using a 14.9-28 bias-ply tire (unloaded overall diameter = 1364 mm with a static loaded radius of 627 mm) is to be used to drive a center pivot irrigation system.

(a) For operation on a soil with a cone index of 600 kN/m^2, the wheel must develop a net tractive force of 6000 N. Estimate to the nearest 1000 N the weight W required to be supported by the wheel to achieve maximum tractive efficiency. (W = 12,000 N, tractive efficiency = 0.720, slip = 14.9%)

(b) The following list gives a portion of the load-inflation table values for the tire:

11,565 N @ 96 kPa

12,588 N @110 kPa

13,344 N @124 kPa

At what inflation pressure should the tire operate? (110 kPa)

(c) For the above conditions, what is the torque required to drive the wheel if the rolling circumference is 4061 mm? (4587 Nm)

(d) If a 15 kW variable speed motor will be used to drive the wheel, what will be the rotational speed of the wheel? (3.27 radians/second = 31.2 rpm) (e) What would be the resulting actual forward travel speed? (1.799 m/s = 6.48 km/h)

13.7 The measurement of traction related variables such as axle torque and wheel slip generally requires special instrumentation. Some agricultural tractors are equipped from the factory with an electronic drive wheel slip readout based on measurements of drive wheel rotational speed and true ground speed. In the absence of such instrumentation, describe how distance measurements can be used for calculating slip in a field situation (see Goering, 1971).

References and Suggested Readings

Abu-Hamdeh, N.H., T.G. Carpenter, R.K. Wood, and R.G. Holmes. 1995. Soil compaction of four-wheel drive and tracked tractors under various draft loads. SAE Paper No. 952098. Warrendale, PA: SAE.

Abu-Hamdeh, N.H., T.G. Carpenter, R.K. Wood, and R.G. Holmes. 1995. Combine tractive devices: effects on soil compaction. SAE Paper No. 952159. Warrendale, PA: SAE.

Armbruster, K., and H.D. Kutzbach. 1991. Combined lateral and longitudinal forces on driven angled tractor tyres. *Journal of Terramechanics* 28(4): 331-338.

ASAE. 1995. General terminology for traction of agricultural tractors, self-propelled implements, and traction and transport devices. ASAE Standard S296.4. *ASAE Standards,* 42nd ed. St. Joseph, MI: ASAE.

ASAE. 1998. Agricultural machinery management data. ASAE Data D497.4. *ASAE Standards,* 45th ed. St. Joseph, MI: ASAE.

ASAE. 1999. Soil cone penetrometer. ASAE Standard S313.3. *ASAE Standards,* 46th ed. St. Joseph, MI: ASAE.

ASAE. 1999. Procedures for using and reporting data obtained with the soil cone penetrometer. Engineering Practice EP542. *ASAE Standards,* 46th ed. St. Joseph, MI: ASAE.

Aubel, T. 1993. FEM-simulation of the interaction between an elastic tyre and soft soil. In *Proceedings of the 11th ISTVS International Conference* (2): 791-802. Lake Tahoe, NV, 27-30 September 1993.

Bashford, L.L., and M.F. Kocher. 1999. Belts vs. tires, belts vs. belts, tires vs. tires. *Applied Engineering in Agriculture* 15(3): 175-181.

Bédard, Y., S. Tessier, C. Laguë, Y. Chen, and L. Chi. 1997. Soil compaction by manure spreaders equipped with standard and oversized tires and multiple axles. *Transactions of the ASAE* 40(1): 37-43.

Bekker, M.G. 1956. *Theory of Land Locomotion.* Ann Arbor, MI: The University of Michigan Press.

Bekker, M.G. 1969. *Introduction to Terrain-Vehicle Systems.* Ann Arbor, MI: The University of Michigan Press.

Book, R.S., and C.E. Goering. 2000. A new traction model for crawler tractors. *Transactions of the ASAE* 43(1): 39-46.

Brixius, W.W. (a.k.a. C.M. Hollis), and R.D. Wismer. 1978. The role of slip in traction. ASAE Paper No. 781538. St. Joseph, MI: ASAE.

Brixius, W.W. (a.k.a. C.M. Hollis). 1987. Traction prediction equations for bias-ply tires. ASAE Paper No. 871622. St. Joseph, MI: ASAE.

Burt, E.C. 1982. Load and inflation pressure effects on tires. *Transactions of the ASAE* 25(4): 881-884.

Charles, S.M., and D.J. Schuring. 1984. An empirical model for predicting the effective rolling radius of agricultural drive tires. ASAE Paper No. 841555. St. Joseph, MI: ASAE.

Crolla, D.A., D.N.L. Horton, and R.M. Stayner. 1990. Effect of tyre modeling on tractor ride vibration prediction. *Journal of Agricultural Engineering Research* 47(1): 55-77.

Crolla, D.A., and A.S.A. El-Razaz. 1987. A review of the combined lateral and longitudinal force generation of tyres on deformable surfaces. *Journal of Terramechanics* 24(3): 199-225.

Dwyer, M.J. 1984. The tractive performance of wheeled vehicles. *Journal of Terramechanics* 21(1): 19-34.

Ellis, R.W. 1977. Agricultural tire design requirements and selection considerations. ASAE Distinguished Lecture Series No. 3. St. Joseph, MI: ASAE.

Esch, J.H., L.L. Bashford, K. VonBargen, and R.E. Ekstrom. 1990. Tractive performance comparisons between a rubber belt track and a four-wheel drive tractor. *Transactions of the ASAE* 33(4): 1109-1115.

Freitag, D.R. 1977. History of wheels for off-road transport. *Journal of Terramechanics* 16(2): 49-68.

Freitag, D.R. 1985. Soil dynamics as related to traction and transport systems. In *Proceedings of First International Conference on Soil Dynamics* (4): 605-629. Auburn, AL, 17-19 June 1985.

Gee-Clough, D. 1985. The special problem of wetland traction and flotation. *Journal of Agricultural Engineering Research* 32: 279-288.

Gill, W.R., and G.E. Vandenberg. 1967. Soil dynamics in tillage and traction. Agriculture Handbook No. 316. U.S. Government Printing Office, Washington, D.C.: USDA-ARS.

Gillespie, T.D. 1992. *Fundamentals of Vehicle Dynamics.* SAE Publication R-114. Warrendale, PA: SAE.

Goering, C.E. 1971. Tractor pulling contest in a teaching laboratory. *Agricultural Engineering* 52(5): 268-269.

Håkansson, I. 1994. Subsoil compaction caused by heavy vehicles–A long-term threat to soil productivity. *Soil & Tillage Research* 29(2-3): 105-110.

Holloway, D.C., W.H. Wilson, and T.J. Drach. 1989. Examination of ATV tire forces generated on clay, grass and sand surfaces. SAE Paper No. 891106. Warrendale, PA: SAE.

Janssen, M.L., and D.J. Schuring. 1985. Structural and performance characteristics of agricultural tires. SAE Paper No. 851090. Warrendale, PA: SAE.

Karafiath, L.L., and E.A. Nawatzki. 1978. *Soil Mechanics for Off-road Vehicle Engineering.* Clausthal, Germany: TransTech Publications.

Kising, A., and H. Göhlich. 1989. Dynamic characteristics of large tyres. *Journal of Agricultural Engineering Research* 43(1): 11-21.

Krick, G. 1969. Radial and shear stress distribution under rigid wheels and pneumatic tires operating on yielding soils with consideration of tire deformation. *Journal of Terramechanics* 6(3): 73-98.

Leviticus, L.I., and J.F. Reyes. 1985. Tractor performance on concrete. *Transactions of the ASAE* 28(5): 1425-1429.

Lines, J.A., and K. Murphy. 1991. The stiffness of agricultural tractor tyres. *Journal of Terramechanics* 28(1): 49-64.

Lines, J.A., and K. Murphy. 1991. The radial damping of agricultural tractor tyres. *Journal of Terramechanics* 28(2-3): 229-241.

Macmillan, R. H. 2003. The Mechanics of Tractor-Implement Performance: Theory and Worked Examples. University of Melbourne ePrints Repository (UMER). The book can be downloaded free of charge from UMER at http://eprints.unimelb .edu.au/archive/00000204 (accessed 30 May 2003).

McKyes, E. 1989. *Agricultural Engineering Soil Mechanics.* New York, NY: Elsevier.

Pitts, M.J., and C.E. Goering. 1979. Modeling soil cone index changes induced by drive wheel traffic. ASAE Paper No. 791552. St. Joseph, MI: ASAE.

Plackett, C.W. 1985. A review of force prediction methods for off-road wheels. *Journal of Agricultural Engineering Research* 31(1): 1-29.

Raper, R.L., B.H. Washington, and J.D. Jarrell. 1999 A tractor mounted multiple probe soil cone penetrometer. *Applied Engineering in Agriculture* 15(4): 287-290.

Reaves, C.A., and A.W. Cooper. 1960. Stress distribution in soils under tractor loads. *Agricultural Engineering* 41(1): 20-21, 31.

SAE. 1994. *Central Tire Inflation Systems–Managing the Vehicle to Surface.* SAE Publication SP-1061. Warrendale, PA: SAE.

Söhne, W. 1958. Fundamentals of pressure distribution and soil compaction under tractor tires. *Agricultural Engineering* 39(5): 276-281, 290.

Söhne, W. 1969. Agricultural engineering and terramechanics. *Journal of Terramechanics* 6(4): 9-30.

Southwell, P.H. 1964. An investigation of traction and traction aids. *Transactions of the ASAE* 7(2): 190-193.

Taylor, J.H. 1973. Lug angle effect on traction performance of pneumatic tractor tires. *Transactions of the ASAE* 16(1): 16-18.

Taylor, J.H. 1974. Lug spacing effect on traction of pneumatic tractor tires. *Transactions of the ASAE* 17(2): 195-197.

Taylor, J.H. 1976. Comparative traction performance of R-1, R-3, and R-4 tractor tires. *Transactions of the ASAE* 19(1): 14-16

Taylor, J.H., G.E. Vandenberg, and I.F. Reed. 1967. Effect of diameter on performance of powered tractor wheels. *Transactions of the ASAE* 10(6): 838-842.

Taylor, J.H., and E.C. Burt. 1975. Track and tire performance in agricultural soils. *Transactions of the ASAE* 18(1): 3-6.

Taylor, J.H., E.C. Burt, and A.C. Bailey. 1976. Radial tire performance in firm and soft soils. *Transactions of the ASAE* 19(6): 1062-1064.

Taylor, J.H., A.C. Trouse, E.C. Burt, and A.C. Bailey. 1982. Multipass behavior of a pneumatic tire in tilled soils. *Transactions of the ASAE* 25(5): 1229-1231, 1236.

Taylor, J.H., and W.R. Gill. 1984. Soil compaction: state-of-the-art report. *Journal of Terramechanics* 21(2): 195-213.

Taylor, J.H., and E.C. Burt. 1987. Total axle load effects on soil compaction. *Journal of Terramechanics* 24(3): 179-186.

Turner, R.J., L.R. Shell, and F.M. Zoz. 1997. Field performance of rubber belted and MFWD tractors in southern Alberta soils. SAE Paper No. 972730. Warrendale, PA: SAE.

Upadhyaya, S.K., et al. 1994. *Advances in Soil Dynamics*, vol. 1. St. Joseph, MI: ASAE.

Upadhyaya, S.K., et al. 2002. *Advances in Soil Dynamics*, vol. 2. St. Joseph, MI: ASAE.

Upadhyaya, S.K., M. Sime, N. Raghuwanshi, and B. Adler. 1997. Semi-empirical traction prediction equations based on relevant soil parameters. *Journal of Terramechanics* 34(3): 141-154.

Vandenberg, G.E., and W.R. Gill. 1962. Pressure distribution between a smooth tire and the soil. *Transactions of the ASAE* 5(2): 105-107.

Voorhes, W.B. 1986. The effect of soil compaction on crop yield. SAE Paper No. 860729. Warrendale, PA: SAE.

Wiley, J.C. 1995. Down to earth breakthrough–new guidelines for radial tractor tire pressures pay off. *Resource* 2(9): 9-12.

Wismer, R.D. 1982. Soil dynamics: a review of theory and application. SAE Paper No. 820656. Warrendale, PA: SAE.

Wismer, R.D., and H.J. Luth. 1974. Off-road traction prediction for wheeled vehicles. *Transactions of the ASAE* 17(1): 8-10, 14.

Wolf, D., I. Shmulevich, and U. Mussel. 1996. Wheel traction prediction on hard soil. *Transactions of the ASAE* 39(4): 1275-1283.

Wong, J.Y. 1989. *Terramechanics and Off-road Vehicles*. New York, NY: Elsevier.

Wong, J.Y. 2001. *Theory of Ground Vehicles*. New York, NY: Wiley.

Wood, R.K., R.C. Reeder, M.T. Morgan, and R.G. Holmes. 1993. Soil physical properties as affected by grain cart traffic. *Transactions of the ASAE* 36(1): 11-14.

Year Book. 2000. Copley, OH: Tire and Rim Assn.

Yong, R.N., E.A. Fattah, and N. Skiadas. 1984. *Vehicle Traction Mechanics*. New York, NY: Elsevier.

Zhang, N., and W. Chancellor. 1989. Automatic ballast position control for tractors. *Transactions of the ASAE* 32(4): 1159-1164.

Zoz, F.M. 1997. Belt and tire tractive performance. SAE Paper No. 972731. Warrendale, PA: SAE.

Zoz, F.M., and W.W. Brixius (a.k.a. C.M. Hollis). 1979. Traction prediction for agricultural tires on concrete. ASAE Paper No. 791046. St. Joseph, MI: ASAE.

Zoz, F.M., and J.C. Wiley. 1995. A theoretical basis for tractor ballasting recommendations. In *Proceedings of the 5th North American Conference of the ISTVS,* 80-87. Saskatoon, Saskatchewan, 10-12 May 1985.

Zoz, F.M., R.J. Turner, and L.R. Shell. 2002. Power delivery efficiency: A valid measure of belt/tire tractive performance. *Transactions of the ASAE* 45(3): 509-518.

Zoz, F.M., and R.D. Grisso. 2003. Traction and tractor performance. ASAE Distinguished Lecture Series No. 27. St. Joseph, MI: ASAE.

CHAPTER 14

CHASSIS MECHANICS

If the agricultural engineer who is specializing in farm power is to make his best contribution to the agricultural engineering profession, if as an agricultural engineer he is to earn and hold a recognized place among other engineers, he must have more than the garage mechanic's conception of the tractor. He must visualize the tractor as a unit and have a clear conception of all forces acting upon it. He must be informed concerning the fundamental laws of mathematics and physics which govern its kinematic and dynamic response to these forces. In no other way will he be able to make the tractor perform its maximum service.

E.G. McKibben, 1927

14.1 Introduction

In this chapter, the farm tractor is used as an example of the application of the principles of statics and dynamics in the analysis of an off-road vehicle's tractive performance, stability, ride, and handling. This chapter provides an introduction to these individual areas while attempting to emphasize their interrelationships.

Because of the introductory nature of this chapter, the two-dimensional analysis of a rear-wheel-driven wheeled tractor is emphasized. The Suggested Readings at the end of the chapter provide additional information on methods of three- as well as two-dimensional analysis.

14.2 Simplifying Assumptions

The following simplifying assumptions apply to the rear-wheel-driven tractor shown in Figures 14.1 and 14.2:

1. The ground surface is assumed to be planar and nondeformable.
2. The motion of the tractor can be analyzed as two-dimensional.
3. Rotational motion of the front wheels is neglected.
4. Aerodynamic forces are neglected.

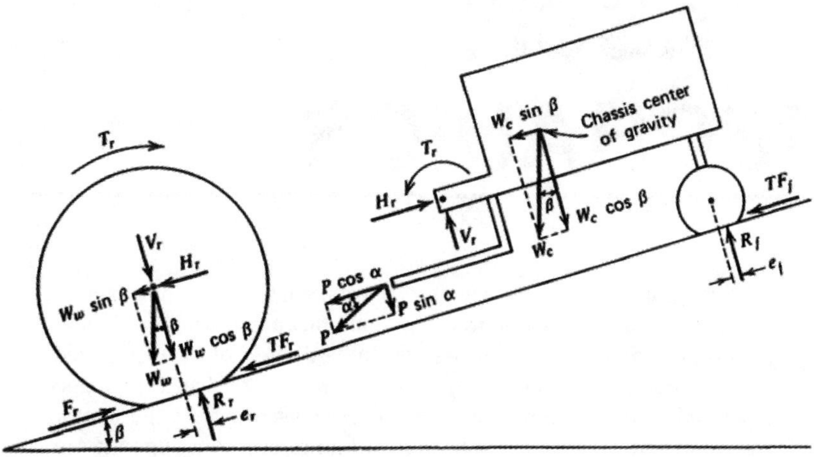

Figure 14.1. Free-body diagrams of the chassis and drive wheels of a rear-wheel-driven tractor.

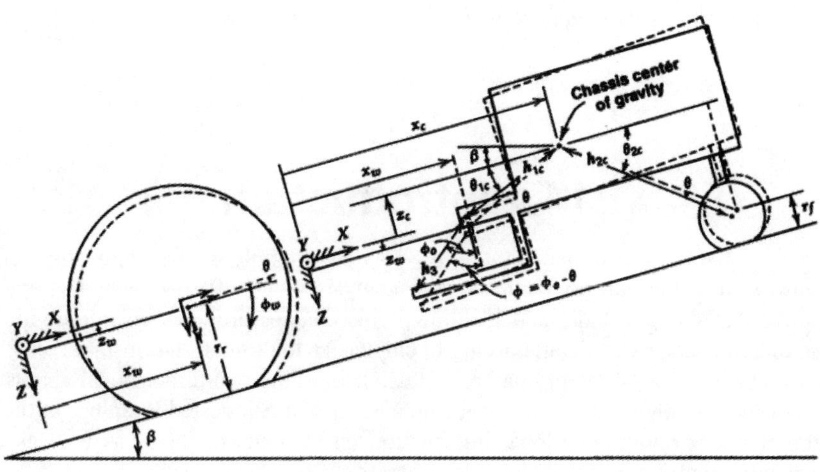

Figure 14.2. Kinematics associated with the planar motion of the chassis and drive wheels. The dashed outline shows the position of the chassis and drive wheels after an angular displacement θ.

14.3 Equations of Motion

Figure 14.1 contains free body diagrams of the two major components of the tractor: the chassis and the rear wheels. The traction mechanics developed in Chapter 13 is used to describe the force systems acting on both the rear drive wheels and the front unpowered wheels. The weights of the rear wheels (W_w) and chassis (W_c) are decomposed into components acting parallel and perpendicular to the ground surface. Similarly the external force P exerted at the chassis drawbar is broken down into a parallel component $P \cos \alpha$ (commonly called the drawbar pull) and a perpendicular component $P \sin \alpha$.

The kinematics used to describe the rear wheel and chassis motions are illustrated in Figure 14.2. The translational motions of the rear wheels and chassis are referenced to the fixed or inertial X-Z coordinate system. Such a coordinate system, in which the positive Z axis points downward, is consistent with the terminology widely used in other areas of vehicle dynamics. The angle of rotation of the drive wheels, ϕ_w, is measured relative to the tractor chassis so that the absolute angular motion of the drive wheels is given by the angle $\phi_w - \theta$ where θ is the pitch angle of the chassis.

Letting m_w be the mass of the rear wheels and summing forces on the rear wheels in the X and Z directions,

$$m_w \ddot{x}_w = F_r - TF_r - W_w \sin\beta - H_r \tag{14.1}$$

$$m_w \ddot{z}_w = V_r + W_w \cos\beta - R_r \tag{14.2}$$

Letting I_{yyw} be the *moment of inertia* of the wheels about the y or lateral axis passing through their *center of gravity* and summing moments about the rear axle center (assumed to be coincident with the center of gravity of the rear wheels),

$$I_{yyw}(\ddot{\phi}_w - \ddot{\theta}) = T_r - (F_r - TF_r)r_r - R_r e_r \tag{14.3}$$

However, in the traction mechanics of Chapter 13, e_r is considered to be equal to $(TF_r/R_r)r_r$, so that Equation 14.3 may be written

$$I_{yyw}(\ddot{\phi}_w - \ddot{\theta}) = T_r - F_r r_r \tag{14.4}$$

Letting m_c be the mass of the chassis and summing forces on the chassis in the X and Z directions,

$$m_c \ddot{x}_c = H_r - W_c \sin\beta - TF_f - P\cos\alpha \tag{14.5}$$

$$m_c \ddot{z}_c = W_c \cos\beta + P\sin\alpha - V_r - R_f \tag{14.6}$$

Letting I_{yyc} be the moment of inertia of the chassis about the y or lateral axis passing through its center of gravity and summing moments about the chassis center of gravity,

$$\begin{aligned} I_{yyc}\ddot{\theta} = {} & T_r + H_r h_{1c}\sin(\theta_{1c}+\theta) - V_r h_{1c}\cos(\theta_{1c}+\theta) \\ & + P\sin\alpha\,[h_3\sin\phi + h_{1c}\cos(\theta_{1c}+\theta)] \\ & - P\cos\alpha\,[h_3\cos\phi + h_{1c}\sin(\theta_{1c}+\theta)] \\ & - TF_f\,[h_{2c}\sin(\theta_{2c}-\theta) + r_f] \\ & + R_f\,[h_{2c}\cos(\theta_{2c}-\theta) + e_f] \end{aligned} \tag{14.7}$$

Equation 14.7 may be simplified, since the traction mechanics of the unpowered front wheels assumes that $e_f = (TF_f/R_f)r_f$.

Equations 14.1, 14.2, and 14.4 are the differential equations of motion governing the three degrees of freedom that the rear wheels possess for general planar motion. Equations 14.5, 14.6, and 14.7 are the corresponding equations for the chassis. However, the constraint that the rear wheels are attached to the chassis implies that the corresponding 6 degrees of freedom (x_w, z_w, ϕ_w, x_c, z_c, θ) are not independent.

Referring to Figure 14.2, the following two equations express the constraint relation:

$$x_c = x_w + h_{1c}\cos(\theta_{1c}+\theta) \tag{14.8}$$

$$z_c = z_w - h_{1c}\sin(\theta_{1c}+\theta) \tag{14.9}$$

The two constraint equations reduce the number of degrees of freedom of the system from 6 to 4 and thus imply that the rear wheel-chassis system (the total tractor) can also be described by four independent differential equations. These four equations can be derived from Equations 14.1, 14.2, 14.3, 14.5, 14.6, and 14.7 by eliminating the internal reactions H_r, V_r, and T_r, using constraint Equations 14.8 and 14.9, and considering the relations between the locations of the centers of gravity of the rear wheels, chassis, and the tractor. After some algebraic manipulation, the equations of motion for the tractor as a whole may be found. The free-body and kinematic diagrams associated with these equations are shown in Figures 14.3 and 14.4.

$$m_t\ddot{x}_t = F_r - W_t\sin\beta - TF_f - TF_r - P\cos\alpha \tag{14.10}$$

$$m_t\ddot{z}_t = W_t\cos\beta + P\sin\alpha - R_r - R_f \tag{14.11}$$

$$\begin{aligned} I_{yyt}\ddot{\theta} = {} & I_{yyw}\ddot{\phi}_w + (F_r - TF_r)[h_{1t}\sin(\theta_{1t}+\theta) + r_r] \\ & - TF_f[h_{2t}\sin(\theta_{2t}-\theta) + r_f] \\ & + R_f[h_{2t}\cos(\theta_{2t}-\theta) + e_f] \\ & - R_r[h_{1t}\cos(\theta_{1t}+\theta) - e_r] \\ & + P\sin\alpha[h_3\sin\phi + h_{1t}\cos(\theta_{1t}+\theta)] \\ & - P\cos\alpha[h_3\cos\phi + h_{1t}\sin(\theta_{1t}+\theta)] \end{aligned} \tag{14.12}$$

The mass of the tractor is m_t ($m_t = m_c + m_w$), while $\ddot{x}_t$ and $\ddot{z}_t$ are the translational accelerations of the tractor center of gravity in the X and Z directions. I_{yyt} is the moment of inertia of the entire tractor about the y or lateral axis passing through the center of gravity of that body.

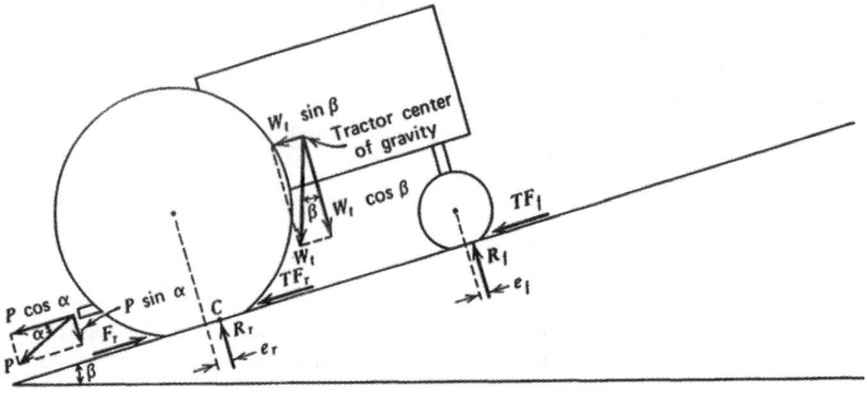

Figure 14.3. Free-body diagram of a rear-wheel-driven tractor.

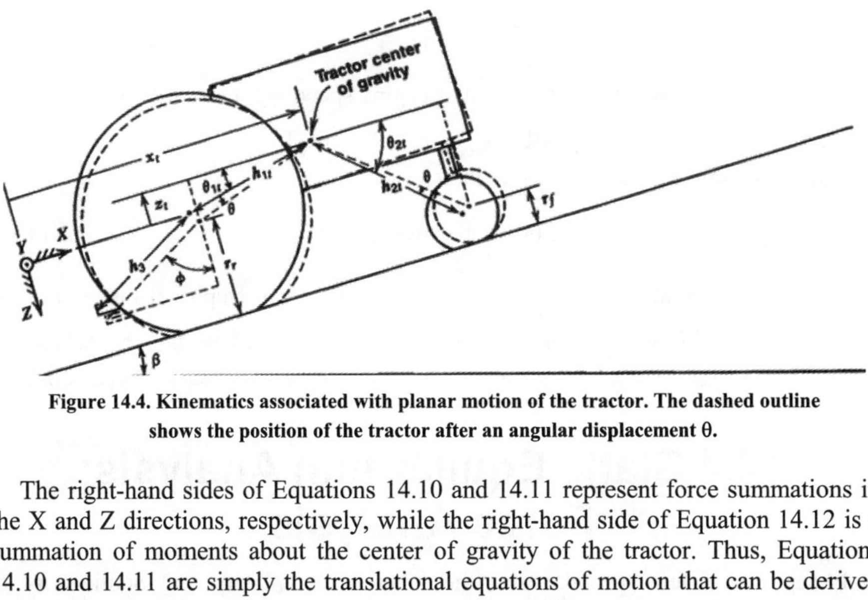

Figure 14.4. Kinematics associated with planar motion of the tractor. The dashed outline shows the position of the tractor after an angular displacement θ.

The right-hand sides of Equations 14.10 and 14.11 represent force summations in the X and Z directions, respectively, while the right-hand side of Equation 14.12 is a summation of moments about the center of gravity of the tractor. Thus, Equations 14.10 and 14.11 are simply the translational equations of motion that can be derived more directly from the free-body diagram of Figure 14.3. Similarly, with the exception of the term $I_{yyw}\ddot{\phi}_w$, Equation 14.12 can be derived directly from a summation of moments about the center of gravity of the tractor. Equations 14.10, 14.11, and 14.12 in combination with Equation 14.4 form the four independent differential equations of motion discussed above.

Equation 14.10 governs the forward translational motion of the tractor. The gross tractive force F_r is seen to be the only force acting to propel the tractor forward. This force must exceed the sum of the motion resistance forces acting at the wheels, the portion of the tractor weight acting parallel to the ground surface, and the drawbar pull in order for the tractor to accelerate forward. If the rear wheels are braked instead of driven, F_r changes direction and acts to decrease the forward speed of the tractor.

As discussed in Chapter 13, F_r may be considered to be equal to the product of the coefficient of gross traction μ_g and the soil reaction R_r. The coefficient of gross traction is, in turn, a function of the slippage s of the rear wheels as well as of tire and soil surface parameters. Using Equation 13.7 in Chapter 13 and the kinematics illustrated in Figure 14.2,

$$s = 1 - \dot{x}_w /(r_r \dot{\phi}_w) \tag{14.13}$$

Equations 14.11 and 14.12 govern the vertical translation and pitch rotation of the tractor. These equations are particularly important for describing tractor rearward overturning and vibration. Equation 14.12 may be simplified by using the relations $e_r = (TF_r/R_r)r_r$ and $e_f = (TF_f/R_f)r_f$.

If we consider that the power delivered to the rear axle is equal to the power output of the engine multiplied by an overall efficiency η, we can write

$$T_r \dot{\phi}_w = \eta T_e \dot{\phi}_e$$

where T_e is the engine torque produced at engine speed $\dot{\phi}_e$. Letting G be the ratio of engine speed to axle speed $(G = \dot{\phi}_e / \dot{\phi}_w)$, then $T_r = \eta G T_e$

Thus, Equation 14.4 can also be written

$$I_{yyw}(\ddot{\phi}_w - \ddot{\theta}) = \eta G T_e - F_r r_r \tag{14.14}$$

For a given throttle position, T_e can be expressed as an empirical function of $\dot{\phi}_e$ and, thus, given G, of $\dot{\phi}_w$. Thus, Equation 14.14 describes how the engine delivers torque to the rear wheels and in turn is loaded by the gross tractive force acting on them.

14.4 Static Equilibrium Analysis: Force Analysis

Except for a few special cases, the inherent complexity and nonlinearity of Equations 14.10, 14.11, 14.12, and 14.14 make their analytical solution difficult, if not impossible. Multi-body dynamics simulation software, which automatically formulates and numerically solves the differential equations for interconnected mechanical systems, is a particularly powerful tool for modeling the dynamic behavior of a wide range of off-road vehicles.

However, much useful information, particularly for evaluating tractor field performance, can be obtained from a static equilibrium analysis. For the static equilibrium situation, $\ddot{z}_t = 0$ and Equation 14.11 becomes

$$W_t \cos\beta + P\sin\alpha - R_r - R_f = 0 \tag{14.15}$$

Similarly, for static equilibrium, $\ddot{\theta} = \ddot{\phi}_w = 0$, and Equation 14.12 could be used to represent the moment equilibrium of the tractor. However, since the sum of moments about any transverse axis must be zero for the tractor to be in static equilibrium, the location of the axis may be chosen to help simplify the resulting moment summation. A convenient axis to use passes through Point C of Figure 14.3. This axis has the advantage that the resulting moment arms of forces F_r, TF_r, R_r, and TF_f all become zero. Then, summing moments about Point C, taking counterclockwise moments as positive, and assuming $\theta = 0$ when the tractor is in static equilibrium,

$$\begin{aligned} &R_f(h_{1t}\cos\theta_{1t} + h_{2t}\cos\theta_{2t} + e_f - e_r) \\ &+ W_t \sin\beta(r_r + h_{1t}\sin\theta_{1t}) - W_t\cos\beta(h_{1t}\cos\theta_{1t} - e_r) \\ &+ P\cos\alpha\,(r_r - h_3\cos\phi) + P\sin\alpha\,(h_3\sin\phi + e_r) = 0 \end{aligned} \tag{14.16}$$

Solving Equation 14.16 for R_f,

$$\begin{aligned} R_f &= \{W_t[(h_{1t}\cos\theta_{1t} - e_r)\cos\beta - (r_r + h_{1t}\sin\theta_{1t})\sin\beta] \\ &- P[(r_r - h_3\cos\phi)\cos\alpha + (h_3\sin\phi + e_r)\sin\alpha]\} \\ &/\{h_{1t}\cos\theta_{1t} + h_{2t}\cos\theta_{2t} + e_f - e_r\} \end{aligned} \tag{14.17}$$

Although useful for computation, the complexity of Equation 14.17 tends to obscure its physical meaning. Using the notation of Figure 14.5, Equation 14.17 may be written

$$R_f = (W_t L_2 - P y_r)/L_1 \tag{14.18}$$

With R_f now known, Equation 14.15 may be solved for R_r

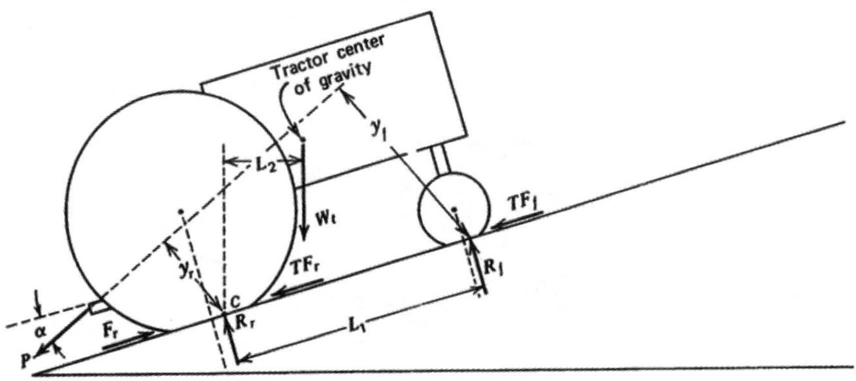

Figure 14.5. Free-body diagram illustrating the calculation of force R_f.

$$\begin{aligned} R_r &= W_t \cos\beta + P\sin\alpha - R_f \\ &= W_t \cos\beta + P\sin\alpha - \frac{W_t L_2}{L_1} + \frac{P y_r}{L_1} \\ &= W_t \cos\beta - \frac{W_t L_2}{L_1} + \frac{P(y_r + L_1 \sin\alpha)}{L_1} \\ &= W_t \cos\beta - \frac{W_t L_2}{L_1} + \frac{P y_f}{L_1} \end{aligned} \tag{14.19}$$

The terms Py_r/L_1 and Py_f/L_1 express changes in the soil reactions R_f and R_r as a result of the drawbar force P. Although there is no actual shift of weight, the changes in the forces R_r and R_f are commonly known as weight transfer.

It would be straightforward to solve Equations 14.17 and 14.19 for R_f and R_r if e_f and e_r did not depend on those forces. However, this is not the case if the equations discussed in Chapter 13 are used to predict the gross tractive and motion resistance forces. In such cases, an iterative numerical solution technique is required, allowing for the simultaneous calculation of all the forces acting on the vehicle.

Figure 14.6, a free-body diagram of a four-wheel-drive tractor, illustrates one such approach. The only difference from the previous analysis is the addition of a gross tractive force, F_f, acting on the front wheels, which are turning at a rotational speed $\dot{\phi}_{wf}$. To simplify the resulting equations, the motion resistance forces TF_f and TF_r are considered to act at the centers of the respective wheels, while the lines of action of the corresponding reaction forces R_f and R_r are shifted to likewise pass through the wheel centers. The net change in the moments acting on the tractor is zero for $TF_i\, r_i - R_i\, e_i = 0$ (i = f, r) when using the traction mechanics of Chapter 13.

Since the traction mechanics applies to individual wheels, let n_f and n_r be the number of tires on the front and rear axle, respectively. Further, let us assume all the front tires are identical as are all the rear tires and that radial-ply tires are being used. Then if W_f and W_r are the normal loads on a single front and a single rear tire

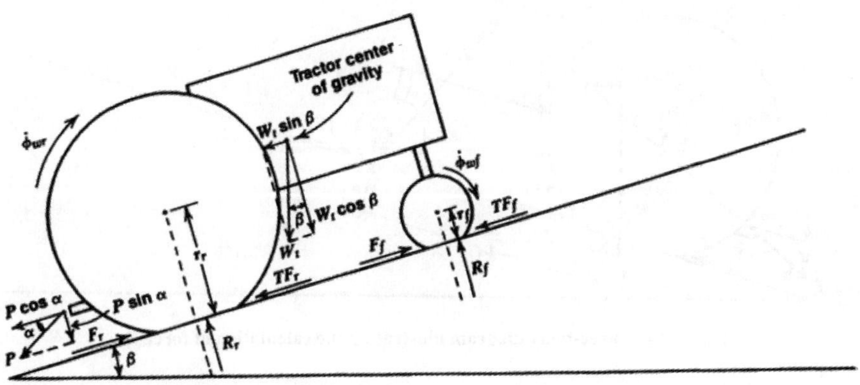

Figure 14.6. Free-body diagram of a four-wheel-drive tractor.

respectively, the forces acting on the tractor must simultaneously be compatible with the following equations:

$$W_f = R_f / n_f \tag{14.20}$$

$$W_r = R_r / n_r$$

$$\delta_f = f(W_f) \quad \text{(The tire deflections are functions of the normal loads on the tires.)}$$

$$\delta_r = g(W_r)$$

$$B_{nf} = \left(\frac{CI_f b_f d_f}{W_f}\right)\left(\frac{1+5\frac{\delta_f}{h_f}}{1+3\frac{b_f}{d_f}}\right)$$

$$B_{nr} = \left(\frac{CI_r b_r d_r}{W_r}\right)\left(\frac{1+5\frac{\delta_r}{h_r}}{1+3\frac{b_r}{d_r}}\right)$$

$$TF_f = \rho_f R_f = \left(\frac{0.9}{B_{nf}} + 0.0325 + \frac{0.5 s_f}{\sqrt{B_{nf}}}\right) R_f$$

$$TF_r = \rho_r R_r = \left(\frac{0.9}{B_{nr}} + 0.0325 + \frac{0.5 s_r}{\sqrt{B_{nr}}}\right) R_r$$

$$R_f = \{W_t \cos\beta h_{1t} \cos\theta_{1t} - W_t \sin\beta(r_r + h_{1t}\sin\theta_{1t}) - P\cos\alpha(r_r - h_3\cos\phi) - P\sin\alpha h_3 \sin\phi - TF_f r_f - TF_r r_r\} / \{h_{1t}\cos\theta_{1t} + h_{2t}\cos\theta_{2t}\}$$

$$R_r = W\cos\beta + P\sin\alpha - R_f$$

$$F_f - TF_f + F_r - TF_r - W_t \sin\beta - P\cos\alpha = 0$$

$$F_f = [0.88\,(1 - e^{-0.1B_{nf}})\,(1 - e^{-9.5 s_f}) + 0.0325] R_f$$

$$F_r = [0.88\,(1 - e^{-0.1B_{nr}})\,(1 - e^{-9.5 s_r}) + 0.0325] R_r$$

Thus far, our system consists of 13 equations in 14 unknowns (R_f, R_r, W_f, W_r, δ_f, δ_r, B_{nf}, B_{nr}, TF_f, TF_r, F_f, F_r, s_f, s_r). To complete the solution, an additional equation involving s_f and s_r is needed. Given the slippage, Equation 14.13 may be rearranged to calculate the forward velocity of the wheel. Since the forward velocities of the front and rear wheels are equal,

$$r_f\dot{\phi}_{wf}(1-s_f) = r_r\dot{\phi}_{wr}(1-s_r) \qquad \text{so that} \qquad s_f = 1-[(r_r\dot{\phi}_{wr})/(r_f\dot{\phi}_{wf})](1-s_r)\,.$$

Assuming a mechanical transmission connects an engine running at speed $\dot{\phi}_e$ to the front and rear axles with speed ratios of $G_f = \dot{\phi}_e/\dot{\phi}_{wf}$ and $G_r = \dot{\phi}_e/\dot{\phi}_{wr}$, then $s_f = 1-[(r_rG_f)/(r_fG_r)](1-s_r)$.

With G_f and G_r (or their ratio, G_f/G_r) known, Equations 14.20 can be used to determine the forces acting on the tractor. As Roscoe Pershing noted in SAE Paper No. 710525 (1971), "The solution of this set of equations will yield the unique set of operating forces that the actual vehicle solves routinely."

14.5 Static Equilibrium Analysis: Maximum Achievable Drawbar Pull

The drawbar pull, P cos α, that can be continuously sustained by a tractor is, of course, an important factor in determining the tractor's productivity. The maximum drawbar pull that can be developed may be limited by one of three factors: *stability*, traction, or power. Although it is impossible to control the situations in which a tractor may be operated, the maximum drawbar pull should be limited by traction or power availability.

14.5.1 Stability

Through the weight transfer effect, the drawbar force P may be large enough to cause force R_f to become zero and thus endanger the stability of the tractor. The drawbar force P_s required for this situation to occur can be found from Equation 14.18 by letting $R_f = 0$.

$$P_s = W_tL_2/y_r \tag{14.21}$$

Thus, P_s may be increased by increasing the weight of the tractor, W_t, by moving forward the center of gravity of the tractor (increasing L_2), or by lowering and/or moving forward the drawbar hitch point (decreasing y_r).

The generation of sufficient lateral forces for steering the tractor also depends on the value of R_f. Thus from the standpoint of steering control, the maximum value of P should be somewhat less than P_s.

14.5.2 Traction

The maximum achievable drawbar force P may also be limited by the tractive conditions of the surface on which the tractor is operating. As mentioned above, the gross tractive coefficient is a function of both drive wheel slippage and tire and soil parameters. As discussed in Chapter 13, the gross tractive coefficient μ_g versus slippage s relation has the following functional form for radial-ply tires

$$\mu_g = F/R = 0.88\,(1 - e^{-0.1B_n})\,(1 - e^{-9.5s}) + 0.0325 \qquad (14.22)$$

where the dimensionless parameter B_n incorporates tire and soil parameters as well as the vertical load on the drive wheel. The exponential nature of Equation 14.22 indicates that the maximum value of μ_g obtainable is $\mu_{max} = 0.88\ (1 - e^{-0.1Bn}) + 0.0325$.

For a given drawbar force P, the above set of equations may be used to calculate the complete set of forces acting including the gross tractive forces. Dividing the gross tractive force F_i by the normal load R_i determines the required gross tractive coefficient μ_{gi}. If μ_{gi} is greater than μ_{max}, it is obvious that the required tractive coefficient cannot be obtained under the given operating conditions.

If the required tractive coefficient is less than but close to the value of μ_{max}, and if sufficient engine power is available, the given drawbar force may be developed, but the slippage of the drive wheels may be so great that operation under such conditions would be impractical.

14.5.3 Power

Assuming that sufficient traction is available, a check should be made to determine if sufficient engine torque is available for developing the given drawbar force. Starting with the required gross tractive forces, the engine torque T_e required becomes

$$T_e = \frac{1}{\eta}\left(\frac{F_f r_f}{G_f} + \frac{F_r r_r}{G_r}\right) \qquad (14.23)$$

If T_e is less than the maximum torque that can be produced by the engine, the engine has sufficient torque capacity for the tractor to pull the given load.

14.5.4 Operating Conditions

To complete the analysis of the tractor operating conditions for a given drawbar pull, the engine's torque-speed relation may be used to estimate the steady-state engine speed $\dot{\phi}_e$ at which torque T_e may be produced. The resulting wheel speeds, $\dot{\phi}_{wi}$, are then

$$\dot{\phi}_{wi} = \dot{\phi}_e / G_i \quad (i = f, r) \qquad (14.24)$$

Since the front and rear wheel slips s_i have already been determined, the forward velocity of the drive wheels determines the forward velocity of the tractor:

$$\dot{x}_w = r_f \dot{\phi}_{wf}(1 - s_f) = r_r \dot{\phi}_{wr}(1 - s_r) \qquad (14.25)$$

The procedure just outlined can be used to predict the steady-state operating conditions for an off-road vehicle operating under a given drawbar loading. These operating conditions also might be used as the initial conditions for a dynamic analysis.

Powering the front wheels of a rear-wheel-driven tractor will reduce the slippage of the rear wheels, resulting in the tractor being able to pull the given loading at a higher forward travel speed, thus increasing its productivity. Since the rear wheels of a four-wheel-drive tractor usually follow in the tracks left by the front wheels, the rear wheels will usually encounter higher strength soil. This can be accounted for in the above analysis by allowing the cone index to vary from front to rear and providing an additional equation to estimate the relation between the front and rear cone index values.

14.6 Longitudinal Stability

There are several situations in which a tractor may be in danger of rearward overturning. The equations associated with the free-body diagrams of Figure 14.1 are particularly useful for the analysis of these situations. In such situations where the front wheels have left the ground, forces R_f and TF_f will be zero. By using *D'Alembert's principle,* the tractor chassis may be considered to be in static equilibrium, allowing moments to be summed about any point. D'Alembert's principle is applied by adding two fictitious forces, often called inertial forces, acting at the chassis center of gravity: $m_c\ddot{x}_c$ in the –X direction and $m_c\ddot{z}_c$ in the –Z direction. In addition, a fictitious clockwise moment $I_{yyc}\ddot{\theta}$ is applied to the chassis. Now, with the tractor in the assumed static equilibrium state, moments may be summed about the rear axle, thus eliminating the presence of the internal reaction forces V_r and H_r from the resulting equation. Using Equations 14.8 and 14.9 to express $\ddot{x}_c$ and $\ddot{z}_c$ in terms of $\ddot{x}_w$, $\ddot{z}_w$, $\ddot{\theta}$, and $\dot{\theta}^2$, the resulting moment equation may be written

$$\begin{aligned}(I_{yyc} + m_c h_{1c}^2)\ddot{\theta} = T_r &- W_c h_{1c}\cos(\theta_{1c} + \beta + \theta) \\ &- P\cos\alpha\, h_3 \cos(\phi_o - \theta) \\ &+ P\sin\alpha\, h_3 \sin(\phi_o - \theta) \\ &+ m_c \ddot{z}_w h_{1c}\cos(\theta_{1c} + \theta) \\ &+ m_c \ddot{x}_w h_{1c}\sin(\theta_{1c} + \theta)\end{aligned} \qquad (14.26)$$

Equation 14.26 indicates that the parallel component, P cos α, of the drawbar force exerts a moment tending to resist positive rotation of the chassis while an overturning moment is exerted by the perpendicular component, P sin α. One situation that has led to many rearward overturns is the application of the drawbar force P at a point near or above the rear axle. In such a situation, the angle of inclination α of the drawbar force with the ground surface may also be fairly large. As a result, the moment arm, $h_3 \cos(\phi_0 - \theta)$, of the parallel component may be small or even negative, while the perpendicular component P sin α is increased. As Equation 14.26 indicates, such a situation makes the development of a positive angular acceleration $\ddot{\theta}$ more probable, and thus the chances for a rearward overturn are increased.

Even though sufficient traction and power are available to develop a drawbar force in excess of P_s, the drawbar force required to cause the front wheels to just lose contact with the ground, the tractor may not be in danger of overturning if the drawbar load is properly hitched to the tractor. Equation 14.26 and Figure 14.2 indicate that the moment arm, $h_3 \cos(\phi_0 - \theta)$, of the parallel drawbar force component, $P \cos \alpha$, tends to increase as the tractor rotates counterclockwise through a positive angle θ. Simultaneously, the moment arm, $h_3 \sin(\phi_0 - \theta)$, of the perpendicular component, $P \sin \alpha$, tends to decrease as θ increases. Thus, for a properly hitched load, rotation of the tractor chassis may tend to stabilize the tractor, thus providing an explanation for situations in which a tractor pulling a drawbar load may be in equilibrium with the front wheels of the tractor off the ground.

Another situation in which a rearward overturn may take place occurs when a post or log is chained to the rear drive wheels in an effort to free a tractor that has become immobilized in soft soil. If the post or log prevents the rear wheels from turning when the tractor clutch is released, sufficient rear axle torque T_r may be developed to overturn the tractor.

Equation 14.26 indicates that the moment arm, $h_{1c} \cos(\theta_{1c} + \beta + \theta)$, of the chassis weight decreases as the rotation angle θ increases. In addition, this moment arm is also decreased by operation on a slope and is influenced by the height of the chassis center of gravity. Thus, the longitudinal stability of the tractor may be increased by adding weight to the tractor chassis in such a manner as to move forward and lower the chassis center of gravity. Such an action would also probably have the favorable effect of increasing the moment of inertia of the chassis.

When no drawbar force P is applied, a tractor becomes statically unstable when the angle $(\theta_{1c} + \beta + \theta)$ reaches 90°. However, in a dynamic sense, the tractor may become unstable at a considerably smaller angle. At this angle, the angular velocity $\dot{\theta}$ may be sufficient to allow the tractor to become statically unstable even though the rear axle torque T_r and drawbar force P may be reduced to zero. In such a situation where T_r, P, $\ddot{x}_w$, and $\ddot{z}_w$ are assumed to be zero, Equation 14.26 becomes

$$(I_{yyc} + m_c h_{1c}^2)\ddot{\theta} = -W_c h_{1c} \cos(\theta_{1c} + \beta + \theta) \tag{14.27}$$

Equation 14.27 can be integrated using the identities $\ddot{\theta}\, d\theta = \dot{\theta}\, d\dot{\theta}$ and $d\theta = d(\theta_{1c} + \beta + \theta)$. Thus, by multiplying both sides of Equation 14.27 by $d\theta$,

$$(I_{yyc} + m_c h_{1c}^2)\ddot{\theta}\, d\theta = -W_c h_{1c} \cos(\theta_{1c} + \beta + \theta)\, d\theta$$

$$(I_{yyc} + m_c h_{1c}^2)\dot{\theta}\, d\dot{\theta} = -W_c h_{1c} \cos(\theta_{1c} + \beta + \theta)\, d(\theta_{1c} + \beta + \theta) \tag{14.28}$$

Both sides of Equation 14.28 may now be integrated once appropriate limits are chosen. Assume that for a given angle of rotation θ_s, we desire to find the angular velocity $\dot{\theta}_s$ just sufficient to cause the chassis to become statically unstable. Then the appropriate limits are $\theta_{1c} + \beta + \theta = \theta_{1c} + \beta + \theta_s$ when $\dot{\theta} = \dot{\theta}_s$, and $\theta_{1c} + \beta + \theta = \pi/2$

when $\dot{\theta} = 0$ (the tractor becomes statically unstable just as the angular velocity $\dot{\theta}$ becomes zero).

Integrating Equation 14.28 with the above limits results in

$$(I_{yyc} + m_c h_{1c}^2)\dot{\theta}_s^2 / 2 = W_c h_{1c}(1 - \sin[\theta_{1c} + \beta + \theta_s]) \tag{14.29}$$

Equation 14.29 may be used to estimate the angular velocity $\dot{\theta}_s$ required for rotation of the tractor chassis to the point of static instability starting from a given angle of rotation θ_s. Equation 14.29 may also be derived from the work-energy theorem of mechanics.

14.7 The Tractor as a Two-Degree-of-Freedom Vibratory System

The ride motions of tractors have historically been associated with their *bounce* (vertical translation) and *pitch* motions. Considerable insight into these motions can be obtained from linearization of Equations 14.11 and 14.12. Assume that the tractor is traveling at a constant forward speed ($\ddot{x}_t = \ddot{\phi}_w = 0$) on level ground ($\beta = 0$), under no drawbar load ($P = 0$) and that the forces R_r and R_f pass through the rear and front wheel centers respectively ($e_r = e_f = 0$). Note in Equation 14.12 that the moment arms of the forces F_r, TF_f, and TF_r are equal to each other. Further from Equation 14.10, $F_r - TF_f - TF_r = 0$. Under these conditions, Equations 14.11 and 14.12 become

$$m_t \ddot{z}_t = W_t - R_r - R_f \tag{14.30}$$

$$I_{yyt}\ddot{\theta} = R_f h_{2t} \cos(\theta_{2t} - \theta) - R_r h_{1t} \cos(\theta_{1t} + \theta) \tag{14.31}$$

The tires, which represent the only suspension element besides a seat suspension on conventional tractors, can be idealized as parallel combinations of a linear spring and damper as shown in Figure 14.7. The spring and damping rates are the sums of the corresponding rates for each individual tire. For example, K_r is the sum of the spring rates of the rear tires. See Figures 13.4 and 13.5 in Chapter 13 for more information on the spring and damping properties of off-road tires.

The time histories of the ground displacements encountered by the rear and front tires are represented by z_{gr} and z_{gf} respectively. These would normally be thought of as functions of the forward travel of the tractor. However, through use of the constant forward velocity of the tractor, z_{gr} and z_{gf} may be converted into functions of time.

When the tractor is in static equilibrium ($\theta = z_t = 0$), forces R_r and R_f take the values R_{rs} and R_{fs} respectively, where

$$R_{rs} = W_t h_{2t} \cos\theta_{2t} / (h_{1t} \cos\theta_{1t} + h_{2t} \cos\theta_{2t}) \tag{14.32}$$

$$R_{fs} = W_t h_{1t} \cos\theta_{1t} / (h_{1t} \cos\theta_{1t} + h_{2t} \cos\theta_{2t}) \tag{14.33}$$

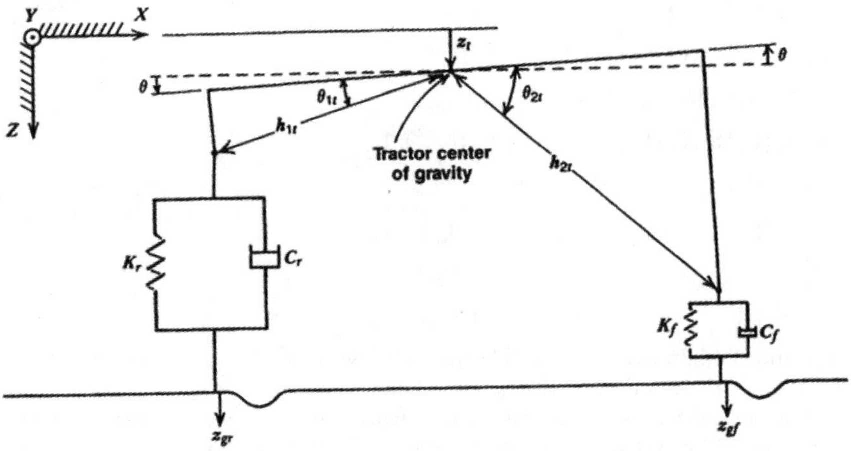

Figure 14.7. Representation of the tractor as a two-degree-of-freedom spring-mass-damper system.

For the displacements shown in Figure 14.7,

$$R_r = R_{rs} + K_r[z_t + h_{1t}\sin(\theta_{1t}+\theta) - h_{1t}\sin\theta_{1t} - z_{gr}] + C_r[\dot{z}_t + h_{1t}\dot{\theta}\cos(\theta_{1t}+\theta) - \dot{z}_{gr}] \quad (14.34)$$

$$R_f = R_{fs} + K_f[z_t + h_{2t}\sin(\theta_{2t}-\theta) - h_{2t}\sin\theta_{2t} - z_{gf}] + C_f[\dot{z}_t - h_{2t}\dot{\theta}\cos(\theta_{2t}-\theta) - \dot{z}_{gf}] \quad (14.35)$$

After substituting Equations 14.34 and 14.35 into Equations 14.30 and 14.31, utilizing the static equilibrium equations ($W_t - R_{rs} - R_{fs} = 0$ and $R_{fs}\, h_{2t} \cos\theta_{2t} - R_{rs}\, h_{1t} \cos\theta_{1t} = 0$), using the trigonometric identities for the sine and cosine of the sum and difference of two angles, and assuming that since θ is considered to be small, $\sin\theta \approx \theta$, $\cos\theta \approx 1$, and that second-order terms such as $\theta\dot{\theta}$ may be neglected, Equations 14.30 and 14.31 may be linearized, yielding

$$\ddot{z}_t + K_1\dot{z}_t + K_2\dot{\theta} + K_3 z_t + K_4\theta = (K_r z_{gr} + K_f z_{gf} + C_r\dot{z}_{gr} + C_f\dot{z}_{gf})/m_t \quad (14.36)$$

$$\ddot{\theta} + K_5\dot{z}_t + K_6\dot{\theta} + K_7 z_t + K_8\theta = [(K_r z_{gr} + C_r\dot{z}_{gr})(h_{1t}\cos\theta_{1t}) - (K_f z_{gf} + C_f\dot{z}_{gf})(h_{2t}\cos\theta_{2t})]/I_{yyt} \quad (14.37)$$

where

$K_1 = (C_f + C_r)/m_t$

$K_2 = (C_r h_{1t}\cos\theta_{1t} - C_f h_{2t}\cos\theta_{2t})/m_t$

$$K_3 = (K_f + K_r)/m_t$$
$$K_4 = (K_r h_{1t} \cos\theta_{1t} - K_f h_{2t} \cos\theta_{2t})/m_t$$
$$K_5 = (C_r h_{1t} \cos\theta_{1t} - C_f h_{2t} \cos\theta_{2t})/I_{yyt}$$
$$K_6 = (C_f [h_{2t} \cos\theta_{2t}]^2 + C_r [h_{1t} \cos\theta_{1t}]^2)/I_{yyt}$$
$$K_7 = (K_r h_{1t} \cos\theta_{1t} - K_f h_{2t} \cos\theta_{2t})/I_{yyt}$$
$$K_8 = (K_f [h_{2t} \cos\theta_{2t}]^2 + K_r [h_{1t} \cos\theta_{1t}]^2 - [R_{fs} - K_f z_{gf} - C_f \dot{z}_{gf}][h_{2t} \sin\theta_{2t}] - [R_{rs} - K_r z_{gr} - C_r \dot{z}_{gr}][h_{1t} \sin\theta_{1t}])/I_{yyt}$$

Equations 14.36 and 14.37 define the response of the linearized system to terrain excitation. If the terrain excitation is random in nature, a frequency analysis of the time histories of the system response will usually show prominent peaks at the natural frequencies of the system, since the damping provided by the tires is usually relatively small. The natural frequencies of the system are important in seat suspension design, since the natural frequency of the seat suspension must be somewhat less than the major frequency input to the seat if the suspension is to attenuate the input.

For free vibration ($z_{gf} = z_{gr} = 0$), the natural frequencies may be determined by setting the damping rates C_f and C_r to zero. Equations 14.36 and 14.37 then become

$$\ddot{z}_t + K_3 z_t + K_4 \theta = 0 \qquad (14.38)$$

$$\ddot{\theta} + K_7 z_t + K_8 \theta = 0 \qquad (14.39)$$

Assume that a periodic solution exists for Equations 14.38 and 14.39 of the form

$$z_t = Z \sin \omega t \qquad (14.40)$$

$$\theta = \Theta \sin \omega t \qquad (14.41)$$

Substituting Equations 14.40 and 14.41 into Equations 14.38 and 14.39 and simplifying,

$$(K_3 - \omega^2) Z + K_4 \Theta = 0 \qquad (14.42)$$

$$K_7 Z + (K_8 - \omega^2) \Theta = 0 \qquad (14.43)$$

Equations 14.42 and 14.43 have a nontrivial solution only if the determinant of the coefficient matrix formed from these equations is zero. Thus,

$$(K_3 - \omega^2)(K_8 - \omega^2) - K_4 K_7 = 0$$

or

$$\omega^4 - (K_3 + K_8)\,\omega^2 + (K_3 K_8 - K_4 K_7) = 0$$

Using the quadratic formula to solve for ω^2:

$$\omega^2 = \frac{(K_3 + K_8) \pm \sqrt{(K_3 + K_8)^2 - 4(K_3K_8 - K_4K_7)}}{2} \tag{14.44}$$

The two values of ω^2 given by Equation 14.44 lead to the two natural frequencies of the system.

Each natural frequency is associated with a corresponding mode of vibration defined by the ratio Z/Θ obtainable from either Equation 14.42 or 14.43:

$$\frac{Z}{\Theta} = \frac{K_4}{\omega^2 - K_3} = \frac{\omega^2 - K_8}{K_7} \tag{14.45}$$

The mode of vibration may be illustrated by assuming an arbitrary value for the amplitude Θ. The amplitude Z may then be calculated from Equation 14.45. The two amplitudes may then be combined with a scaled drawing of the tractor to indicate the mode of vibration as shown in Figure 14.8. In Figure 14.8, the mode of vibration is

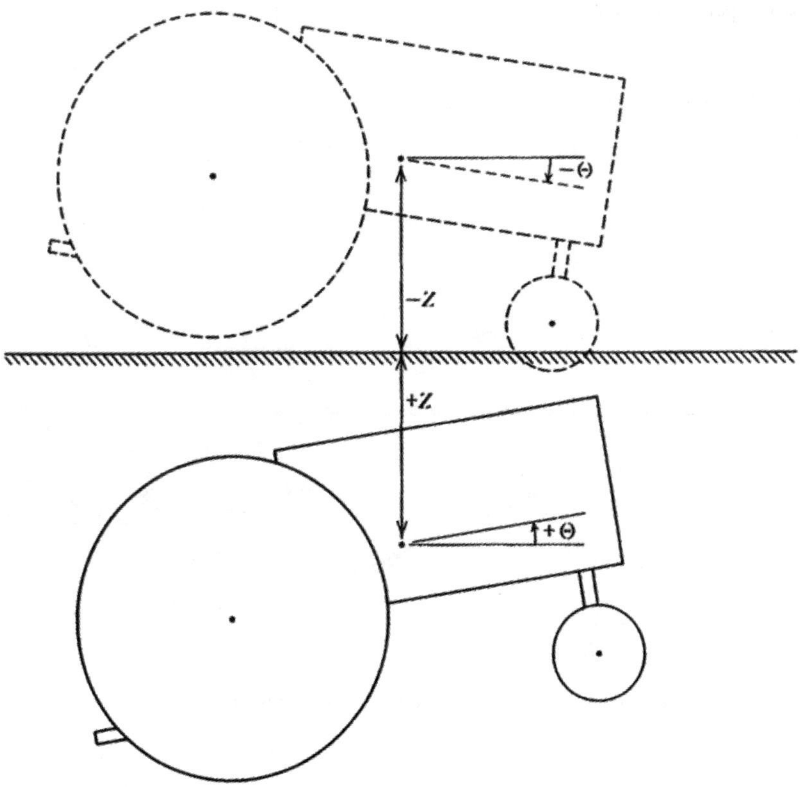

Figure 14.8. Method of illustrating a mode shape.

represented by what might be described as a double exposure photograph. The solid line drawing of the tractor represents one extreme position of the mode of vibration (sin ωt = 1), while the dashed line drawing illustrates the other extreme (sin ωt = –1). Note in general the vibration modes will contain both bounce and pitch motion.

Of course, the amplitude of the motion indicated by a mode shape drawing such as Figure 14.8 has no relation to the amplitude of motion that would result if that mode were excited experimentally or in the field. However, the mode shape drawing does provide a valuable insight into the type of motion associated with each of the natural frequencies of the system.

The natural frequencies of an off-road vehicle are important both in the field and on the road. For an unsuspended vehicle, the bounce and pitch motions exited by rough terrain will occur primarily at the natural frequencies. If a four-wheel-drive vehicle is pulling a drawbar load, the possibility exists for additional excitation of a bounce/pitch mode through variations in the tractive forces acting on the tires. In the right circumstances, this excitation can result in a severe bounce/pitch vibration termed *power hop.* Finally, tire and/or wheel out-of-roundness can excite vehicle bounce/pitch motion as the vehicle travels on the road during a transport operation. For farm tractors, this vibration, termed *road lope,* is becoming of increasing concern as transport speeds are increased.

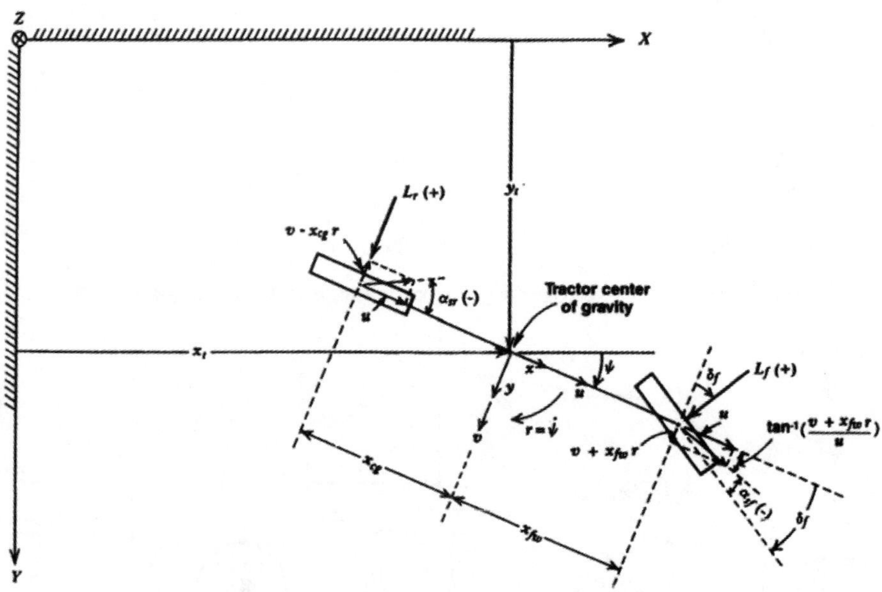

Figure 14.9. Free-body diagram for transient handling model illustrating front and rear slip angles.

14.8 Transient and Steady-State Handling

Figure 14.9 illustrates a simplified model that can be used to describe tractor transient handling. The wheels on the front and rear axles have been replaced by an equivalent single wheel at each axle having lateral force properties equivalent to the total of the wheels on that axle.

The model has two degrees of freedom, the lateral translational velocity, v, of the center of gravity and the yaw angular velocity, r, of the tractor about its center of gravity. The forward velocity, u, of the center of gravity is assumed to be constant while the steer angle, δ_f, of the front wheels is assumed to be a function of time.

Recalling from Chapter 13 that the lateral forces are functions of the slip angles of the tires, Figure 14.9 illustrates the slip angles of the front, α_{sf}, and rear, α_{sr} , wheels. From that figure,

$$\alpha_{sf} = \tan^{-1}((v + x_{fw} r)/u) - \delta_f$$

$$\alpha_{sr} = \tan^{-1}((v - x_{cg} r)/u)$$

The equations of motion for the tractor of Figure 14.9 are most easily expressed in terms of the vehicle coordinate system. In such a system, the lateral acceleration of the center of gravity is $\dot{v} + ur$. Thus,

$$m_t(\dot{v} + ur) = L_f \cos\delta_f + L_r \tag{14.46}$$

Letting I_{zzt} be the moment of inertia of the tractor about the vertical (or z) axis through its center of gravity,

$$I_{zzt}\dot{r} = (L_f \cos\delta_f)x_f - L_r x_{cg} \tag{14.47}$$

In transport situations, the steer and slip angles will be small so that the above equations can be linearized yielding

$$m_t(\dot{v} + ur) = -C_{\alpha f}\left[\frac{v + x_{fw} r}{u} - \delta_f\right] - C_{\alpha r}\left[\frac{v - x_{cg} r}{u}\right] \tag{14.48}$$

$$I_{zzt}\dot{r} = -C_{\alpha f}\left[\frac{v + x_{fw} r}{u} - \delta_f\right]x_{fw} + C_{\alpha r}\left[\frac{v - x_{cg} r}{u}\right]x_{cg} \tag{14.49}$$

Note that $C_{\alpha f}$ and $C_{\alpha r}$ represent the total cornering stiffness at the front and rear axles, respectively.

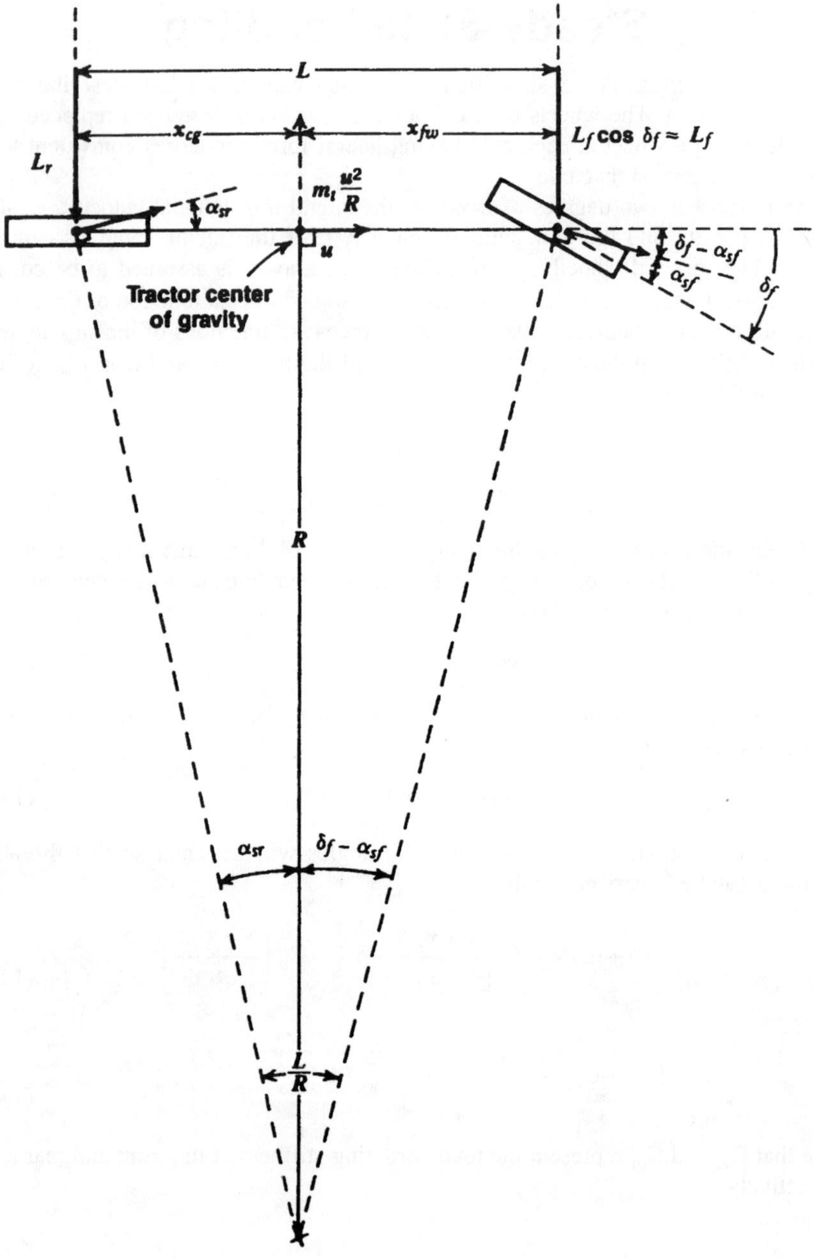

Figure 14.10. Tractor making a steady-state turn of radius R at a constant forward speed u.

Let us now turn to an analysis of a steady-state turning situation. In Figure 14.10, the tractor shown is traveling on level ground at a constant forward velocity u in such a manner that the center of gravity of the tractor is traversing a circle of radius R. Note that although both the front and rear slip angles are negative (resulting in positive lateral forces), the absolute values of the angles are used in the geometric analysis.

From Figure 14.10, $L/R = \alpha_{sr} + \delta_f - \alpha_{sf}$ or $\delta_f = L/R + \alpha_{sf} - \alpha_{sr}$. But, for this steady-state situation, $\dot{v} = \dot{r} = 0$ and $r = u / R$. Thus, with δ_f small, Equations 14.46 and 14.47 become:

$$m_t(\dot{v} + ur) = m_t u^2 / R = L_f + L_r \tag{14.50}$$

$$I_{zzt}\dot{r} = 0 = L_f x_{fw} - L_r x_{cg} \tag{14.51}$$

From Equation 14.51, $L_r = L_f (x_{fw} / x_{cg})$. Substituting in Equation 14.50,

$$m_t u^2 / R = L_f (1 + x_{fw} / x_{cg}) = L_f \frac{(x_{cg} + x_{fw})}{x_{cg}} = L_f \frac{L}{x_{cg}}$$

Thus,

$$L_f = \frac{x_{cg}}{L} m_t \frac{u^2}{R} = \frac{x_{cg}}{L} \frac{W_t}{g} \frac{u^2}{R}$$

But, $x_{cg}W_t/L = W_f$ is the portion of the tractor weight, W_t, statically supported at the front axle so that $L_f = (W_f/g)(u^2/R)$. Similarly, $L_r = (W_r/g)(u^2/R)$ where W_r is the portion supported at the rear axle. Then

$$L_f = \frac{W_f}{g} \frac{u^2}{R} = C_{\alpha f} \alpha_{sf}$$

$$L_r = \frac{W_r}{g} \frac{u^2}{R} = C_{\alpha r} \alpha_{sr}$$

$$\alpha_{sf} = \frac{1}{C_{\alpha f}} \frac{W_f}{g} \frac{u^2}{R}$$

$$\alpha_{sr} = \frac{1}{C_{\alpha r}} \frac{W_r}{g} \frac{u^2}{R}$$

Thus,

$$\delta_f = \frac{L}{R} + \frac{u^2}{gR}\left(\frac{W_f}{C_{\alpha f}} - \frac{W_r}{C_{\alpha r}}\right) \tag{14.52}$$

Equation 14.52 is the fundamental equation describing the steady-state handling behavior of a road vehicle. The quantity L/R is termed the *Ackerman steer angle* and is the front wheel steer angle required for the vehicle to move in a turn of radius R at low speed.

For higher speeds, the term $(W_f /C_{\alpha f} - W_r/C_{\alpha r})$, called the *understeer coefficient*, determines the steering behavior of the vehicle. If this term is zero, the steer angle required to negotiate a turn of radius R is independent of the forward speed u. Such a vehicle would be termed as having neutral steering. If the understeer coefficient is positive, the wheel steer angle must be increased with increasing speed. This condition is termed *understeering* and is the situation desired on most road vehicles. Finally, if the understeer coefficient is negative, the wheel steer angle δ_f must be decreased as the speed u increases. Such a condition is called *oversteering*.

For an oversteering vehicle, the speed at which the required steer angle δ_f becomes zero is termed the critical speed, u_c. From Equation 14.52, $u_c^2 = - Lg/(W_f /C_{\alpha f} - W_r/C_{\alpha r})$. By examining the eigenvalues of Equations 14.48 and 14.49, it can also be shown that an oversteering vehicle is directionally unstable above the critical speed. Compared to on-road vehicles, off-road vehicles may experience considerable changes in their static axle loads with the potential for resulting changes in their steering behavior. In addition, the development of substantial tractive or braking forces while steering has the effect of reducing the lateral forces affecting the maneuverability of the off-road vehicle.

14.9 Lateral Stability in a Steady-State Turn

The geometric configuration of tricycle tractors, combined with their ability (aided by individual brakes on the drive wheels) to make sharp turns at moderately high travel speeds, can result in a lateral overturning situation. Figure 14.11 illustrates a tricycle tractor in the steady-state circular turn analyzed in the previous section.

Assuming that the lateral tire forces are sufficient to generate the assumed acceleration, D'Alembert's principle may be applied by assuming the presence of a force, $m_t u^2/R$, acting at the center of gravity and in a direction opposite to the lateral acceleration of the center of gravity. The tractor may now be considered to be in static equilibrium.

Assume that the forward speed u of the tractor is gradually increased as the center of gravity traverses the circle of radius R. The lateral tire forces required to sustain this motion will also increase and will create a moment about the center of gravity tending to lift the right front and right rear tires off the ground.

(a)

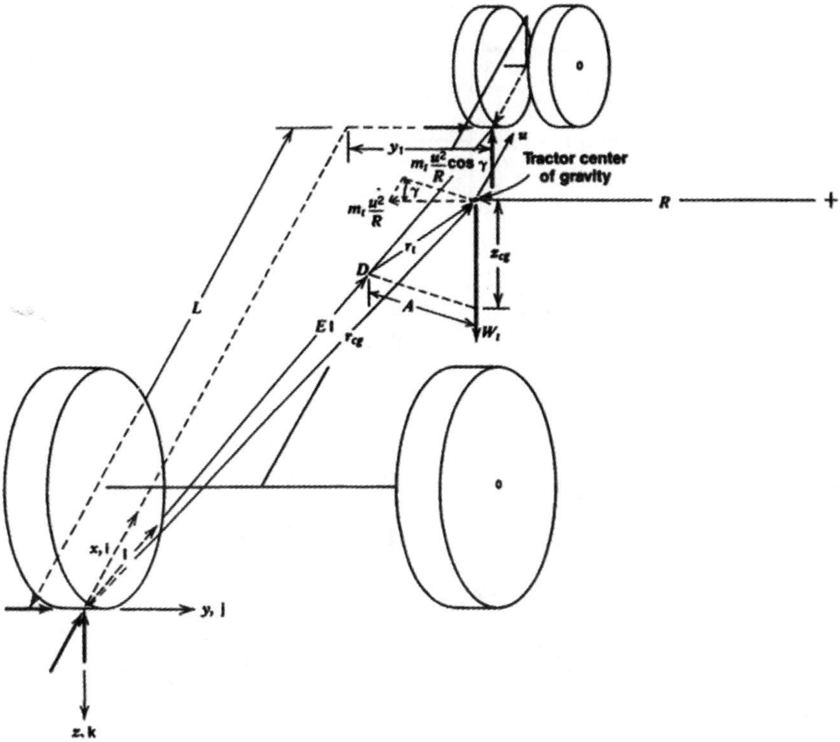

(b)

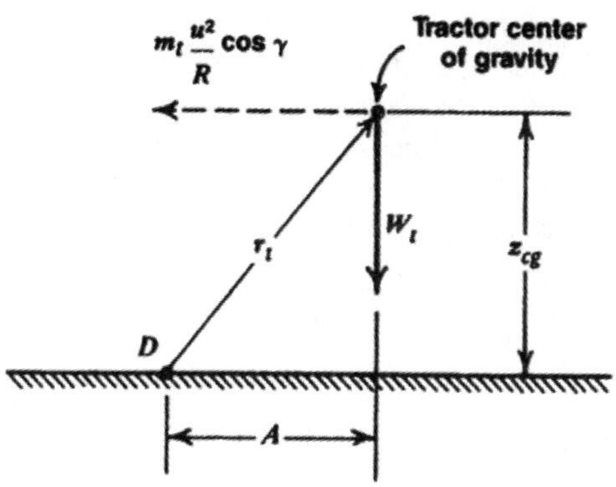

Figure 14.11. (a) Free-body diagram of a tractor about to overturn laterally as a result of making a steady-state turn. (b) Plane containing the tipping motion of the tractor (continued on next page).

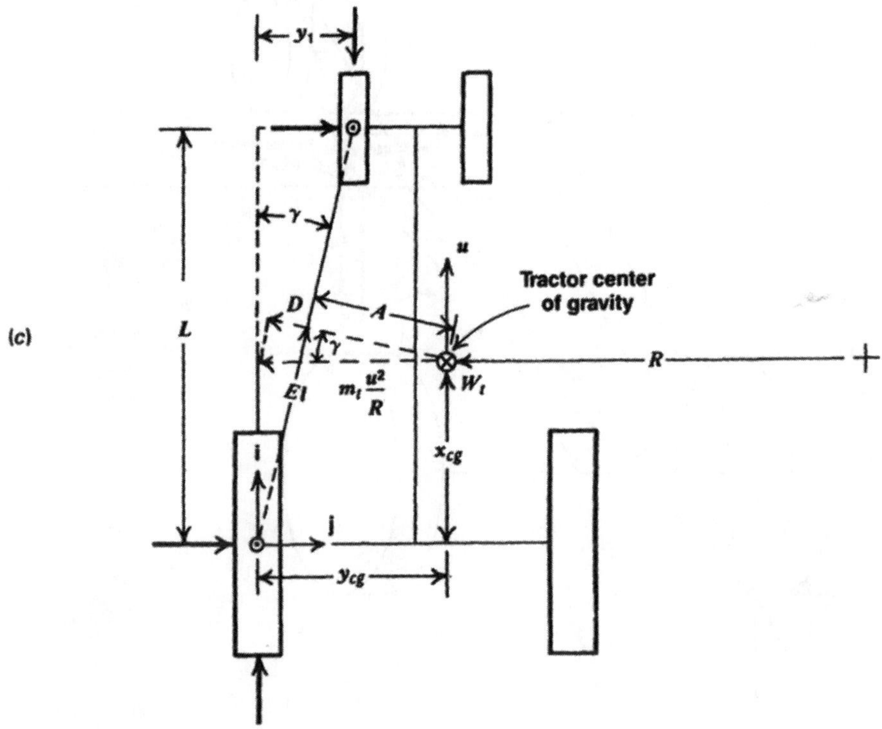

Figure 14.11 (continued from previous page). (c) Top view of tractor.

Assume that the right front and rear tires have just lost contact with the ground when the forward speed u_s is attained. The tractor is now on the verge of overturning about an axis connecting the ground contact points of the left rear and front tires. An analytical relation for u_s can be determined by summing moments about this tipping axis. Assuming the tire force components all act through the tipping axis, a summation of moments about that axis produces

$$m_t \frac{u_s^2}{R} \cos\gamma\, z_{cg} - W_t A = 0$$

Thus,

$$u_s = \sqrt{\frac{gAR}{z_{cg} \cos\gamma}} \tag{14.53}$$

where g is the acceleration of gravity. γ, the angle between the assumed force $m_t u^2/R$ and the tipping plane, is given by $\tan^{-1}(y_1/L)$. Equation 14.53 indicates u_s is decreased as the moment arm A of the tractor weight is decreased or as the height z_{cg} of the center of gravity is increased.

To use Equation 14.53, a relation for A must be determined from the tractor geometry. Let the xyz coordinate system and associated unit vector system **i**, **j**, and **k** (shown in Figure 14.11) have its origin at the point on the ground surface directly beneath the center of the left rear wheel. Relative to this coordinate system, the tipping axis is defined by the unit vector **l**.

$$\mathbf{l} = (L/B)\mathbf{i} + (y_1/B)\mathbf{j} \qquad (14.54)$$

where L and y_1 locate the point directly beneath the left front wheel center where the corresponding tire forces are assumed to act and $B = \sqrt{L^2 + y_1^2}$.

The plane containing the motion of the center of gravity as it rotates about the tipping axis intersects the tipping axis at Point D. The vector from the origin of the xyz system to Point D is denoted E**l**. Given the vector $\mathbf{r}_{cg} = x_{cg}\mathbf{i} + y_{cg}\mathbf{j} - z_{cg}\mathbf{k}$ locating the center of gravity of the tractor ($x_{cg} = h_{1t}\ \cos\theta_{1t}$, $z_{cg} = r_r + h_{1t}\sin\theta_{1t}$), it follows that E may be found by finding the component of $\mathbf{r}_{cg}$ in the **l** direction. This can be done by taking the dot or scalar product of the $\mathbf{r}_{cg}$ and **l** vectors:

$$E = \mathbf{r}_{cg} \bullet \mathbf{l} = (x_{cg}L + y_{cg}y_1)/B \qquad (14.55)$$

The vector $\mathbf{r}_t$ from Point D to the center of gravity of the tractor can now be found using the relation $\mathbf{r}_t = \mathbf{r}_{cg} - E\mathbf{l}$.

$$\mathbf{r}_t = (x_{cg} - EL/B)\ \mathbf{i} + (y_{cg} - Ey_1/B)\ \mathbf{j} - z_{cg}\ \mathbf{k}$$

The value of A is just the component of $\mathbf{r}_t$ in the direction lying in the ground surface and perpendicular to the tipping axis. Thus,

$$A = \sqrt{(x_{cg} - EL/B)^2 + (y_{cg} - Ey_1/B)^2} \qquad (14.56)$$

The addition of a front-end loader to a tricycle tractor may considerably reduce the lateral stability of the tractor in a turning situation. If a load is transported in the bucket of the loader with the bucket raised, the center of gravity of the tractor-loader combination will probably be both raised (increasing z_{cg}) and moved forward (decreasing A) as compared to the center of gravity location of the tractor alone. Thus the value of u_s for the loader equipped tractor may be considerably less than for the tractor alone.

14.10 Center of Gravity Determination

In the above analyses, the location of the center of gravity of the tractor was assumed to be known. Since most tractors are composed of many comparatively irregularly shaped parts, it is difficult to analytically find the center of gravity of a tractor still in the design stage. However, several methods exist for experimentally determining the center of gravity location of an assembled tractor. The locations determined experimentally are then often of use in estimating the center of gravity location of a new tractor design.

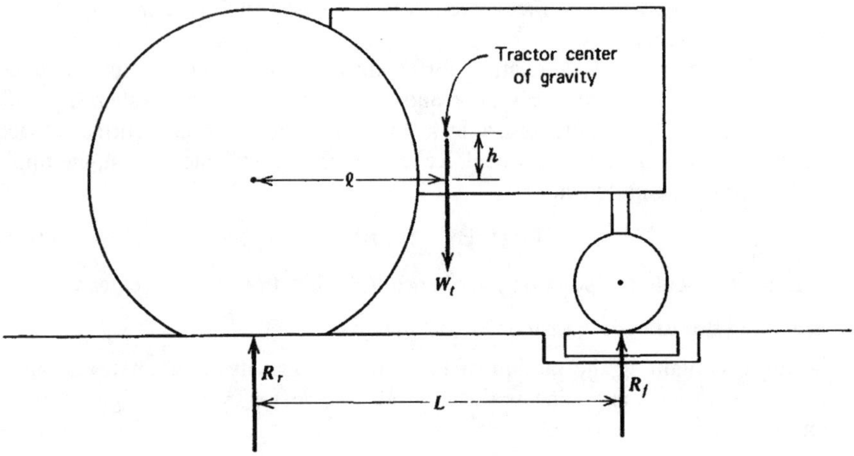

Figure 14.12. Determination of the longitudinal location of the center of gravity using the weighing method.

Only the weighing method for center of gravity location will be discussed in this section. Several other methods are discussed in the Suggested Readings at the end of the chapter.

Given the tractor weight W_t and wheelbase $L = h_{1t} \cos \theta_{1t} + h_{2t} \cos \theta_{2t}$, the longitudinal location $\ell = x_{cg} = h_{1t} \cos\theta_{1t}$ of the center of gravity may be found by placing the front axle of the tractor on a scale as in Figure 14.12 to determine force R_f. Then,

$$\ell = R_f L / W_t \tag{14.57}$$

As a check, the reaction R_r may be found and the distance $L - \ell$ determined.

Since most tractors are approximately symmetrical about the vertical plane perpendicular to the axles and passing midway between the wheels, the lateral location of the center of gravity will normally be quite close to this plane. Given the wheel tread setting, the lateral location may be found by weighing one of the sides of the tractor. Weighing the other side can serve as a check.

The measurements required for finding the longitudinal and lateral location of the center of gravity are straightforward. However, the determination of the height, $h = h_{1t} \sin \theta_{1t}$, is considerably more difficult. Again the weighing method can be used with either the front or rear axle elevated. Figure 14.13 illustrates the geometry involved in the following derivation. Here the symbols r_f and r_r refer to the static loaded radius of the front and rear tires respectively. Summing moments about the rear axle,

$$\ell' = R_f' L' / W_t$$

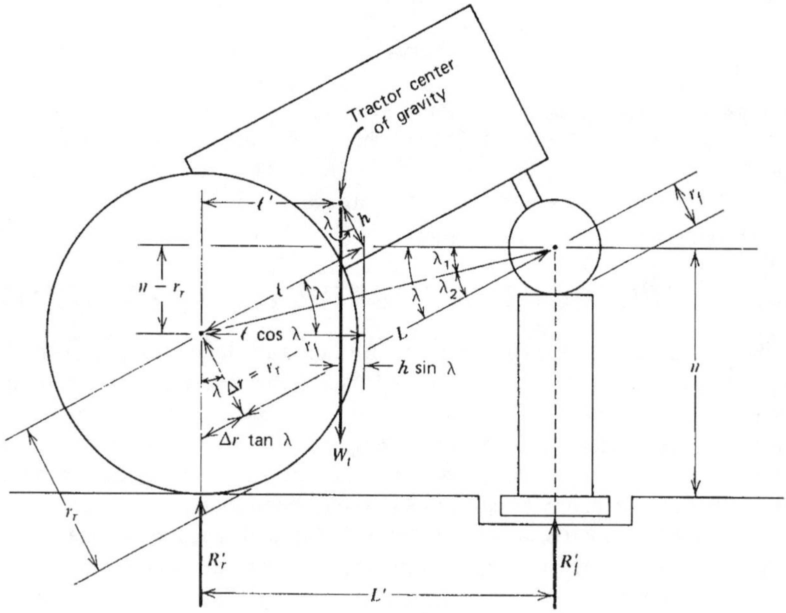

Figure 14.13. Determination of the vertical location of the center of gravity using the weighing method.

But from the geometry of Figure 14.13,

$$\ell' = \ell \cos\lambda - h \sin\lambda$$

Thus,

$$R'_f L' / W_t = \ell \cos\lambda - h \sin\lambda \tag{14.58}$$

But,

$$L' = (L + \Delta r \tan\lambda)\cos\lambda \tag{14.59}$$

Substituting Equation 14.59 into Equation 14.58, dividing by $\cos\lambda$, and solving for h,

$$h = \frac{W_t \ell - R'_f L}{W_t \tan\lambda} - \frac{R'_f \Delta r}{W_t} \tag{14.60}$$

The angle λ is the only quantity in Equation 14.60 that cannot be directly measured. However, $\lambda = \lambda_1 + \lambda_2$, where

$\tan\lambda_1 = (n - r_r)/L'$

$\tan\lambda_2 = \Delta r/L$

To eliminate the need for measuring L′ for use in calculating tan λ_1,

$$L' = \sqrt{L^2 + (\Delta r)^2 - (n - r_r)^2}$$

Two assumptions implicitly made in the preceding analysis should be pointed out: (1) the tires are assumed rigid and (2) fluid shifts occurring in the fuel, coolant, and oil compartments when the tractor is tilted are ignored. For accurate measurement of h, the front wheels should be elevated by an amount sufficient to obtain a significant difference between the reactions R_f and R'_f.

14.11 Moment of Inertia Determination

In the preceding analyses, the moments of inertia of the chassis, rear wheels, and complete tractor were assumed to be known about transverse axes passing through the center of gravity of each of these bodies. As was the case for the center of gravity location, the many and often irregularly shaped parts comprising a tractor make it difficult to analytically calculate the moments of inertia of a tractor about the longitudinal, transverse, or vertical axes passing through its center of gravity.

The moments of inertia about the transverse (pitch) and longitudinal (roll) axes passing through the center of gravity of the tractor may be measured using the setup in Figure 14.14. The tractor plus the supporting sling form a compound pendulum that will oscillate with a period T given by

$$T = 2\pi\sqrt{\frac{I_o}{WR_o}} \tag{14.61}$$

where

I_o = moment of inertia of tractor plus sling about pivot O

W = weight of tractor plus sling

R_o = distance between pivot O and the center of gravity of the tractor plus sling

Measurements of the distance R_o and the period of oscillation allow the calculation of I_o.

Often the weight and moment of inertia of the sling can be neglected with respect to the corresponding quantities for the tractor. If so, the moment of inertia of the tractor about its center of gravity, I_t, can be computed using the parallel axis theorem:

$$I_t = I_o - m_t R_o^2 \tag{14.62}$$

For accurate measurement of I_t, the distance R_o should be made as small as practical.

The yaw moment of inertia may be measured using a trifilar or quadrifilar pendulum. For such a measurement, the tractor is supported either directly or on a platform by three or four vertical cables. By knowing the length of the cables and measuring the period for small oscillations of the tractor about the vertical, the yaw moment of inertia may be calculated.

By measuring the moments of inertia with the tractor tilted instead of level, it is possible to calculate the products of inertia. In many cases, the tractor will be very

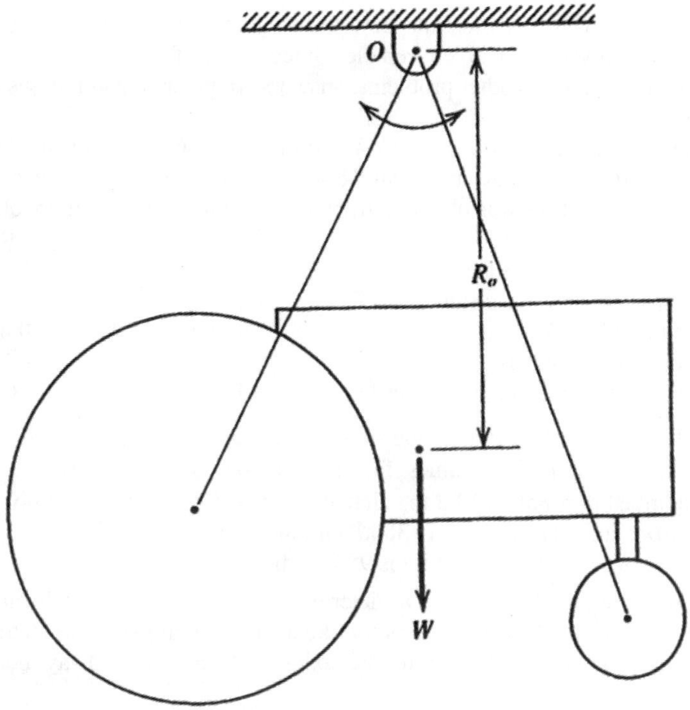

Figure 14.14. Determination of the pitch moment of inertia using the pendulum method.

nearly symmetric about the xz plane. In such a case, the only nonzero product of inertia will be that associated with the x and z axes.

14.12 Summary

The basic principles of statics and dynamics provide the basis for an analysis of the tractive performance, ride, handling, and stability of an off-road vehicle. Because of the need to negotiate sloping terrain and/or pull draft loads, tractive performance is important for both mobility and power delivery efficiency. Ride vibration, especially if the vehicle is unsuspended, may limit vehicle speed and hence productivity. Vehicle maneuverability is important to enhance productivity and negotiate terrain obstacles. The provision of adequate ground clearance can result in higher vehicle center of gravity locations with resulting lowered slope and transport stability.

Homework Problems

The following problems illustrate how the subject matter of the chapter may be applied to the analysis of a given vehicle. Since results from one problem may be needed as input data to the other problems, answers are given in parentheses at the end of each problem.

The vehicle to be analyzed is a 120 kW farm tractor equipped with single 14.9R30 radial tires on the front axle and dual 18.4R42 radial tires on the rear axle. The following data on the tires was obtained from a tire manufacturer's handbook:

	14.9R30	18.4R42
Static Loaded Radius (mm)	632.0	846.0
Rolling Circumference (mm)	4251.0	5601.0
Rated Load (kN)	20.789	30.889
Section Width (mm)	386.0	475.0
Overall Unloaded Diameter (mm)	1417.0	1862.0

Except for Problem 14.1 (where the static loaded radius values should be used), use the rolling radius numerical values for the symbols r_f and r_r. Use the empirical relationship given in Problem 13.3 to calculate the tire deflection δ_i for a given vertical load. Both front and rear tires use a 138 kPa inflation pressure.

Assume the acceleration of gravity is 9806 mm/sec^2.

14.1 The weighing method is used to determine the center of gravity location of the tractor. Because of tractor symmetry, the center of gravity is known to lie in the vertical plane perpendicular to the axles and passing midway between the wheels.

(a) With the tractor level, the weights supported by the rear (R_r) and front (R_f) axles are measured. The following values are recorded: R_r = 55.2177 kN, R_f = 44.6858 kN. The wheelbase of the tractor, L, is 2675.0 mm. Find the longitudinal location, ℓ, of the center of gravity of the tractor. (ℓ = 1196.5 mm)

(b) With the center of the front wheels raised to n = 1600 mm above the level surface used in (a), the weight supported by the rear axle, R'_r, is measured as 59.1881 kN. Find the height h of the center of gravity of the tractor above the rear axle center. (h = 191.3 mm)

14.2 To determine the pitch moment of inertia of the tractor about its center of gravity, the tractor is supported as shown in Figure 14.14. The lengths of the supporting cables connecting the axles to the pivot are adjusted so that the tractor is level in the equilibrium position. Because the distance R_o between the pivot and the tractor center of gravity is difficult to measure directly, the vertical distance between the pivot and the rear axle center is measured and found to be 2200 mm. The tractor is set into oscillation and the elapsed time for several cycles recorded. The average period is determined as 3.694 seconds. Assuming the weight and moment of inertia of the supporting cables are negligible, find the pitch moment of inertia I_{yyt} of the tractor about its center of gravity.

(I_{yyt} = 28,256 kg m^2)

14.3 One of the four rear wheels of the tractor is removed and found to weigh 3.510 kN. Because of its symmetry, the center of gravity of the wheel is assumed to be coincident with the center of the wheel. The moment of inertia of the wheel about a transverse axis through its center of gravity is measured using the pendulum method and found to be 120.1 kg m^2.

(a) Show analytically that the center of gravity of the chassis lies on the extension of the line connecting the center of the rear axle and the center of gravity of the tractor. That is, show $\theta_{1c} = \theta_{1t}$. *Hint*: Recall that the center of gravity of a body made of several parts can be found if the center of gravity locations and weights of the individual parts are known. The tractor is such a body composed of the chassis and rear wheels.

(b) Find the weight of the chassis (W_c), the location of the center of gravity of the chassis ($h_{1c} \cos \theta_{1c}$ and $h_{1c} \sin \theta_{1c}$), and the moment of inertia (I_{yyc}) of the chassis about its center of gravity. *Hint*: Recall that, by using the parallel axis theorem, the moment of inertia of a body made of several parts can be found if the moments of inertia of the parts about their centers of gravity are known in addition to the locations of the individual part centers of gravity relative to the center of gravity of the complete body.

(W_c = 85.8635 kN; $h_{1c} \cos \theta_{1c}$ = 1392.2 mm; $h_{1c} \sin \theta_{1c}$ = 222.6 mm; I_{yyc} = 25,329.7 kg m^2)

14.4 (a) Determine the steady-state drawbar pull, $P_s \cos \alpha$, required to just lift the front wheels of the tractor off the ground ($R_f = 0$) when the tractor is operating on level ground. The drawbar force P applied to the tractor is inclined downward from the horizontal at an angle α of 15°. Assume $TF_r/R_r = 0.0325$ in the relation $e_r = (TF_r/R_r)r_r$. The drawbar force is applied 907 mm behind and 337 mm below the rear axle center. ($P_s \cos \alpha$ = 144.864 kN)

(b) If the gross tractive coefficient μ_g has the form

$$\mu_g = F/R = 0.88\,(1 - e^{-0.1B_n})(1 - e^{-9.5s}) + 0.0325,$$

show the tractor cannot develop the tractive force required for developing this drawbar pull.

14.5 Suppose the drawbar hitch point used in Problem 14.4 is raised by 300 mm above the standard location.

(a) What drawbar pull is now required to just lift the front wheels off the ground when the tractor is operating on level ground? The inclination of the drawbar force with the horizontal is maintained at 15°. ($P_s \cos \alpha$ = 105.540 kN)

(b) Show that sufficient traction is available to pull this load and calculate the resulting slippage of the drive wheels. The tractor is operating on a firm soil for which the cone index CI is 1750 kN/m^2. Remember that the traction and motion resistance equations apply to single wheels. (slip = 0.293 = 29.3%)

(c) The linearized full throttle torque (T_e) – speed ($\dot{\phi}_e$) relation for the tractor engine can be represented by the following equations where the engine torque T_e is in Nm and the engine speed $\dot{\phi}_e$ in revolutions per minute (rpm):

$$T_e = 2.0\,(2400 - \dot{\phi}_e) \qquad 2150 \text{ rpm} \le \dot{\phi}_e \le 2400 \text{ rpm},\ T_e \le 500 \text{ Nm}$$

$$T_e = 500 + 0.4\,(2150 - \dot{\phi}_e) \qquad \dot{\phi}_e \le 2150 \text{ rpm},\ T_e \ge 500 \text{ Nm}$$

The tractor has fifteen forward speeds for which the gear ratios, G_r, are (the subscript indicates the gear):

$G_1 = 316.55$	$G_5 = 126.36$	$G_9 = 73.02$	$G_{13} = 40.34$
$G_2 = 221.11$	$G_6 = 111.34$	$G_{10} = 64.35$	$G_{14} = 28.84$
$G_3 = 182.88$	$G_7 = 96.66$	$G_{11} = 55.85$	$G_{15} = 23.31$
$G_4 = 145.60$	$G_8 = 84.13$	$G_{12} = 49.92$	

The efficiency of power transmission, η, between the engine and the axles is assumed to be independent of the gear used and equal to 0.85. Show that sufficient engine torque is available to pull the load in third gear and calculate the resulting engine and axle speeds. ($\dot{\phi}_e = 1827.2$ rpm, $\dot{\phi}_{wr} = 1.046$ rad/s)

(d) Using the drive wheel slip calculated in (b), what is the forward speed of the tractor? ($\dot{x}_w = 0.659$ m/s).

(e) What are the engine, axle, and drawbar powers?

(engine: 120.38 kW, axle: 102.32 kW, drawbar: 69.60 kW)

(f) What is the rear axle tractive efficiency? (0.680 = 68.0%)

(g) What is the power delivery efficiency? (0.578 = 57.8%)

14.6 (a) Assume the tractor is operating in two-wheel-drive mode and is exerting a drawbar pull P cos α of 30 kN while operating in seventh gear on a level surface. The drawbar force P is inclined downward at an angle α of 15° to the horizontal and is applied at the standard drawbar location. The tractor is operating on a firm soil with a cone index of 1750 kN/m^2. Assume

$$TF/R = \left(\frac{0.9}{B_n} + 0.0325 + \frac{0.5s}{\sqrt{B_n}}\right)$$

in the relations $e_r = (TF_r/R_r)r_r$ and $e_f = (TF_f/R_f)r_f$. Assume the cone index for the soil encountered by the rear wheels is the same as for the soil encountered by the front wheels. Determine the resulting forces, the drive wheel slip, and the resulting forward speed of the tractor. Note an iterative equation solving technique will be required to solve for the forces.

($R_f = 34.178$ kN, $R_r = 73.762$ kN, $TF_f = 1.590$ kN, $TF_r = 3.488$ kN, $F_r = 35.078$ kN, $s_r = .074 = 7.4\%$, $\dot{x}_w = 1.977$ m/s)

(b) If the tractor is operating under the same conditions as in (a) but on a slope of 10°, what is the resulting forward speed of the tractor? ($\dot{x}_w = 1.657$ m/s).

14.7 Assume the front wheels of the tractor are powered and that the now four-wheel-drive tractor is operating under the same conditions as those of Problem 14.6(a). The ratio $G_f / G_r = 0.759$ for all the transmission gears. (If you were setting this ratio, what would be a logical way to determine it?) Determine the resulting forces, the drive wheel slippages and the resulting forward speed of the tractor.

($R_f = 34.197$ kN, $R_r = 73.743$ kN, $TF_f = 1.682$ kN, $TF_r = 3.361$ kN, $F_f = 11.088$ kN, $F_r = 23.955$ kN, $s_f = 0.0425 = 4.25\%$, $s_r = 0.0425 = 4.25\%$, $\dot{x}_w = 2.043$ m/s)

14.8 The tractor has become immobilized while traveling under no drawbar load on soft ground. The rear wheels have dug into the ground to such an extent that the tractor is on an effective slope β of 15°. In an effort to free the tractor, the operator chains a post in front of the rear wheels and releases the clutch with the transmission in first gear. With the rear wheels restrained from moving ($\ddot{x}_w = \ddot{z}_w = \ddot{\phi}_w = 0$), the tractor begins to tip backward. When the angle of rotation θ reaches 20°, the operator depresses the clutch. At this time, the angular velocity $\dot{\theta}$ is 1.6 rad/s. Is the tractor in danger of overturning?

(Since the angular velocity, $\dot{\theta}_s$, required to bring the tractor to the point of static instability is 1.31 rad/s, the tractor is in danger of overturning.)

14.9 (a) Use the load-deflection relation given in Problem 13.3 to estimate the vertical spring rates of the tires at their static deflections. Estimate the natural frequencies and illustrate the modes of vibration of the tractor considering it as a linear two-degree-of-freedom system. ($K_f = 353$ kN/m, $K_r = 379$ kN/m. Natural frequencies are 1.68 and 2.45 Hz. The corresponding mode shapes are shown in Figure 14.15 where the dashed lines represent the undisturbed tractor.)

(b) Neglecting slip, at what forward speeds will a once per revolution out-of-roundness of the tire-wheel assemblies excite the two natural frequencies? (front: 7.13 and 10.42 m/s, rear: 9.40 and 13.73 m/s)

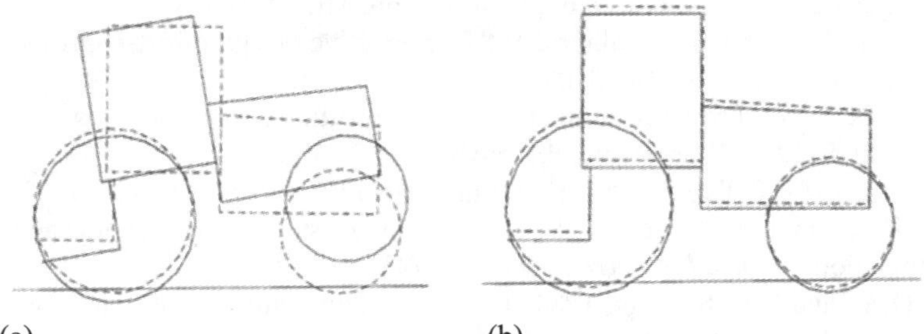

(a) (b)

Figure 14.15. Modes of vibration corresponding to the (a) 1.68 Hz and (b) 2.45 Hz natural frequencies of the tractor. The dashed outline represents the static equilibrium position of the tractor.

14.10 (a) If the cornering stiffnesses of a single front and a single rear tire are 176 and 158 kN/radian, respectively, determine the understeer coefficient.

(0.0396 radians)

(b) Will the tractor tend to understeer or oversteer? (understeer)

14.11 The tractor is put into a steady-state turn such that the center of gravity of the tractor traverses a circle of radius 7 m. Assume the tractor is of the tricycle type and has a front wheel tread setting of 0.5 m and a rear wheel tread setting of 2 m.

(a) If the forward speed of the tractor is 6.2 m/s, is the tractor in danger of turning over sideways? Assume the tractor is traveling on a firm surface on which the tires can develop the lateral forces required by the specified turn.

(No. $u_s = 6.49$ m/s)

(b) A front-end loader weighing 4000 N is attached to the tractor. A 5000 N load is to be carried in the loader bucket while the bucket is raised to its maximum height. In this position, the center of gravity of the loader plus the bucket load is estimated to be 2 m ahead of and 1.75 m above the rear axle. Is the loader-equipped tractor in danger of overturning in the turn described in (a)?

(Yes. $u_s = 6.05$ m/s)

References and Suggested Readings

Bashford, L.L., G.R. Woerman and G.J. Shropshire. 1985. Front wheel assist tractor performance in two and four-wheel drive modes. *Transactions of the ASAE* 28(1): 23-29.

Chisholm, C.J. 1979. A mathematical model of tractor overturning and impact behavior. *Journal of Agricultural Engineering Research* 24: 375-394.

Claar II, P.W., W.F. Buchele and P.N. Sheth. 1980. Off-road vehicle ride: review of concepts and design evaluation with computer simulation. SAE Paper No. 801023. Warrendale, PA: SAE.

Clay, R., and P. Hemingway. 2001. Engineering tractors for higher speeds. ASAE Distinguished Lecture Series No. 25. St. Joseph, MI: ASAE.

Coddington, R.C., and G.P. Kalhorn. 1987. Interactive design of off-highway vehicle suspensions. SAE Paper No. 870960. Warrendale, PA: SAE.

Cornell, C.R. 1981. Dynamic simulation of a hydrostatically propelled vehicle. SAE Paper No. 811253. Warrendale, PA: SAE.

Crolla, D.A. 1981. Off-road vehicle dynamics. *Vehicle System Dynamics* 10: 253-266.

Crolla, D.A., and F.D. Hales. 1979. The lateral stability of tractor and trailer combinations. *Journal of Terramechanics* 16(1): 1-22.

Crolla, D.A., and H.B. Spencer. 1984. Tractor handling during control loss on sloping ground. *Vehicle System Dynamics* 13: 1-17.

Davis, D.C., and G.E. Rehkugler. 1974. Agricultural wheel-tractor overturns–Part I: Mathematical model. Part II: Mathematical model verification by scale-model study. *Transactions of the ASAE* 17(3): 477-488, 492.

Dufner, D.L., and T.E. Schick. 2002. New level of seat performance. *Resource* 9(12): 7-8.

Dwyer, M.J. 1970. The braking performance of tractor-trailer combinations. *Journal of Agricultural Engineering Research* 15(2): 148-162.

Dwyer, M.J., and G. Pearson. 1976. A field comparison of the tractive performance of two and four-wheel-drive tractors. *Journal of Agricultural Engineering Research* 21(1): 77-85.

Gee-Clough, D., M. McAllister, G. Pearson and D.W. Everden. 1978. The empirical prediction of tractor-implement field performance. *Journal of Terramechanics* 15(2): 81-94.

Gerke, F.G., and D.L. Hoag. 1981. Tractor vibrations at the operator's station. *Transactions of the ASAE* 24(5): 1131-1134.

Gillespie, T.D. 1992. *Fundamentals of Vehicle Dynamics.* SAE Publication R-114. Warrendale, PA: SAE.

Goering, C.E., and W.F. Buchele. 1967. Computer simulation of an unsprung vehicle. *Transactions of the ASAE* 10(2): 272-280.

Göhlich, H. 1991. Tractor ride dynamics. In *Progress in Agricultural Physics and Engineering,* Chapter 3. New York, NY: C.A.B. International.

Hilton, D.J., and P. Moran. 1975. Experiments in improving tractor operator ride by means of a cab suspension. *Journal of Agricultural Engineering Research* 20: 433-448.

Horton, D.N.L., and D.A. Crolla. 1984. The handling behavior of off-road vehicles. *International Journal of Vehicle Design* 5(1-2): 197-218.

Hunter, A.G.M. 1991. Stability of agricultural machinery on slopes. In *Progress in Agricultural Physics and Engineering,* Chapter 4. New York, NY: C.A.B. International.

Itoh, H., A. Oida and M. Yamazaki. 1999. Numerical simulation of a 4WD-4WS tractor turning in a rice field. *Journal of Terramechanics* 36(2): 91-115.

Kim, K.W., and G.E. Rehkugler. 1987. A review of tractor dynamics and stability. *Transactions of the ASAE* 30(3): 615-623.

Koch, J.A., W.F. Buchele and S.J. Marley. 1970. Verification of a mathematical model to predict tractor tipping behavior. *Transactions of the ASAE* 13(1): 67-72, 76.

Larson, D.L., D.W. Smith and J.B. Liljedahl. 1976. The dynamics of three-dimensional tractor motion. *Transactions of the ASAE* 19(1): 195-200.

Lines, J., M. Stiles and R. Whyte. 1995. Whole body vibration during tractor driving. *Journal of Low Frequency Noise and Vibration* 14(2):87-104.

Liu, J., and P.D. Ayers. 1999. Off-road vehicle rollover and field testing of stability index. *Journal of Agricultural Safety and Health* 5(1): 59-72.

Lohmann, W.E., and P.D. Fotsch. 1994. Flexible body dynamic simulation of a large mining truck. SAE Paper No. 941117. Warrendale, PA: SAE.

MacInnis, D.D., W.E. Cliff and K.W. Ising. 1997. A comparison of moment of inertia estimation techniques for vehicle dynamics simulation. SAE Paper No. 970951. Warrendale, PA: SAE.

Macmillan, R. H. 2003. The Mechanics of Tractor-Implement Performance: Theory and Worked Examples. University of Melbourne ePrints Repository (UMER). The book can be downloaded free of charge from UMER at http://eprints.unimelb.edu.au/archive/00000204 (accessed 30 May 2003).

Matthews, J. 1973. The measurement of tractor ride comfort. SAE Paper No. 730795. Warrendale, PA: SAE.

McConville, J.B., and J.C. Angell. 1984. The dynamic simulation of a moving vehicle subject to transient steering inputs using the ADAMS computer program. ASME Paper No. 84-DET-2. New York, NY: ASME.

McCormick, E. 1941. Some engineering implications of high speed farming. *Agricultural Engineering* 22(5): 165-167.

McKibben, E.G. 1927. The kinematics and dynamics of the wheel type farm tractor. *Agricultural Engineering* 8(1-7): 15-16, 39-40, 43, 58-60, 90-93, 119-122,155-160,187-189.

Mitchell, B.W., G.L. Zachariah and J.B. Liljedahl. 1972. Prediction and control of tractor stability to prevent rearward overturning. *Transactions of the ASAE* 15(5): 838-844.

Owen, R.H., and J.E. Bernard. 1982. Directional dynamics of a tractor-loader-backhoe. *Vehicle System Dynamics* 11:251-265.

Pershing, R.L. 1971. Simulating tractive performance. SAE Paper No. 710525. Warrendale, PA: SAE.

Pershing, R.L., and R.R. Yoerger. 1964. Steady state behavior of tractors on roadside slopes. SAE Paper No. 903B. Warrendale, PA: SAE.

Pershing, R.L., and R.R. Yoerger. 1969. Simulation of tractors for transient response. *Transactions of the ASAE* 12(5): 715-719.

Plaskitt, R.J., and C.J. Musiol. 2002. Developing a durable product. ASAE Distinguished Lecture Series No. 26. St. Joseph, MI: ASAE.

Queiroz, D.M., H.G. Gibson, G.W. Krutz, K. Haghighi, J.M. Starkey and G.E. Miles. 1995. Steering performance analysis of rubber tracked tractors using mechanical system simulation. ASAE Paper No. 951324. St. Joseph, MI: ASAE.

Rackman, D.H., and D.P. Blight. 1985. Four-wheel-drive tractors – a review. *Journal of Agricultural Engineering Research* 31: 185-201.

Raney, J.P., J.B. Liljedahl and R. Cohen. 1961. The dynamic behavior of farm tractors. *Transactions of the ASAE* 4(2): 215-218, 221.

Rehkugler, G.E. 1982. Tractor steering dynamics–simulated and measured. *Transactions of the ASAE* 25(6): 1512-1515.

Rummer, R.B. 1989. Computer simulation of ride dynamics for the design of forest machines. *Transactions of the ASAE* 32(1): 12-16.

Sack, H.W. 1956. Longitudinal stability of tractors. *Agricultural Engineering* 37(5): 328-333.

Schwanghart, H. 1978. Tipping and overturning behavior of a tractor. *Landtechnik* 33(11): 488-492.

Slättengren, J. 2000. Utilization of ADAMS to predict tracked vehicle performance. SAE Paper No. 2000-01-0303. Warrendale, PA: SAE.

Smith, D.W. 1977. Computer simulation of tractor ride for design evaluation. SAE Paper No. 770704. Warrendale, PA: SAE.

Smith, D.W. 1984. The influence of drawbar position on tractor rearward stability. ASAE Paper No. 84-1560. St. Joseph, MI: ASAE.

Smith, D.W. 1986. The influence of fore-aft center of gravity location on the slope performance of two-wheel-drive vehicles. ASAE Paper No. 861572. St. Joseph, MI: ASAE.

Smith, D.W., and J.B. Liljedahl. 1972. Simulation of rearward overturning of farm tractors. *Transactions of the ASAE* 15(5): 818-821.

Smith, D.W., J.V. Perumpral and J.B. Liljedahl. 1974. The kinematics of tractor sideways overturning. *Transactions of the ASAE* 17(1): 1-3.

SAE. 1981. Vehicle dynamics terminology. SAE Recommended Practice J670e. Warrendale, PA: SAE.

SAE. 1986. Classification of agricultural wheeled tractors for vibration tests of operator seats. Recommended Practice J1386. Warrendale, PA: SAE.

Song, A., B.K. Huang and H.D. Bowen. 1989. Simulating a powered model wheel-tractor on soft ground. *Transactions of the ASAE* 32(1): 2-11.

Spencer, H.B. 1978. Stability and control of two-wheel-drive tractors and machinery on sloping ground. *Journal of Agricultural Engineering Research* 23: 169-188.

Stayner, R.M., T.S. Collins and J.A. Lines. 1984. Tractor ride vibration simulation as an aid to design. *Journal of Agricultural Engineering Research* 29: 345-355.

Steinbruegge, G.W. 1969. Improved methods of locating centers of gravity. *Transactions of the ASAE* 12(5): 681-684, 689.

Stikeleather, L.F. 1981. Operator seats for agricultural equipment. ASAE Distinguished Lecture Series No. 7. St. Joseph, MI: ASAE.

Unruh, D.H. 1971. Determination of wheel loader static and dynamic stability. SAE Paper No. 710526. Warrendale, PA: SAE.

Wiley, J.C., B.E. Romig, L.V. Anderson and F.M. Zoz. 1992. Optimizing dynamic stability and performance of tractors with radial tires. ASAE Paper No. 921586. St. Joseph, MI: ASAE.

Winkler, C.B. 1973. Measurement of inertial properties and suspension parameters of heavy highway vehicles. SAE Paper No. 730182. Warrendale, PA: SAE.

Wong, J.Y. 2001. *Theory of Ground Vehicles.* New York, NY: John Wiley and Sons.

Worthington, W.H. 1949. Evaluation of factors affecting the operating stability of wheel tractors. *Agricultural Engineering* 30(3-4): 119-123, 179-183.

Zoz, F.M., and R.D. Grisso. 2003. Traction and tractor performance. ASAE Distinguished Lecture Series No. 27. St. Joseph, MI: ASAE.

CHAPTER 15

HUMAN FACTORS AND SAFETY

15.1 Introduction

Design of off-road vehicles should include an understanding of the operator-machine interface, knowledge of human factors engineering, and integration of safety into the design. When these factors are properly incorporated in design, the operator can perform complex tasks with efficiency, safety, and a minimum of fatigue. In general, *human factors* include such items as *riding comfort*, *visibility*, location and arrangement of *controls*, ease of operating controls, *thermal comfort*, and *noise control*. A typical workspace control center for an off-road vehicle is shown in Figure 15.1.

Safety in the overall design of an off-road vehicle is a major consideration by industry. Design teams and individual designers employ the concept of "Safety through Design" (Ahlschwede and Klotzbach, 2000), which is "the integration of hazard analysis and risk assessment methods early in the design and engineering stages and the taking of actions necessary so that the risks of injury or damage are at an acceptable level." This concept is based on the principles of

- Avoiding, eliminating, or reducing the probability of a hazard related incident occurring;
- Minimizing the severity of harm or damage should an incident occur; and
- This is most effectively performed when designing new equipment.

The Safety through Design Process involves five steps:

1. Review design proposal (scope)
2. Identify known safety requirements; e.g. standards, regulations, standard safety features

Figure 15.1. Workspace control center for an off-road vehicle. Reprinted with permission from Deere and Company.)

3. Review the design concept, schematic and logic.
   (a) Identify potential hazards assuming perfect parts (revise concept as appropriate)
   (b) Identify potentially hazardous failure modes
   (c) Risk assessment
   (d) Hazard treatment
4. Hardware review
   (a) Detail design hazard analysis
   (b) Manufacturing process hazard analysis
   (c) Risk assessment
   (d) Hazard treatment
5. Tracking and verification

Designing for safety is not a simple process. In addition to the above, engineers must be knowledgeable in vehicle function, engineering standards, engineering practices, human factors principles, epidemiological data, user feedback, government regulations, laws, and court decisions.

This chapter provides an introduction to principles of human factors that are essential in the design of an off-road vehicle operator's workspace. The reader interested in further exploration is directed to the suggested readings provided at the end of the chapter.

15.2 The Operator-Machine Interface

When a person operates an off-road vehicle, the sensing, decision-making, and muscular powers of the operator are joined to an engineering system. The operator uses hearing, sight, and touch to interpret inputs and to interface with the control-instrumentation components to achieve the desired output of the tractor. Human engineering data defining acceptable ranges for environmental factors, noise, and vibration must be incorporated in the design to ensure operator safety and comfort. This is a challenging task because many gaps exist in human engineering data. Although significant progress has been made over the years, additional research, testing, and experience for off-road vehicles is needed to develop more useful and reliable levels of design technology.

Regarding design of the operator's workplace, Hansson et al. (1970) stated:

> Safety, comfort and convenience should be considered in the design, location and construction of the operator's work place. The work place should be located on the machine so that visibility in the driving position is good without requiring the operator to work in an awkward, tiring position. Levers, pedals and instruments should be conveniently and logically located and the work place should fit both tall and short operators. In addition, the operator should be able to change his working position easily and the work area should be free of sharp edges and obstructions such as transmission cases.

Important tools of the designer are engineering standards, engineering practices, recommendations, and engineering data. Such information is published by the American Society of Agricultural Engineers (ASAE), and is found in the *ASAE Standards,* which is published annually. The Society of Automotive Engineers (SAE) also has a number of standards for agricultural tractors and other off-highway vehicles. Several of the standards are published jointly by ASAE and SAE. The International Standards Organization (ISO), too, has standards relative to agricultural tractors. Current standards are cited in this text, but the reader is advised to consult the most recent publications for revisions and additions.

The reader may want to consult *Human Factors Design Handbook* by Woodson (1992), *Human Factors in Engineering and Design* by Sanders and McCormick (1987). In addition, the *Journal of Applied Ergonomics* is an excellent source of published research.

15.3 Operator Exposure to Environmental Factors

Off-road vehicles are used under varied geographic and climatic conditions. Direct exposure to temperature, humidity, wind, thermal radiation, dust, and chemicals is encountered. Table 15.1 defines comfort and bearable zones for four of these parameters as they apply to humans. Temperature, humidity, ventilation, and thermal radiation zones are interrelated. Permissible guideline limits for dust concentrations on

Table 15.1. Environmental zones for selected parameters.

Environmental Parameter	Comfort Zone		Bearable Zone	
	Lower Limit	Upper Limit	Lower Limit	Upper Limit
Temperature, °C	18	24	-1	38
Humidity, % RH	30	70	10	90
Ventilation, m^3/min	00.37	00.57	00.14	00.14
Ultraviolet radiation	Unknown		Unknown	

Source: Dreyfus, 1960, p. 159.

combines in the Netherlands is 15 mg/m^3 (Zander, 1972). Noren (1985) reported peak dust concentrations of 577 mg/m^3 occurring during soil tillage operations. Mean values were 146 mg/m^3. Tractor cabs that were pressurized at 50 Pa and fitted with air filters of "fine" quality reduced the dust concentration to a mean value of 24.7 mg/m^3 in the operator's breathing zone.

Greater use of pesticides in agriculture and forestry and increased human protection requirements suggest the importance of a tractor cab as a protective device. Akesson et al. (1974) measured the health hazards to workers applying certain chemicals commonly used in California agriculture. In pesticide application operations, tractor operators should use personal protective equipment that meets EPA requirements. An option is an enclosed cab that provides respiratory and skin protection designed to meet ANSI/ASAE S525-1.1 and ANSI/ASAE S525-2 (ANSI/ASAE, 1999) standards for environmental air quality.

Design of an operator enclosure for a tractor should include thermal comfort, ride comfort, noise protection, air quality, and rollover protection.

15.4 Spatial, Visual, and Control Requirements of the Operator

Functional considerations in design of the workplace for an off-road vehicle operator are visibility, easy access to controls, and clearance. Easy access to all controls in the workplace and cab are necessary so the operator can achieve the desired machine performance and productivity. Primary visibility requires that the operator can look outside the cab in any direction, both near and far. Secondary visibility is needed to monitor instruments and controls in the workplace and cab. Clearance at various locations is necessary to provide access to and from the workplace. Proper workplace dimensions in relation to the seat are important for ease in grasping and operating controls. These factors are related to the operator's anthropometric and biomechanical characteristics.

15.4.1 Anthropometric Data

Anthropometric or human physical data are used to properly design the operator's workplace to meet visibility and clearance requirements. Human physical data for adults obtained from SAE Recommended Practice J833 MAY89 (SAE 1999) is shown Figures 15.2 and 15.3. Table 15.2 shows selected human physical data for adults,

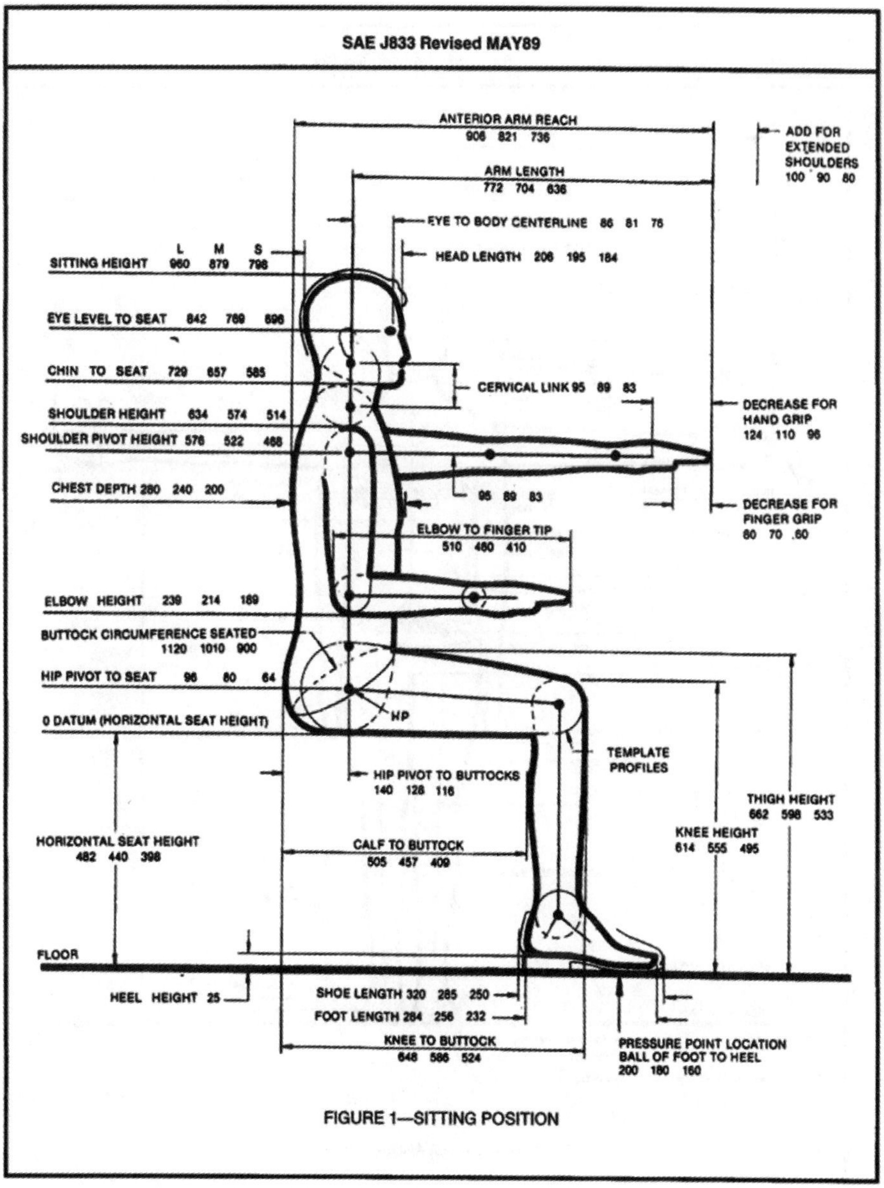

Figure 15.2. Human dimensions for sitting position. (Reprinted with permission from SAE J833 MAY89 © 1989 Society of Automotive Engineers, Inc.)

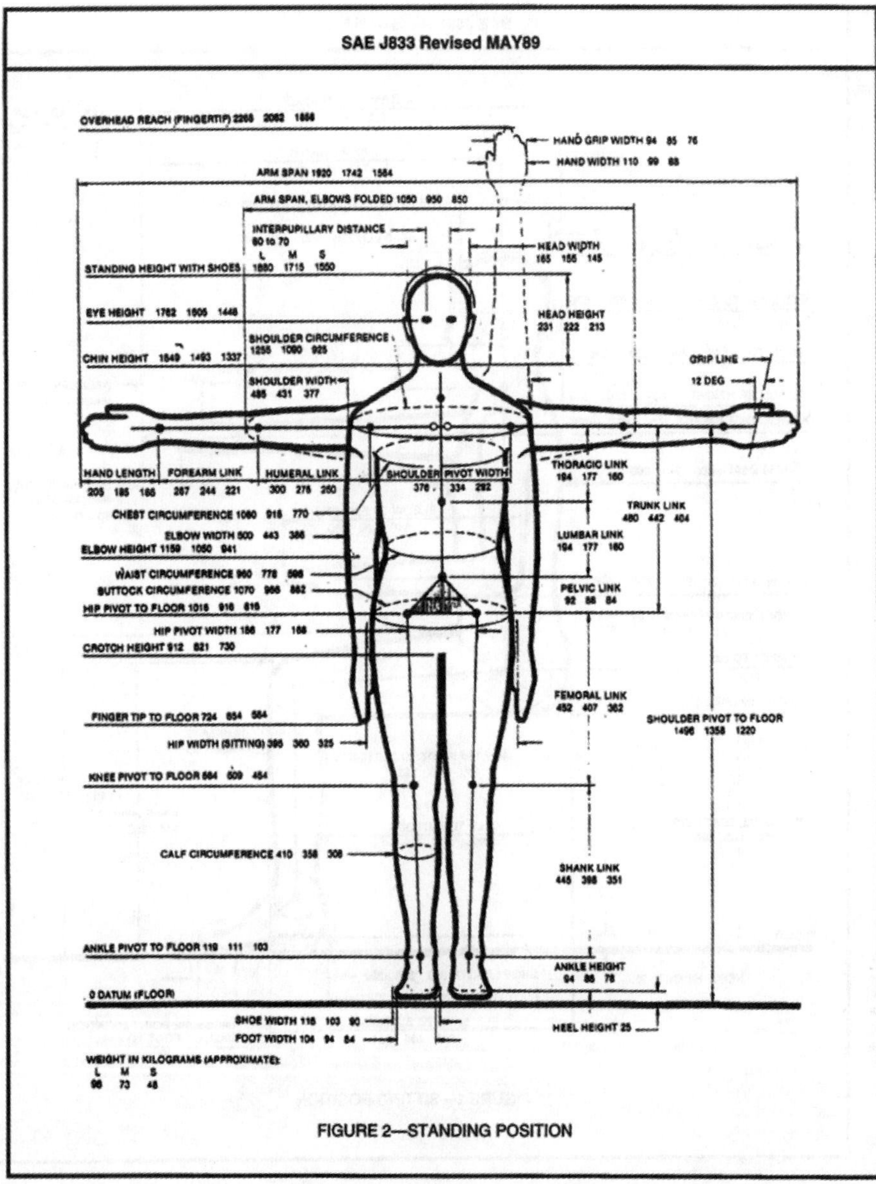

Figure 15.3. Human dimensions for standing position. (Reprinted with permission from SAE J833 MAY89 © 1989 Society of Automotive Engineers, Inc.)

Table 15.2. Selected human physical dimensions for adults.

Body Measurement	Dimension, mm		
	5th Percentile	50th Percentile	95th Percentile
Standing height with shoes	1550	1715	1880
Sitting height, normal	798	879	960
Knee height sitting	495	555	614
Horizontal seat height	398	440	482
Elbow sitting height	189	214	239
Thigh height sitting	533	598	662
Knee to buttock length	524	586	648
Buttock to calf length	409	457	505
Elbow-to-elbow width	386	443	500
Hip width sitting	325	360	395
Hip pivot to horizontal seat height	64	80	96
Hip pivot to floor (leg length)	816	916	1016
Arm span	1564	1742	1920
Hip pivot width	168	177	188
Shoe width	90	103	116
Foot width	84	94	104
Weight, kg	48	73	98

Source: SAE Standard J833 MAY89 (SAE, 1999)

Notes:

1. 5th percentile data is for female population.
2. 95th percentile data is for male population.
3. African descents can have 2% longer arms and 4% longer legs than dimension provided.
4. Oriental descents can have 7% shorter arms and 10% shorter legs than dimensions provided.

extracted from the figures. It is general practice to use the 95th and 5th percentile values in workplace design. A minimum dimension in some cases should be based on an upper percentile value, such as establishing clearance in doors and escape hatches. On the other hand, maximum dimensions are established from lower percentiles to ensure that a short functional arm can reach a control or steering wheel.

15.4.2 Operator Space Envelope and Access

The *operator space envelope* dimensions are defined in SAE Standard J154 JUN92 "Operator Space Envelope Dimensions for Off-Road Machines" (SAE, 1999). The purpose is to define minimum normal operating space around a clothed operator for seated and standing situations. Figure 15.4 shows the minimum operator space envelope dimensions for sitting in an enclosure. Figure 15.5 shows the minimum operator space envelope for standing in an enclosure. SAE Recommended Practice J185 JUN88 "Access Systems for Off-Road Machines" (SAE, 1999) establishes criteria for access systems primarily to aid in minimizing accidents and injury to personnel getting on, off, or moving about while servicing or preparing to operate off-road machines. Data for steps, handrails, and handholds are included.

SAE J154 Revised JUN92

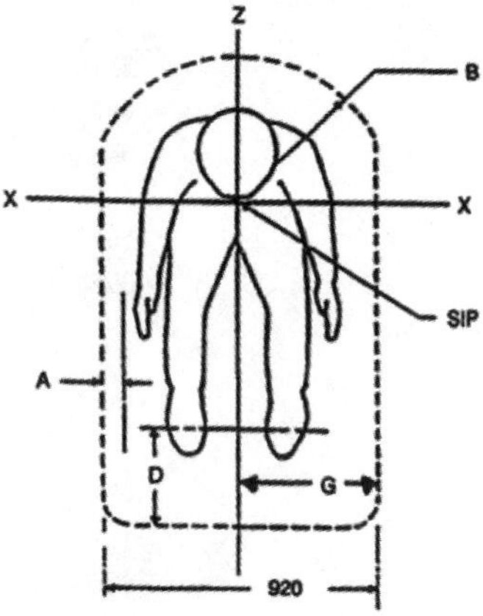

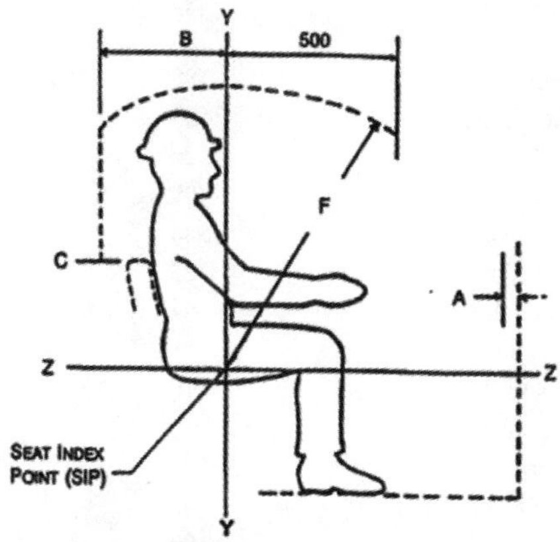

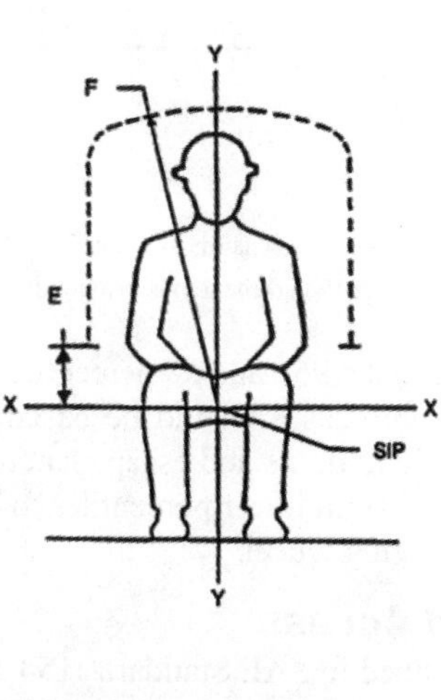

Minimum Operator Enclosure Dimensions

A = 50 mm clearance between enclosure and controls (add 25 mm for cold weather clothes).
B = 400 mm plus half of the horizontal seat adjustment dimension. See paragraph 3.3.4.
C = Terminus of rearward clearance is horizontal plane tangent to top of seat back.
D = 150 mm clearance from enclosure to centerline of the foot control. 400 mm clearance from heel rest for treadle pedals. 50 mm clearance between shoe and enclosure for suspended pedals.
E = Enclosure minimum width requirement terminates at a horizontal plane 150 mm above SIP or at contact with the undeflected top surface of the arm rest, if provided.
F = 1050 mm. See paragraphs 3.3.1.
G = 460 mm. See paragraphs 3.3.2 and 3.3.3.

Notes:

1. Dimensions (in mm) are minimum requirements.
2. Controls are positioned in closest proximity to enclosure when determining minimum control clearances.
3. Maximum radius at the intersection of enclosure's internal walls with each other and with enclosure ceiling is 250 mm.
4. Seat Index Point is considered equivalent to the H-Point as defined in SAE J1163.
5. Outline is not intended to imply the shape of the enclosure.
6. The minimum operator space envelope dimensions are referenced from the SIP even when the seat is offset or angled with respect to the enclosure centerline.

X-X: Horizontal transverse axis through SIP
Y-Y: Vertical axis through SIP
Z-Z: Longitudinal axis through SIP

FIGURE 1—MINIMUM OPERATOR SPACE ENVELOPE DIMENSIONS FOR SITTING ENCLOSURE (95TH PERCENTILE OPERATOR, SEE SAE J833)

Figure 15.4. Minimum operator space envelope dimensions for sitting enclosure for the 95th percentile operator. See SAE J833. (Reprinted with permission from SAE J154 JUN92 © 1992 Society of Automotive Engineers, Inc.)

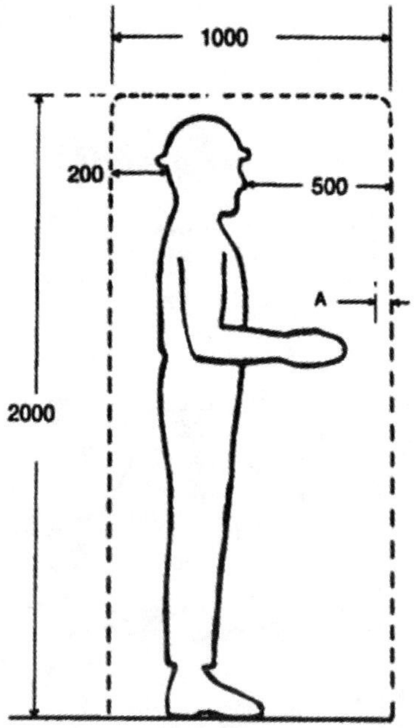

A—50 mm clearance between enclosure and controls (add 25 mm for cold weather clothes).

1. All notes for sit down enclosure (Figure 1) apply to standing enclosure.
2. Enclosure width dimension is same as sit down enclosure width (Figure 1).
3. Clearance between foot controls and standing enclosure is same as sit down enclosure clearance (Figure 1)

Figure 15.5. Minimum operator space envelope dimensions for standing enclosure for the 95th percentile operator. See SAE J833. (Reprinted with permission from SAE J154 JUN92 © 1992 Society of Automotive Engineers, Inc).

15.4.3 Operator Seating

Proper location of the operator seat in the workplace is essential for operator comfort and efficiency. In Figure 15.4 a key reference point, the *seat index point* (SIP) is identified. In addition to being a reference point for several minimum operator enclosure dimensions, it is also the key reference point for locating the operator seat in the workplace and cab. Figure 15.6, from SAE Standard J1163 AUG97 "Determining Seat Index Point" (SAE, 1999), shows a side view of a seat with location of the SIP. Dimension A is a key dimension and it is desirable to establish a range of values for A so that a vertical seat adjustment can be incorporated in the operator seat suspension mechanism.

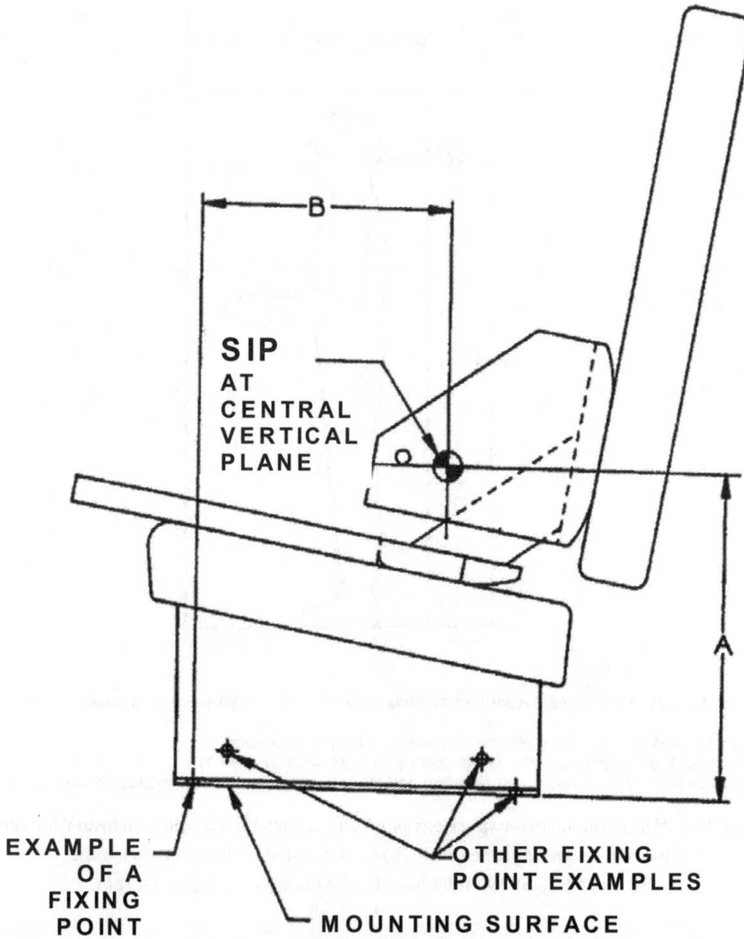

Figure 15.6. Seat index point dimensions. (Reprinted with permission from SAE J1163 AUG97 © 1997 Society of Automotive Engineers, Inc.)

SAE Standard J899 DEC88 "Operator's Seat Dimensions for Off-Road Self-Propelled Work Machines" (SAE, 1999), provides seat dimensions and adjustments for the design of the operator's seat. Figure 15.7 shows the operator's seat dimensions. All seat dimensions and adjustments are referenced to the SIP. Table 15.3 shows the seat dimensions and adjustments. Nominal values are included wherever possible and their use will generally provide maximum ergonomic benefit. The nominal values are based on the body dimensions of SAE J833 and will accommodate operator sizes from the 5th percentile female through the 95th percentile male of that document. Dimensions or adjustments other than those in the standard may be used if they will provide equivalent or greater ergonomic benefit. Consideration may be given for specific controls or operator requirements such as smaller worldwide population body dimensions.

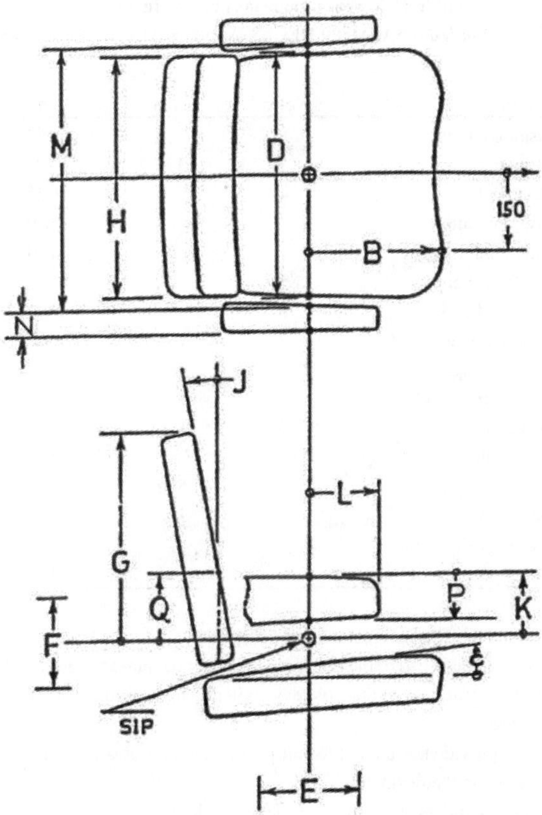

Figure 15.7. Operator's seat dimensions for off-road self-propelled work machines. See Table 15.3 for dimensions. (Reprinted with permission from SAE J899 DEC98 © 1988 Society of Automotive Engineers, Inc.)

Data in Table 15.3 are useful to properly design the operator's seat. Seat adjustments should provide a nominal 150 mm of fore and aft movement and a nominal 75 mm of vertical movement. The seat cushion supports the weight of the thighs and upper body. To insure comfort, the seat cushion angle from horizontal should be 5°, 10°, and 15° for minimum, nominal, and maximum, respectively. If an adjustment is provided, it should be ±3° and ±5°, respectively, for minimum and nominal. Likewise, for comfort, the back cushion angle from vertical should conform to the same angles as for the horizontal. Purcell (1980) suggests that the angle between the backrest and seat should be approximately 95°, with a minimum of 90°. The backrest must also provide lumbar support of 145, 130, and 115 mm for minimum, nominal, and maximum, respectively. Armrests are necessary to provide support and comfort to the shoulders. Also, they must be adjustable vertically to accommodate varying sizes of persons for freedom of movement of the forearm and elbow when operating hand controls.

Table 15.3. Seat dimensions and adjustment.
(Reprinted with permission from SAE J899 DEC88 © 1988 Society of Automotive Engineers, Inc.)

Item	Description	Notes	Maximum,[2] mm	Nominal,[1] mm	Minimum, mm
B	Seat cushion length		315	265	215
C	Seat cushion angle	3	15°	10°	5°
	Adjustment	4	–	± 5°	± 3°
D	Seat cushion width		–	500	430
E	Fore and aft adjustment	5	–	150	100
F	Vertical adjustment	5	–	75	60
G	Back cushion height	6	–	400	150
H	Back cushion width	7	–	500	300
J	Back cushion angle	8	15°	10°	5°
	Adjustment	4	–	± 5°	± 3°
K	Armrest height	9	160	140	95
L	Armrest length	190		140	90
M	Armrest lateral spacing	550	500	450	
N	Armrest width		–	75	50
P	Armrest depth		–	100	50
Q	Lumbar support	10	145	130	115

Notes:

1. A value which might have the broadest general acceptance.
2. Where no maximum value is listed, the nominal value may be increased arbitrarily.
3. Angle of the top of the seat base of the SIP device after being positioned and weighted using the SAE J1163 procedure.
4. Angle adjustment, if provided, is about the mid-position, not necessarily a latched position.
5. Adjustment values are total adjustments.
6. Where free swing of shoulders and arms over the top of the back is necessary or appropriate for visibility when traveling in the reverse or when controlling rear-mounted implements, the minimum back cushion height should be 300.
7. Where free aft swing of the elbows is desired, the maximum width should be 330.
8. Measure the angle of the centerline of the back cushion. If a lumbar support is provided, it should be set at the mid-range position, and the back angle measured on centerline of the back cushion above the lumbar support. For back cushions with the lumbar support, the allowable angles of Table 15.3 may be increased by 5 deg or more.
9. Armrests attached to the seat should move with the vertical and horizontal seat adjustments. The ability to adjust the armrests vertically to the maximum-minimum values of (K) is desirable. (K) is measured to the top of the pad (if used) at the lateral centerline of the armrest.
10. Radius of curvature of the lumbar support in the vertical plans should be nominally 300 with a minimum of 150.

15.4.4 Operator Controls

Procedural efficacy in operator workplace design is the logical arrangements of task elements to enhance operator performance and minimize error. Proper association between controls and displays along with logical grouping of instruments and controls is important in a compatible man-machine system. A number of standards and engineering practices are used to achieve proper location of operator controls and instruments. SAE Standard J898/ISO 6682 OCT94 "Control Locations for Off-Road Work Machines" (SAE, 1999), defines positions for convenient placement of hand- and foot-operated controls derived from the overlapping reach capability of large and small operators. The large operator approximates the 95th percentile male, and small operator approximates the 5th percentile female U.S. Population. For ethnic group data

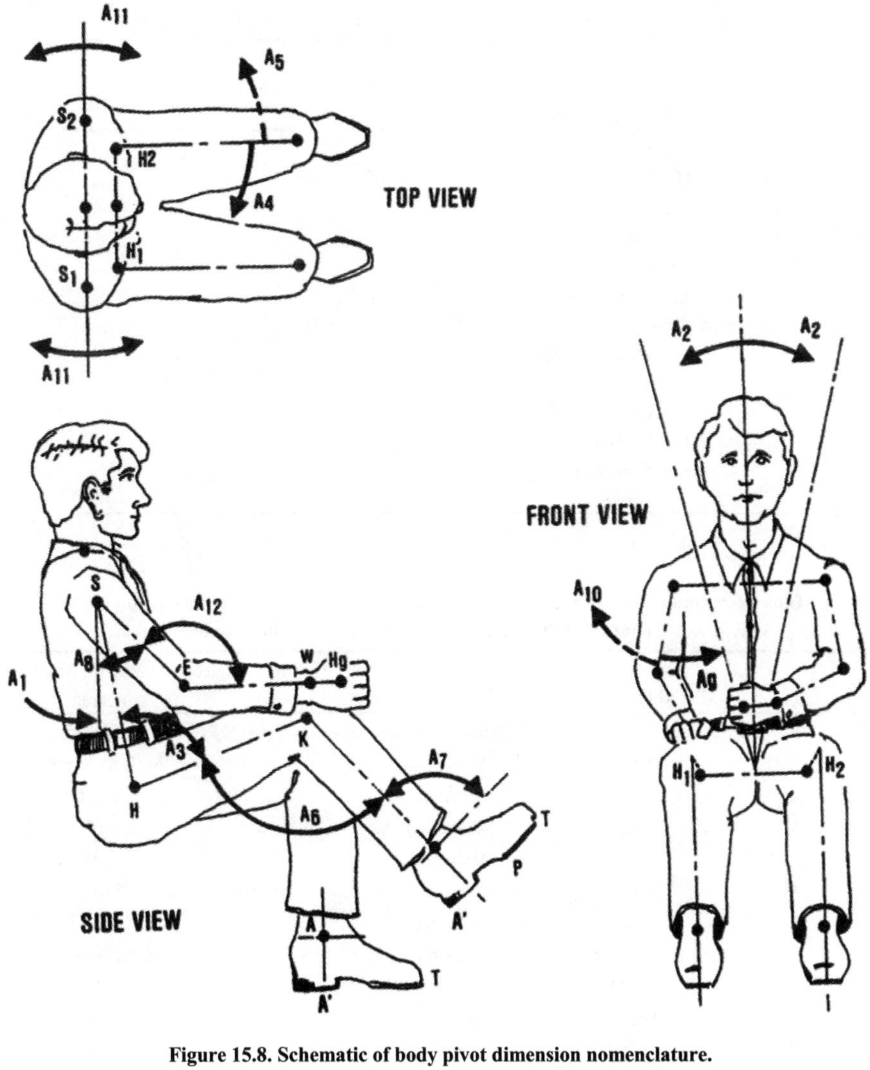

Figure 15.8. Schematic of body pivot dimension nomenclature. See Tables 15.4 and 15.5 for dimensions. (Reprinted with permission from SAE J898/ISO 6682 © 1994 International Organization for Standardization.)

adjustments see SAE J925. Figure 15.8 is a schematic of body pivot dimension nomenclature. Tables 15.4 and 15.5 are a summary of the dimensions associated with Figure 15.8. Purcell (1980) suggests the knee angle should be 110° to 120° for comfortable pedal operation, with a minimum ankle angle of 90°.

Table 15.4. Summary of body pivot dimensions. (Reprinted with permission from SAE J898/ISO 6682 © 1994 International Organization for Standardization.)

Item	Body Elements	Large Operator, 95th Percentile Male, mm	Small Operator, 5th Percentile Female, mm
SH	Shoulder-hip	480	396
HK	Hip-knee	452	372
KA	Knee-ankle	445	367
AA^1	Ankle-shoe sole	119	98
A^1P	Ankle-pedal	150	124
SE	Shoulder-elbow	300	247
EW	Elbow-wrist	267	220
Eh_g	Elbow-hand grasp	394	325
A^1T	Ankle-tow (When $A_1 = 90°$)	243	200
H_1H_2	Hip-hip (lateral)	185	125
S_1S_2	Shoulder-shoulder (lateral)	376	310

Table 15.5. Summary of range of movement angles. (Reprinted with permission from SAE J898/ISO 6682 © 1994 International Organization for Standardization.)

Item	Angle (Right-Side Joint)	Movement	Comfort Angle, degrees	Maximum Angle, degrees
A_1	Seat-Back Angle	Aft to vertical	10	5 to 15
A_2	Trunk	Abduction	0	20 (Left or Right)
A_3	Hip	Flexion	75 to 100	60 to 110
A_4	Hip	Abduction	10	10
A_5	Hip	Abduction	22	30
A_6	Knee	Flexion adduction	75 to 160	75 to 170
A_7	Ankle	Flexion	85 to 108	78 to 115
A_8	Shoulder	Flexion	-35 to 85	-50 to 180
A_9	Shoulder	Adduction	20	20
A_{10}	Shoulder	Abduction	70	120
A_{11}	Clavicle	Circumduction	20	20
A_{12}	Elbow	Flexion	60 to 180	45 to 180

Figures 15.9, 15.10, and 15.11 show the zones of comfort and reach for an average (50th percentile) operator, respectively, for side, top, and front views of the workplace. Use of this information allows the designer to properly locate the hand controls and foot controls for comfortable and efficient operation.

ASAE Standard S335.4 DEC98, "Operator Controls on Agricultural Equipment" (ASAE, 1999), provides guidelines for the uniformity of location and direction of motion of operator controls to improve operator efficiency and convenience. The controls considered are brake, clutch, engine speed, ground speed (transmission), lift, and steering. For example,

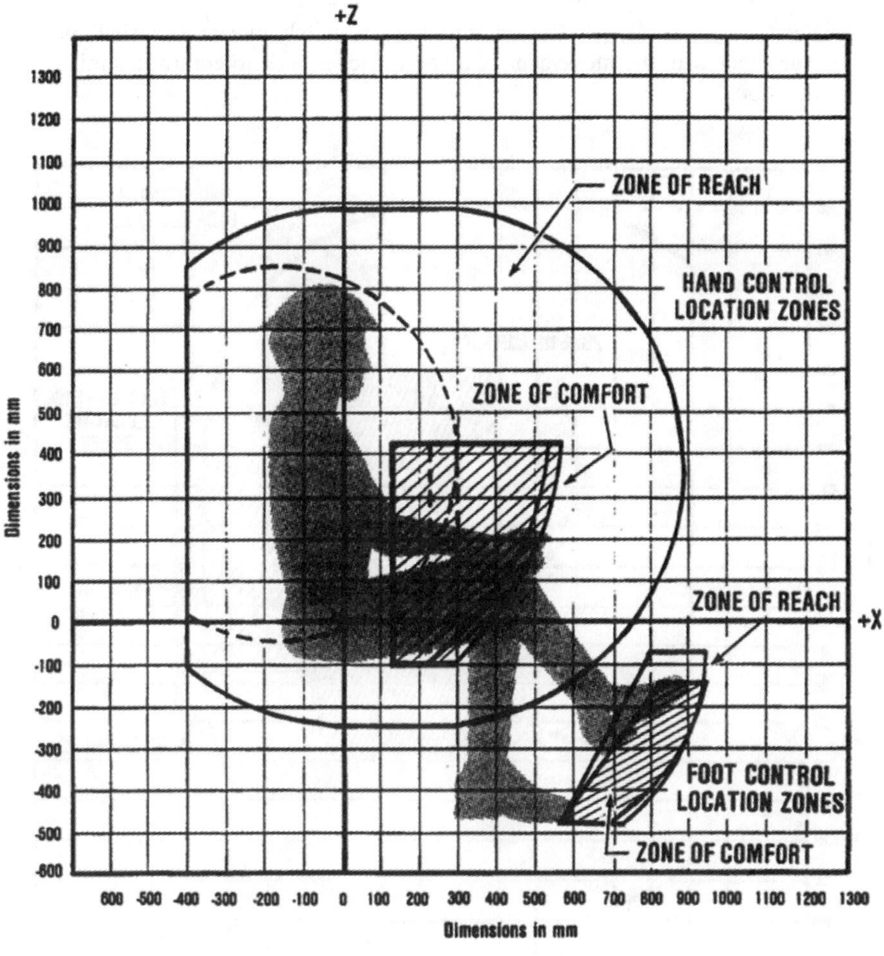

Figure 15.9. Side view of zones of comfort and reach for a average operator with seat adjusted to the mid-range position. (Reprinted with permission from SAE J898/ISO 6682 © 1994 International Organization for Standardization.)

Clutch control 5.1.1. When a foot pedal control is provided, it shall be actuated by the operator's left foot with the direction of motion forward and/or downward for disengagement.

Engine speed control 6.1. The engine speed hand-operated control on agricultural field equipment shall be located to operate with the right hand.

Lift controls for implements or equipment 12.1.3. When a hand control lever is provided the direction of motion shall be generally forward, downward or away from the operator to lower the implement or equipment; and rearward, upward, or toward the operator to raise the implement or equipment.

Electrical controls 10. Rocker and toggle switch movements shall be consistent with the movement of a control lever used to control a similar function.

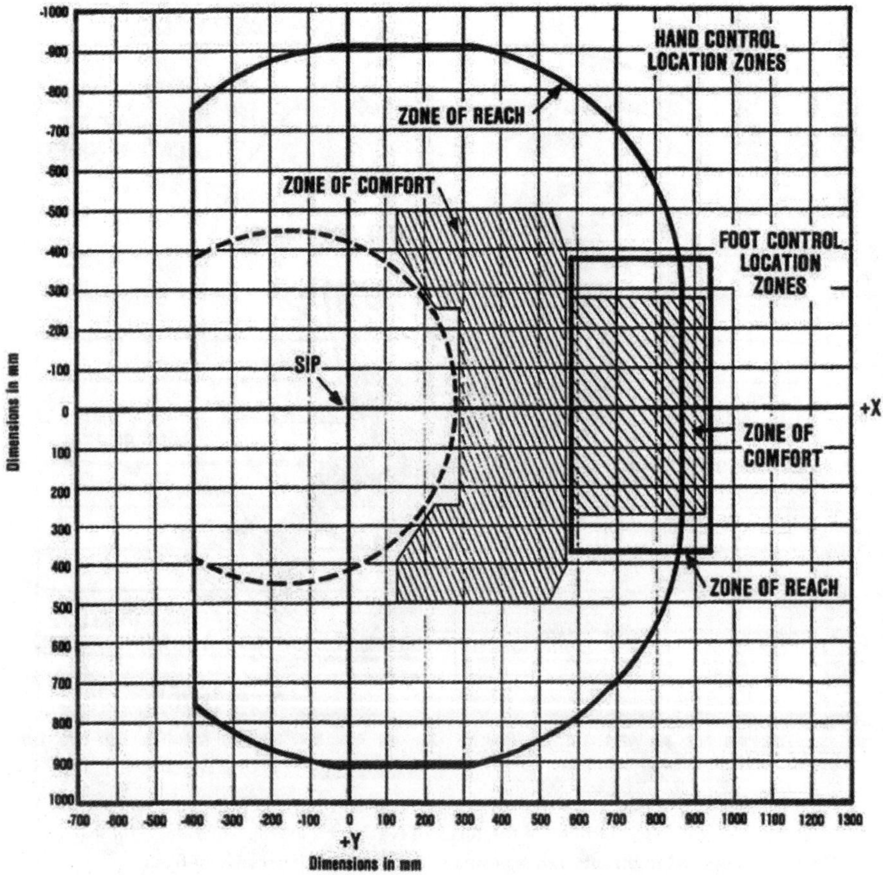

Figure 15.10. Top view of zones of comfort and reach. (Reprinted with permission from SAE J898/ISO 6682 © 1994 International Organization for Standardization.)

Universal symbols for operator controls have been standardized to achieve proper association between controls and displays and are published in the ANSI/ASAE Standard S304.6 DEC95 "Graphical Symbols for Operator Controls and Displays on Agricultural Equipment" (ANSI/ASAE, 1999). A few selected symbols are shown in Figure 15.12.

Instrument face design for readout of analog and digital devices must be quickly readable and understandable. A digital monitor for an off-road vehicle is shown in Figure 15.13.

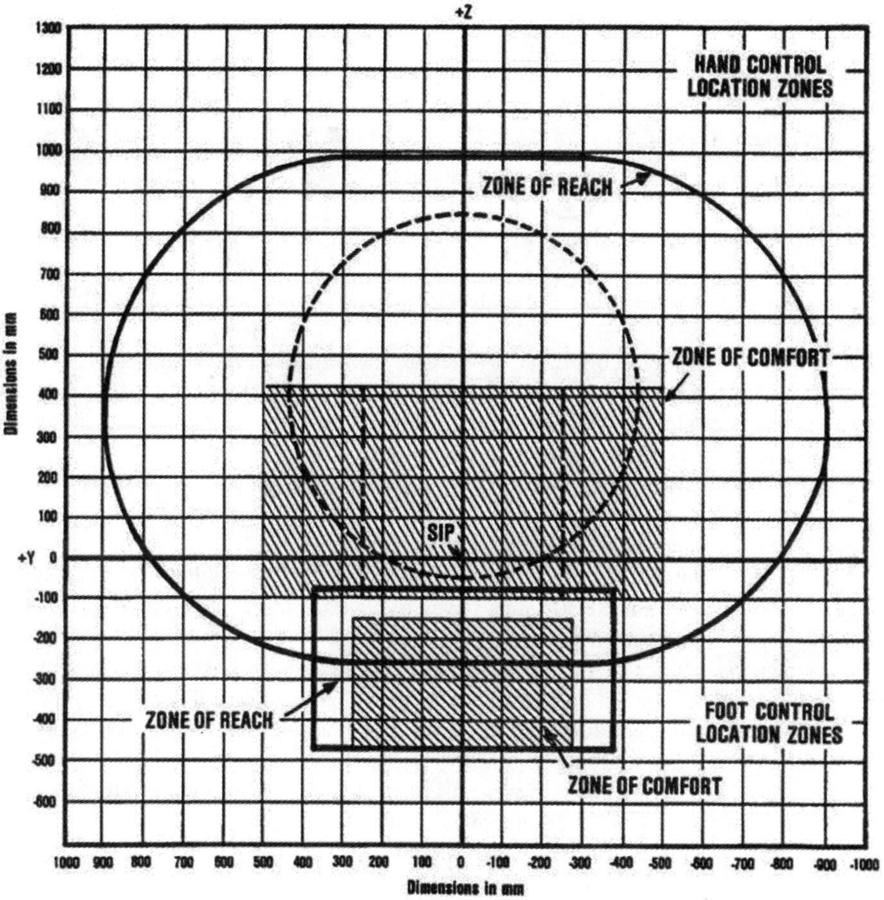

Figure 15.11. Front view of zones of comfort and reach. (Reprinted with permission from SAE J898/ISO 6682 © 1994 International Organization for Standardization.)

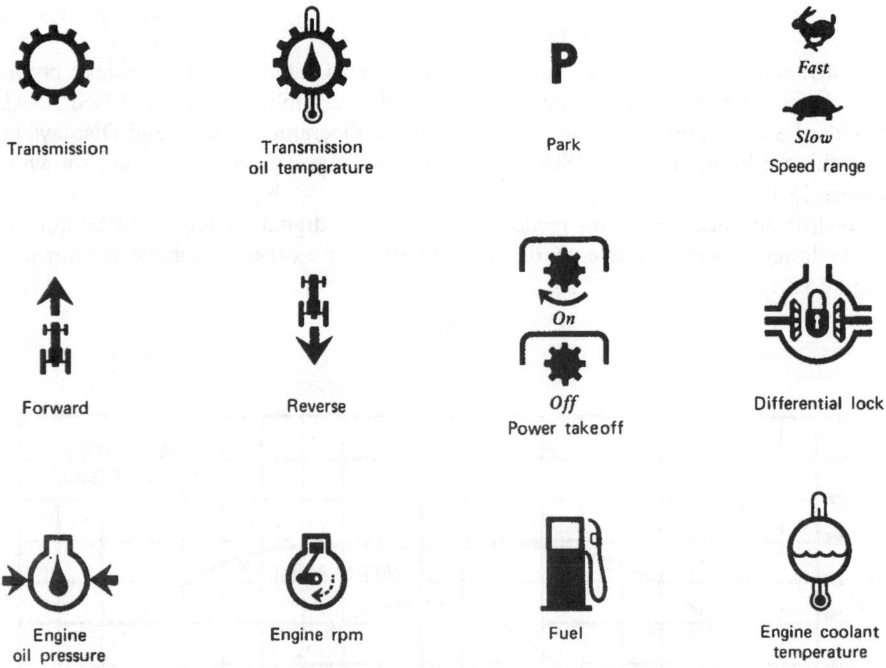

Figure 15.12. Selected universal symbols for operator controls (ANSI/ASAE, 1999).

SAE J209 JAN87 "Recommended Practice, Instrument Face Design and Location for Construction and Industrial Equipment" (SAE, 1999), Figure 15.14, shows the recommended grouping for instruments. Purcell (1980) suggests the following guidelines to instrument design:

- Group similar function instruments or controls on the panel.
- Make related groups of instruments equal in size.
- Design the panel as symmetrically as possible.
- Provide a central zone on the panel for the highest priority instruments. This is the arc lying within the 30° cone for easy eye movement, which is 380 mm wide at 700 mm from the operator's eye.
- Group the gauges in the priority zone horizontally and according to function with engine gauges to the left of the panel center and transmission gauges to the right. The panel center should be marked by either the steering column, or by the tachometer, or by a group of indicator lights.
- Place all remaining instruments on either side of the priority group and keep their relative positions the same on all vehicles in the line.
- Controls on the instrument panel should also follow a standard for the whole line of vehicles.

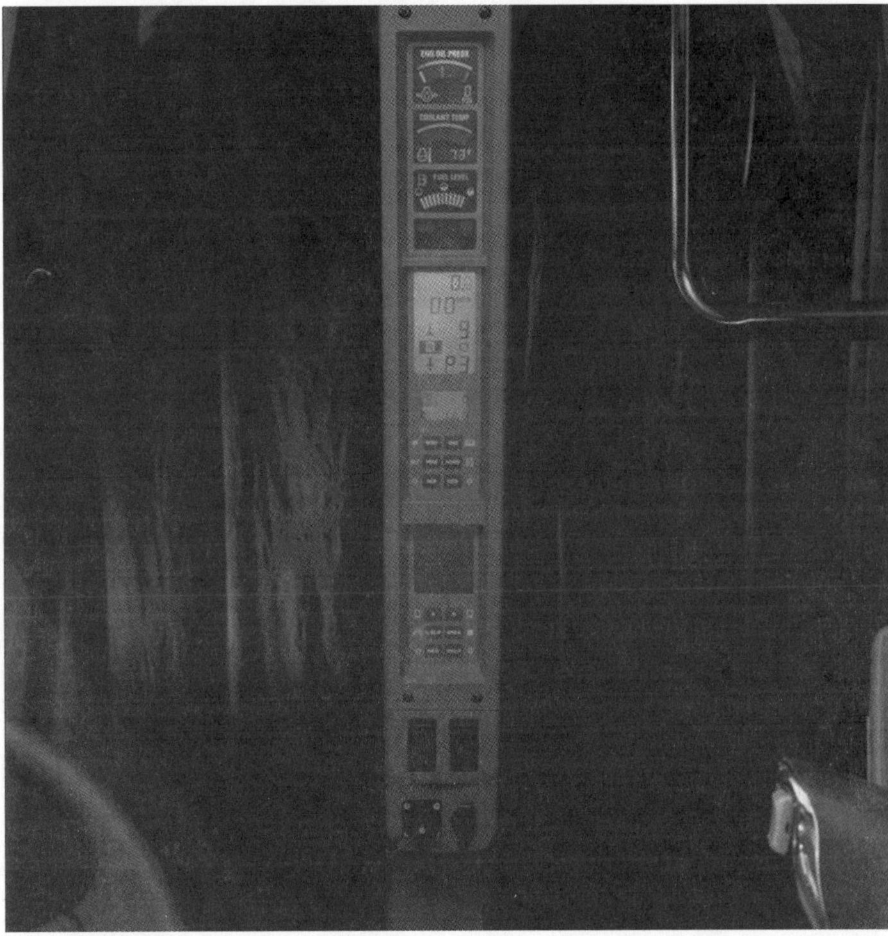

Figure 15.13. Operator's console monitor. (Reprinted with permission from Case IH).

Purcell (1980) also suggest that primary hand controls that are used continuously such as hand throttle, shift levers, and rock shaft control lever should be assigned to the right hand, and the left hand be available for steering at all times. The less frequently used controls such as PTO clutch, parking brake, and differential lock can be allocated to left hand or foot operation.

Use of color-coding of hand controls, as recommended in ASAE Standard EP443 DEC98, "Color Coding Hand Controls" (ASAE, 1999), will aid the operator in identification in various types of controls. Red shall be used only for single-function engine stop controls. Orange shall be used only for machine ground motion controls, such as engine speed controls, transmission controls, parking brakes or park-locks, and independent emergency brakes. Yellow shall be used only for function controls that involve the engagement of mechanisms, such as power take-off. Black or some other

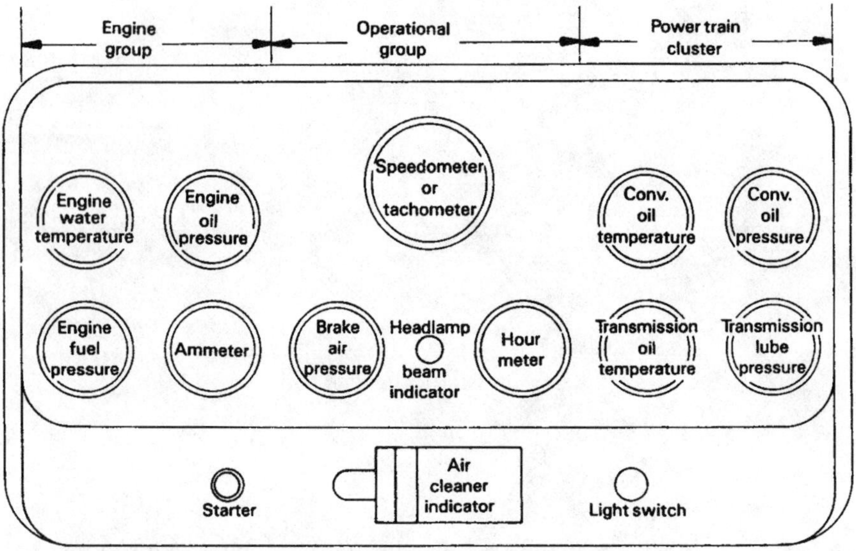

Figure 15.14. Instrument panel grouping. (Reprinted with permission from SAE J209 JAN87 © 1987 Society of Automotive Engineers, Inc.)

dark color to harmonize with décor shall be used for all controls that have positioning and adjusting functions such as steering, hydraulic control, implement hitch, seat adjustment, and machine lighting.

Off-road vehicles require lighting systems that allow safe operation during nighttime operation. ANSI/ASAE Standard S279.10 OCT98 "Lighting and Marking of Agricultural Equipment on Highways" (ASAE, 1999), provides specifications for lighting and marking of agricultural equipment whenever such equipment is operating or is traveling on a highway.

Example Problems

- Determine the design height for a standing operator inside a tractor cab. Refer to Figure 15.3 and Table 15.2 to select the "Standing height with shoes" at the 95th percentile of 1880 mm. Design standing height inside cab is 1880 mm.
- Determine the design range of values for the height (dimension A) of the seat index point (SIP). See Figures 15.2, 15.4, 15.6, and 15.7. Refer to Table 15.2 to select the "horizontal seat height" at the 5th percentile of 398 mm. Then select the "hip pivot to horizontal seat height" at the 5th percentile of 64 mm. Dimension A is thus 398 + 64 for a total of 462 mm. To establish the maximum vertical distance for the SIP from the floor, repeating the above analysis using the 95th percentile data gives 578 mm. The resulting design range is 462 mm to 578 mm with a vertical seat adjustment of 116 mm.

15.5 Thermal Comfort in Operator Enclosures

Thermal comfort is defined in the ASHRAE (1972) *Handbook of Fundamentals* (1972) as the state of mind that expresses satisfaction with the thermal environment. Operator enclosure design must include cab pressurization, air filtration, air movement, heating, cooling, and window defrosting. These factors must be considered in order to provide environmentally clean air with proper air velocity and direction, temperature, and humidity for acceptable human thermal comfort. Design values suggested by Zitko (1977) provide basic guidelines for heating, cooling, air movement, cab pressurization, and air filter properties and are shown in Table 15.6. Cab pressurization is necessary to prevent particulate matter and vapors from entering through leaks into the enclosure. To insure that proper pressurization exists in an enclosure, measurement of cab pressurization should be done as specified in SAE Recommended Practice J1012 JUN93/ISO 3737-1976 (SAE, 1999), "Operator Enclosure Pressurization System Test Procedure."

Table 15.6. Design parameters for thermal comfort in an operator enclosure.

Parameter	Rating or Capacity
Heating	8.2 kW at 66°C and water flow 11.4 L/m
Cooling	7.0 kW at 32°C and 60% relative humidity
Air movement	Three-speed blower rated at 0.236 m^3/s at 50 P_a
Cab pressurization	50 to 100 P_a above outside of cab
Fresh air filter	1.92 m^2 pleated paper, self-cleaning

Source: Zitko, R.F. 1977.

Fanger (1972) was the first to generalize the physiological basis of comfort so that for any activity level (metabolism) and clothing values, it is possible to analytically predict comfort in terms of combinations of air temperature, mean radiant temperature, air humidity, and relative air velocity. Fanger derived the original comfort equation that takes the form

$$F\,(H/A_{Du}, I_{cl}, t_a, t_{mrt}, P_a, v, t_s, E_{sw}/A_{Du}) = 0 \qquad (15.1)$$

where

H/A_{Du} = internal heat production per unit body surface area, kcal/hm^2

A_{Du} = Dubois area, m^2

I_{cl} = thermal resistance of clothing, clo

t_a = air temperature, °C

t_{mrt} = mean radiant temperature, °C

P_a = pressure of water vapor in ambient air,

v = relative air velocity, m/s,

t_s = mean skin temperature, °C

E_{sw}/A_{Du} = heat loss per unit body surface area by evaporation of sweat secretion, kcal/hm^2

From a comfort equation so derived, it is possible to predict any combination of environmental factors (i.e., t_a, t_{mrt}, P_a, and v) that produce a "comfortable" environment for a clothed (clo) person performing any selected activity (H). To use Fanger's approach, use either Fanger's original equation, or use Fanger Comfort Charts, which are computer solutions of his equation.

A tractor operator will expend 60 to 150 kcal/hm^2. Sedentary persons will average 50 kcal/hm^2. Light, medium, and heavy clothing will have values of 0.5, 1.0, and 1.5 clo respectively. On a bright, sunny day, the mean radiant temperature, t_{mrt} can be 5° to 10°C above the cab air temperature t_a. Cab air velocities should be in the 1 to 3 m/s range.

Kaufman et al. (1976) concluded that Fanger's approach to thermal comfort was effective in predicting comfort in a tractor cab for summer conditions in South Dakota. Eriksson and Domier (1975) did work on comfort of subjects in heated tractor cabs. They compared the cab temperature chosen by the subject to a calculated comfortable temperature based on Fanger's comfort equation and found a large spread in the data.

The thermodynamic process of heat exchange between humans and the environment can be described by the general heat balance equation of Gagge et al. (1941).

$$\text{S (storage)} = \text{M (metabolism)} - \text{E (evaporation)} \pm \text{R (radiation)} \pm \text{C (convection)} - \text{W (work accomplished)} \tag{15.2}$$

S is the amount of heat gained or lost. If the body is in a state of thermal balance, S becomes zero. Positive values for storage will cause the mean body temperature to rise; negative values will cause it to fall. It is most convenient to express the preceding terms as power per unit of body surface (W/m^2), since heat exchange is always related in some way to the body's surface area. The metabolic rate for an off-road driver will be in the range of 70 to 174 watts/m^2. Average body surface area of an adult male will be 2 m^2, resulting in 140 to 348 watts. An enclosure environment must be maintained that will balance the metabolism of the operator so that thermal equilibrium is maintained.

An empirical index for human thermal comfort, effective temperature (ET*), was first developed by Houghton and Taylor and in 1972, a new comfort chart (shown in Figure 15.15) was published by ASHRAE (1972) *Handbook of Fundamentals*. The new *effective temperature scale* (ET*) is based on a simple model of human physiological response. Most individuals will be comfortable when the ET* is between 24° and 27°C (Figure 15.15). Note that various combinations of dry- and wet-bulb temperatures satisfy the "comfortable" zone. If the operator enclosure is maintained at a dry-bulb temperature of 26°C and 40% to 50% range of relative humidity, reasonable comfort will result. The operator can make adjustments to meet his or her comfort by changing air speed, air direction, air temperature, and amount of clothing worn. In winter conditions, relative humidity will be lower, in the 10% to 30% range, which calls for a slightly higher air temperature for the desired thermal comfort.

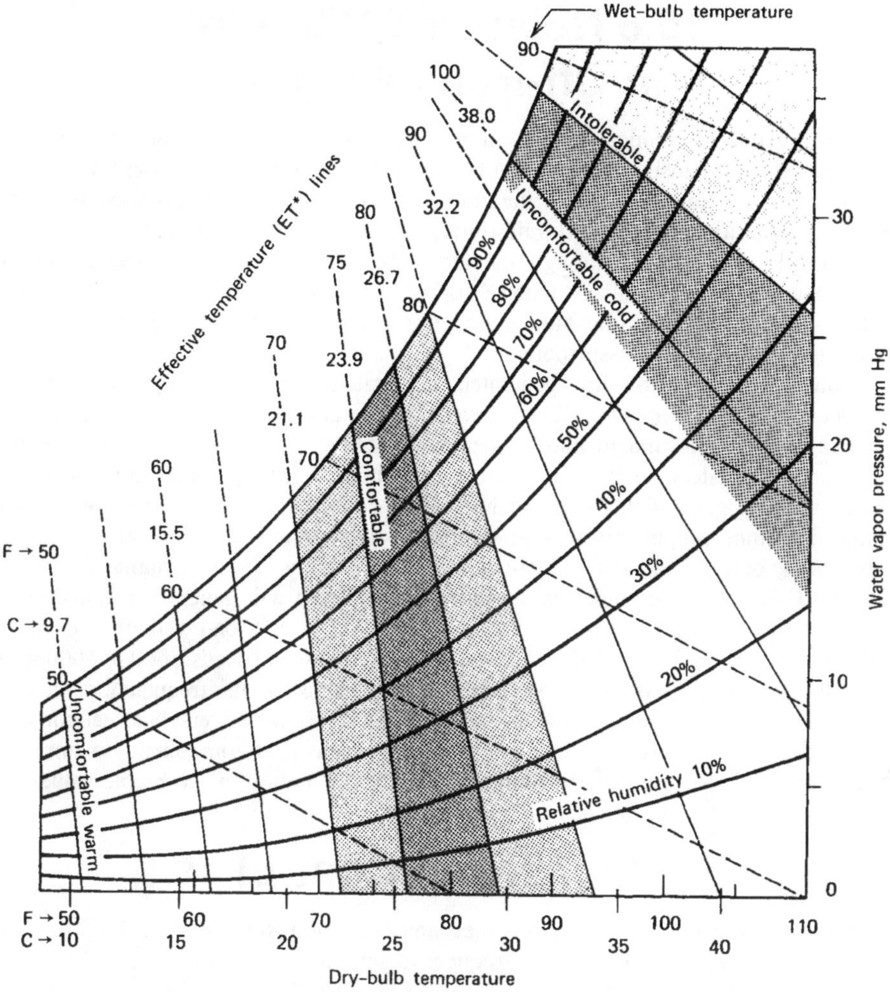

Figure 15.15. New effective temperature scale ET*. (Copyright 1972, American Society of Heating, Refrigerating and Air-Conditioning Engineers, Inc., 1791 Tullie Circle, NE, Atlanta, GA 30329, 404-636-8400, www.ashrae.org. Reprinted by permission from ASHRAE 1972 *Handbook-Fundamentals.*)

The new *effective temperature scale* ET* applies to altitudes from sea level to 2134 m and to the special case for indoor thermal environments in which the mean radiant temperature is nearly equal to dry-bulb air temperature and the air velocity is less than 14 m/min. In applying the new ET* to off-road vehicle enclosures, the mean radiant temperature may be 2° to 5°C greater than or 2° to 20°C less than the dry-bulb, and air velocities may be two to three times greater than 14 m/min. The designer needs to provide some additional heating and cooling capacity to compensate for these effects.

15.6 Noise and Vibration in Off-Road Vehicles

The overall mechanical design of an off-road vehicle, including the operator workplace and enclosure, has a direct bearing on the noise and vibration levels at the operator's site. Noise levels are the summation of rotating and reciprocating parts, structural vibrations, gases, and fluid flows, all of which are transmitted either through the vehicle structure or through the air. Vibration levels are the summation of components that are similar to noise source components plus ground inputs from wheels or tracks. Thus the human operator experiences various levels of sound, vibration, and vehicle chassis motion.

Sound and vibration are generated in rotating, reciprocating, and vibrating components of the vehicle. Fans, blowers, gears, and internal combustion engines generate sound at a fundamental frequency, respectively, of rpm/60 times the number of blades, impellers, teeth, and number of cylinders. Also, components of higher harmonics are generated. Design techniques to control structure and air borne noise must determine proper stiffness of stationary and moving parts and proper use of absorbing, barrier, and damping materials. The hearing range in humans is 16 to 11,000 Hz and a six-cylinder engine rated at 2100 rpm will generate a fundamental sound frequency of 210 Hz. Mechanical vibrations in off-road vehicles lie in the frequency range of 1 to 100 Hz, corresponding to rotational speeds of 60 to 6000 rpm. Ground input vibrations (bumps) to wheels and tracks are in the frequency range of 1 to 10 Hz. Vibrations in the frequency range of 1 to 50 Hz can adversely affect humans. Thus, the engineer has the challenge to design mechanical components to limit the levels of sound energy, vehicle vibration, and chassis motion to acceptable values for operator comfort and efficiency.

15.7 Noise Control

Noise is acoustic sound, which is mechanical vibrations in gaseous, fluid, or solid media. Sound is characterized by frequency, amplitude, and phase. The frequency range of the human ear extends from as low as 16 Hz to as high as 20,000 Hz; however, few adults can perceive sound above 11,000 Hz.

General-purpose sound-measuring instruments are normally equipped with three frequency-weighting scales, A, B, and C. These scales approximate the ear's response characteristics at different sound levels. Nebraska OECD Tractor test data are reported in decibels, using the A weighting scale, and written as dB(A). Design for noise control, however, requires more detailed analysis of sound spectrums. Sound analyzers measure sound levels in bandwidths ranging from octaves to only a few hertz in width.

The official guideline for noise exposure in the United States is based on the Occupational Safety and Health Agency (OSHA) Walsh-Healy Act. Table 15.7 shows current acceptable levels. Note that for exposure eight hours per day, the sound level should not exceed 90 dB(A). For shorter exposure times, higher levels of sound can be

Table 15.7. Occupational Safety and Health Act (OSHA) noise criteria.

Duration per Day, hours	Sound Level, dB(A)[1]
8	90
6	92
4	95
3	97
2	100
1 ½	102
½	105
¼ or less	115

Note: Exposure to impulsive or impact noise should not exceed 140 dB peak sound pressure level.

1. Sound level meter using A-scale, slow response.

tolerated. These criteria were developed for factory worker environments. Application to off-road vehicle operators needs judgment because of intermittent, variable-intensity noise and seasonal variations in daily exposures. Referring to Figure 3.4, Nebraska OECD Tractor Test 1678, tractor with cab, the sound level at 75% load in 7th gear was 73.0 dB(A) at the operator's site. For the bystander test in 12th gear, 82.5 dB(A) was reported. Bystander distance is 7.5 m, measured in a direction normal to the centerline of the travel path.

The predominant noise in off-road vehicles is in the frequency range of 125 to 500 Hz. Reducing noise levels at the operator's site has been accomplished by incorporating sound-control measures in the cab. Sound reaching the operator is structure-borne, airborne, or a combination of both. Structure-borne sound results from vibrations transmitted from the vehicle to the cab through the attachment points. Airborne sound is transmitted through air and enters the operator area through windows, walls, and holes. Rubber interfacing between the mounting bracket and the control platform achieves vibration isolation. The magnitude of isolation is dependent on the transmissibility of the rubber used. The cab floor is treated with sound barrier materials and walls are treated with noise absorption material, which is effective in the 125 to 2000 Hz range.

For tractors without cabs, sound levels measured by Nebraska OECD Tractor Tests at the operator's site will be in the 89 to 98 dB(A) range. For these conditions, the operator should wear protective gear to protect against noise-induced hearing loss.

15.8 Vibration Control

Exposure of the human body to vibrations can result in physiological, psychological, and mechanical effects. Ride vibration intensities are positively correlated with ground speed and often become intolerable as speed increases. Matthews (1972) reported in a survey that the majority of large tractors were found to be operating at less than two-thirds maximum power because of the operator's inability or unwillingness to withstand the ride at full speed. Rossegger and Rossegger

Table 15.8. Pathological spine deformations in different occupations as determined by x-ray examination.

Occupation	% with Spine Deformations	Average Age of Sample (years)
Truck drivers	80.0	--
Tractor drivers	71.3	26
Miners	70.0	51
Bus drivers	43.6	40
Factory workers	43.0	45
Construction workers	37.0	51

Source: Rossegger and Rossegger, 1960.

(1960) conducted research suggesting that ride-induced vibrations do have ill effects on the operator. The results of their work, which emphasizes the importance of proper seating for tractor drivers, are shown in Table 15.8. Of all the occupations listed, only truck drivers had more difficulties with their spines than tractor drivers. As a general rule, the percentage of the normal population with symptoms of spine degeneration increases with age, so if a correction for age were made as shown in Table 15.8, the seriousness of the problem would be even greater.

The Rosseggers also studied the gastrointestinal difficulties of tractor drivers as compared to a control group (Table 15.9). They conclude:

> The investigation has revealed that tractor driving may have considerable ill effects on the health of the operator. This is largely due to the effects of vibration and shocks continuously acting upon the human body and setting up harmful stimuli, and partly to the need to keep the body in a cramped condition and unhealthy posture for long periods. The response of the body in an effort to counteract these effects and to maintain equilibrium imposes an additional strain upon the tractor driver, thus increasing fatigue.
>
> Care should, therefore, be taken in designing tractors, and particularly tractor seats, to reduce vibration and shocks to a minimum by appropriate suspension and shock absorption and to arrange the tractor controls in a manner to insure a comfortable posture and minimum effort.

Effects reported by the Rosseggers are caused by low-frequency vertical vibration, that is, frequencies up to 20 Hz. This low-frequency vibration results in whole-body excitation. Suggs (1973) has observed the 4 to 8 Hz range as being critical. In this range resonance occurs in parts of the human body, producing discomfort. Suggs states:

Table 15.9. Comparison of the percentage of tractor drivers who had stomach difficulties and the control group.

	Tractor Drivers (n = 322)	Control (n = 37)
Stomach complaints	76%	46%
X-ray findings	32%	30%

Source: Rossegger and Rossegger, 1960.

> One of the more interesting resonances, which occur, is that of the viscera. In vertical vibration the viscera resonates at about 4 to 5 Hz and forces the diaphragm at the same frequency. This produces a mechanical pumping action in the chest, which over ventilates the lungs. Carbon dioxide, a component of the plasma acid balance is flushed out of the blood stream and respiratory alkalosis results. The drowsiness associated with alkalosis may be implicated in some tractor accidents.

Suggs further states:

> Hand coordination, foot pedal control, reaction time, visual acuity and tracking are all adversely affected by low-frequency vibration. While these effects in themselves do not constitute a health or safety problem, they do produce a decrement in the ability of the operator to control a vehicle which implies a higher accident probability.

The classic Rossegger and Rossegger (1960) studies represent benchmark work. Studies by Bongers and Boshuizen (1992) provide additional information on whole-body vibration effects on off-road vehicle operators.

Low-frequency vertical vibration is present during normal field operations. Amplitude of vibration, in part, is dependent on roughness of the field. The undamped natural frequencies of wheel tractors commonly lie in the 3 to 5 Hz range. Higher frequency vibration, 30 Hz and up, results in part-body vibration. Although it is not important with regard to whole-body vibration, it is the source of foot and hand-arm excitation. Higher frequency vibrations can be present at the steering wheel, gearshift levers, control levers, and floor panels.

Vibration is an oscillatory motion of a mechanical system, which can be a simple or extremely complex harmonic. Mechanical vibration is characterized by frequency, amplitude, and phase.

For a simple harmonic wave form;

$$X = A \sin \omega t \tag{15.3}$$

where

X = displacement, m

A = peak amplitude, m

ω = frequency of vibration, rad/s

then

$$\dot{X} = \omega A \cos \omega t \tag{15.4}$$

where $\dot{X}$ = velocity, m/s

and

$$\ddot{X} = -\omega^2 A \sin \omega t \tag{15.5}$$

where $\ddot{X}$ = acceleration, m/s^2

The acceleration is proportional to the displacement and to the square of the frequency.

Vibration is measured with an *accelerometer,* which senses $\ddot{X}$ in Equation 15.5 An accelerometer is an electromechanical transducer that produces an output voltage signal proportional to the acceleration to which it is subjected. Readout is normally expressed as acceleration, RMS, which mathematically is

$$\ddot{X}_{RMS} = \frac{1}{\sqrt{2}}\left(\omega^2 A\right) \tag{15.6}$$

It is also convenient to express acceleration in terms of g's. This is achieved by dividing $\ddot{X}$ or $\ddot{X}_{RMS}$ by 9.8 m/s^2. Vibration levels are also expressed in decibels. Equation 15.7 expresses it mathematically with a reference level of 1 m/s^2 RMS.

$$X_{dB} = 20 \log X_{RMS} / 1 \tag{15.7}$$

$$\text{dB re } 1 \text{ m/s}^2$$

where

X_{dB} = vibration acceleration level, dB

X_{RMS} = measured RMS acceleration, m/s^2

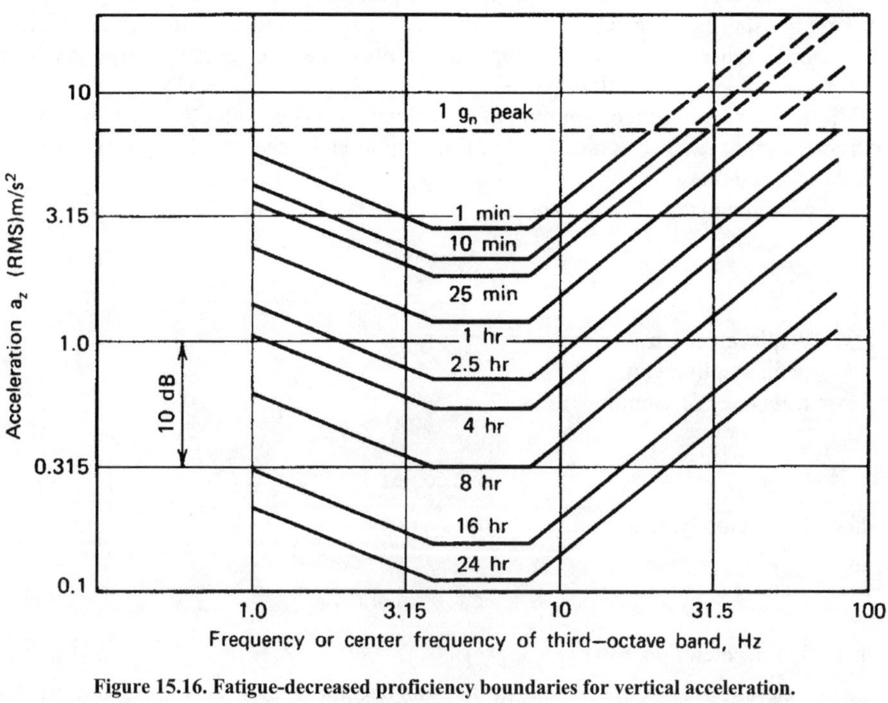

Figure 15.16. Fatigue-decreased proficiency boundaries for vertical acceleration. (Reprinted with permission from ISO 2631© International Organization for Standardization.)

Engineers and scientists continuously work to quantify human vibration tolerance to vehicle ride vibration. One such effort is represented by the International Standards Organization in ISO 2631, "Guide for the Evaluation of Human Exposure to Whole Body Vibrations" (ISO, 1972). Figure 15.16 shows the format of the boundary curves for vertical vibration as experienced by the operator. It is important to note that in the 4- to 8-Hz range, human tolerance to vibration is a minimum.

This discussion has been limited to vertical vibration effects with respect to a seated operator. However, horizontal (lateral and longitudinal) and angular (pitch, roll, and yaw) acceleration inputs are also experienced by the seated operator. Design for vibration isolation and control requires detailed analysis of complex vibrations. Vibration analyzers (similar to sound analyzers) are used to determine vibration levels in bandwidths ranging from octaves to only a few hertz.

15.8.1 Operator Seating

An off-road vehicle can be considered as a spring-mass-damper system vibration with four degrees of freedom (DOF). Displacement and acceleration inputs to the base of an operator's seat will consist of vertical, lateral, longitudinal, roll, pitch, and yaw components. The research of Raney et al. (1961), Pradko and Lee (1968), and Janeway (1975) shows that the predominant vibration motion of a wheel tractor is vertical (Z_s) and that the seated operator is more sensitive to vertical acceleration.

If we consider only vertical motion imparted to the base of the operator's seat, the 2 DOF system described in Section 14.6 Longitudinal Stability applies. Referring to Figure 14.7, the vertical motion imparted to the tractor seat is approximately

$$Z_{input} = Z_t + R\theta \quad (15.8)$$

where

Z_{input} = vertical motion imparted to base of seat, m

Z_t = vertical motion of tractor center of gravity, m

θ = rotation or pitch about tractor center of gravity, radians

R = longitudinal distance from operator's seat to the tractor center of gravity, m

For small θ, the term $R\theta$ represents the vertical motion resulting from the pitch motion of the tractor.

Then

$$\ddot{Z}_{input} = \ddot{Z}_t + R\ddot{\theta} \quad (15.9)$$

where

$\ddot{Z}_{input}$ = vertical acceleration imparted to base of operator's seat, m/s^2

$\ddot{Z}_t$ = vertical acceleration of tractor center of gravity, m/s^2

$R\ddot{\theta}$ = vertical acceleration resulting from pitch, m/s

A simple model of a seat and suspension is shown in Figure 15.17. For suspension-type seats the cushion is mounted to a guided linkage incorporating some type of spring and damper (shock absorber) along with travel limit stops. Friction is also

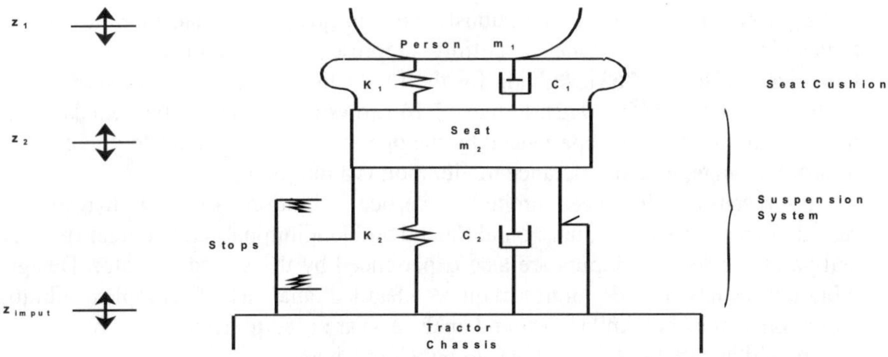

Figure 15.17. Schematic of tractor operator's seat.

inherent. The seat suspension system is a vibratory mechanical system subjected to the tractor chassis input, $\ddot{Z}_{input}$. Seat motion is a function of seat spring rate, mass, damping, and frequency content of the input vibration. For a given input vibration $\ddot{Z}_{input}$ to the seat base, the seat develops a corresponding output vibration.

We simplify the system in Figure 15.17 to a rigid mass, $M = m_1 + m_2$, coupled to the tractor chassis via a linear spring K_2 and a viscous damper C_2, which is a second order single DOF linear differential equation,

$$M\ddot{Z}_2 + C_2\dot{Z}_2 + K_2 Z_2 = \ddot{Z}_{input} \tag{15.10}$$

where $\ddot{Z}_{input}$ is steady-state and $R\ddot{\theta}$ is zero.

Solution of Equation 15.10 and development of Equations 15.11 and 15.12 is derived by Den Hartog (1956). The ratio of output vibration intensity to input vibration intensity is called the *transmissibility*, as

$$\text{Transmissibility} = \frac{\text{output vibration intensity}}{\text{input vibration intensity}} \tag{15.11}$$

The relationship between the output (seat) vibration for a given input (off-road vehicle chassis) vibration as a function of frequency is known as the transfer function or the frequency response function of the system, as

$$T = \left[\frac{1 + 4\zeta^2 \left(\frac{\omega_t}{\omega_s}\right)^2}{\left[1 - \left(\frac{\omega_t}{\omega_s}\right)^2\right]^2 + 4\zeta^2\left(\frac{\omega_t}{\omega_s}\right)^2} \right]^{0.5} \tag{15.12}$$

where

T = transmissibility

ς = *damping ratio*

ω_t = tractor chassis frequency, rad/s

ω_s = undamped *natural frequency* of the seat, rad/s

Equation 15.12 is shown in Figure 15.18. For a clearer understanding of Equation 15.12 some additional details are as follows:

$$\varsigma = c/c_c \tag{15.13}$$

where

$c = C_2$ = seat suspension damping rate, N·s/m

c_c = critical damping rate, N·s/m

where

$$c_c = 2M\omega_s \tag{15.14}$$

where

M = mass of seat and operator, kg

ω_s = undamped natural frequency of the operator seat, rad/s

$$\omega_s = \sqrt{\frac{k}{M}} \tag{15.15}$$

where

$k = K_2$ = spring rate of seating system, N/m

M = mass of seat and operator, kg

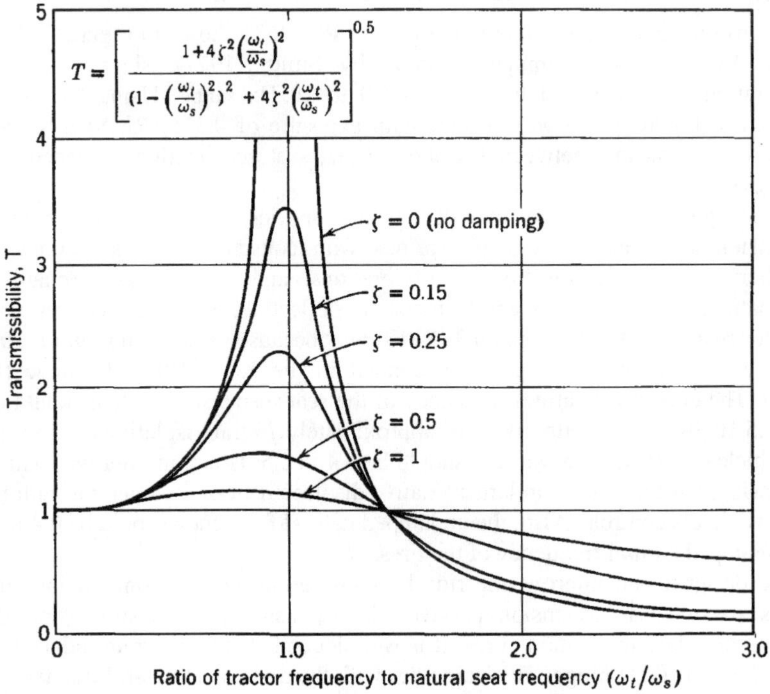

Figure 15.18. Transmissibility as a function of frequency ratio for various amounts of viscous damping.

For proper seat-suspension design, values of C_2 and K_2 must be judiciously selected so that the transmissibility is less than unity, which occurs at ω_t/ω_s ratios of greater than 1.414. Note that the frequency ω_t of the input vibration must be known in order to design a seat-suspension system for the vehicle. Vehicle frequency will be in the 3 to 10 Hz range. Because occasionally a ω_t/ω_s ratio near unity will be experienced, the importance of damping should be apparent. Transmissibility in the range of 0.4 to 0.5 is considered satisfactory. Finally, it should be pointed out that transmissibility is independent of input amplitude for linear systems.

The true transmissibility of a seat is much more complex than Equation 15.12 indicates because of nonlinear cushion characteristics and nonlinearities in the suspension such as inherent friction and limit stops. In practice, the transmissibility of a seat is determined by computer model simulation of the seat suspension system and laboratory testing.

Once the seat transmissibility is determined, the ride quality can be related to vehicle ride criteria. For example,

$$a_z = (\text{transmissibility})(\ddot{Z}_{\text{input}}) \tag{15.16}$$

where

a_z = vertical acceleration, (RMS) m/s^2

$\ddot{Z}_{\text{input}}$ = RMS value of the acceleration defined in Equation 15.9, m/s^2

a_z corresponds to the acceleration criteria of ISO 2631 shown in Figure 15.16. Figure 15.19, from a simulation study by Smith (1977), shows vertical seat acceleration for a wheeled tractor in a 1.0 to 20 Hz range. Using the eight-hour duration in Figure 15.16 with a maximum exposure of 0.315 (RMS) m/s^2, Smith's simulation shows that between 1.5 and 2.7 Hz, seat acceleration exceeded the ISO 2631 level of 0.315.

Basic guidelines govern overall design for ride comfort. For passive seat suspension, components used are springs with constant or variable rates, shock absorbers to dampen low-frequency vibrations, and torsion bar mechanisms or pneumatic cylinders to compensate for static deflection. Vertical acceleration at the operator seat should be less than 0.3 (RMS) m/s^2 because wheeled agricultural tractors have vertical acceleration frequency content in the 1.0 to 20.0 Hz range (Figure 15.19). The undamped natural frequency of the seat suspension system should be less than 1.5 Hz, with maximum travel of approximately 76 mm. Isolation of the cab from the vehicle chassis transmissibility should be 0.4 to 0.5. Isolation improves with softer suspension, but this results in larger relative displacement of the operator with respect to the vehicle's controls. Also, the undamped natural frequency should be 0.4 to 0.5 of the undamped natural frequency of the tires.

The designer can improve the ride by considering various combinations of four factors: passive seat suspension, passive cab suspension, primary suspension of front and/or rear axles, and suitable tires. It is well documented that terrain-induced bumps are in the 1 to 10 Hz range. This coincides with the undamped natural frequency of an agricultural tractor, which is in the 2 to 4 Hz range, thus amplified vertical motion at

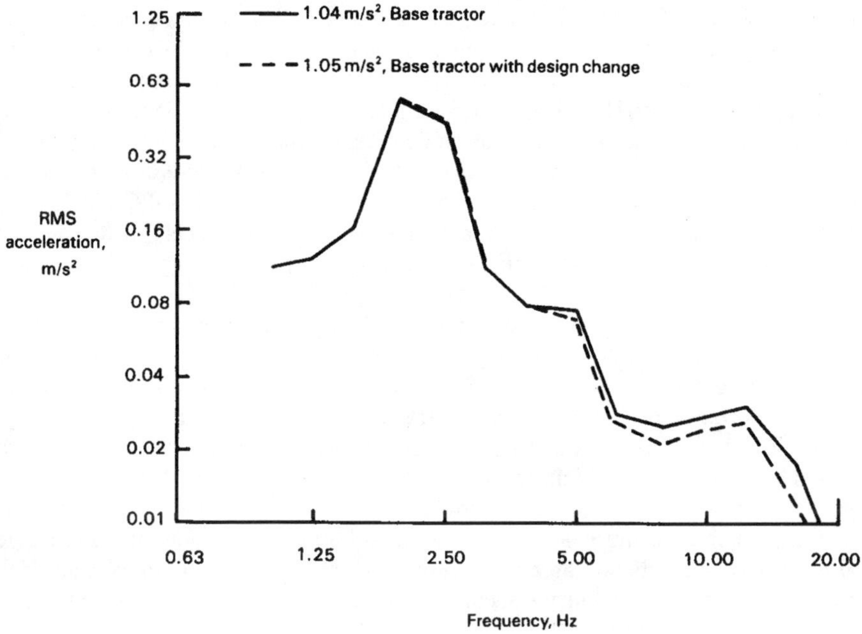

Figure 15.19. Third-octave spectra of the vertical seat acceleration. (From Smith, 1977.)

the operator's site is likely. The operator also experiences lateral and longitudinal vibrations in the 1.5 to 10 Hz range that are lower in magnitude than the vertical vibration. The largest diameter tires possible should be used for improved ride comfort. This will help the tire better negotiate obstacles because there is more time to raise the center of the axle to the bump height. Also, a large tire will not follow rough terrain exactly and will act as a filter.

Example Problems

- A vertical acceleration measured at the floor of a cab was 18 Hz. What is the frequency in Rad/s? 18 cycles/s × 2 π Rad/cycle = 36 π Rad/s.
- If the damping ratio of the seat suspension is 0.5, what is the transmissibility T for a $\omega_t / \omega_s = 2.0$? Using Figure 15.18, T is approximately 0.8.
- If the measured acceleration at the cab floor is 2.0 m/s^2, what level will the operator experience? Output vibration = input vibration × transmissibility. 0.8 × 2.0 m/s2 = 1.6 m/s^2.

Computer modeling is used to study ride performance as affected by vehicle geometry and spring, mass, and damping parameters. Simulations provide the user with relative comparisons as design parameters are varied, thus indicating where the largest improvements can be made. Smith (1977) developed a three-dimensional mathematical model that included 13 DOF. Lagrangian dynamics were used to derive the differential equations of motion. Random terrain profiles generated by the National Institute of Agricultural Engineering (NIAE) (Matthews, 1966) for smooth and rough test tracks at 12 km/h were used as the excitation for the model developed. Results of the simulation indicated a vertical motion frequency of 1.99 Hz of the chassis and cab. The longitudinal motion of the chassis and cab had a fundamental frequency of 2.22 Hz, with higher order resonance at 4.75 and 10.17 Hz. Also, lateral motion of the chassis and cab occurred at frequencies of 1.21, 1.51, 2.67, 5.97, 6.48, and 12.4 Hz. Because it is difficult to correlate acceleration amplitude levels with operator-perceived ride comfort, the simulation was largely used for providing a relative, rather than an absolute, evaluation of the effect of a design change.

Prasad et al. (1995) reviewed the progress made in reducing tractor ride vibration at the operator site. Several models for tires, seat suspension, and cab suspension are discussed. Progress in determining subjective response to whole body vibration is also reviewed. To date, the low-frequency vibrations caused by ground inputs combined with the low, natural undamped frequency of the vehicle cause vibration levels at the operator's seat that are too high. Since tractors and other off-road vehicles do not have suspension systems, isolation of this low-frequency vibration is extremely difficult. One manufacturer has introduced a tractor with a limited suspension through linkages and hydraulic cylinders that absorb some of the vertical and lateral movement of the front axle. Another draft linkage member is used to absorb some of the fore-to-aft movement of the rear axle.

Ahmed and Goupillon (1997) predicted the ride vibration of an agricultural tractor. He developed equations of motion similar to Equation 15.10, but with 19 DOF compared to the one DOF used in Equation 15.10. Elements of the model included chassis and front axle with seven DOF, cab with six DOF, and seat suspension with six DOF. To test the model, a vibration test was carried out on the tractor on a standard ISO-5008 smooth track. Vertical, lateral, and longitudinal accelerations of the chassis center of gravity were measured and compared with predicted values. Good correlations between predicted and measured values occurred for the vertical and lateral vibrations. However, the longitudinal correlations did not agree until the lateral stiffness of the tires was reduced by 40%. The next level of simulation was with a cab and seat suspension added to the chassis. Good correlation between the predicted and measured values occurred for the vertical acceleration of the seat base. However, good correlations between the lateral and longitudinal vibrations at the seat base were not obtained. The authors concluded this was due to inadequate data for some of the dynamic characteristics of the elements in the simulation. Figure 15.20 shows the measured and predicted vertical vibrations of the base of the seat. Note a measured acceleration of about 1 m/s^2 and predicted of 0.9 m/s^2 at 2.5 Hz. These figures are consistent with other findings. Wan and Schimmels (1997) did a computer simulation for optimal seat suspensions based on minimizing the computer simulated subjective response of an average human. He used a four DOF linear lumped parameter model for assessing human exposure to vertical whole body vibration. This work gives insight to the defining the range of design values for the seat damping (C_2) and stiffness (K_2).

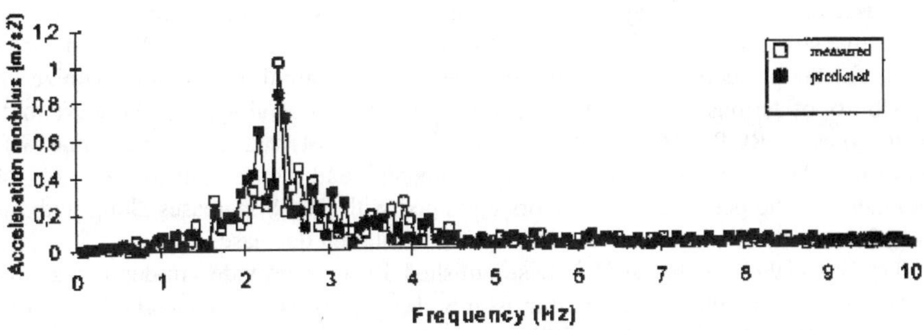

Figure 15.20. Vertical acceleration of base of seat. (Source: Ahmed and Goupillon, 1997.)

During the 1990s, tractor manufacturers introduced vehicles that included one track on each side of the tractor. Compared to wheels, the rubber tracks increased draft capability and provided a better ride for the operator. One manufacturer introduced a tractor in which a small rubber track system replaced each of the wheels on a four wheel drive tractor.

Duncan and Wegscheid (1982) developed an off-road vehicle motion simulator to determine effects of vibration, noise, and task complexity on operator and system performance. The simulator was used to improve suspension systems for seats and operator enclosures. Six DOF with motion in the x, y, z, roll, pitch, and yaw coordinates were possible. Ride programs drive the simulator from data files made from recordings of vehicle accelerations collected from field operations. Data files constructed from vehicle computer simulations also drive the simulator.

Other research done in seat-suspension design and ride comfort include Bell (1982), Claar II and Sheth (1980), Claar II et al. (1980), Janeway (1975), Matthews (1973), Rakheja and Sankar (1984a, 1984b), Stikeleather and Suggs (1970), Stikeleather (1973, 1976, 1981), and Young and Suggs (1973).

15.9 Roll-Over Protection

Hard et al. (1999) reported findings on work-related fatalities in the agricultural production sector using two national occupational fatality surveillance systems during 1990-1995. Data from the National Traumatic Occupational Fatalities (NTOF) surveillance from 1990 to 1993 and Census of Fatal Occupational Injuries (CFOI) surveillance from 1992 to 1994 were collected and analyzed. The NTOF reports fewer fatalities because juvenile deaths under age 16 are not counted and also due to differences in protocol. The NTOF reported 19.5 deaths per 100,000 workers over the four-year period and CFOI reported 25.5 deaths per 100,000 workers over three years. This compares to the overall total occupational fatality rate of 4.4 per 100,000 (NTOF) and 5 per 100,000 (CFOI) workers. The leading source of fatal injury for the agricultural production sector identified by both national surveillance systems was farm tractors, which accounted for 46% (NTOF) and 53% (CFOI) of the fatalities. Non-highway overturn was the most frequent cause of tractor fatalities.

The preceding figures emphasize the importance of roll-over protection and seat belt use. In 1985 all major tractor manufacturers voluntarily adopted roll-over protective structures (ROPS) and seatbelts as standard equipment. A ROPS is a cab or frame for the protection of operators of wheeled agricultural tractors to minimize the possibility of serious operator injury resulting from accidental upsets. There are three basic types of ROPS: the two-post frame, the four-post frame, and the frame with enclosure. With an enclosed cab, environmental and noise control can also be included. As the percentage of tractors equipped with ROPS increases along with the use of seat belts, fatality rates caused by overturn should decrease.

Teaford (1993) in his ASAE Distinguished Lecture provides in-depth resource information for engineers concerning ROPS. Testing of ROPS is specified in ASAE Standard S383.1 DEC98 (ASAE 1999). Tests consist of either a static or dynamic rear or side loading in the laboratory. After either the static or dynamic load test, the same ROPS is subjected to a static crush to verify the effectiveness of the deformed ROPS in supporting the tractor in an upset position. There is also a field upset test both to the rear and side, to verify the effectiveness of the protective system under actual dynamic conditions (this test can be omitted under certain conditions). A temperature-material requirement will also be met if the ROPS passes either the static or dynamic test at a metal temperature of -18°C or below.

The engineering design of ROPS is beyond the scope of this text. Finite element methods are used to identify and predict the mode and sequence of yielding that takes place during a roll-over situation. It also aids in analysis of welds and joints.

Templeton and Strong (1998), in their ASAE Distinguished Lecture, provide additional insights to vehicle cab design.

15.10 Chapter Summary

Design of an off-road vehicle such as a tractor should include an understanding of the operator-machine interface, knowledge of human factors engineering, and integration of safety into the design. When these factors are properly incorporated into the design, the operator can perform complex tasks with efficiency, safety, and a minimum of fatigue.

When a person operates an off-road vehicle such as a tractor, the sensing, decision-making, and muscular powers of the operator are joined to an engineering system. The operator uses hearing, sight, and touch to interpret inputs and to interface with the control-instrumentation components to achieve the desired output of the tractor.

Functional considerations in design of the workplace for an off-road vehicle operator are visibility, easy access to controls, and clearance. Easy access to all controls in the workplace and cab are necessary so the operator can achieve the desired machine performance and productivity.

Procedural efficacy in operator workplace design is the logical arrangements of task elements to enhance operator performance and minimize error. Proper association between controls and displays along with logical grouping of instruments and controls is important in a compatible man-machine system.

Thermal comfort is defined as the state of mind that expresses satisfaction with the thermal environment. Operator enclosure design must include cab pressurization, air filtration, air movement, heating, cooling, and window defrosting.

The overall mechanical design of an off-road vehicle, including the operator workplace and enclosure, has a direct bearing on the noise and vibration levels at the operator's site. Exposure of the human body to vibrations can result in physiological, psychological, and mechanical effects. Ride vibration intensities are positively correlated with ground speed and often become intolerable as speed is increased. Low-frequency vertical vibration is present during normal field operations. Amplitude of vibration, in part, is dependent on roughness of the field. The undamped natural frequencies of wheel tractors commonly lie in the 3 to 5 Hz range. Higher frequency vibration, 30 Hz and up, results in part-body vibration.

There are basic guidelines governing overall design for ride comfort. For passive seat suspension, components used are springs with constant or variable rates, shock absorbers to dampen low-frequency vibrations, and torsion bar mechanisms or pneumatic cylinders to compensate for static deflection.

Farm tractors were the leading source of agricultural fatal injury identified by two national occupational fatality surveillance studies during 1990-1995, accounting for 46% and 53% of the fatalities. Non-highway overturn was the most frequent cause of tractor fatalities. These figures emphasize the importance of roll-over protective structures (ROPS) and seat belt use.

Homework Problems

15.1 Why is the 5th percentile for females and 95th percentile for males used in dimensions for adults?

15.2 In the first Example Problem why was the 95th percentile of males used?

15.3 SAE J833 Standard specifies the minimum standing height in a cab (see Figure 15.5). How does the design height of 1880 mm from the first Example Problem compare with the standard? Why the difference?

15.4 Rework the first Example Problem and determine the standing height for African decent (See Table 15.2). How much headroom is now available using SAE J833?

15.5 In the second Example Problem why were the 5th percentile female dimensions used?

15.6 In the second Example Problem the resulting design height of the seat index point (SIP) from the floor of the operator's platform was 462 mm minimum and 578 mm maximum which results in a vertical adjustment of 116 mm. Referring to SAE J899, Figure 15.7, and Table 15.3, what dimension applies to the vertical adjustment and how does it compare to the design dimensions? Should the design dimensions be changed?

15.7 Design an operator's seat (see Figure15.7 and Table 15.3) for the following conditions:

(a) Minimum operator enclosure width of 920 mm.

(b) Operator must control rear-mounted implements. Specify design dimensions for the:

(1) Seat cushion; length B, width D, angle C, and angle adjustment.

(2) Back cushion; height G, width H, angle J, and angle adjustment.

(3) Armrest; height K, length L, lateral spacing M, width N, and depth P.

(4) Lumbar support Q.

(5) Adjustments; vertical adjustment F (use your answer from Problem 15.6), fore and aft E, and armrest vertical.

For each dimension selected, give brief description why chosen i.e., max, nom, use of information in Figures 15.2 and 15.3, etc.

15.8 How does the vertical armrest height adjustment you determined in Problem 15.7, compare with the elbow height for a human in sitting position? See Table 15.2 and in Figure 15.2, assume the hip pivot (HP) and the seat index point (SIP) are the same point.

15.9 Using Figure 15.9, determine the x and z coordinates for a "zone of comfort" for foot controls. Reference them to the SIP point.

15.10 Using Figure 15.10, determine the x and y coordinates for a "zone of comfort" for foot controls. Reference them to the SIP point.

15.11 Using Figure 15.11, determine the y and z coordinates for a "zone of comfort" for foot controls. Reference them to the SIP point.

15.12 Design floor pedal location and width of pedal for:

(a) Left foot pedal that operates the clutch of vehicle, forward/reverse travel.

(b) Right foot—two pedals, the right pedal that operates the right wheel brake and the left pedal that operates the left wheel brake. Determine the "touch point" P for each pedal as measured horizontally from the seat index point (SIP) and the vertical distance from the floor of the workplace in mm. Use X as the horizontal axis and Z as the vertical axis. See Figures 15.3 and 15.8, Tables 15.2, 15.4, and 15.5. SIP = 500 mm above floor, $A_1 = 0°$, $A_3 = 90°$, $A_6 = 120°$, and $A_7 = 90°$. Determine the lateral location of each "touch point" P from the longitudinal centerline of the floor of the workplace (i.e., center line of steering column). Make a sketch (horizontal plane) and locate the centerlines of the steering column and pedals. On the centerlines locate SIP, H, P (for each pedal), and width of each pedal. Also include the height of P above floor.

Check your design by

(a) comparing with your answers to Problems 15.9, 15.10, and 15.11, and

(b) taking appropriate measurements from an existing vehicle and comparing to your design values. What adjustments would you make to your design?

15.13 Using Figure 15.15, define two sets of dry-bulb temperature and relative humidity that would be in the comfortable zone for

(a) summer conditions and

(b) winter conditions. How would you propose to make measurements inside a cab to relate cab conditions to the comfort criteria of Figure 15.15?

15.14 An accelerometer mounted to the waist of a seated off-road vehicle operator records a vertical RMS acceleration of 1.5 m/s^2 at 8 Hz. What is the most likely source of this vibration? Would it be desirable for the operator to experience this level of vibration? Why? What is the amplitude of this vibration in millimeters?

15.15 If the RMS acceleration in Problem 15.14 was decreased by 10 dB because of improved seat design, what could be the resulting RMS acceleration? How many hours could the operator tolerate this level of vibration?

15.16 Design a seat suspension for seat and operator mass of 90 Kg, transmissibility of 0.315, vehicle chassis frequency of 4 Hz, and static spring deflection of seat and operator of 10 cm.

(a) Determine the spring rate, k in N/m for the seat suspension.

(b) Determine undamped natural frequency of the seat.

(c) Determine the damping (shock absorber) c in N/ms for seat suspension.

(d) Determine the decibel reduction in RMS acceleration that the seat suspension provides.

(e) Using ISO 2631 criteria, how long should an off-road vehicle operator be exposed to a vehicle chassis input of 1 RMS m/s^2 for the seat suspension you designed?

(f) If the vehicle chassis frequency decreases, will the seat suspension be more or less effective? Why?

References and Suggested Readings

Ahlschwede, B., and K. Klotzbach. 2000. Integrating safety into the design of agricultural equipment. ASAE 2000 Agricultural Equipment Technology Conference. Kansas City, MO, 23-25 February.

Ahmed, O.B., and J.F. Goupillon. 1997. Predicting the ride vibration of an agricultural tractor. *Journal of Terramechanics* 34(1): 1-11.

Akesson, N.B., et al. 1974. Health hazards to workers from application of pesticides. ASAE Paper No. 741007.

ANSI/ASAE. 1999. Lighting and marking of agricultural equipment on highways. ANSI/ASAE Standard S279.10 OCT98. *ASAE Standards,* 46th ed. St. Joseph, MI: ASAE.

ANSI/ASAE. 1999. Graphical symbols for operator controls and displays on agricultural equipment. ANSI/ASAE Standard S304.6 DEC95. *ASAE Standards,* 46th ed. St. Joseph, MI: ASAE.

ANSI/ASAE. 1999. Agricultural cabs—Environmental air quality. Part 1: Definitions, test methods, and safety practices. ANSI/ASAE Standard S525-1.1. *ASAE Standards,* 46th ed. St. Joseph, MI: ASAE.

ANSI/ASAE. 1999. Agricultural cabs–Environmental air quality. Part 2: Pesticide vapor filters–Test procedure and performance criteria. ANSI/ASAE Standard S525-2. *ASAE Standards,* 46th ed. St. Joseph, MI: ASAE.

ASAE. 1999. Color coding hand controls. ASAE Standard EP443 DEC98. *ASAE Standards,* 46th ed. St. Joseph, MI: ASAE.

ASAE. 1999. Operator controls on agricultural equipment. ASAE Standard S335.4 DEC98. *ASAE Standards,* 46th ed. St. Joseph, MI: ASAE.

ASAE, 1999. Roll-over protective structures (ROPS) for wheeled agricultural tractors. ASAE Standard S383.1 DEC98. *ASAE Standards,* 46th ed. St. Joseph, MI: ASAE.

ASHRAE. 1972. *Handbook of Fundamentals.* Atlanta, GA: ASHRAE.

Bell, L.H. 1982. *Industrial Noise Control Fundamentals and Applications,* 1st Ed. New York, NY: Marcel Dekker, Inc.

Bongers, P.M., and H.C. Boshuizen. 1992. Back disorders and exposure to whole-body vibration. Thesis summary. *Clinical Biomechanics* 7(3): 185-186.

Claar II, P.W., and P. Sheth. 1980. Off-road vehicle ride: review of concepts and design evaluation with computer simulations. SAE Paper No. 801023. Warrendale, PA: SAE.

Claar II, P.W., W.F. Buchele, and S.J. Marley. 1980. Agricultural tractor chassis suspension system for improved tractor ride comfort. ASAE Paper No. 801565. St. Joseph, MI: ASAE.

Den Hartog, J.P. 1956. *Mechanical Vibrations,* 4th Ed. New York, NY: McGraw-Hill Book Co.

Dreyfus, H. 1960. *The Measure of Man*, p. 159. New York, NY: Whitney Library Design.

Duncan, J.R., and E.L. Wegscheid. 1982. Off-road vehicle simulation for human factors research. ASAE Paper No. 821610. St. Joseph, MI: ASAE.

Eriksson, H., and K.W. Domier. 1975. Heating and ventilating of tractor cabs. ASAE Paper No. 751516. St. Joseph, MI: ASAE.

Fanger, P.O. 1972. *Thermal Comfort: Analysis and Applications in Environmental Engineering.* New York, NY: McGraw-Hill Book Co.

Gagee, A.P., A.C. Burton, and H.C. Bazett. 1941. A practical system of units for the description of the heat exchange of man with his environment. Science 94: 428-430.

Hansson, J.E., L. Sjoflot, and C.W. Suggs. 1970. Matching the farm machine to the operator's capabilities and limitations. *Implement & Tractor* (21 August).

Hard, D.L., J.R. Myers, K.A. Snyder, V.J. Casini, L.L. Morton, R. Cianfrocco, and J. Fields. 1999. Identifying work-related fatalities in the agricultural production sector using two national occupational fatality surveillance systems, 1990-1995. *Journal of Agricultural Safety and Health* 5(2): 155-169.

ISO. 1972. Guide for the evaluation of human exposure to whole body vibration. ISO 2631. Geneva, Switzerland: ISO.

Janeway, R.N. 1975. Human vibration tolerance criteria and applications to ride evaluation. SAE Paper No. 750166. Warrendale, PA: SAE.

Kaufman, K.R., P.K. Turnquist, and R.N. Swanson. 1976. Physiological responses and thermal comfort of subjects in a tractor cab. ASAE Paper No. 761577. St. Joseph, MI: ASAE.

Matthews, J. 1966. Ride comfort for tractor operators. IV. Assessment of the ride quality of seats. *Journal of Agricultural Engineering Research* 2(1): 44-57.

Matthews, J. 1972. The ergonomics of tractor design and operation. In *Proceedings of the XVI CIOSTA Congress,* Wageningen, The Netherlands: Wageningen University Press.

Matthews, J. 1973. The measurement of tractor ride comfort. SAE Paper No. 730795. Warrendale, PA: SAE.

Noren, O. 1985. Dust concentrations during operations with farm machines. ASAE Paper No. 851055. St. Joseph, MI: ASAE.

Pradko, F., and R. Lee. 1968. Analysis of human vibration. *SAE Transactions* 77. Warrendale, PA: SAE.

Prasad, N., V.E. Tewari, and R. Yadav. 1995. Tractor ride vibration–A review. *Journal of Terramechanics* 32(4): 205-219.

Purcell, W.F.H. 1980. The human factor in farm and industrial equipment design. ASAE Distinguished Lecture Series No.6. St. Joseph, MI: ASAE.

Rakheja, S., and S. Sankar. 1984a. Suspension designs to improve tractor ride: I. Passive seat suspension. SAE Paper No. 841108. Warrendale, PA: SAE.

Rakheja, S., and S. Sankar. 1984b. Suspension designs to improve tractor ride: II. Passive cab suspension. SAE Paper No. 841108. Warrendale, PA: SAE.

Raney, J.P., J.B. Liljedahl, and R. Cohen. 1961. The dynamic behavior of farm tractors. *Transactions of the ASAE* 4(2): 215-218, 221.

Rossegger, R., and S. Rossegger. 1960. Health effects of tractor driving. *Journal of Agricultural Engineering Research* 5(3): 241.

SAE. 1999. Operator space envelope dimensions for off-road machines. SAE Standard J154 JUN92. *SAE Handbook,* vol. 3. Warrendale, PA: SAE.

SAE. 1999. Access systems for off-road machines. SAE Recommended Practice J185 JUN88. *SAE Handbook*, vol. 3. Warrendale, PA: SAE.

SAE. 1999. Human physical dimensions. SAE Recommended Practice J833 MAY89. *SAE Handbook*, vol. 3. Warrendale, PA: SAE.

SAE. 1999. Control locations for off-road work machines. SAE Standard J898/ISO 6682 OCT94. *SAE Handbook*, vol. 3. Warrendale, PA: SAE.

SAE. 1999. Operator's seat dimensions for off-road self-propelled work machines. SAE Standard J899 DEC88. *SAE Handbook,* vol. 3. Warrendale, PA: SAE.

SAE. 1999. Determining seat index point. SAE Standard J1163 AUG97. *SAE Handbook,* vol. 3. Warrendale, PA: SAE.

SAE. 1999. Instrument face design and location for construction and industrial equipment. SAE Recom-mended Practice J209 JAN87. *SAE Handbook,* vol. 3. Warrendale, PA: SAE.

SAE. 1999. Operator enclosure pressurization system test procedure. SAE Recommended Practice J1012 JUNE93/ISO 3737-1976. *SAE Handbook,* vol. 3. Warrendale, PA: SAE.

Sanders, M.S., and E.J. McCormick. 1987. *Human Factors in Engineering and Design,* 6th Ed. New York, NY: McGraw-Hill Book Co.

Smith, D.W. 1977. Computer simulation of tractor ride for design evaluation. SAE Paper No. 77074. Warrendale, PA: SAE.

Stikeleather, L.F. 1973. Evaluating the vibration of shock isolation qualities of operator seats for construction machinery. SAE Paper No. 730823. Warrendale, PA: SAE.

Stikeleather, L.F. 1976. Review of ride vibration of shock isolation qualities of operator seats for construction machinery. SAE Paper No. 730823. Warrendale, PA: SAE.

Stikeleather, L.F. 1981. Operator seats for agricultural equipment. ASAE Distinguished Lecture Series No. 7. St. Joseph, MI: ASAE.

Stikeleather, L.F., and C.W. Suggs. 1970. An active seat suspension system for off-road vehicles. *Transactions of the ASAE* 13(1): 99.

Suggs, C.W. 1973. Agricultural machinery noise and vibration levels in comparison to human comfort and safety limits. ASAE Paper No. 73524. St. Joseph, MI: ASAE.

Teaford, W.J. 1993. Roll-over protective structures (ROPS) for wheeled agricultural tractors. ASAE Distinguished Lecture Series No. 18. St. Joseph, MI: ASAE.

Templeton, D.J., and R.W. Strong. 1998. Essential design considerations for today's tractor cab and a challenge for the future. ASAE Distinguished Lecture Series No. 22. St. Joseph, MI: ASAE.

Wan, Y., and J.M. Schimmels. 1997. Optimal seat suspension design based on minimum "simulated subjective response." *Journal of Biomechanical Engineering* 119: 409-415.

Woodson, W.E. 1992. *Human Factors Design Handbook,* 2nd Ed. New York, NY: McGraw-Hill Book Co.

Young, R.E., and C.W. Suggs. 1973. Seat suspension system for isolation of roll and pitch in off-road vehicles. *Transactions of the ASAE* 16(5): 876.

Zander, J. 1972. Ergonomics in machine design (a case-study of the self-propelled combine harvester). Wageningen, The Netherlands: Wageningen University Press.

Zitko, R.F. 1977. Control center design concepts series 86 tractor. ASAE Paper No. 771049. St. Joseph, MI: ASAE.

INDEX